CHARLES DARWIN,
GEOLOGIST

CORNELL UNIVERSITY PRESS • ITHACA AND LONDON

CHARLES DARWIN, GEOLOGIST

Sandra Herbert

Cornell University Press gratefully acknowledges receipt of a subvention from the University of Maryland, Baltimore County, which aided in bringing this book to publication.

First published 2005 by Cornell University Press

Printed in the United States of America

Design by Scott Levine

Library of Congress Cataloging-in-Publication Data

Herbert, Sandra.
Charles Darwin, geologist / Sandra Herbert..
p. cm.
Includes bibliographical references and index.
ISBN 13: 978-0-8014-4348-0 (cloth : alk. paper)
ISBN-10: 0-8014-4348-2 (cloth : alk. paper)
1. Darwin, Charles, 1809–1882. 2. Geologists—Great Britain—Biography. 3. Geology—Great Britain—History. I. Title.

QE22.D27H47 2005
551′.092—dc22

2005002690

Cornell University Press strives to use environmentally responsible suppliers and materials to the fullest extent possible in the publishing of its books. Such materials include vegetable-based, low-VOC inks and acid-free papers that are recycled, totally chlorine-free, or partly composed of nonwood fibers. For further information, visit our website at www.cornellpress.cornell.edu.

Cloth printing 10 9 8 7 6 5 4 3

For Jim, my loving husband and ready companion

CONTENTS

ILLUSTRATIONS

Figures

Plates

PREFACE

In the public mind the name of Charles Darwin is firmly attached to the idea of evolution, and properly so. Yet in his working life as a scientist Darwin had wider interests. He pursued a variety of subjects, among which, prominently, was geology. His geological works are marked with lasting though partial triumph, keen insight bordering on intuition, and reversals of opinion.

In the nineteenth century geology attracted persons of imagination, like Darwin, because of its promise of knowledge of the distant past. Inquiring minds also found the breadth of its subject matter stimulating. The study of stratigraphy and the processes involved in forming strata, of fossils and the reconstruction of the history of life, of volcanoes, earthquakes, and petrology all fell within its purview. On the practical side, geological science went hand in hand with mineral surveying and thus promised economic benefit and national advantage.

As a young man Darwin was known both popularly and professionally as a geologist. In February 1859 the Geological Society awarded him its highest honor, the Wollaston Medal, for his signal contributions to the field. In reading the citation for the medal, Charles Lyell noted Darwin's studies on the growth of coral reefs, on evidence for the modern elevation of Chile and the repeated elevation of the Andes, on South American geology generally, on the distribution of boulders in Britain, and on fossil barnacles.[1]

In November 1859, eight months after receiving the Wollaston Medal, Darwin published *On the Origin of Species by Means of Natural Selection*. At once the book secured him an international reputation and overshadowed everything that had come before it in his life as a scientist. The book has had a curious relationship to the history of geology. On the one hand, it brought the evolutionary perspective into a permanent relationship with the field: thereafter every geologist had to be for or against an evolutionary interpretation of the fossil record. Yet Darwin's argument in the *Origin* did not position geology prominently. One would not guess from the book alone the key role played by geology in Darwin's arrival at his theory of evolution, nor the complexity of the relation of the book to the field.

Following publication of the *Origin,* Darwin's reputation as a biologist superseded his renown as a geologist. This imbalance was redressed in 1909 at celebrations honoring the centenary of his birth. John Wesley Judd, who had become Darwin's geological confidant after Lyell died, wrote on Darwin the geologist. Archibald Geikie, the foremost geologist of his day, devoted his Rede Lecture at the University of Cambridge to the subject.[2]

Geikie and Judd, young men in 1859, remembered publication of the *Origin* as an epoch-making event. Geikie recalled its effect on younger geologists such as himself. Both geologists emphasized Darwin's debt to Lyell, and, despite Lyell's slow acceptance of the idea of evolution, regarded the *Origin* as of a piece with Lyell's *Principles of Geology* (1830–1833).[3]

As we approach 2009, the bicentenary of Darwin's birth, it is timely to look again at Darwin as a geologist. While those of us writing today cannot speak from personal knowledge of Darwin and his contemporaries, we write with other advantages. The lasting significance of the tradition of scientific exploration, along with the political context that supported it, is better appreciated, and with the passage of time, the structure of the field of geology stands in greater relief than it could have to Darwin and his contemporaries.

On the biographical side, details of Darwin's early education have emerged, revealing the importance to his later work of chemical experiments performed as a boy in a home laboratory—continuing the tradition of his Lunar Society forebears—and the benefits to him of his Edinburgh years under Robert Grant and Robert Jameson as well as his Cambridge years under John Stevens Henslow and, slightly later, Adam Sedgwick. The *Beagle* voyage still looms large, but it is now clear that Darwin was moving toward a distinguished career in science quite apart from that opportunity. The role of Lyell remains of central importance, but the influence of other authors—among them Alexander von Humboldt, John Herschel, George Poulett Scrope, Henry De la Beche, William Buckland, Richard Owen, Louis Agassiz, John Phillips, and Andrew Crombie Ramsay —is more easily recognized now that the extent and direction of Darwin's reading are understood.

The primary sources for this knowledge are Darwin's manuscripts, including notes and correspondence, the great preponderance of which have come into the public domain since the time of Geikie and Judd. These resources have also made it possible to trace Darwin's development as a geologist from his earliest field studies, through his authorship of a number of theoretically bold essays during the voyage, to his published work. His ambition for a "simple" geology stands revealed, his theories and speculations on a variety of other topics identified, and the intensity of his involvement with contemporary geological opinion—as in the debate over glaciation—better appreciated. As one studies Darwin's career closely, the *Origin*'s geological moorings become increasingly apparent.

The temporal boundaries of this book extend from Darwin's earliest interest in geological subjects to the publication of the *Origin* in 1859, with occasional reference to later developments. During this period, a number of geological questions were raised and resolved fairly quickly within the geological community while other issues remained unsettled, rendering a strictly chronological, year-by-year approach to the subject inadequate. There is thus a sinuous quality to the narrative of this book: issues appear and, if not resolved (as happened with discussion of the "diluvium"), reappear in later chapters.

To aid the reader in following the narrative, I have provided numerous visual aids and quoted at length from texts that are hard to come by. Persons referred to in the text are identified on first mention by their full names and by birth and death dates if known. After first mention, full names are used only if necessary to avoid confusion between persons of identical surnames or if the narrative thread would seem to require it.

Chapters 1 and 2 recount Darwin's education as an early-nineteenth-century geologist against the background of the growth of geology as a science. Chapter 3 focuses on the narrower topic of specimens. Chapter 4, an intermezzo, considers the aesthetic appeal of the experience of nature afforded by a geological perspective. Chapter 5 describes Darwin's maturation as a geologist during the voyage of H.M.S. *Beagle*. Chapter 6 examines the "Genesis and geology" dispute as it played out for Darwin. Chapters 7 and 8 analyze Darwin's ideal of a simple geology, and the challenges to that ideal. Chapters 9 and 10 situate Darwin's species theory in the context of his geology.

One subject remains to be addressed: how I first became interested in the subject of Darwin as a geologist. As a graduate student I scrutinized Darwin's manuscripts in search of the "origins of the 'Origin,'" but as I read through the material from the *Beagle* voyage, I was struck by his remarkable and, to me, unanticipated focus on geology. The point crystallized as I read one of his small notebooks deposited at Down House, his former home. In that notebook he wrote both of a simple geology based on the vertical movements of the earth's crust and of what might be true if species do change. This book started there.

I express gratitude to the late George Pember Darwin, to William Huxley Darwin, and to the Syndics of Cambridge University Library for permission to publish selections from the manuscripts of Charles Darwin.

My research for this book was generously supported by the National Endowment for the Humanities, the John Simon Guggenheim Memorial Foundation, the National Science Foundation, the Smithsonian Institution, and the University of Maryland Baltimore County (UMBC). I thank these institutions for their support.

Scholarship occurs in a collegial setting, and I am indebted to many individuals. Between them Robert K. Webb and David R. Oldroyd read all of the chapters in the book. Their rich, sharp, but always tactful comments kept me on course. Similarly, Frederick Burkhardt provided continuing assistance throughout the writing of this book. Chapter 3 on specimens benefited from the direct help of three persons: most important, Graham Chinner gave me access to the collection of Darwin's geological specimens at Cambridge and patiently explained much about them as he went along, as well as reading drafts of the chapter; Steven Laurie took time to help me go through a portion of that collection; and Sally E. Newcomb served as my resource on matters petrological here at home. For assistance with the Welsh materials, discussed in chapters 1 and 8, I am grateful to Kenneth Addison, Peter Lucas, and Michael B. Roberts. Few places on earth have attracted the devotion of residents and visitors as has Wales, Darwin being no exception, and my correspondents on Wales have been generous and helpful. Stephen Brush, Lindley Darden, Ernst Mayr, David Norman, and Laszlo Takacs also kindly read other chapters and offered highly informed and pointed corrections and additions. I am greatly indebted to each of them. Richard Darwin Keynes provided key assistance at a number of points, and I thank him.

Over the years the informal band of scholars working on Charles Darwin, often convening in the tearoom at Cambridge University Library, has proved a source of edification and camaraderie. While I cannot name all the members of this society, I cannot fail to mention Paul H. Barrett, Gillian Beer, Peter Bowler, Janet Browne, Anne Burkhardt, Sheila Ann Dean, Mario Di Gregorio, Peter J. Gautrey, Nick Gill, John C. Greene, Joy Harvey, Jonathan Hodge, Thomas Junker, David Kohn, William M. Montgomery, James Moore, Perry O'Donovan, Stephen Pocock, Duncan M. Porter, Robert J. Richards, Frank H. T. Rhodes, Marsha L. Richmond, Michael Ruse, Anne Secord, James A. Secord, and Sydney Smith.

In similar fashion I cannot fail to mention individuals, many of them associated with the History of Science Society, the History of Earth Sciences Society, the International Commission on the History of Geological Sciences, Cambridge University Library, the Geological Society of Washington, the Library of Congress, the U.S. Geological Survey, the Smithsonian Institution, and my university, with whom I have conversed about the questions discussed in this book, or who have otherwise aided my research: Michele L. Aldrich,

Toby A. Appel, Joseph L. Arnold, Cathy Barton, Robert Basila, John Biebel, Stella Brecknell, Eric Brown, Joe D. Burchfield, Harold Burstyn, Joe Cain, Susan Faye Cannon, Thomas Carroll, Constance Carter, Wendy Cawthorne, Anne Chapman, Warren I. Cohen, Gretchen Cooke, Ronald E. Doel, Jon Eklund, Gregory Estes, Bernard S. Finn, Gary L. Fitzpatrick, Cyril Galvin Jr., Janet B. Garber, Maureen Gautrey, Gerald L. Geison, Stanley Goldberg, Gregory A. Goode, Thalia Grant, Mott T. Greene, James S. Grubb, Martin Guntau, Sharon K. Hanley, Anthony P. Harvey, Pamela M. Henson, Brian Huber, David L. Hull, Jane Insley, John W. Jeffries, Randy Kloetzli, William Lauffer, Linda Lear, Elizabeth Leedham-Green, Alan E. Leviton, Rod Long, Roy MacLeod, Bruce Martin, Iain McCalman, Eileen McLellan, Everett Mendelsohn, David Mitch, Joseph F. Mulligan, Kevin Omland, Margaret Pamplin, Paul N. Pearson, Jessica Pfeifer, Valerie Pinsky, Philip F. Rehbock, Jayne Ringrose, Shirley A. Roe, Marc Rothenberg, Jean-Paul Schaer, Stephen Schneckler, Andrew Sclater, James Sharp, Tom Simkin, Phillip Reid Sloan, Michael M. Sokal, Phillip Sokolove, Alan Swanson, Nora Tamberg, Joseph N. Tatarewicz, Kenneth L. Taylor, John Thackray, Hugh Torrens, Stephen Toulmin, Yoram Unguru, Ezio Vaccari, Godfrey Waller, Deborah J. Warner, Victor Wexler, Matthew A. White, Leonard G. Wilson, Patrick Wyse-Jackson, Ellis Yochelson, and Barbara York.

I am grateful to the editors at Cornell University Press, particularly Roger M. Haydon, for their efforts. Comments on the manuscript by an anonymous reader for the Press were insightful and valuable. Susan Jacobs and Elizabeth Kight proofread and Mary Babcock copyedited the manuscript with care. Faye Adams and Carla Ison provided loyal secretarial support at UMBC.

As always in geology, visual aids are essential. I wish to express my appreciation to Richard Keynes and Mark Smyth for permission to reproduce paintings by Conrad Martens, to Kenneth Addison for permission to reproduce a photograph of Cwm Idwal, to Dudley Simons for photographs of the Darwin specimens at the Sedgwick Museum, to Peter Grant for permission to reproduce one of his maps, to English Heritage for photographs of Darwin's clinometer and hammer, to Cambridge University Press for permission to reproduce drawings from the Darwin correspondence, and to Cambridge University Library, the Geological Society of London, and the Library of Congress for permission to reproduce materials from their collections. I wish also to thank Joseph School of the UMBC Cartographic Services for a number of maps and drawings.

Some material in this book has appeared earlier. A portion of chapter 2 first appeared in *The Darwinian Heritage*. Early drafts for chapters 5 and 6 were published in the *British Journal for the History of Science* and in *Religion and Irreligion in Victorian Society,* respectively. A portion of chapter 8 first appeared in *Eclogae Geologicae Helvetiae.* I thank publishers for permission to use this material in another setting.

This book is dedicated with gratitude to my husband Jim, whose love of the history of ideas has come to embrace geology. We both thank our daughters Kristen and Sonja, who have shared in our adventures. I also recall with pleasure the quiet, scholarly home provided by my parents, Emrick and Dorothy Swanson, for my sister Joann and me. At one time or another the whole lot of us found ourselves in Cambridge, and enjoyably so. From the more distant past I recall with respect the labor of my grandfathers at the Chicago steel mills, metallurgy being at the applied end of the earth sciences. Our generation has had easier work, but one hopes it is life enhancing in its own way.

S. H.

CHARLES DARWIN,
GEOLOGIST

CHAPTER 1

"I A GEOLOGIST"

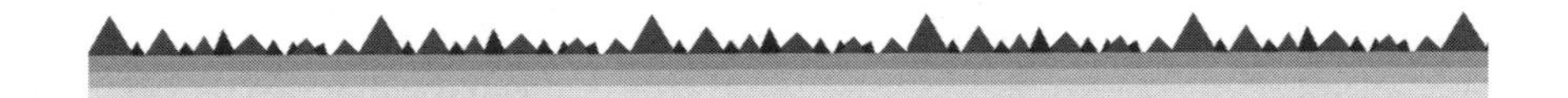

Youth

In August 1838, at the age of twenty-nine, Charles Darwin (1809–1882), who had already returned from his five-year passage aboard H.M.S. *Beagle,* recorded his earliest memories. That summer he was pondering the implication for man of his newly adopted theory of transmutation, and his examination of the subject extended to himself. In the few paragraphs he wrote in this early "*Life,*" he reflected on the character of these memories. He recalled the moment, before he was four, of receiving an injury, and of a visit to the seaside when he was four and a half. He commented that, unlike those of his younger sister Catherine (1810–1866), "All my recollections seem to be connected most closely with self.—"[1] Also unlike Catherine, he remembered the impression external scenery made on him. He recalled his experiences at Mr. Case's school when he was eight and a half.

> I was very timid by nature. I remember I took great delight at school in fishing for newts in the quary pool.—I had thus young formed a strong taste for collecting, chiefly seals, franks & but also pebbles & minerals,—one which was given me by some boy, decided this taste.—I believe shortly after this or before I had smattered in botany, & certainly when at Mr Case's school I was very fond of gardening, & invented some great falsehoods about being able to colour crocuses as I liked.—At this time

> I felt strong friendships for some boys.—It was soon after I began collecting stones, ie when 9 or 10 I distinctly recollect the desire I had of being able to know something about every pebble in front of the Hall door—it was my earliest—only geological aspiration at that time.—

And, further:

> I do not remember any mental pursuits excepting those of collecting stones &c.—gardening, & about this time often going with my father in his carriage, telling him of my lessons, & seeing game & other wild birds, which was a great delight to me.—I was born a naturalist.—[2]

Darwin had located in his schooldays the roots of his mature self. In that same summer of 1838, in July, he had considered privately the general subject of the "pleasure of imagination." He believed that trains of imaginative thought varied among people. For example, he suggested that "an agriculturist, in whose mind supply of food was [an] evasive & ill defined thought would receive pleasure from thinking of the fertility.—" In contrast, he chose to characterize himself this way: "I a geologist have illdefined notion of land covered with ocean, former animals, slow force cracking surface &c truly poetical."[3] The lad who had wanted to know something about every pebble in front of the Hall door had become the adult geologist and naturalist.

Fortunately for the historian, Darwin intended his early "*Life*" as a serious exercise, and he did not spare himself. He commented on the self-referential nature of his early memories, and compared them to Catherine's—she seemed "to recollect scenes, where others were chief actors.—" In this regard he seems to lament the paucity and trivial nature of his memories surrounding his mother's death when he was eight and a half. He remarked,

> I was in these days a very great story teller,—for the pure pleasure of exciting attention & surprise. . . . I scarcely ever went out walking without saying I had seen a pheasant or some strange bird, (natural History taste). these lies, when not detected, I presume excited my attention, as I recollect them vividly,.—not connected with shame, though some I do,—but as something which by having produced great effect on my mind, gave pleasure, like a tragedy.—

Later in his life, when he was ten, he recalls himself not fabricating but "at that time very passionate, (when I swore like a trooper) & quarrelsome,—the former passion has I think nearly wholly, but slowly died away." Darwin's early vanity would funnel into a strong ambition, storytelling would be checked by the disciplines of school and maturity, and rough edges would be smoothed into a deferential politeness that some have connected to his motherless adolescence. Still, the dominant note in his early recollection is that of enjoyment. In his autobiographical sketch of 1838 he remembered in Wales in 1819 "a certain shady green road (where I saw a snake) & a waterfall with a degree of pleasure, which must be connected with the pleasure from scenery, though not directly recognized as such.—"[4] This was the part of his psychology that fed most directly into his work as a geologist.

Darwin's early autobiographical exercise holds other indications of his situation in 1838. He remarked that he remembered the "illumination after the Battle of Waterloo, & the militia exercising, about that period, in the field opposite our House.—"[5] Born in 1809, he would have been six at the time of Waterloo. As was clear to Darwin writing in 1838, the world had become more open to peaceful exploration by the English since their victory, for they had defeated all rivals for domination of the seas. Darwin's own travel aboard the *Beagle* was a case in point. Other indications of Darwin's situation were his references to schooling, where his reading and associations set the stage for his future work. As he recalled in a later autobiography, it was learning about the "wonders of the world" that drew him on:

> Early in my school-days a boy had a copy of the *Wonders of the World,* which I often read and disputed with other boys about the veracity of some of the statements . . . ; and I believe this book first gave me a wish to travel in remote countries, which was ultimately fulfilled by the voyage of the *Beagle.*[6]

The table of contents for *Wonders of the World* lists the following:

> Giants' Causeway.—The Geysers: or, boiling springs.—The Great Wall of China.—The Great Kentucky Cavern.—The Pyramids of Egypt.—Mount Vesuvius.—Breakwater at Plymouth.—The Andes.—The Ruins of Palmyra.—The Falls of Niagara.—Mont Blanc.—The Steam Engine.—[7]

The book closes by citing "Dr. [Erasmus] Darwin"—Charles's paternal grandfather—on the steam engine:

> Soon shall thy arm, unconquer'd steam, afar
> Urge the slow barge, or drive the rapid car;
> Or on wide waving wings expanded bear
> The flying chariot thro' the fields of air.[8]

Darwin's school was private and single sex. (There was no universal free public education in England at that time.)[9] Of a piece with his schooling were the material circumstances of his family, which allowed for the variety of his early experiences, at the seaside, in Wales, and in his birthplace of Shrewsbury, where his father's large house stood high above the river Severn. Then, too, the practice of collecting is enhanced if one has adequate opportunities for storage and display. These he had from childhood. A "room of one's own" can be used to store one's treasures. It can also serve, as it did for Darwin, to provide a quiet place for writing and reflection. Opportunity for peaceful exploration and the room to store his collections were part of Darwin's situation as much as his desire to collect and to know, and his enjoyment of scenery.

Yet there have undoubtedly been other young boys, or girls, with an inborn curiosity and talent as strong as Darwin's and similarly advantaged in a material sense. True, few would have been so advantaged in their family heritage. Charles's grandfathers Erasmus Darwin (1731–1802) and Josiah Wedgwood (1730–1795) were members of the Lunar Society of Birmingham, which also included the chemist Joseph Priestley (1733–1804).[10] Charles's father, Robert Waring Darwin (1766–1848), a physician (M.D., Leiden, 1785) and a Fellow of the Royal Society (1788), was quieter in his scientific interests, but well educated—as a young man he had studied under the great chemist Joseph Black (1728–1799) at Edinburgh University.[11] Charles's elder brother, Erasmus Alvey Darwin (1804–1881), also actively pursued scientific interests as a young man, particularly in chemistry and mineralogy. While at school, Erasmus had set up a laboratory in the toolhouse of the garden at the family home in Shrewsbury, where he performed chemical experiments, assisted by his younger brother Charles.[12] Erasmus also served as the literal forerunner to his younger brother. He preceded Charles to the University of Cambridge (where he, unlike Charles, took an M.B. degree) and to Edinburgh University to pursue medical studies; however, Charles attended the universities in reverse order, starting at Edinburgh in 1825 at age sixteen. Erasmus also preceded his younger brother Charles as a freethinker.[13]

These circumstances combine to suggest that Charles Darwin began his adult life from a high platform of familial achievement. Combined with a sense, gained at Edinburgh, that he would inherit substantial property from his father, he began his scientific life with numerous advantages.[14] Yet, taken together, they do not account for his particular course. To offer an explanation for Darwin's career in geology, one has to add to his biography and to the account he offered in his "*Life*." There Darwin offered a glimpse of himself as a pebble-gathering boy, but clearly other elements beyond curiosity were at work.

Three circumstances—all propitious—underlay Darwin's choice of geology. First, the political situation in England was sufficiently stable to allow for the leisured pursuit of scientific knowledge. Second, and relatedly, the tradition of scientific exploration was well established in England, and Europe generally. And, third, by the 1830s, when Darwin grew into adulthood, the field of geology also was well established, and Darwin's way into it, if not obvious, was at least well signposted. In a period when disciplinary lines were in flux—as between chemistry and mineralogy, or between mineralogy and geology—the example set by mentors also figured. All these factors complemented Darwin's native curiosity and talent. We will consider each one in turn.

Politics

On 18 June 1815 Napoleon's army met its final defeat, thus ending twenty-three years of war in Europe. The peace left Britain, as one of the victors, in a strong position, both absolutely, by virtue of its wealth, and relatively, in comparison with the power of other nations, particularly at sea. Indeed, from 1815 to the outbreak of war in 1914, Britain was more powerful than any other single nation in shaping world affairs. For a talented and ambitious man, like Darwin, born into wealth at the beginning of the period of British hegemony, whatever he chose to do in life would be enhanced by the position of his country.

During the voyage of H.M.S. *Beagle*, from December 1831 to October 1836, Darwin was fully cognizant of British power. The simple fact of the ship's mission spoke to the point. A principal purpose was to chart the southernmost tip of South America, thus completing work begun on the *Beagle*'s first voyage of 1826–1830. The hydrographic department of the Royal Navy sponsored the mission. Over time, from its founding in 1795, that office was increasingly active in sending out surveying expeditions: nine were in progress in 1818, twelve in 1823, nineteen in 1838.[15] Under Francis Beaufort (1774–1857), who became hydrographer in 1829, "officers of the naval Surveying Service [were] at sea in every ocean and charting nearly every remote and distant shore."[16] That no interference to this goal would be forthcoming from Spanish authorities in South America was a consequence of the Napoleonic wars. As a historian of Latin America has written,

> The revolutions for independence in Spanish America were sudden, violent and universal. When, in 1808, Spain collapsed under the onslaught of Napoleon, she ruled an empire stretching from California to Cape Horn, from the mouth of the Orinoco to the shores of the Pacific, the site of four viceroyalties, the home of seventeen million people. Fifteen

> years later she retained only Cuba and Puerto Rico, and new nations were already proliferating.[17]

At the same time, influences rivaling those of Spain were welcomed in the region. Thus, in the 1830s, a British presence was permitted in what had been a protectionist Spanish domain. As an example, the value of British goods imported into Argentina went from £369,000 in 1812 to £1,104,000 in 1824.[18] Prior to the late 1820s the only British hydrographic survey in South America was Beaufort's 1807 survey of the river Plate, the Río de la Plata, on whose southern bank lies the city of Buenos Aires.[19] Over the next decade, the British Hydrographic Office increased its attention to South America, with H.M.S. *Beagle*'s two expeditions leading the way. These surveys were done with the permission of local authorities. This open and usually hospitable attitude toward British nationals during the period was extended to a young naturalist leading his own small party on inland excursions: Darwin found the Spanish- and Portuguese-speaking residents of the ports he visited generally helpful. He took weapons with him on the voyage nonetheless. (Two pistols, one large, one small, and a "life preserver"—a short section of cable fitted with metal ends—are on display at Down House Museum in Kent, Darwin's home from 1842 until his death.)

Darwin's attitude toward British power was appreciative. Frequently during the voyage he expressed his sense of national pride. To his sister Catherine he wrote from Hobart, Tasmania, in February 1836,

> It is necessary to leave England, & see distant Colonies, of various nations, to know what wonderful people the English are.—It is rather an interesting feature in our Voyage, seeing so many of the distant English Colonies.—Falklands Island, (the lowest in the scale), 3 parts of Australia: Is^{d} of France, the Cape.—St Helena, & Ascension—[20]

Nonetheless, as a supernumerary passenger aboard a naval vessel he only rarely involved himself in the day-to-day operations of the ship or in the use of force. Early in the voyage he helped handle the sails coming into harbor at Rio de Janeiro. He was pleased to be part of a common effort and, as a shipmate later recalled, "enjoyed the fun of it."[21] But soldiering held no appeal. In August 1832 he wrote to a Cambridge colleague from Montevideo:

> At Buenos Ayres, a shot came whistling over our heads; it is a noise I had never before heard, but I found I had an instinctive knowledge of what it meant. The other day we landed our men here & took possession at the request of the inhabitants of the central fort. We Philoso-

phers do not bargain for this sort of work and I hope there will be no more."[22]

The day of the landing, Darwin had chosen to return to the ship at sunset "as I had a bad headache."[23]

Darwin kept a similar measured distance from political events at home. The year 1831, when his voyage began, also saw the Reform Bill introduced into the House of Commons; it became the Reform Act in June 1832. This law enlarged the electorate to more than 650,000, and by virtue of its provisions "took a long step towards substituting individuals for interests as the basis of representation."[24] Darwin's family at home was aware of his distance from current politics. His ordinarily reticent brother Erasmus wrote enthusiastically to him in 1832 of the possible consequences "now that we have got the Reform bill," but concluded on the regretful note, "I have written you all this politics tho' I suppose you are too far from England to care much about it. Politics wont travel." Certainly Charles was not as touched as were his brother and a cousin, William Darwin Fox (1805–1880), who worried in the wake of the Reform Bill, "Party Spirit runs now very high indeed—. . . . For some days we certainly were on the very verge of Revolution." Still, in his own limited way, Charles remained a Whig loyalist, affirming to his mentor John Stevens Henslow (1796–1861), "I would not be a Tory, if it was merely on account of their cold hearts about that scandal to Christian Nations, Slavery.—"[25]

The England to which Darwin returned in 1836 was wealthy and changed. Peace had brought prosperity that, though unequally shared, was apparent to all. Yet toward national politics Darwin displayed something like the same disparity of appreciative interest and aloofness that he had aboard ship. He observed more than he engaged. The democratically minded Chartist reform movement that swept England in the late 1830s had strongholds in the midlands, where his family owned substantial property, and in London, where he lived in the years immediately following the voyage. But Darwin did not comment on it. Nor did he express fear in the face of public demonstrations in its favor. He pursued his own affairs.[26] On more particular political subjects, those touching government and his own science, Darwin was an engaged actor. From the initial arrangements for the *Beagle* voyage, to the negotiation for the £1000 grant from the Treasury for publication of the zoological results from the voyage, and throughout a lifetime of activity in the scientific societies, Darwin pursued his ends single-mindedly.[27] That he was so focused on his own scientific pursuits was striking, even to himself, and later in life, he wondered if it had been altogether a good thing. He regretted that he had not behaved more philanthropically.[28] But his devotion to science both gave his life consistency and limited his involvement with other spheres of life, including the political.

Scientific Exploration

Evolution is so dazzling an idea that it can obscure the tradition of scientific exploration that gave it birth. Yet if one looks to the objects that inspired the idea—*Megatherium* bones, the Galápagos mockingbirds—one realizes that these are not European in origin. They were found by Europeans elsewhere and brought home for interpretation. While the immediate circumstances of their discovery may have had the feel of accident (for example, the Galápagos archipelago was not a required stop for the *Beagle*), the objects were observed in the course of systematic searches made from particular points of view. In this context, Darwin's discoveries, like those of other European naturalists, resulted from forethought.

The historian who has argued most succinctly for the intentional aspect of discovery is William H. Goetzmann. He has suggested that exploration should be viewed as an ongoing process, an activity, rather than as a sequence of discrete discoveries. He has noted, "The words 'exploration' and 'discovery' are most often and most casually linked in the popular imagination simply as interchangeable synonyms for 'adventure.' But exploration is something more than adventure, and something more than discovery. . . . It is purposeful. It is the seeking." In Goetzmann's view, exploration is the "result of purpose or mission. As such, it is an activity, which to a very large degree, is 'programmed' by some older center of culture."[29] Bruno Latour has expressed a similar idea in his suggestion that successive missions of exploration result in "cycles of accumulation" of findings, which are formed into scientific knowledge at "centres of calculation" at home.[30] An illustration of Latour's point is the fact, noted by Darwin, that the geologist George Greenough (1778–1855) recorded data from travelers on his "Map of the World" held at the Geological Society of London.[31]

The tradition of scientific exploration that came to include Darwin had its great inspiration in the work of Captain James Cook (1728–1779), whose three voyages both settled great questions and provided a model for work to come, even half a century after Cook's death. Indeed, for some, Cook was held within living memory well into the new century; his widow lived until 1835 and was buried in Cambridge. Unlike the European voyagers of the fifteenth and sixteenth centuries, Cook, and several other eighteenth-century explorers, were sent to gather the treasure of scientific knowledge rather than precious goods. They were to seek knowledge of many varieties: astronomical, geographical, natural historical, and cultural. Further, it was to be shared among nations rather than hoarded. Sales of charts were permitted to foreign nationals, astronomical data were combined, and natural historical objects exchanged. The scientific explorers themselves could be granted free range in the world, regardless of the politics of their countries. Both the Americans and the French protected Cook, inspiring Joseph Banks (1743–1820) to suggest it as axiomatic that the "science of two Nations may be at Peace while their politics are at war."[32] Scientific expeditions also tended to unite, temporarily, opposing factions at home. In February 1791, with the French Revolution in full swing, the National Assembly successfully petitioned the king to send out an expedition in search of the lost voyage that had been led by Jean François de Galaup, comte de Lapérouse (1741–1788).[33] In Britain, when naval surveying vessels went out, parties across the board could rejoice in their accomplishments. Such expeditions brought knowledge and potential advantage without entailing the enormous costs of war. The *Beagle* voyage of 1831–1836 was a relatively modest venture, one of many naval surveying voyages, but FitzRoy and Darwin sensed that they were acting in the tradition of Cook.

A similarly enlightened attitude pertained to the object of interest: nature on a global scale. A reified Nature was celebrated and her laws appealed to as the foundation for science and society. This was the Enlightenment vision.[34] On the practical level, there came to be conventions for sharing the new knowledge with the public of all nations. Narratives of scientific travel developed into a popular and established genre. Museums were founded or enlarged to receive the incoming specimens. More cerebral results, such as measurements of latitude and longitude, were accommodated in newly drawn charts, which were sold commercially. The benefits of this new knowledge were thus rendered directly to sponsors of the work, nations or individuals, and to the world. Ideally, there was a two-tier layering of benefits. Reflecting the survival of such a rationale, an English official wrote in 1885 of the work of H.M.S. *Beagle:*

> In 1825, the best charts of the South American coasts which had been made by Spain, or by Portugal, were found inadequate, and it was then that France and England undertook to survey those shores *for the benefit of the world*. The French examined the coasts of Brazil, the English those of Patagonia, Tierra del Fuego, Chile and Peru.[35]

National self-interest and the interest of the whole world were claimed to be in concert.

Finally, the experience of a long and arduous voyage would likely prove to be indelible in the life of the traveler. This was obviously the case for Cook, whose three voyages dominate his biography and who died during the third voyage. Equally truly, if less dramatically, this was also true for Darwin. As his older sister Caroline (1800–1888) said presciently on his arrival back in England, "Now we have him really again at home I intend to begin to be glad he went [on] this expedition & now I can allow he has gained happiness & interest for the rest of his life."[36]

The achievements of Cook's three voyages need to be recounted both because they suggested to his successors in the Admiralty what might be accomplished and because they indicate to us how mixed were the motives, even on scientific voyages. Cook's first voyage in 1768, aboard the *Endeavour*, was directed toward observing the transit of Venus (3 June 1769), an operation that, in concert with about 150 others around the world, would improve the accuracy of the measurement of the solar parallax and thereby knowledge of the distance from the earth to the sun, a quantity that has been sought since antiquity.[37] However, alongside the astronomical charge, originating from the Royal Society of London and effected by the Admiralty, went the government's "secret instructions" to "search for a continent down to latitude 40° south and, if none was found, to go to New Zealand (discovered by Tasman in 1642)."[38] Thus began the British penetration of the eastern portion of what was to be called Australia.[39] As Harry Woolf has put it, "Interest in the voyage to the South Seas was not limited to the transit of Venus alone, for one of the greatest ambitions of the young King (George III was twenty-eight in 1768) was to achieve the exploration of the unknown areas of the globe."[40] The *Beagle* voyage of 1831–1836 operated at a level of reduced scientific interest as compared to Cook's first voyage, though perhaps the substitution of Charles Darwin for Cook's naturalist Joseph Banks might be regarded as at least an equal trade. For Darwin, Banks was a presence. In 1833 he wrote home,

> My first introduction to the notorious Tierra del F was at Good Success Bay & the master of ceremonies was a gale of wind.—This place was visited by Capt. Cook; when ascending the mountains, which caused so many disasters to Mr Banks I felt that I was treading on ground which to me was classic.—[41]

Following the success of his first voyage, Cook was allowed to plan a second. On this second voyage, he chose to prove or disprove the existence of the long-rumored *Terra Australis incognita*. As John Beaglehole has written of Cook, "Plunging south from the Cape of Good Hope and sailing east, he circumnavigated the world, utterly destroying the ancient hypothesis of a great southern continent, and reached latitude 71°10′."[42]

At a philosophical level, there was an affinity between the motivation for Cook's second voyage and the young Darwin's intellectual enterprise. Cook had been testing the notion of a southern continent. The idea had struck geographers as likely since the sixteenth century, in part because the earth's symmetry demanded it: "for in the absence of this tremendous mass of land what, asked Mercator, was there to prevent the world from toppling over to destruction amidst the stars?"[43] In 1756 Charles de Brosses again asked, "It is not possible, that there is not in such a vast sea, some immense continent of solid land south of Asia, capable of keeping the globe in equilibrium in its rotation."[44] Mercator and de Brosses aside, the question of the symmetry of the earth, in regard to the distribution of continents and oceans, was fundamental to Darwin's geology. Whereas Cook in his second voyage had disproved the existence of *Terra Australis incognita* in its expected place in the Pacific basin, John Biscoe (?–1848) in February 1831 identified the continent of Antarctica (the opposite of the Arctic, the *Anti-Arktikos*). Darwin himself remarked, "It is certain that a new continent has been discovered somewhere far South. Perhaps we may be sent in search.—"[45] Still, a satisfactory scientific explanation had yet to be given for the present-day distribution of oceans and continents on the earth's surface. The global perspective of such inquiries remained fresh for Darwin half a century after Cook's second voyage.

Cook's third voyage of 1776–1778 was undertaken in search of a northwestern passage from the Pacific across the North American continent to the Atlantic. It had less relevance for Darwin's experience aboard the *Beagle* than had the first two voyages. Even so, the discovery of Hawaii during the voyage represented a further expansion of European knowledge of the Pacific. As Bernard Smith has written, Cook "helped to make the world one world; not a harmonious world, as the men of the Enlightenment had so ardently hoped, but an increasingly interdependent one."[46] Cook's death in 1779 at the hands of native Hawaiians, while he was in the act of recovering a stolen ship's boat, also represented the perils of travel and of communication across cultures. As a footnote to current scholarly interests, the interpretation of Cook's death has become problematic for anthropologists.[47] To a lesser degree, the contacts between the *Beagle*'s company and the natives of Tierra del Fuego have been subject to renewed interest for historians.[48]

The value of enlightened travel, which Cook's voyages had demonstrated, was clear to contemporaries of several nations. The French expedition (1785–1788) of Lapérouse was sent out with Cook's achievements as its background. An artifact of French intent is the beautiful set of three maps made for the voyage, "showing the projected itinerary for the voyage as well as tracings of the principal European voyages hitherto undertaken"—Cook's voyages being prominent.[49] Louis XVI's own set of these maps, backed in blue silk, is extant. The Spanish expedition (1789–1794) of Alessandro Malaspina was made in a similar grand tradition of enlightenment. While a report of the voyage was not published on Malaspina's return to Spain owing to political circumstances, the voyage itself was highly successful.[50] Alexander von Humboldt (1769–1859) was an important person connecting Cook's tradition to Darwin. It was his account of his 1799–1804 journey to the Americas that stimulated in Darwin a desire to travel.[51] Humboldt had traveled earlier with Georg Forster (1754–1794), a naturalist who had been on Cook's second voyage.[52] Further, Humboldt credited Banks, then president of the Royal Society of London, with assistance in preserving the geological specimens Humboldt had collected. Banks, amid the "political agitations of Europe, has unceasingly labored to strengthen the ties by which are united the scientific [persons] of all nations."[53] Banks the botanist admired Erasmus Darwin's verses; for his part the potter Josiah Wedgwood produced a portrait of Cook in blue jasper.[54] There were thus numerous links between the Cook legacy and Charles Darwin. Indeed, Darwin rather combined aspects of Cook and Humboldt. Like the Cook voyages, the *Beagle* journey was an official naval, primarily seagoing, affair. Like Humboldt as a traveler, Darwin investigated the land and negotiated terms, even on the ship, that allowed him to act with much of the freedom of a private person. Also, like Humboldt, Darwin's desire to travel was strong and separable from destination. Before Humboldt settled on the Americas, he had considered several destinations, including Egypt, India, and the South Pole.[55] Similarly, and fundamentally, Darwin was ready for science, adventure, and travel; South America represented the opportunity of the moment.

In turning to discuss the hold Alexander von Humboldt had on the imagination of the young Darwin, one must consider the avenue of that influence since the two men did not meet until after Darwin's voyage. As Nicolaas Rupke has argued, Humboldt's fame during his life, and since, displays a number of disparate aspects. One must therefore specify which Humboldt is being invoked: the explorer, the founder of physical geography, the statesman, the publicist for European economic expansion, the universalist, and even, more recently, the gay liberal.[56] For Darwin, the Humboldt who mattered was the Humboldt who was author of the narrative of his five-year travels in the Americas. This Humboldt was the explorer and the physical geographer. However, there was a decidedly liberal frame to his travel narrative. Helen Maria Williams (1761–1827), an English woman of liberal sympathies, had translated Humboldt's work from French. A supporter of the Girondin faction during the French Revolution, Williams lived in France from 1790 until 1793 when she was imprisoned. Both before and after the revolution she kept an influential salon in Paris where Humboldt and Aimé Bonpland (1772–

1858), the botanist and Humboldt's traveling companion, were among her guests. Earlier, while still in England, she was also acquainted with the poet Anna Seward (1747–1809), Erasmus Darwin's neighbor and friend, which adds to the sense of the Darwins participating in an intellectual circle of wide diameter. Humboldt's narrative thus arrived on the scene in England with its intellectual pedigree on display.[57] Moreover, Humboldt's book—the *Personal Narrative of Travels to the Equinoctial Regions of the New Continent during the Years 1799–1804*—had ample currency in Darwin's circle. His neighbor in Shropshire, Sarah Owen (dates?), alluded casually to its title; Robert FitzRoy (1805–1865), commander of the *Beagle* (1828–1836), had his own copy; the Cambridge geologist Sedgwick recommended it; and Henslow presented him with a copy before the *Beagle*'s departure.[58] Darwin in turn recommended the book to his Cambridge contemporaries, his cousin Fox and Thomas Campbell Eyton (1809–1880).[59] Darwin also copied sections—it is not known which ones—from Humboldt's description of the first stop on his voyage, the Canary Islands, to read to several friends, to inspire them with the desire to join him in travel.[60] While Darwin had the wish to see the tropics, he enticed his friends with the report that "according to Humboldt Teneriffe is a very pretty specimen."[61]

From reading Humboldt's *Personal Narrative,* one can see what drew Darwin to him, and also where it may have led him. Humboldt described with painstaking accuracy the scenery of the places he visited and infused his descriptions with his own passionate sensibility. As W. H. Brock has said, Humboldt's writings "conveyed urban Britons on a magic carpet to places of extraordinary colour, beauty and strangeness."[62] Humboldt also combined lush description with science, not only in botany, but also, more to Darwin's liking, in geology. Finally, he alternated attractively between discussion of the local and the general.

A few examples of Humboldt's style should make the point. First, consider how sympathetically Humboldt presented his own desire to travel. He wrote,

> From my earliest youth I had felt an ardent desire to travel into distant regions, which Europeans had seldom visited. This desire is the characteristic of a period of our existence, when life appears an unlimited horizon, and when we find an irresistible attraction in the impetuous agitations of the mind, and the image of positive danger."[63]

(He assured the reader that a trip departing from Spain to the Canaries and thence to South America is "often less dangerous than crossing one of the great lakes of Switzerland."[64]) Whatever the dangers, however, Humboldt encouraged the reader to believe that the trip would be worthwhile. In a beautifully sustained passage, he described his own wonder at first seeing the stars of the Southern Hemisphere. Placing himself in the position of the universal traveler, he wrote that the traveler "feels he is not in Europe, when he sees the immense constellation of the Ship, or the phosphorescent clouds of Magellan, arise on the horizon. The heaven, and the earth, every thing in the equinoctial regions, assumes an exotic character." When he finally saw the "Cross of the south" appearing "from time to time between the clouds," Humboldt was moved to conclude, "If a traveller may be permitted to speak of his personal emotions, I shall add, that in this night I saw one of the reveries of my earliest youth accomplished."[65]

Of climbing the volcanic peak at Teneriffe, in the Canaries, Humboldt would write,

> As the temperature diminished, the Peak became covered with thick clouds. The approach of night interrupts the play of the ascending current, which, during the day, rises from the plains towards the high regions of the atmosphere; and the air, in cooling, loses its capacity of suspending water. A strong northerly wind chased the clouds; the moon at intervals, shooting across the vapors, exposed its disk on a firmament of the darkest blue; and the view of the volcano threw a majestic character over the nocturnal scenery. Sometimes the Peak was entirely hidden from our eyes by the fog, at others, it broke upon us in terrific nearness; and, like an enormous pyramid, threw its shadow over the clouds rolling beneath our feet.[66]

Such passages inspire a desire to travel. All the characteristic Humboldtian elements are there: the presence of the traveler as witness, the vivid description of the scene, and the search for what causes may have produced the sight. But Humboldt could swiftly change his tone from reverie to analysis. After concluding his narrative of his excursion to the top of the peak, he turned his attention to the geological structure of the volcano. He suggested questions such as the following:

> Is the conical mountain of a volcano entirely formed of liquified matter, heaped together by successive eruptions; or does it contain in its centre a nucleus of primitive rocks covered with lavas, which are these same rocks altered by fire? . . . Has the central nucleus of volcanoes been heated in its primitive position, and raised up, in a softened state, by the force of the elastic vapours, before these fluids communicated, by means of a crater, with the external air? What is the substance, which, for thousands of years, keeps up this combustion, which is sometimes slow, and

> at other times so active? Does this unknown cause take place in secondary rocks lying on granite?[67]

He regretted that more progress had not been made on these questions, but such ignorance only increased "the desire of the traveller, to see with his own eyes."[68] The peak of Teneriffe was the first active volcano Humboldt had visited; in South America he was to see more. But the scientific interest with which he described the sight would be characteristic of Darwin as well.

On a final point, in the first volume of his *Personal Narrative* Humboldt struck a theme that would resonate with Darwin in several ways that will become apparent in the later chapters of this book. Humboldt's own predilection was for finding large scale connections. As he put it in the introduction, his goal in travel was the straightforward one of making known the countries he had visited, but more than that, "to collect such facts as are fitted to elucidate a science, of which we have possessed scarcely the outline, and which has been vaguely denominated *natural history of the world, theory of the Earth,* or physical *geography*. The last of these two objects seemed to me the most important."[69] Common to these goals, and underlying his science, was the belief that the processes of nature must be viewed from the perspective of the entire world. As he put it,

> It is by isolating facts, that travellers, on every other account respectable, have given birth to so many false ideas of the pretended contrasts, which Nature offers in Africa, in New Holland, and on the ridge of the Cordilleras. The great geological phaenomena are subject to the same laws, as well as the forms of plants and animals. The ties which united these phaenomena, the relations which exist between such varied forms of organized beings, are discovered only when we have acquired the habit of viewing the Globe as a great whole; and when we consider in the same point of view the composition of rocks, the forces which alter them, and the productions of the soil, in the most distant regions.[70]

This "habit of viewing the Globe as a great whole" was to become ingrained in Charles Darwin in the course of the voyage. In his openness to the study of "complex interrelationships of the physical, the biological, and even the human" he was Humboldtian.[71] Darwin's interpretation of a global perspective differed from Humboldt's—Humboldt's isothermal lines were not for Darwin—but there was an underlying unity of vision between the two men.[72]

For Darwin and Humboldt the story regarding influence quickly came full circle. In the summer of 1839, Darwin sent Humboldt a copy of his *Journal of Researches* from the voyage. Darwin's accompanying letter has not been located, but Humboldt's reply is known. No young author could hope for more than what Humboldt gave Darwin:

> You told me in your kind letter, that when you were young, the manner in which I studied and depicted nature in the torrid zones contributed towards exciting in you the ardour and desire to travel in distant lands. Considering the importance of your work, Sir, this may be the greatest success that my humble work could bring. Works are of value only if they give rise to better ones.[73]

Humboldt then proceeded to praise the work of Charles's grandfather Erasmus and to enumerate particular points of interest in Darwin's *Journal:*

> Your work is remarkable for the number of new and ingenious observations on the geographical distribution of organisms, the physiognomy of plants, the geological structure of the earth's crust, the ancient oscillations, the influence of that unusual littoral climate which unites Cycads, hummingbirds, and parrots with forms found in Lapland, on the perpetually green and damp vegetation of *paramos* [a high bleak plain] at sea level, on primeval bones, the possibility of feeding the great pachyderms in the absence of luxuriant vegetation, the ancient cohabitation of animals which are now separated by enormous distances, on the origin of coral islands and the marvellous uniformity of their progressive construction, on the phenomena of glaciers descending to the sea, on the frozen earth covered with plants, on the reason for the absence of forests, on the action of earthquakes and their effects on the surrounding air. . . .
>
> You see, Sir, that I like going over the principal points on which you have enlarged and corrected my views.[74]

Humboldt also characterized Darwin's book as "one of the most remarkable works, that in the course of a long life, I have had the pleasure to see published."[75]

Darwin was hesitant at the strength of the praise ("even a young author cannot gorge such a mouthful of flattery—"), but replied to Humboldt with such information on the coldness of seawater at the Galápagos Islands as would interest him.[76] In his reply Darwin also referred to having copied favorite passages from Humboldt, who, in his own letter, had listed the page numbers of his favorite pages in Darwin's book. Thus, we have close readings by two authors of each other's work. From it Darwin had gained inspiration and approach; Humboldt, the satisfaction of a kindred spirit. Thereafter, as an added benefit, Darwin, when he chose, could put Humboldt's prestige behind his requests for information, as in prodding Henslow toward describing the plant collection from the Galápagos brought home aboard the *Beagle*.[77]

Uncontroversially, this much we have suggested so far: that the voyage of *Beagle* of 1831–1836 should be placed within the tradition of scientific exploration exemplified by James Cook, and that Darwin himself was inspired to visit the tropics by the writings of Alexander von Humboldt. We must now consider the particular mission of the ship on this voyage and Darwin's circumstances within it and, briefly, the broader relations of both to imperial science.

The Admiralty set the mission of the ship's voyage. In his instructions to FitzRoy of 11 November 1831, Francis Beaufort, hydrographer, concluded by referring to the expedition as "devoted to the noblest purpose, the acquisition of knowledge."[78] There were two distinct parts to the *Beagle*'s charge. As Darwin put it in describing the mission to his cousin Fox in a letter written on 19 September 1831,

> The expedition, under the command of Cap FitzRoy, is fitted out principally for completing a survey of the S. parts of S America: The western shores of these parts have been well done by Cap King, under whom Fitzroy went out second in command. We accordingly shall principally work on the Eastern coast of Patagonia from Rio de Plata [*sic*] to St[s] of Magellan.—The second object is to ascertain the longitudes of several places, more accurately than they are at present, & to carry a series of them round the world.—[79]

The "Memorandum" from Beaufort to FitzRoy spelled out the particulars. The longitude of Rio de Janeiro was to be determined anew since authorities were not in agreement, and "as all our median distances in South America are measured from thence, it becomes a matter of importance to decide between these conflicting authorities." More detailed instructions were listed. Then Beaufort turned to the principal task: "To the southward of the Rio de la Plata, the real work of the survey will begin." Again, more detailed instructions are given, with especial attention to the east coast of Tierra del Fuego, and an encouragement, if possible, to survey the Falkland Islands. Beaufort allowed that all "these several operations may probably be completed in the summer of 1833–34." (In fact, the *Beagle* took a few months longer, leaving Tierra del Fuego for the western coast of the subcontinent in June 1834.) Beaufort also had various survey requirements for the west coast. He directed that "the survey should be continued to Coquimbo, and indefinitely to the northward, till that period arrives when the Commander must determine on quitting the shores of South America altogether." Beaufort continued, "If he should reach Guayaquil, or even Callao, it would be desirable he should run for the Galapagos, and if the season permits, survey that knot of islands."[80] H.M.S. *Beagle*'s expert survey of the Galapágos Islands is reproduced in Figure 1.1.

The farther Pacific portion of the circumnavigation was treated perfunctorily in Beaufort's instructions. "But whatever route may be adopted, it should conduct her to Tahiti, in order to verify the chronometers at Point Venus, a point which may be considered as indisputably fixed by Captain Cook's and by many concurrent observations." Further, the *Beagle* was to call at Port Jackson (Sydney). And, if circumstances allowed, he might also "look at the Keeling Islands, and settle their position."[81]

The offhand tone of several of Beaufort's remarks regarding the *Beagle*'s work in the Pacific suggests only that it was not the main purpose of the voyage. Viewed from the perspective of Pacific exploration, the *Beagle*'s accomplishments in the Pacific region were part of a tidying-up operation that lasted from the end of Cook's voyages to the *Challenger* expedition of 1872–1876.[82] By the time the *Beagle* set sail, the outlines of what geographers term the "Pacific basin" were known to the European world. The work that had gone into this effort had been done over a period of almost three hundred years, beginning with Magellan's entrance into the Pacific in 1520. The eastern boundary to the Pacific—the west coast of the Americas—was known earliest, having been identified by the Spanish. The southwestern boundary of the Pacific basin—Indonesia, the Philippines, and parts of the northern coast of Australia—was identified in the sixteenth and seventeenth centuries, by the Dutch among others. The northern and western boundaries of the Pacific were known from Cook's third voyage and by the efforts of Russian explorers. The southern portion of the Pacific was known last, not fully until the 1830s and 1840s. Particularly important here was the work done by French explorers in the reef-dotted area of Polynesia, by the British and French in Australia, and by various nations, including the United States, in Antarctica.

Aaron Arrowsmith's "Chart of the Pacific Ocean," issued in 1798, is a tangible indication of the newly emerging understanding of the geography of the region. Several features of Arrowsmith's work are significant. His chart emphasized accuracy: unknown portions of coastlines were left undefined (as the outline of Australia in Figure 1.2 suggests), routes of previous explorations were shown, natural historical information—birds seen, languages spoken—was included, and the chart was revised repeatedly to reflect current geographical knowledge. One can imagine Darwin aboard the *Beagle* with a current version of Arrowsmith's chart spread out before him, absorbing its information and comparing his own progress to that of previous voyagers. The actual chart is so large (188 × 243 cm) that, if reproduced in miniature, its

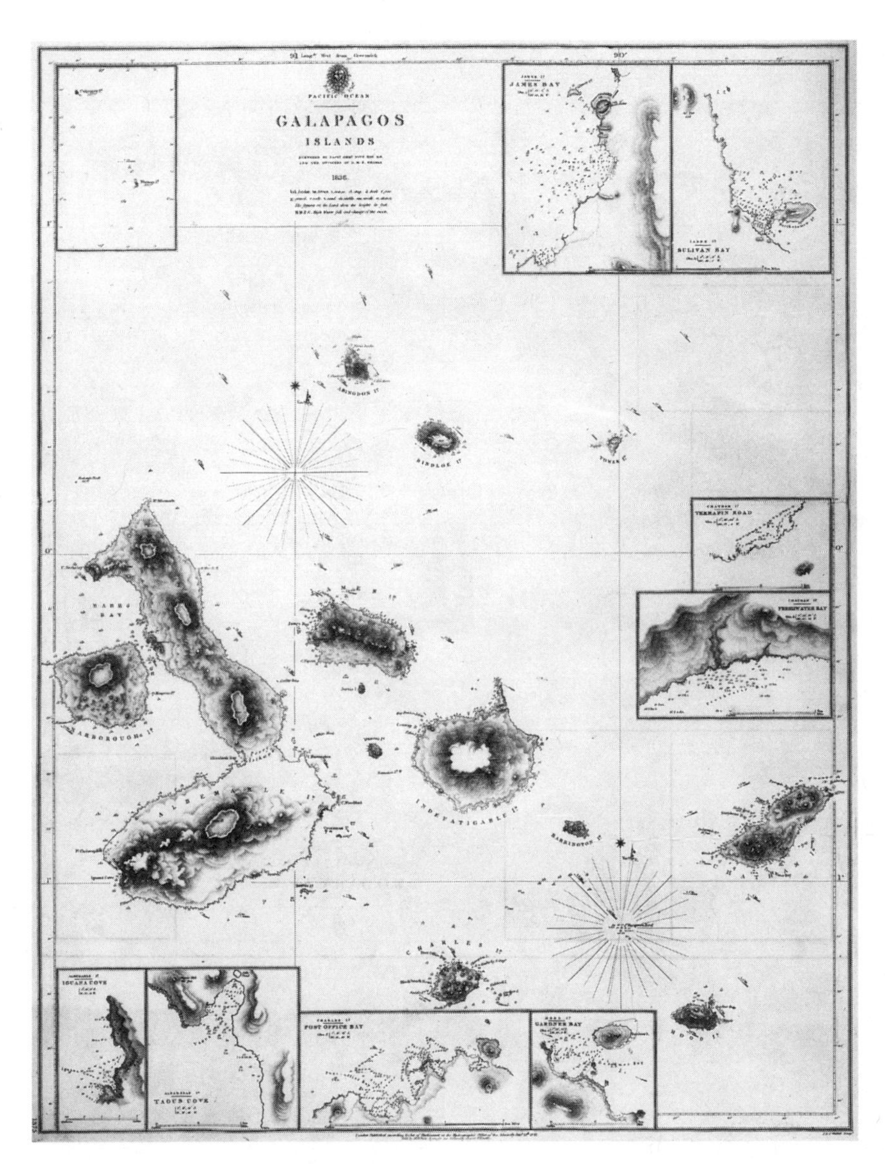

Figure 1.1. FitzRoy's chart of the Galápagos Islands. Courtesy of the Library of Congress.

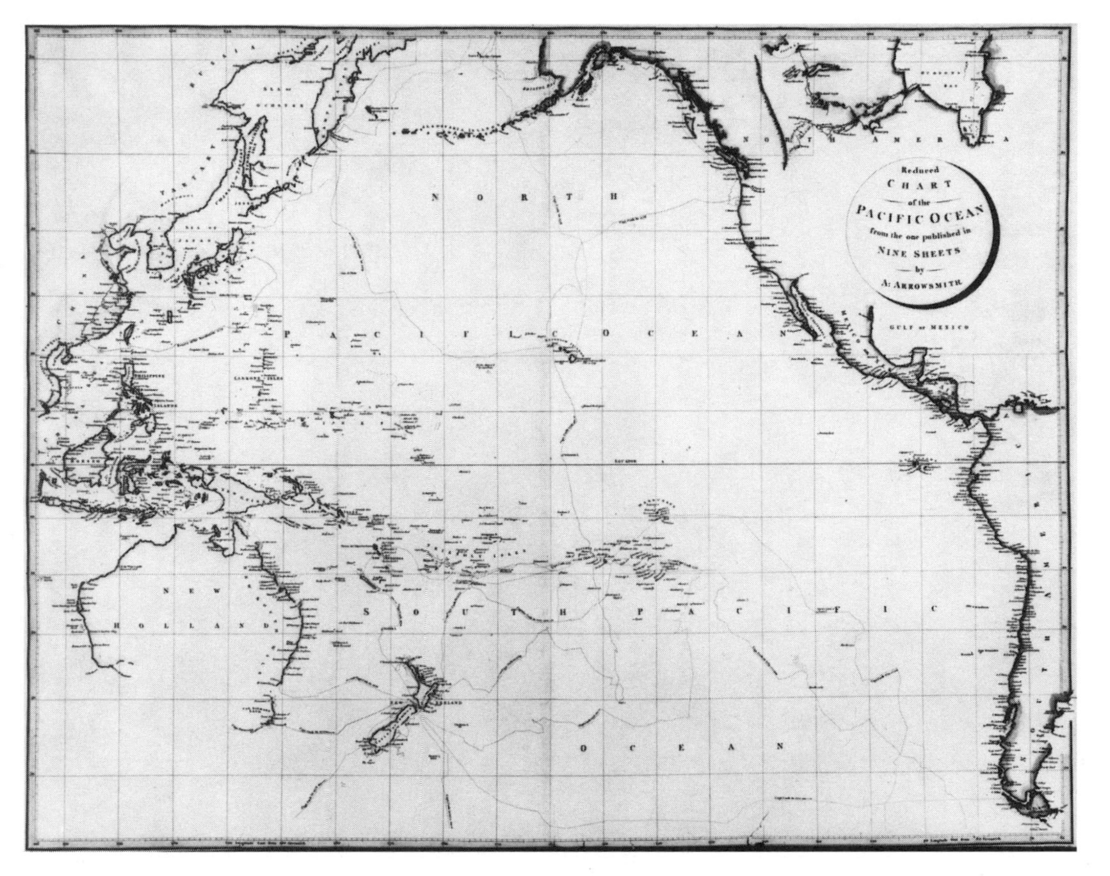

Figure 1.2. Arrowsmith's chart of the Pacific Ocean. Courtesy of the Library of Congress.

detail would be illegible. However, Arrowsmith published a reduced version of the chart, with some of the information from the full-scale version, that conveys the impact of the original. This chart is shown in Figure 1.2.[83]

Beaufort's "Memorandum" is also a useful guide to the practice of the naval surveyor's craft and to the extensive series of scientific observations made during the course of the voyage. Directions are given on such matters as drawing: "a few strokes of a pen will denote the extent and direction of the several slopes much more distinctly than the brush." The practice of naming newly discovered places is discussed, with agreement that "officers and crews . . . have some claim on such distinction," which such names on current maps for Tierra del Fuego as "Isla Wickham" and "Isla Darwin" may document. FitzRoy was required to send back reports "from time to time" to the Hydrographic Office "so that if any disaster should happen to the Beagle, the fruits of the expedition may not be altogether lost." He was also directed to compile navigational instructions for all places visited, a traditional and essential duty. Information was requested on rock types in the shoals in eastern Tierra del Fuego, with the thought that the rock of origin might be related to tidal patterns. Similarly, and here one detects Beaufort's status as a member of the Geological Society of London, information was requested pertaining to the formation of coral reefs. The author of the "Memorandum" was obviously fully aware of recent conjectures by Charles Lyell (1797–1875) concerning the origin of atolls.[84]

The key scientific information to be gathered involved the reading of the ship's chronometers. Following the successful construction of chronometers by John Harrison (1693–1776) in the eighteenth century, longitude could be established by comparing local solar time, obtained by astronomical measurement, with Greenwich time recorded on the ship's chronometers, each hour of difference corresponding to fifteen degrees of longitude. Beaufort required, however, that astronomical means should also be used to check the accuracy of chronometers, that is, by determination of the moon's position relative to the stars. In addition, magnetic readings were to be taken, "not only the magnetic angle, but the dip, intensity, and diurnal variation." The goal of such measurements was "inferring the magnetic curves."[85] Adding further work for officers and crew, meteorological registers were to be kept. Such would include barometer and thermometer readings. The temperature of the sea at the surface was also suggested to be of use in determining the current.

A confluence of chart-making and pursuit of scientific knowledge occurred during the tenure of Francis Beaufort as hydrographer from 1829 to 1855.[86] While the instructions for the *Beagle* were systematic and scientific in a nautical sense, an immediately earlier voyage had held even greater promise. H.M.S. *Chanticleer*, sent out by the Admiralty in 1828, was to determine the "true figure of the earth, and the law of the variation of gravity in different points of its surface."[87] As described after the voyage by the astronomer Francis Baily (1774–1844), the method employed in studying the variation of gravity was to measure "the difference in the number of vibrations made in a mean solar day by the different pendulums at the various stations."[88] Both in its association with scientific societies (the Royal, the Geological, and the Astronomical) and in the charge that measurements be taken "at the nearest practicable point to the antipodes of London," the voyage stood in the tradition established by Cook.[89] Gathering longitude and magnetic measurements worldwide was also part of the *Chanticleer*'s mission. The journey was cut short, however, by the accidental death by drowning of the captain, Henry Foster, on 5 February 1831, while he was at work measuring the differences of longitude on the isthmus of Panama. The *Chanticleer* returned to England by an Atlantic route, arriving in Falmouth on 17 May 1831. It had not completed the mission of circumnavigation. In a sense, H.M.S. *Beagle,* commissioned on 4 July 1831, was able to finish a portion of the work that the more scientifically ambitious mission of the H.M.S. *Chanticleer* could not. On its voyage of 1831–1836, the *Beagle* became the "first full-scale attempt made by the British to plot the course of an entire circumnavigation by marine chronometers."[90] With his marine chronometers, FitzRoy had achieved impressive results. As he wrote with justifiable pride, "The whole chain [of measurements] exceeds twenty-four hours by about thirty-three seconds of time."[91] In accuracy of measuring longitude, that translated to less than a sixth of a degree of arc over the earth's circumference of 360°.[92]

Complementary to this pattern of activity, but distinct from it, were Darwin's own endeavors. By the terms of his negotiations prior to the voyage, he retained the right to quit the vessel when he chose and to retain ownership of his collections. On the official ship's list he appears as a "supernumerary."[93] Because of his independent status, he was able to set his own research agenda. Overall he spent only eighteen months of the five-year voyage aboard ship; for the rest he was on land.[94] In this he was quite like Humboldt, and for both it was family money that gave them freedom. Humboldt inherited money from his mother's estate; Darwin drew on his father's bank account. Darwin's drafts en route amounted to just under £1200 and were in addition to the £600 it took to outfit him for the voyage.[95] Eighteen hundred pounds sterling was then a substantial sum. A single gentleman, with free housing, might live modestly at that time in London for about £200 per annum.[96] Five years on the *Beagle* thus cost Darwin's father about what nine years of modest support in London would have required.

To what extent Darwin's position aboard ship was an irregular one has been the subject of academic discussion. The position of surgeon-naturalist would have been a more regular arrangement, and the person, Robert McCormick (1800–1890), appointed to that position on the *Beagle* soon left ship once he realized that Darwin had supplanted him as naturalist.[97] But in keeping with an amateur and individualistic tradition in British science, other arrangements had also been made in the past (as Cook's voyages would attest), so that Darwin's position aboard ship might be considered a regular sort of irregularity. (Similarly, in regard to collections, instructions varied among voyages as to rights of ownership.)[98] While he was solely responsible for his pursuits, Darwin did not usually work alone. Officers might accompany him on walks, contribute to his collections, or assist him in other ways. For more directed help he turned to his servant, Syms Covington (1816?–1861), whom he hired early in the voyage at a cost of about £60 per annum.[99] While on land, Darwin was often in contact with British residents, some of whom had scientific interests. One instance of a man who befriended him was the diplomat Woodbine Parish (1796–1882), chargé d'affaires at Buenos Aires. Finally, and importantly, he trusted local authorities and guides for advice and assistance at all levels. As he said of the traveling naturalist in his *Journal of Researches,* "He will discover, how many truly goodnatured people there are, with whom he never before had, or ever again will have any further communication, who yet are ready to offer him the most disinterested assistance."[100]

Before leaving the subject of scientific exploration in regard to the voyage of the *Beagle,* something needs to be said on the thorny question of the relationship between science and imperialism.[101] While the terms "imperialism" and "imperialistic" were not ones that the men of the *Beagle* would have used of themselves—indeed, the phrases probably would have drawn a puzzled stare—in the twentieth century the terms have been widely employed.[102] They also carry with them grand intimations that are appropriate to geology.

In discussing imperialism, it is helpful to distinguish among motive, policy, and effect. "Motive" refers to individuals and is a psychological term. "Policy" refers to the concerted planning undertaken by institutions, even if under the direction of an individual. An "effect" is the result of action. Effect can be short or long term, intended or unintended.

Thus, Darwin's motive, which was his desire for scientific adventure and travel, is sufficient to account for his presence on the ship. This is so, even though his presence served other men's motives, such as the desire of FitzRoy for companionship. (The terms of FitzRoy's arrangement with Darwin, premised, as it was, on Darwin's being allowed maximum freedom on shore precluded much economic benefit. FitzRoy had suggested at the end of the first voyage that a geologist be aboard ship to search for metals in Tierra del Fuego.[103] In fact, Darwin never acted as prospector.)

The policy governing the actions of H.M.S. *Beagle* was designed by the Hydrographic Office under Beaufort. The short-term effect of the voyage was chiefly the intended one of increased geographical knowledge. The long-term effect is more difficult to assess, particularly with regard to the question of imperialism.[104]

If one takes a definition of imperialism that emphasizes the role of the state in "extending or maintaining . . . direct or indirect political control over any other inhabited territory," then the voyage can be regarded as an imperial venture in only a limited, though genuine, sense.[105] There was no policy for annexation, and it did not occur. The *Beagle* voyage was hardly part of the sort of land grab that was characteristic of late-nineteenth-century European imperialism. Unlike Cook's first voyage, there was not even a secret instruction that would lead to a new colony. Only one adventure in South America could properly be called imperialistic in the direct sense. By its presence, the *Beagle* was in the position of supporting a reassertion of British claims to the Falkland Islands, where the flag had been raised in January 1833 by H.M.S. *Clio* and H.M.S. *Tyne,* but had been challenged by an Argentinean protest in June of that year.[106] The *Beagle* called on the Falklands in March–April 1833. Thus, it played a role in maintaining a British claim to the Falklands. After leaving South America, the ship called at points already claimed. Australia, for example, was clearly already in the British orbit. Simply by making calls, the ship was, however, maintaining and thereby strengthening ties already established. Interestingly, the Falkland Islands, like many other oceanic islands, were valued as safe harbors more than as potential lands for settlement. In Beaufort's instructions to FitzRoy, for example, he referred to the desirability of rigorously surveying the Falkland Islands "as they are the frequent resort of whalers, and as it is of immense consequence to a vessel that has lost her masts, anchors, or a large part of her crew, to have a precise knowledge of the port to which she is obliged to fly."[107]

But the situation that developed in the Falklands illustrates the gradation between formal and informal, direct and indirect control to which Ronald Robinson and John Gallagher called attention.[108] Thus, for example, whereas the surveyors aboard the *Beagle,* on both its voyages, took only their maps and charts home with them, British sheep raisers were able to come into Tierra del Fuego forty years later, in the 1870s, to settle, with the accurate maps to guide them. Further, British missionaries had a long-time, though intermittent, presence in Tierra del Fuego, dating from the time of the *Beagle*. Thus, the British established trade, settlement, and some degree of control over limited areas, even in the inhospitable lands of the South Atlantic. As Europeans settled in Tierra del Fuego later in the century, the indigenous population declined sharply in number. Surveying voyages obviously contributed to this pattern of change.

By way of comparison to Darwin's experience, both on the *Beagle* and thereafter, one may consider two other Victorians, Henry Thomas De la Beche (1796–1855) and Roderick Murchison (1792–1871), whose geological work in relation to British expansion has been described by Robert Stafford and James Secord.[109] De la Beche was the first director-general of the Geological Survey of Great Britain, established in 1835 as a branch of the Ordnance Survey. Murchison became director-general in 1855, on De la Beche's death. While De la Beche's primary concern was Great Britain, he also arranged for survey work in various British colonies, as indicated in a table provided by Stafford. However, it was Murchison, building on De La Beche's work, who made geology "an agent of British overseas expansion."[110] Further, Murchison sought to put geology to the service of the search for gold. In contrast, Darwin was not involved in official surveys—they had not yet been founded—and he paid almost no attention to searching for metals. Overall, in comparing the careers of De la Beche, Murchison, and Darwin, one sees the strengths of education and institutional attachment: De la Beche and Murchison were educated in a military setting and during portions of their careers served as government scientists. Unlike them, Darwin was university educated and came into geology from that relatively cloistered environment. Darwin's career was as an author. So education, with its associated power to influence motivation, and employment, with the opportunities it offered to form policy, separate Darwin from the other two men. On the question of the effect of their lives and work, however, the differences among the men narrow; there were communalities. Darwin was carried along—literally—by British expansion; the name "Darwin" was left behind at points on charts assembled at the Admiralty Hydrographic Department.[111] His name was thus a marker for British expansion.

For many geologists there was a link between their science and political developments, in the intrinsic requirements of the science of geology itself. Large-scale theorizing requires comparison among rocks of all nations. As Adam Sedgwick (1785–1873) put it in his 1831 presidential address to the Geological Society of London, "The foundations on which we build are so widely parted, that we require nothing less than a free range through all the kingdoms of the earth."[112] Sedgwick had in mind the peace that Europe had experienced in the previous fifteen years, which had permitted so much progress in geology. Peace in Europe also permitted its geologists to work overseas. Striking a similar chord, the next year, in 1832, at the second meeting of the British Association for the Advancement of Science, William Daniel

Conybeare (1787–1857) called for his fellow geologists to look "to India and to Australia . . . no less than to America" for the evidence that would "lay the foundation" for "an enduring and general geological theory." To this charge Conybeare added the instruction that

> England, the mistress of such vast and remote portions of the globe, seems peculiarly called upon to take the lead in this task. And the increased attention to scientific pursuits, now diffusing itself among her military and naval classes,—one of the most favourable characteristics of the age,—promises to supply her every day with observers more and more competent to achieve this honourable duty.[113]

An imperative toward free access could, of course, slide into an imperative toward direct political control. Geologists themselves were aware of this. In writing to Roderick Murchison, the most expansionist of the geologists, the anatomist and paleontologist Richard Owen (1804–1892), after reflecting on the similarity of the jaw of an extinct crocodile collected at a distant site to that of an ancient British form, could jest, "Would this go at all towards establishing a prior claim to the territory and justifying the occupation?"[114] The site was Sind, now in Pakistan, but which in the 1840s was becoming part of British India.

Owen's tongue-in-cheek remark pointed to questions of ownership. With regard to specimens, there were then no limitations on export in the areas touched by Darwin's itinerary. Objects of natural historical interest could be bought, dug, found, labeled, and shipped legally. Substantially all of the material collected during the *Beagle* voyage remains to this day in British hands. Some of Darwin's own specimens are presently on display at the Natural History Museum in London; others are housed in several research-oriented collections. Within geology, the paleontological specimens are also at the Natural History Museum; the majority of the rock specimens are at the Sedgwick Museum at Cambridge.

Yet in the lands where Darwin collected, there was no attempt to leave a permanent British institutional presence. The area now called Argentina, which is a treasure house of vertebrate fossils, is a case in point. Indeed, later in the century a strong tradition of collection and display of vertebrate fossils was established by Argentinean paleontologists, particularly at the Museo de La Plata, in its association with the Universidad Nacional de La Plata, and at the Museo Argentino de Ciencias Naturales in Buenos Aires.[115] In regard to paleontological institutions, then, nothing developed in South America as it did in Greece regarding classical archeology. Presently in Greece, for example, one sees the signs not only of a recovered classical antiquity but also of the institutional presence of nineteenth-century scholarship that originated outside Greece. The British have responsibility for excavations at Knossos, the Americans for the Agora, the French at Delphi, the Germans at Olympia, and so on.

But knowledge is not only specimens. Overall, the cargo of *Beagle* was knowledge, which included but was not limited to specimens. How did Darwin's pattern of research, particularly in geology, fit into the picture, as one tries to relate the quest for scientific knowledge to the extension of political control? Like other geologists, including De la Beche and Murchison, he had need of access to materials worldwide. Two cases were pertinent for Darwin: first, the worldwide distribution of coral reefs, and, second, the dating of strata worldwide by means of study of their fossils. The first enterprise, begun during the voyage, became one of Darwin's key contributions to geology proper. It was tied into naval interests in the sense that coral reefs were frequently the cause of shipwrecks. In 1770 Cook's ship *Endeavour* was nearly destroyed by running aground onto Australia's Great Barrier Reef; in 1827 wreckage from Lapérouse's expedition was raised from a coral reef in the Solomon Islands.[116] Knowing the kinds of reefs, their distribution, and the causes of their formation was thus of practical utility to a maritime nation. But the question of coral reef formation was also of interest in purely geological terms, for an understanding of the present circumstances of reef formation could be applied toward interpreting the presence of remains of coral reefs in the geological record. Darwin's work on reef distribution was utterly dependent on scientific travel. His own interest in the subject was piqued during the *Beagle* voyage. Equally important, only a small part of his work was based on his own observations. The major portion of the information that he used to compose his great map of coral formations derived from the reports of voyagers (Plate 1). The notes to his map run to fifty pages; the chart itself was borrowed from one published in 1835 by the French Dépôt général de la Marine.[117] The reports from commercial as well as naval ventures were useful to Darwin as well: on the Persian Gulf, for example, he relied on charts "lately published on a large scale by the East Indian Company."[118]

On the second score, Darwin's departure on the *Beagle* coincided with the calls by Sedgwick and Conybeare for comparative geological work throughout the world. Darwin viewed with respect the expectations of those at home. He wrote while collecting at Bahía Blanca, which lies southwest of Buenos Aires, in September and October 1832:

> I have been this particular in describing these beds, in which the organic remains occurred.—for the comparison of formations in different parts of the world, which contain animals of equal grade in the chain of nature, seems at present to be much wanted in Geology.—[119]

This instruction was borne out in his collecting and reporting technique throughout the voyage, for he was careful to note at which place and in which stratum he found specimens. Altogether his efforts would combine with similar efforts by other collections to produce increasingly detailed knowledge of the geological record worldwide. Insofar as British influence enabled him, and others, to do this work by securing access to sites, geological science followed the flag. Sometimes, also, help was more concrete, in the form of potentially useful information. While in Peru, Darwin noted that he received a list of exports of saltpeter from Iquique during the years 1831–1834 from Belford Wilson (1804–1858), the British consul general at Lima, who "allowed me to copy this document.—"[120]

Stafford has used the phrase "science as a servant of national expansion" in relation to the work of Murchison.[121] Equally can one speak of national presence, of various degrees, as operating in support of geological science. Darwin worked in an era of colonial expansion by Britain. But, alongside a colonializing, imperializing impulse went geological imperatives, unique to the science but congruent with Enlightenment valuation of knowledge of the world. There was a duality to this vision: one worldwide, one sprung from British soil. This was perhaps inevitable, for every ship must be launched from some port. For the science of geology, those who surveyed the world often came with prior experience with British geology. In turn, the subdivisions within the geological column worldwide would bear the signs of British expansion during the nineteenth century.[122] Darwin's indebtedness to his Admiralty sponsors eventually bore a tangible expression when he turned as author to write the section on geology for the *Manual of Scientific Enquiry.*[123]

Signposts to a Career

What, then, of geological science? What were the signposts that led Darwin to the subject? Clearly there was more to his affiliation than mere curiosity. There is no reason to doubt the story he told in his autobiography, recounted earlier, of wanting to know something of every pebble in front of the Hall door. However, by his own account, he was also curious regarding other aspects of nature, and the reasons why rocks took precedence over insects and birds during the *Beagle* voyage had more to do with education, and a sense of opportunities in the field, than taste and talent. Taste and talent directed him toward natural history, rather than the mathematical sciences, but within the vast domain of natural historical disciplines, there were choices.

What directed Darwin toward geology in particular was his experience at Cambridge under Henslow. The Latin root of "to educate" is *educere*—to draw out—and Henslow drew out Darwin. He did this practically, for it was he who brought the opportunity of the *Beagle* appointment to Darwin, having passed it up for himself. He also did it intellectually, for, being knowledgeable in many aspects of natural history but having a sense of which fields within in it were most ripe for advancement, he particularly redirected Darwin's attention toward geology. This redirection only occurred, however, during Darwin's last year at Cambridge, 1831. The Darwin residue we have of this education is his letter home in April of that year, in which he wrote that "Henslow & other Dons give us great credit for our plan [to visit the tropics]; Henslow promises to cram me in geology."[124] In a larger sense there is residue of Henslow's mentorship in the deference Darwin showed toward him throughout the voyage, worrying at its end that "I look forward with no little anxiety when Henslow, putting on a grave face, shall decide on the merits of my notes. If he shakes his head in a disapproving manner: I shall then know that I had better at once give up science, for science will have given up on me.—For I have worked with every grain of energy I possess.—"[125] The relationship between Henslow and Darwin is one of the more famous in the history of pedagogy. At the university today one still hears the phrase "no Henslows, no Darwins." The relationship was long-standing and close. As early as 1823, when Charles was only fourteen, his elder brother Erasmus wrote to him from Cambridge, praising Henslow's lectures on mineralogy, adding that "this is his first course so that he will have improved by ye time you come up."[126]

When Charles did "come up" to the university five years later, in 1828, Henslow was his teacher formally in lecture (now it was botany, rather than mineralogy), a guide on field excursions, and his sometime tutor. (Darwin's tutor at his own college—Christ's College—was John Graham [1794–1865]. However, in a contemporary letter Darwin noted that he was being tutored for his final examinations by Henslow, a Fellow of St. John's College.[127]) Although Darwin had benefited greatly from his association with Robert Grant and others at Edinburgh University, where he had been enrolled from 1825 to 1827, he had shown no inclination for medicine. His father then sent him to Cambridge to prepare for the Anglican clergy, which required a degree from an English university. Henslow himself was ordained and demonstrated the compatibility of a clerical vocation with the pursuit of science. On a number of grounds, then, there was a match between teacher and student, and an unusual degree of intimacy developed between the two men while Darwin was at Cambridge.

Henslow frequently invited Darwin to dinner and to at-home gatherings. He advised his student on manners as well as academic matters. As the *Beagle* was to depart, and sensing the potential conflicts between Darwin's "own polished mind" and what he might experience aboard ship, Henslow warned Darwin against unfairly judging his new "comrades" on the basis of their "rough surface."[128] For his part, on leaving Cambridge, Darwin felt free to ask Henslow to forward to him a "Stilton Cheese; fit for eating pretty soon.—"[129] Henslow's and Darwin's bond remained strong until Henslow's death in 1861. In the previous year Henslow performed his last significant service to his student. He ensured that Darwin's just-published *Origin of Species* received a fair hearing by the manner in which he chaired the celebrated session of the British Association of the Advancement of Science at which its conclusions were debated.[130]

Consider now the import of Henslow's offer to Darwin to "cram" him in geology in the spring of 1831. Darwin had passed his examinations for the B.A. degree on 22 January 1831 but was required to remain in residence at the university for two further terms before receiving his degree.[131] The 1831 Lent term ran from 13 January through 25 March (the Friday after Palm Sunday); the Easter term ran from 13 April (the eleventh day after Easter) to 8 July (the Friday after the first Tuesday in July).[132] With his examinations over, Darwin was a free man, with more time for reading and planning than before. During the Lent term, he read the recently published *Preliminary Discourse on the Study of Natural Philosophy* by John Herschel, a Cambridge graduate.[133] The book's title may be slightly misleading to the present-day reader since the phrase "natural philosophy" has an antique ring. William Whewell (1794–1866), also a Cambridge man, substituted the word "sciences" in his nearly contemporary work.[134] If Herschel's treatise had been entitled *Preliminary Discourse on the Study of Science,* it would strike a present-day reader more the way it struck Darwin.

Darwin's examinations for his university degree had consisted of six parts: Homer, Virgil, Euclid, arithmetic and algebra, two works by William Paley, and one by John Locke.[135] Against this traditional background, Herschel's essay represented the refreshingly modern spirit of contemporary natural science, and was read by Darwin as an invitation to contribute to the same. Herschel's short book has since become a standard reference in works on scientific method, partly because he spoke in favor of both inductive and deductive approaches: "the successful process of scientific inquiry demands continually the alternate use of both the *inductive* and the *deductive* method."[136] Not surprisingly, Herschel also left ample room for the forming of hypotheses and theories. Darwin's frequent use of the terms "theory" and "hypothesis" in his later work is consistent with Herschel's usage.

Three other features of Herschel's book were of importance to Darwin in 1831. First, the book provided a digest of recent work in science, primarily in the "inanimate" or physical sciences. Herschel included geology among these. He ranked the science next to astronomy in dignity, reached to it for prime examples (as of a *vera causa* to explain the presence of seashells found in strata high above the sea), and took up a number of questions raised by Lyell in the first volume of his newly published *Principles of Geology,* including climatic change and species extinction.[137] Herschel's second theme was the sympathetic manner in which he suggested the value of science to society. He emphasized, "Between the physical sciences and the arts of life there subsists a constant mutual interchange of good offices."[138] He cited numerous instances of practical benefits of knowledge: quinine used in the treatment of malaria, inoculation for smallpox, and the like. In his view such advances were part of the spread of civilization, by which he meant European civilization. Repeatedly, Herschel wrote in favor of the worldwide increase of science and civilization, moving hand in hand. In Latour-like language ("elements of calculation," "variety of combinations"), he described the need for accurate physical data gathered worldwide and with national support.[139] He also wrote favorably of worldwide searches on behalf of geology:

> . . . the same zeal which animates our countrymen on their native shore accompanies them in their sojourns abroad, and has already begun to supply a fund of information respecting the geology of our Indian possessions, as well as of every other point where English intellect and research can penetrate.[140]

(The nativism inherent in the phrase "English intellect" rings false. Herschel's father, the astronomer William Herschel, was born in Hanover, and John Herschel continued to maintain the family's German connections.) In 1833 Herschel would himself journey to the Cape Colony at the Cape of Good Hope in order to survey the stars visible from the Southern Hemisphere[141]; there he and Darwin would meet in 1836. Both men were on a Herschelian progress, science carried forward by the movement of civilization.

The third aspect of the *Preliminary Discourse* that mattered to Darwin was Herschel's encouragement for others, and not only the few, to pursue science. In Herschel's rather democratic view,

> There is scarcely any well-informed person, who, if he has but the will, has not also the power, to add something essential to the general stock of knowledge, if he will only observe regularly and methodically some particular class of facts which may most excite his attention, or which his situation may best enable him to study with effect.[142]

(Emphasis on what the "well-informed" person might do must have struck a reassuring note to Darwin in February 1831. Unlike Herschel, Sedgwick, and Henslow, Darwin had not pursued the bachelor of arts honors degree at Cambridge, with its mathematical tripos examination, but had taken the ordinary or pass degree.[143]) Herschel then went on to instance meteorology and geology as subjects that should appropriately be cultivated.

The impact of Herschel's book on Darwin was considerable. In February 1831 he advised his cousin Fox regarding the book: "get it directly."[144] Later in life, in his autobiography, Darwin would join Herschel's *Preliminary Discourse* with Humboldt's *Personal Narrative* by saying that reading them "stirred in me a burning zeal to add even the most humble contribution to the noble structure of Natural Science."[145] Herschel's book shared with Humboldt's a worldwide perspective, or, as Herschel put it, "that complete acquaintance with our globe as a whole, which is beginning to be understood by the extensive designation of physical geography."[146] Further, both books, the one by praise and the other by example, accented the role of the scientific traveler. Herschel wrote without reference to Humboldt, but his point of view complemented Humboldt's. During the winter term, Darwin began devising a way to visit the Canary Islands, acting in emulation of Humboldt. In addition to reading and planning while at Cambridge in 1831, Darwin attended Henslow's botany lectures during the Easter term, as he had already done in 1829 and 1830. He was by now Henslow's prodigy, somewhat to the annoyance of his fellow students.[147]

Thus, by mid-1831, though only twenty-two years old, Darwin was more than a novice in natural history. Among other pursuits, he had during his school years, together with his older brother, kept a cabinet of mineralogical specimens. Speaking of his collecting he later wrote, "With respect to science [while at school], I continued collecting minerals with much zeal, but quite unscientifically—all that I cared for was a new *named* mineral, and I hardly attempted to classify them." In those years his older brother Erasmus was already taking a serious interest in chemistry at Cambridge, reporting back to Charles with suggestions for improving the laboratory at home.[148] Four years later, during his second year at Edinburgh, in 1826–1827, Darwin attended the extracurricular course given by Robert Jameson (1774–1854) in natural history. Edinburgh did not have the charm for Darwin that Cambridge did later: he was burdened with medical studies that had little appeal, and instruction was in a large lecture setting rather than the more varied, albeit expensive, Cambridge method centered on individual tutorials.

In his autobiography Darwin was harsh in his judgment of the value of Jameson's lectures, writing, "The sole effect they produced on me was the determination never as long as I lived to read a book on Geology or in any way to study the science."[149] Darwin went on to describe a particularly unhappy memory of Jameson lecturing on a trap-dyke—a dyke of igneous rock—in the Salisbury Crags in Edinburgh, ascribing it to sediments filled in from above, "adding with a sneer that there were men who maintained that it had been injected from beneath in a molten condition."[150] Darwin's reference was to Jameson as an exponent of the "Neptunist" views of Abraham Gottlob Werner (1749–1817). Werner had credited deposition from aqueous solution (hence, "Neptunist") rather than the earth's interior heat (hence, "Plutonist") in accounting for various phenomena. On the interpretation of trap-dykes, Jameson's views were on the wane by the 1820s; the influential *Outlines of the Geology of England and Wales* by Conybeare and William Phillips (1775–1828) cited Sedgwick as declaring the "evidence" of the "igneous formation" of the trap of Northumberland "as complete."[151] Oxford's Charles Daubeny (1795–1867), also writing in the 1820s, attempted to capture the range of opinion by a "geological thermometer" he had drawn up (Figure 1.3).[152] The top of the scale is labeled the "Plutonic Region," the middle the "Volcanic Region," the bottom the "Neptunian Region." Use of actual thermometers was, by then, on the rise in Britain, and the familiar device served for an apt metaphor.

Of course, Darwin, writing in his autobiography, was "right": subsequent opinion supports the igneous interpretation of trap-dykes in the Salisbury Crags. However, there is some background missing in his account. Darwin's grandfather Erasmus had offered an interpretation of the earth that emphasized igneous forces operating in the earth's interior. This point will be considered in the next chapter. Further, as James Secord has shown, the tradition deriving from the work of James Hutton (1726–1797), of which Erasmus Darwin was a part, was well represented in his grandson's Edinburgh education in the person of the Thomas Charles Hope (1766–1844). In his first year at Edinburgh Charles Darwin had attended Hope's lectures on chemistry in which Huttonian views of earth processes were taught as against those of Werner.[153] Thus, Charles received a balanced presentation of geological views at Edinburgh, and presumably he was persuaded of the Vulcanist viewpoint. However, as Secord and others have demonstrated, Darwin received from Jameson a grounding in the fundamentals of mineralogy, as well as some pointers in field geology.[154] Darwin's own copy of Jameson's *Manual of Mineralogy* is one of the most heavily annotated in his library, bearing more than twenty-five-hundred words of comment.[155] Darwin appears to have used Jameson's *Manual* while in Edinburgh, possibly while seated in lecture, and certainly while at the Edinburgh museum of which Jameson was keeper. Thus, for example, one of Darwin's annotations in the *Manual* reads, "Case 2nd of the Museum begins here."[156]

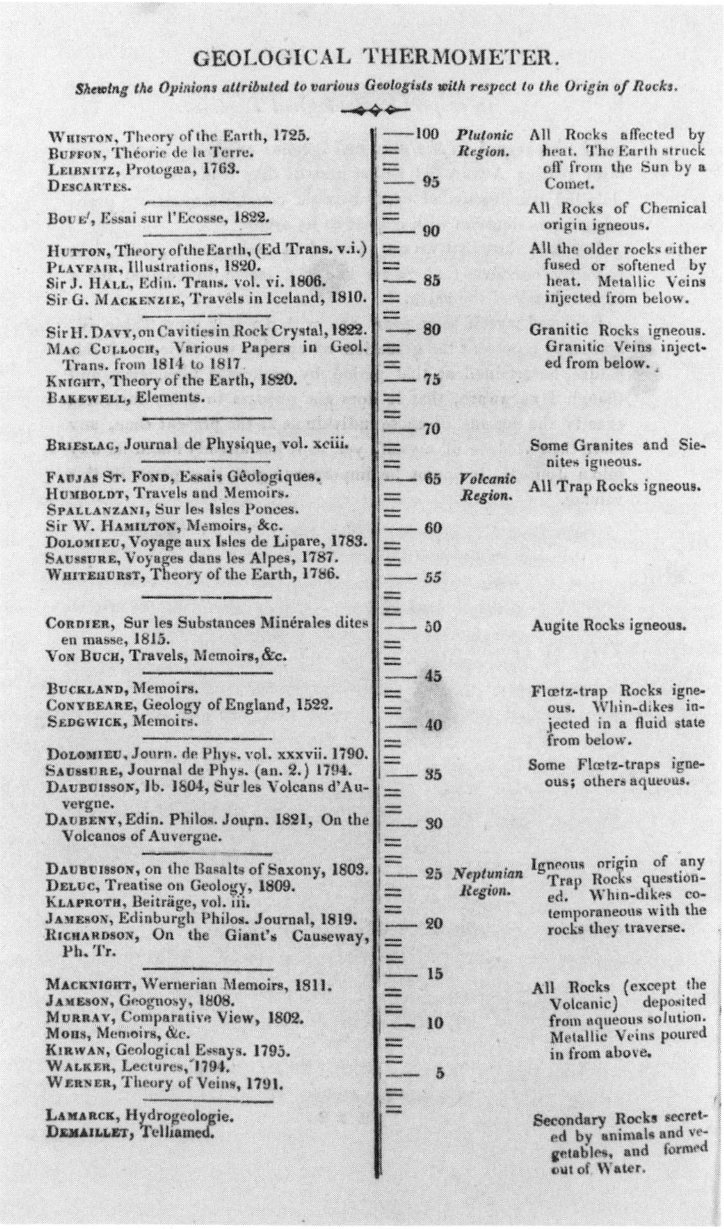

GEOLOGICAL THERMOMETER.

Shewing the Opinions attributed to various Geologists with respect to the Origin of Rocks.

Geologists	Scale	Region	Opinions
Whiston, Theory of the Earth, 1725. Buffon, Théorie de la Terre. Leibnitz, Protogæa, 1763. Descartes.	100 95	*Plutonic Region.*	All Rocks affected by heat. The Earth struck off from the Sun by a Comet.
Boue', Essai sur l'Ecosse, 1822.	90		All Rocks of Chemical origin igneous.
Hutton, Theory of the Earth, (Ed Trans. v.i.) Playfair, Illustrations, 1820. Sir J. Hall, Edin. Trans. vol. vi. 1806. Sir G. Mackenzie, Travels in Iceland, 1810.	85		All the older rocks either fused or softened by heat. Metallic Veins injected from below.
Sir H. Davy, on Cavities in Rock Crystal, 1822. Mac Culloch, Various Papers in Geol. Trans. from 1814 to 1817. Knight, Theory of the Earth, 1820. Bakewell, Elements.	80 75		Granitic Rocks igneous. Granitic Veins injected from below.
Brieslac, Journal de Physique, vol. xciii.	70		Some Granites and Sienites igneous.
Faujas St. Fond, Essais Géologiques. Humboldt, Travels and Memoirs. Spallanzani, Sur les Isles Ponces. Sir W. Hamilton, Memoirs, &c. Dolomieu, Voyage aux Isles de Lipare, 1783. Saussure, Voyages dans les Alpes, 1787. Whitehurst, Theory of the Earth, 1786.	65 60 55	*Volcanic Region.*	All Trap Rocks igneous.
Cordier, Sur les Substances Minérales dites en masse, 1815. Von Buch, Travels, Memoirs, &c.	50		Augite Rocks igneous.
Buckland, Memoirs. Conybeare, Geology of England, 1522. Sedgwick, Memoirs.	45 40		Flœtz-trap Rocks igneous. Whin-dikes injected in a fluid state from below.
Dolomieu, Journ. de Phys. vol. xxxvii. 1790. Saussure, Journal de Phys. (an. 2.) 1794. Daubuisson, Ib. 1804, Sur les Volcans d'Auvergne. Daubeny, Edin. Philos. Journ. 1821, On the Volcanos of Auvergne.	35 30		Some Flœtz-traps igneous; others aqueous.
Daubuisson, on the Basalts of Saxony, 1803. Deluc, Treatise on Geology, 1809. Klaproth, Beiträge, vol. iii. Jameson, Edinburgh Philos. Journal, 1819. Richardson, On the Giant's Causeway, Ph. Tr.	25 20	*Neptunian Region.*	Igneous origin of any Trap Rocks questioned. Whin-dikes cotemporaneous with the rocks they traverse.
Macknight, Wernerian Memoirs, 1811. Jameson, Geognosy, 1808. Murray, Comparative View, 1802. Mohs, Memoirs, &c. Kirwan, Geological Essays. 1795. Walker, Lectures, 1794. Werner, Theory of Veins, 1791.	15 10 5		All Rocks (except the Volcanic) deposited from aqueous solution. Metallic Veins poured in from above.
Lamarck, Hydrogeologie. Demaillet, Telliamed.			Secondary Rocks secreted by animals and vegetables, and formed out of Water.

Figure 1.3. Daubeny's geological thermometer. By permission of the Syndics of Cambridge University Library.

Jameson's book is, in a sense, two books in one. Its full title tells the story: *Manual of Mineralogy: Containing An Account of Simple Minerals, and Also A Description and Arrangement of Mountain Rocks.* In the first part of the book (the "account of simple minerals") Jameson classified rocks according to the system established by Friedrich Mohs (1773–1839). (This is perhaps what Darwin referred to when he wrote that he did not "classify" his childhood rock collection.) Jameson's approach was comprehensive (gases were treated as well as solids) and systematic. For each solid, the crystalline form, the degree of hardness (on an established scale, with "1" being talc and "10" diamond), and the specific gravity were listed.[157] In addition, its geognostic (geological) and geographical occurrences were listed, along with its utility to humans. The information is encyclopedic, modern in its chemistry, and fascinating in its inclusion of natural historical particulars (for example, which buildings were made out of which marble). However, if presented in lectures in the catalogue-like form of the book, the material could well have struck a student as tedious.

The second, and shorter, part of Jameson's book was essentially geological and directed, as was the first part, toward students. It bears the subtitle "Description and Arrangement of Mountain Rocks." The phrase "mountain rocks" is used to indicate "those mineral masses of which the greater portion of the crust of the Earth is composed," including the five classes of "Primitive, Transition, Secondary, Alluvial and Volcanic."[158] There is brief discussion of structural points, as in Jameson's statement, "On a general view, we say the globe is composed of formations; formations of beds; beds of strata; and such strata as are slaty, of layers, or slates."[159] In the same passage, Jameson also suggests what must be looked for in assessing strata in a series of definitions, as, for example, on the meaning the term "dip":

> *The dip* is the point of the compass towards which the stratum inclines. If we know the dip, the direction is given, because it is always at right angles to it. Thus, if a stratum dip to the east, its direction must be north and south; if it dip to the north its direction must be east and west. But we cannot infer the dip from the direction, because a stratum, whose direction continues the same, may dip in opposite directions; thus, a stratum ranging from north to south may dip either to the east or to the west.[160]

(Later the term "strike" would substitute for Jameson's phrase "direction of strata.") Jameson's book also has a brief discussion of fossil petrifactions.[161] What is noticeably lacking in the book are geological maps, or, indeed, diagrams of any kind. It has a colorless, featureless appearance, noticeable in comparison to the well-illustrated books that were becoming associated with geology. Still, Jameson clearly exposed to Darwin the vocabulary and facts of mineralogy, presented against a background of museum specimens. Further, he introduced some of the concepts of the practicing field geologist. What attendance in Jameson's large lecture class did was to inform Darwin with regard to the subject; what it did not do was to recruit him to its pursuit. That happened at Cambridge.

Cambridge did not represent the first nor the last time that favorable circumstances would combine with Darwin's ambition to grant him a second chance. Before he matriculated at Cambridge, he had to refresh the knowledge of Greek and Latin he had gained at school.[162] Possibly there was also some review of knowledge acquired at Edinburgh after his examinations at Cambridge. While a student at Edinburgh, Darwin's own extracurricular researches had led him to the study of marine invertebrates, a convenient subject given Edinburgh's location by the sea.[163] At Cambridge Darwin's field collecting had been directed chiefly toward insects. Attendance at Henslow's botanical lectures had not produced a corresponding interest in plant collecting. But before he had been professor of botany, Henslow had been professor of mineralogy at the university. He had established his own reputation as a geologist with his magnificent study of the island of Anglesey, which is separated from the mainland of Wales by the Menai Strait.[164] In 1822 Sedgwick, who held the Woodwardian chair in geology, accepted nearly one thousand specimens from Henslow's Anglesey fieldwork into the Woodwardian collection.[165] Through his close association with Sedgwick, Henslow remained abreast of progress in the subjects of mineralogy and geology, even after his own research interests and teaching responsibilities shifted to botany. In short, he was well informed, and he had Darwin's confidence. Thus, when in 1831 Henslow offered to "cram" his student in geology, one can be sure that Henslow thought the subject important, and particularly appropriate for a young man seeking to travel. In notes made a few years later, Darwin wrote, "In the Spring Henslow *persuaded* me to think of geology."[166]

Some of the details of this cramming, or persuasion, are known precisely; some are not. It is not certain, for example, how much contact Darwin had with Sedgwick while in residence at the university. Sedgwick completed what Darwin called his "eloquent and interesting" lectures before the end of March, and Darwin's interest in geology was aroused only in the spring.[167] In his autobiography Darwin said he did not attend Sedgwick's lectures ("Had I done so I should probably have become a geologist earlier than I did"), but two contemporaries, writing in the 1880s, remember him in connection with them.[168] Possibly Darwin attended a few of the lectures at the tail end of Sedgwick's lecture course in 1831, after the examinations in January. In what was then the Cambridge system, there would have been no reason not to do so since attendance at lectures was voluntary. There were no examinations attached to specific lecture courses. More importantly, however, a lecture hall was not the only venue for instruction. Sedgwick directed students in the context of field excursions and was present at Henslow's Friday evening gatherings of scientific men in Cambridge. Therefore, Darwin, who was introduced to Sedgwick by Henslow with the intention of acquainting him with geology, had some exposure to the subject at Cambridge through Sedgwick. Fortunately, printed outlines of Sedgwick's lectures exist, so that his orientation toward the subject is known.[169]

Was it odd for Henslow to propose geology to Darwin so late in his career at the university? Had there been a degree program in the subject, it surely would have been. But even Sedgwick's lectures were for one term only. Without an established way into the subject, each major British geologist of the period followed an individual path. Sedgwick regarded himself as little prepared in geology when in 1818 he assumed a professorship in the subject, though immediately afterward he began to remedy that deficiency by beginning serious fieldwork.[170] University studies provided only partial training, and then only for some people.

During this period those interested in geology at Oxford and Cambridge were working to establish a place for it within the curriculum of their institutions. They sought alliances with more established areas of study and recruited able students to their disciplines by lecture, demonstration, and field study. Nicolaas Rupke has described the work of William Buckland (1784–1856) at Oxford to ally geology with classical and historical scholarship.[171] In parallel fashion Brian Dolan has shown how space was made for mineralogy at Cambridge by the efforts of Edward Daniel Clarke (1769–1822), in 1808 the first holder of the university's chair in that subject. Clarke argued persuasively to large audiences that mineralogical analysis was necessary to interpret the physical remains of Greek antiquity.[172] Mineralogy and, by extension, geology were thus pulled into the curriculum by virtue of their utility to classical studies.

At Cambridge in the 1830s geology was better established than a decade earlier, yet it was still without a prescribed sequence of courses. There was thus the difficulty of attracting new practitioners, which made the question of prior preparation negotiable. In 1831 Henslow and Sedgwick saw in Charles Darwin a possible recruit to a discipline of immense potential. Darwin was young, physically robust, educated to a Cambridge standard, and financially secure. Not unlike Charles's father, Henslow and Sedgwick would have seen the only danger to a career for Charles as coming from that scourge of their moral reckoning—idleness, a condition to which Charles's elder brother Erasmus was indeed succumbing. (Sedgwick made this point in 1835 of Darwin, after the *Beagle* voyage was well underway: "There was some risk of his turning out an idle man, but his character will be now fixed."[173]) Further, Charles Darwin's earlier experiences would not be wasted in the new field of geology. He came to Sedgwick with a stock of knowledge: some chemistry and mineralogy, algebra and geometry, as well as natural historical knowledge generally. (The benefit of imported knowledge would have been clear to Sedgwick. His mathematical training at Cambridge—he was fifth Wrangler on graduation in 1808—gave him greater facility with structural geology.[174]) In addition to his own knowledge, Darwin had family tradition to call on. If the Darwin side of the family stood for medicine and university degrees, the Wedgwood side of the family had produced science in an innovative experimental amateur tradition. In the course of canal-building to supply transport for the goods from his pottery, grandfather Josiah had produced an illustration of geological strata in a private letter of 1767.[175] More publicly known would have been such Wedgwood activities as using a pyrometer for estimating high temperatures, one such instrument being part of Erasmus's and Charles's boyhood laboratory.[176] Sedgwick would have seen in Darwin someone worth taking on.

Darwin wrote to Henslow on 11 July, "I have not heard from Prof: Sedgwick, so I am afraid he will not pay the seven [Severn] formations a visit.—I hope & trust you did your best to urge him:—"[177] Clearly Darwin was already an enthusiast. As he wrote in the same letter to Henslow, "As yet I have only indulged in hypotheses; but they are such powerful ones, that I suppose, if they were put into action but for one day, the world would come to an end.—"[178] To Charles Thomas Whitley (1808–1895), his friend from Shrewsbury and Cambridge and a participant in reading holidays in Wales, he wrote the next day,

> I am now mad about Geology & I daresay I shall put a plan which I am now hatching, into execution sometime in August, viz of riding through Wales & staying a few days at Barmouth on my road.—[179]

When he wrote to Whitley, who was then seaside at Barmouth, Darwin had in mind combining a number of activities—the pursuit of his newly found science, the possible visit of Sedgwick, and a holiday riding through the mountains of North Wales. Very probably Darwin was also happy to report to the mathematically minded Whitley that (as he had said of himself three years earlier) he, whose "noddle [*sic*] is not capacious enough to retain or comprehend Mathematics," had found a suitable science, to augment his "Beettle [*sic*] hunting & such things," and was thereby not "infirm of purpose."[180]

In his 11 July 1831 letter to Henslow, Darwin referred to his having received a clinometer, an instrument for measuring inclinations from the horizontal, used by geologists for measuring vertical angles, usually of dip. It had been made especially for him, seemingly after Henslow's design, by the firm of George and John Cary, the London instrument makers. To become familiar with its use, Darwin had "put all the tables in my bedroom, at every conceivable angle & direction[.] I will venture to say I have measured them as accurately as any Geologist going could do.—"[181] Darwin's clinometer included a compass, used for measuring direction of strata, or strike (Figure 1.4).

Thus prepared for the field, Darwin set out to explore the environs of Shrewsbury. Michael B. Roberts has published the notes for one of these excursions.[182] It was to Llanymynech Hill, sixteen miles northwest of Shrewsbury. On the southern face of the hill is a limestone quarry, which provided excellent exposures on which to measure the direction of strata by the compass, and their inclination with respect to the horizon using the clinometer. As Roberts's analysis indicates, Darwin was still relying on terminology ("direction" and "inclination") that he had learned from Jameson. Sedgwick's terminology—"strike" rather than "direction," "dip" rather than "inclination"—was not used. (The term "strike" derived from the usage of German miners. In his earlier notebooks, Sedgwick used the word "range," but by the early 1830s he used "strike."[183] Henslow used the term "direction" in his Anglesea article.) What is most impressive about the excursion to Llanymynech is that it shows Darwin struggling in the field to gain competence in a discipline in which he was a neophyte. Also, in mid-July 1831 Darwin attempted to make a geological map of Shropshire ("dont find it so easy as I expected"), and several of his attempts have survived, one of which includes a set of notes (Figure 1.5).[184] Thus, both in field exploration and in recording geological information on a map, Darwin was endeavoring to make his way.

By the end of July 1831, after he finished some writing, Sedgwick did indeed make his own plans for the summer field season. And, as he promised to Henslow, he took Darwin along for the start of his work in early August. Unbeknownst to Sedgwick and Darwin, the trip would play a role in larger events for each of them. For Darwin it would provide a valuable introduction to fieldwork at a high level of complexity. When the *Beagle* offer came to Darwin unexpectedly at the end of August, he knew that he had been briefly but well trained in the field. For Sedgwick his season of work from August through early October in the contorted older rocks of Wales was the beginning of several years' labor that would establish what he came to call the "Cambrian system." In short, both Sedgwick and Darwin were on the point of moving from ordinary to extraordinary careers.

Existing field notes from both men and some letters allow us to interpret

Figure 1.4. Darwin's clinometer. © English Heritage Photo Library.

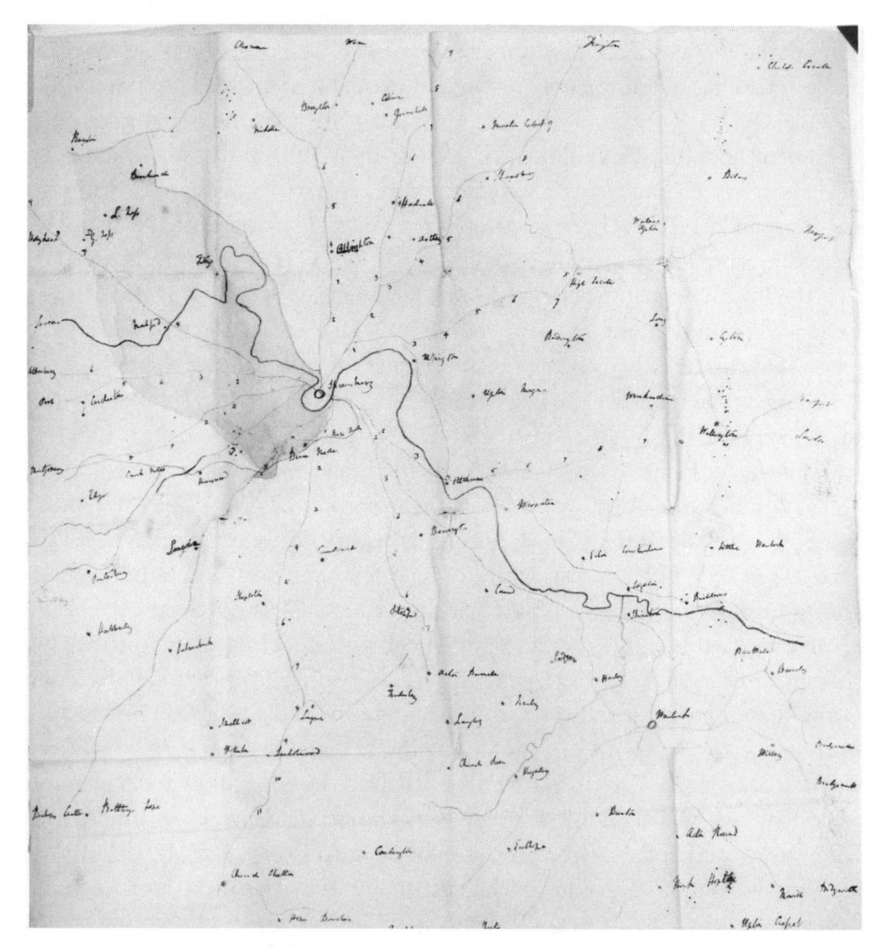

Figure 1.5. Darwin's 1831 geological map of Shropshire. By permission of the Syndics of Cambridge University Library.

the Sedgwick-Darwin excursion. These contemporary documents are important because in his later writing, and especially his autobiography, Darwin wrote retrospectively of what he and Sedgwick failed to see in North Wales—the clear signs of glacial action.[185] In August 1831, the glacial hypothesis had not yet been advanced, and Sedgwick, the master, and Darwin, the apprentice, interpreted the terrain without the benefit of that theory.

In August 1831 Darwin brought with him to Wales prior experience in the countryside, for Wales, lying close to his childhood home of Shrewsbury, was the site of numerous earlier excursions taken for the sake of pleasure. Wales offered mountains and the sea, and, with the Welsh language still spoken in many parts, it was sufficiently foreign as to seem remote from everyday life. (Peter Lucas has identified ten known and probable visits by Darwin to Wales prior to 1831.)[186] Darwin also came to the field with the usual geological equipment, including a geological hammer (its weight in dispute) and his recently purchased clinometer. From an offhand remark by Sedgwick, it appears that Darwin may have been accompanied for at least part of the trip by a "man," presumably a servant or assistant.[187] Darwin also brought to Sedgwick the hospitality of his father's house, for Sedgwick stayed at the Mount, the Darwin family home, en route to Wales. Darwin's agenda, however, was open. He wanted only to become experienced in operating in the field.

Sedgwick's agenda for the later part of the summer of 1831 was settled. It was to explore the older rocks in North Wales. As he put it in a letter to Murchison, "I spent . . . two days at Shrewsbury, and finally entered North Wales on the 5th of August. . . . As the greywacké hills continued in cloud I crossed to the vale of the Clwyd, hoping at least to do some work among the secondaries."[188] His interest in the older rocks was part of a larger scheme among English geologists to establish the order of strata in their own country and to correlate these strata with those abroad. The geological maps of William Smith (1769–1839), and then of Greenough, provided the commonly drawn basis for interpretation. Conybeare and Phillips (1775–1828) reproduced a version of Greenough's map in their *Outlines of the Geology of England and Wales*, Part I (Figure 1.6).[189] Darwin himself owned a portable version of this map (Figure 1.7). By 1828 there had been agreement that Sedgwick would collaborate with Conybeare to produce Part II, devoted to "A Description of the older rocks, throwing the transition and primitive classes together."[190] As Greenough's map indicated, Wales was among the areas of Britain with older rocks, and as Conybeare remarked, "Wales is the most unknown, and from all its local circumstances the most difficult."[191] Sedgwick's work in Wales was thus firmly set as part of a larger objective within the discipline. Henslow's work in Anglesey had been part of this same endeavor.

The way in which Darwin worked with Sedgwick is known from their later accounts and from contemporary field notes. After leaving Darwin's family home on 5 August, they moved north and west from Shrewsbury toward the sea. Darwin recalled that "after a day or two [Sedgwick] sent me across the country in a line parallel to his course, telling me to collect specimens of the rocks, and to note the stratification. In the evening he discussed what I had seen."[192] Darwin's field notes contain clues as to the geology he was practicing. They are especially useful in assessing his work since he was sometimes traveling with Sedgwick and sometimes without. Darwin's notes from his August trip suggested how quickly he was learning; they are of a different or-

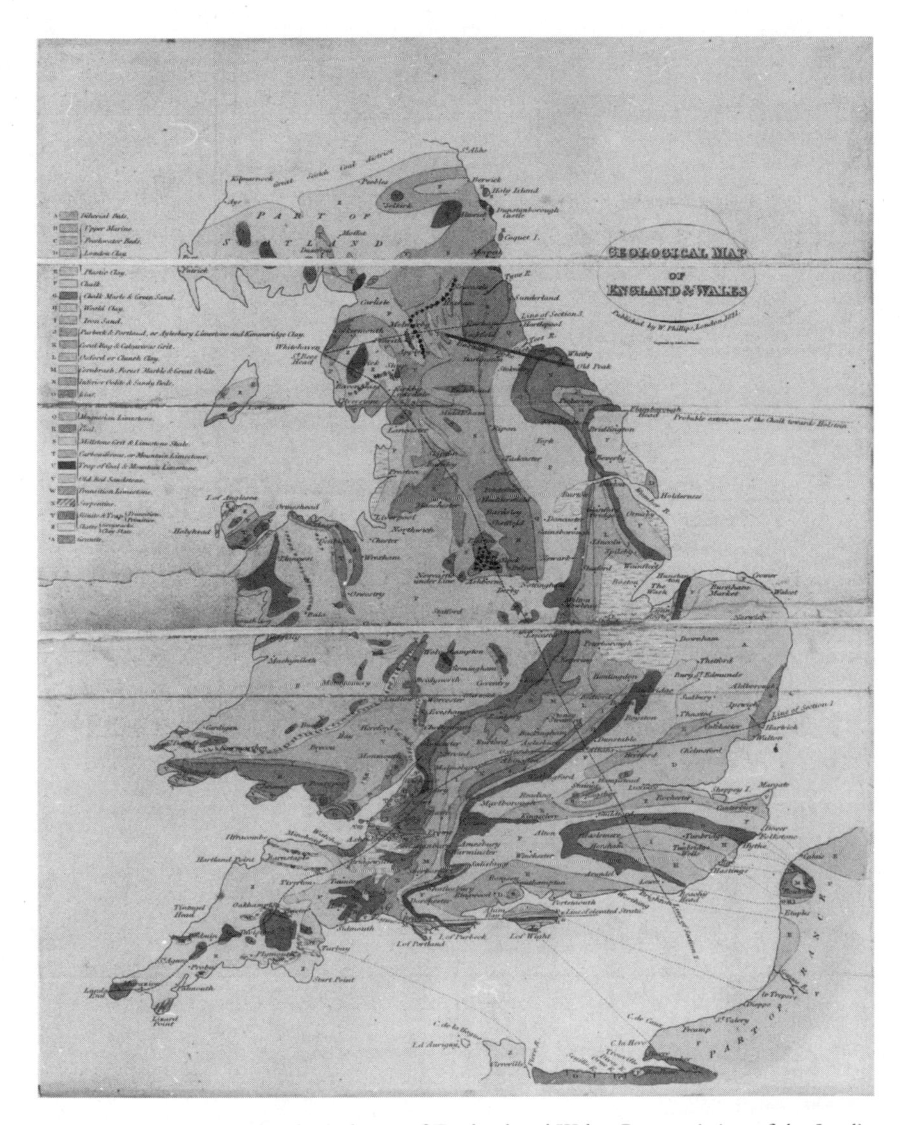

Figure 1.6. Greenough's geological map of England and Wales. By permission of the Syndics of Cambridge University Library.

der of sophistication from his notes on his early fieldwork at Llanymynech.[193] In the first page of his field notes Darwin used Sedgwick's preferred term "striking."[194] He also used Buckland's short-lived English geological term "diluvium," meaning of diluvial formation, to describe "great irregular masses of sand, loam, and coarse gravel," supposedly formed by a catastrophic flood.

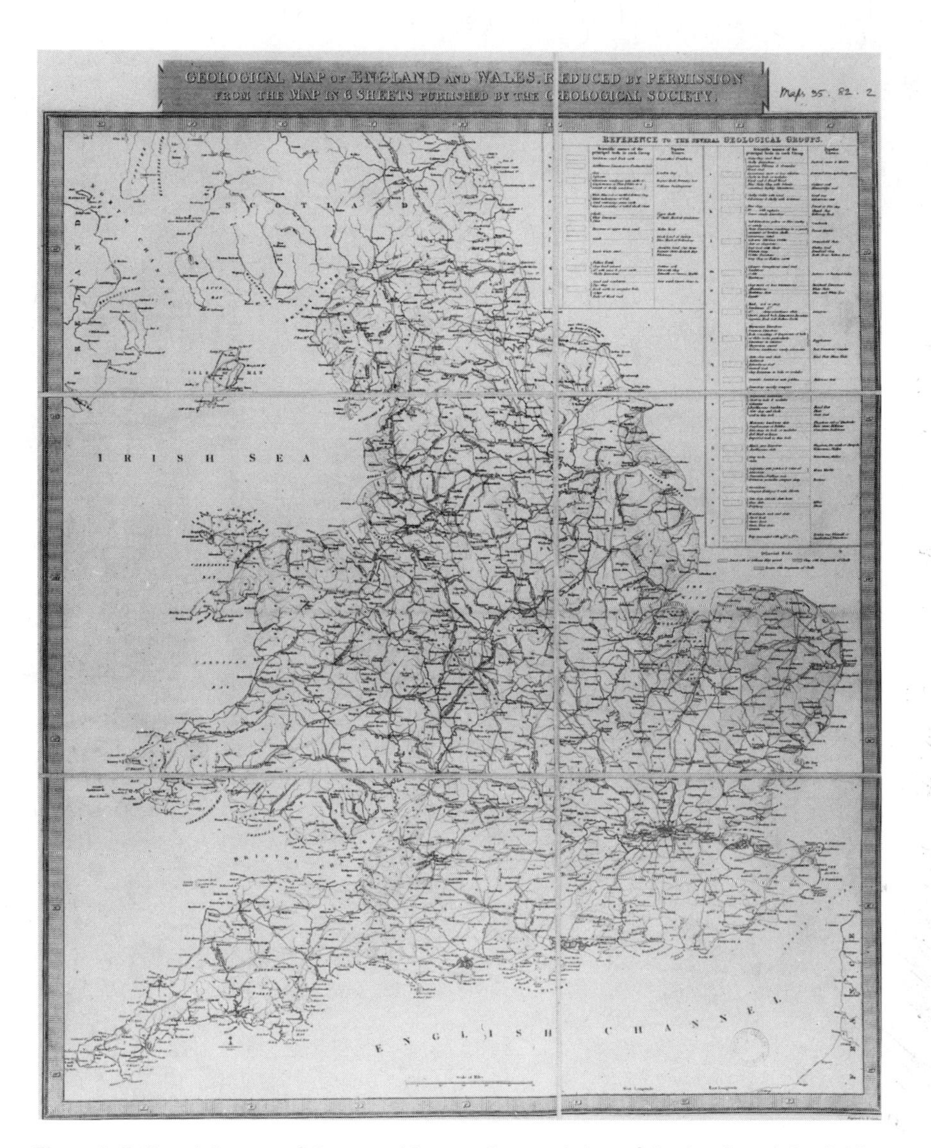

Figure 1.7. Darwin's copy of Greenough's map. By permission of the Syndics of Cambridge University Library.

Darwin's vocabulary marked his knowledge of the ideas of the English school of geologists.[195] On questions of large-scale interpretation of the terrain, Sedgwick's influence is also apparent. In his notes Darwin used the phrase "planes of cleavage"; Sedgwick had introduced to the literature a distinction between bedding and cleavage that was not yet current in geology as a whole.[196] (Henslow may also have made Darwin aware of this point.) As Secord has shown, Darwin's search for signs of the layer of the English strata characterized as "Old Red Sandstone" was prompted by Sedgwick's interest.[197] The formation was marked on Greenough's map, but Sedgwick was doubtful. Working on his own near Abegele, Darwin noted the absence of the formation. Writing to Murchison in September, Sedgwick commented,

> The Old Red all round by Orm Head &c. &c. is a pure fiction. At least I can't see a trace of it. There is not a particle of it between Denbigh and the Isle of Anglesea. There are, however, some red beds (which may pass for Old Red for want of better) in a ravine west of Ruthin [comment on which also appears in Darwin's field notes], and in one or two places near Llangollen under the Mountain Limestone escarpment.[198]

Sedgwick's judgment held: the next version of the geological map of the area, in 1839, showed no Old Red Sandstone.[199] Later, Darwin was uncertain how much his work had aided Sedgwick. Recalling the tour, he wrote,

> In the evening [Sedgwick] discussed what I had seen; and this of course encouraged me greatly, and made me exceedingly proud; but I now suspect that it was done merely for the sake of teaching me, and not for anything of value which I could have told him.[200]

Whatever Darwin's individual contribution, he could see that important issues were at stake in Sedgwick's work. As a last point, it is interesting to see Darwin's high spirits reflected in his geological notes at their very end, when he was clearly on his own. Puzzling out whether he was looking at a case of real or apparent dip of strata, he declared that one of certain beds of porphyry had the "apparent dip to S by W, & which if a true dip would have cut through the beds of slate which is absurd ∴ not true dip."[201] The word "absurd"—an exuberant cry of conclusion—is pure Darwin. The underlying geological meaning to Darwin's remark is that true dip must be determined perpendicular to the strike. If the strike is uncertain, only an apparent dip may be measured.

Sedgwick and Darwin set off from Shrewsbury on 5 August; it is not known when they separated. Did they remain together only until 11 August, with Sedgwick entering Anglesey on 12 August alone? Or, as Peter Lucas and Michael Roberts have suggested, did Darwin accompany Sedgwick through Anglesey?[202] Anglesey, a flat largely agricultural island, forms a scenic contrast to the mountains of North Wales. It provides an ideal plateau from which to view their beauty. Anglesey is connected to the mainland by the Menai Bridge, designed by Thomas Telford (1757–1834). (Of the bridge Emma Darwin [née Wedgwood] [1808–1896] would later write to Fox, “Charles tells me you once carried a toad over the Menai Bridge for fear he should come to a bad end.—”[203]) It is not difficult to imagine Darwin crossing over that bridge with Sedgwick in August 1831. While there is no unambiguous textual evidence available at the moment to document the point, Sedgwick’s goal of studying Anglesey’s geology with Henslow’s memoir in hand would surely have appealed to Darwin as well. As Lucas and Roberts have pointed out, Darwin referred to Anglesey rocks in his own geological notes from the *Beagle* voyage, and though it is possible that his knowledge came from examining the Anglesey specimens held in Cambridge, it is easier to imagine that his knowledge came from examining the Anglesey rocks in situ. Further, as Lucas remarked, “if not with Sedgwick, where was he?”[204]

By 23 August, when he arrived in Barmouth for a holiday with his friends, Darwin finished his part of the geological tour, sending on conclusions to Sedgwick, who was still in the field. That letter has not been located. However, Sedgwick’s response is extant, and it comments on Darwin’s observations. Darwin’s next letter is not extant, but clearly it contained the news that Darwin was to sail aboard the *Beagle*. Sedgwick’s letter in reply itself tells the tale. Now the order of business had become first, the voyage, and second, Welsh geology. Sedgwick responded to what had clearly been Darwin’s request for advice:

> I really dont know what to say about books—N°. 1 Daubeny. N°. 2. a book on geology—D’aubuissions work is one of the best tho’ full of Wernerian nonsense.—I don’t think Bakewell a bad book for a beginner—For *fossil shells* what is to be done?—Go to the Geological Society and introduce yourself to Mr Lonsdale as my friend & fellow traveller & he will counsel you—Humboldts personal narrative you will of course get—He will at least show the right spirit with w^{h}. a man should set to work.—[205]

Only after his work of advising Darwin on books for the *Beagle* did Sedgwick turn to describe the results of his own researches in Wales subsequent to Darwin’s departure. Their paths had separated. Yet, while they never became regular correspondents, the lessons Sedgwick gave Darwin in the field had prepared him for work that he was to do abroad. Sedgwick was the second remarkable teacher Darwin benefited from at Cambridge. Darwin remained permanently in his debt.

For the last third of the year 1831, from the first week of September onward, Darwin prepared for the voyage. Among other tasks, this involved purchasing books. By the time the *Beagle* set sail on 27 December he had assembled a substantial geological library. In addition, Robert FitzRoy had taken with him "all voyages," meaning a wide selection of the narratives of other travelers.[206] Thus, Darwin left with a fair slice of the literature of the field ready at hand. To capture something of the structure of that field is the task of the next chapter of this book. But, before doing so, it would be useful to take stock of Darwin's position in the year of his departure.

In January 1831 Darwin completed his examinations for his university degree. On 12 February he celebrated his twenty-second birthday. He spent most of the first half of the year 1831 in Cambridge, reading and preparing for a life of science combined with travel, and, possibly, with a clerical vocation. When the opportunity of the *Beagle* voyage arose in early September, his preparations for scientific work accelerated but did not mark a radical change from the nature of his activities in the immediately preceding months, except in that they entailed practical details of consultation, purchase, and packing. Had the voyage not come, opportunities for travel and for the pursuit of natural history, including geology, would have remained. There was still to have been a voyage, albeit a lesser one, in the future. Moreover, the scientific life of London, as, possibly, a Fellow of the Geological Society, remained to be tasted. The year 1831 was thus one of transition, but less abruptly a change in intellectual direction than the prospect of a circumnavigation might have suggested. Darwin was already exploring the possibilities of a life in science. A life in the field as a geologist was on the horizon.

CHAPTER 2

GEOLOGY

In my younger days, honorary awards meant a great deal. They were reassuring pats-on-the-back, and they helped to satisfy my ambitions. . . . But I agree with what my brother Jan said when he received the Nobel Prize in economics, "Such distinctions ought to be given to fields of research, rather than to individuals."

NIKO TINBERGEN, QUOTED IN ELIZABETH HALL, "ETHOLOGY'S WARNING"

A Field of Research

Geology in the nineteenth century was a triumphant field. Its achievements were many: the identification and sequencing of the earth's principal strata, an understanding of numerous processes affecting the formation of the earth's topography, and the reconstruction, in outline, of the history of life on the earth.[1] I argue in this book that the field of geology can be usefully regarded as a kind of grid against which Darwin's early work is measured. This is a particularly strong imperative for Darwin because his reputation remains as a biologist. Writing some years ago, the philosopher of science Thomas Kuhn commented,

> The literature on evolution, in the absence of adequate histories of the technical specialties which provided Darwin with both data and prob-

> lems, is written at a level of philosophical generality which makes it hard to see how his *Origin of Species* could have been a major achievement, much less an achievement in the sciences.[2]

In the thirty years since Kuhn's critical comment, many scholars have addressed the problem of the intellectual context for Darwin's work on evolution, but much remains to be done to link Darwin's entire body of work, including evolution, to the various scientific disciplines that, as Kuhn justly remarked, provided him with "both data and problems."

To do that for geology, we must consider first what one means by the "field." There are two aspects to such description: conveying a sense of how geology was distinguished from other aspects of science, and identifying its elements. Timing is also significant, for fields change. We will take the period from the early 1830s to the mid-1840s, which stretches from Darwin's entrance into the field in 1831 to the time when he published his geological trilogy from the voyage: *The Structure and Distribution of Coral Reefs* (1842), *Volcanic Islands* (1844), and *Geological Observations on South America* (1846).[3] Through 1846 Darwin also published nineteen papers, or notices, on geological topics, a number of them presented initially to the Geological Society of London.[4] After the mid-1840s, Darwin's day-to-day work went in less geological directions.

This slice of time encompassed the early years in the history of geology. The field took recognizable shape toward the end of the eighteenth century, and at the turn of the nineteenth century the term "geology" came into circulation.[5] By 1840 the periodization of earth history was understood in a recognizably modern way.[6] Thus, Darwin's own work as a geologist took place when the budding field was establishing its character. The young science and young researcher complemented one another. Darwin's work played off developments in geology at large in creative ways.

Materials for mapping Darwin's career against the growth of the field are readily at hand. Especially valuable are materials that reveal the interface between the man and the field. In this chapter we will use two such sources. One is the substantial collection of books on geology that Darwin took with him aboard the *Beagle*. From his autobiography, we might get the impression that he picked up volume 1 of Charles Lyell's *Principles of Geology* and all was light.[7] But although Darwin did experience reading Lyell as an epiphany, he in fact had with him on the *Beagle* voyage many of the current geological texts in English and some of the literature that existed in French. Darwin's sense of the field was gathered from his entire reading list, not from Lyell's book alone. His library aboard the *Beagle* was also of unusual importance because his geological training at home was relatively brief and because during the voyage he had to assume primary responsibility for his own continuing education.

The second source for plotting Darwin's trajectory against the field of geology is the records of the professional association to which he belonged after his return from the *Beagle* voyage, the Geological Society of London. In particular, the annual presidential addresses from the society grant an entrance into the "conversation of the people who counted."[8] The addresses reviewed what each president perceived as progress in the field for that year. Hence, they give a sense of how the participants in the field felt it was developing year by year, and thus serve as a backdrop for studying an individual career.

The *Beagle*'s Library

A young naturalist setting out as the lone "philosopher" on what would become a five-year voyage presents a romantic image. To complement this image it is necessary to keep in mind that this particular naturalist, Charles Darwin, was traveling on a ship that had a well-stocked and well-run library. Somewhere between 245 and 275 volumes are known to have been on board, and the library rules governing their lending on ship have been published.[9] (Rule eight reads, "Books are never on any account to be taken out of the Vessel.") Books were lent to officers as well as to such supernumeraries as Darwin. They were stored in the small poop cabin in which Darwin also happened to sleep (Figure 2.1).

The library was not static. Mail could be received at an established port, such as Buenos Aires or Montevideo, or it could be delivered to British naval stations. For example, the British arranged to send a "ship to the Falkland Islands, with an Officer & party of soldiers to act as Governor" in early 1834.[10] There was thus a permanent station in the Falklands to which mail might be delivered. Darwin did indeed receive a number of books when the *Beagle* called there in 1834. Some books he requested, as in a letter of May 1833 where he asked his sisters to send several books from the home library, including those by John Playfair (1748–1819) and George Poulett Scrope (1797–1876).[11] Other readings were chosen for him. Henslow wrote to Darwin on 31 August 1833: "I am now in possession of your letter of last April, which has stirred me up to send you off a few books which I thought might interest you, & I have (or rather *shall*) write to your Brother to recommend one or two more."[12] The books that concern us here were the ones dealing with geology, because they played an important role in Darwin's continuing "cramming" in the science.[13] As he put it to Henslow midway through the voyage, he had learned his geology in a "curious fashion."[14]

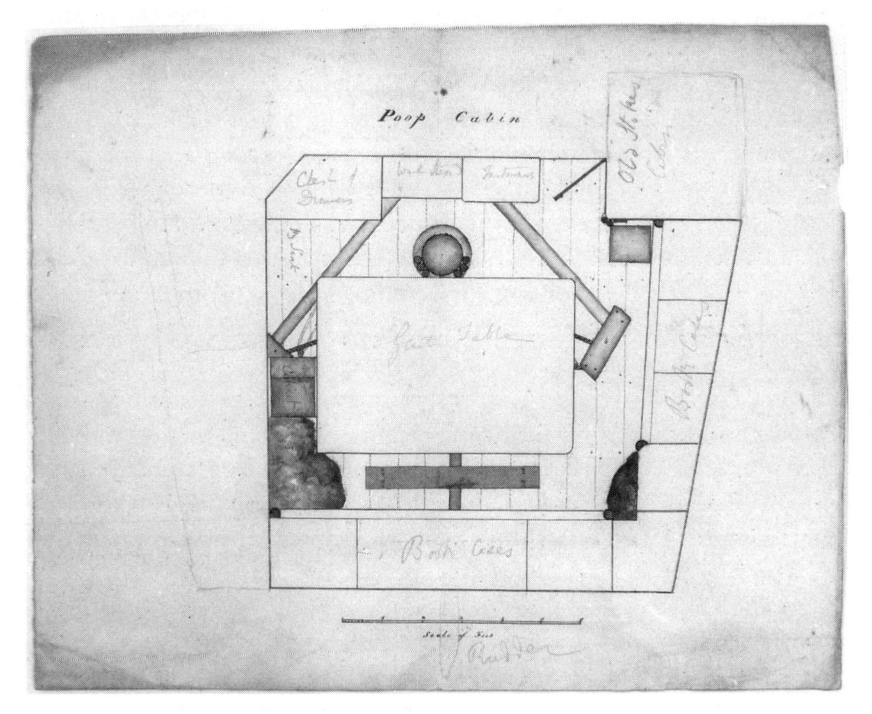

Figure 2.1. Poop cabin of H.M.S. *Beagle*. By permission of the Syndics of Cambridge University Library.

As stated earlier, to define a field of science, one has to ask what sets it apart as a distinct area of inquiry. For geology, that question had been answered by the late eighteenth century. Geology was to concern itself with the crust of the earth.[15] Thus, William Phillips began his *Elementary Introduction to the Study of Mineralogy* stating that "the investigation of the structure of the earth belongs to the science of Geology." He continued,

> In speaking of the earth and of our knowledge of its nature, it is essential that the limited extent of that knowledge should always be had in remembrance. We are acquainted with it, only to a very inconsiderable depth; and when it is recollected that, in proportion to the bulk of the earth, its highest mountains are to be considered merely as unimportant inequalities of its surface, and that our acquaintance does not extend in depth, more than one-fourth of the elevation of these mountains above its general level, we shall surely estimate our knowledge of the earth to be extremely superficial; that it extends only to its crust.[16]

Phillips went on to say that by using the word "crust" he did not imply the hollowness of the globe, "for of this we know nothing."[17] To the same point, the article on "Geology" in the sixth edition of the *Encyclopædia Britannica,* also aboard ship, referred to the subject as the "part of natural history which treats of the internal structure of the earth, as far as we have been able to penetrate below its surface."[18] The French geologist Alexandre Brongniart (1770–1847), quoted in a work that was again part of the *Beagle*'s library, fully and elegantly defined the focus of the young science: "It is now very generally agreed that the end of positive Geology is to be able to understand, as exactly, and completely as possible, the nature and structure of the crust of the globe, and to discover if general and constant laws have governed this structure."[19]

From this focus on the crust, certain things followed. A respectful deference developed toward what might be known only by conjecture. To convey this sense of proportion, De la Beche employed a diagram in his Sections and Views (Figure 2.2). De la Beche's text accompanying the plate is worth quoting in extenso:

> The object of the principal figure in this Plate is to show the proportion which the elevation of the highest mountains bears to the radius of the earth. That the most elevated peaks of the Himalah are insignificant protuberances will be at once observed; and it may be asserted, that from figures of this nature a more definite idea of the relative importance of things is obtained than from pages of description. To one who looks at such a diagram, it will be obvious that slight and unequal contractions of the mass of the earth would produce changes of the surface, which we should consider important; and it may occur to him that mere thermometrical differences beneath the earth's crust might be sufficient to raise whole continents above the level of the sea, or plunge them beneath it.
>
> A line has been drawn representing a depth of one hundred miles below the level of the ocean, in order to show that great disruptions of our planet's surface might take place, and might be produced by causes acting at that depth without the mass of the world being much affected by it.[20]

De la Beche's diagram presumes an interest in the interior of the earth but does not attempt to represent its composition or structure. How different this figure is from the cross section of the earth provided by Erasmus Darwin![21] There the crust of the earth and the interior of the earth are imaginatively de-

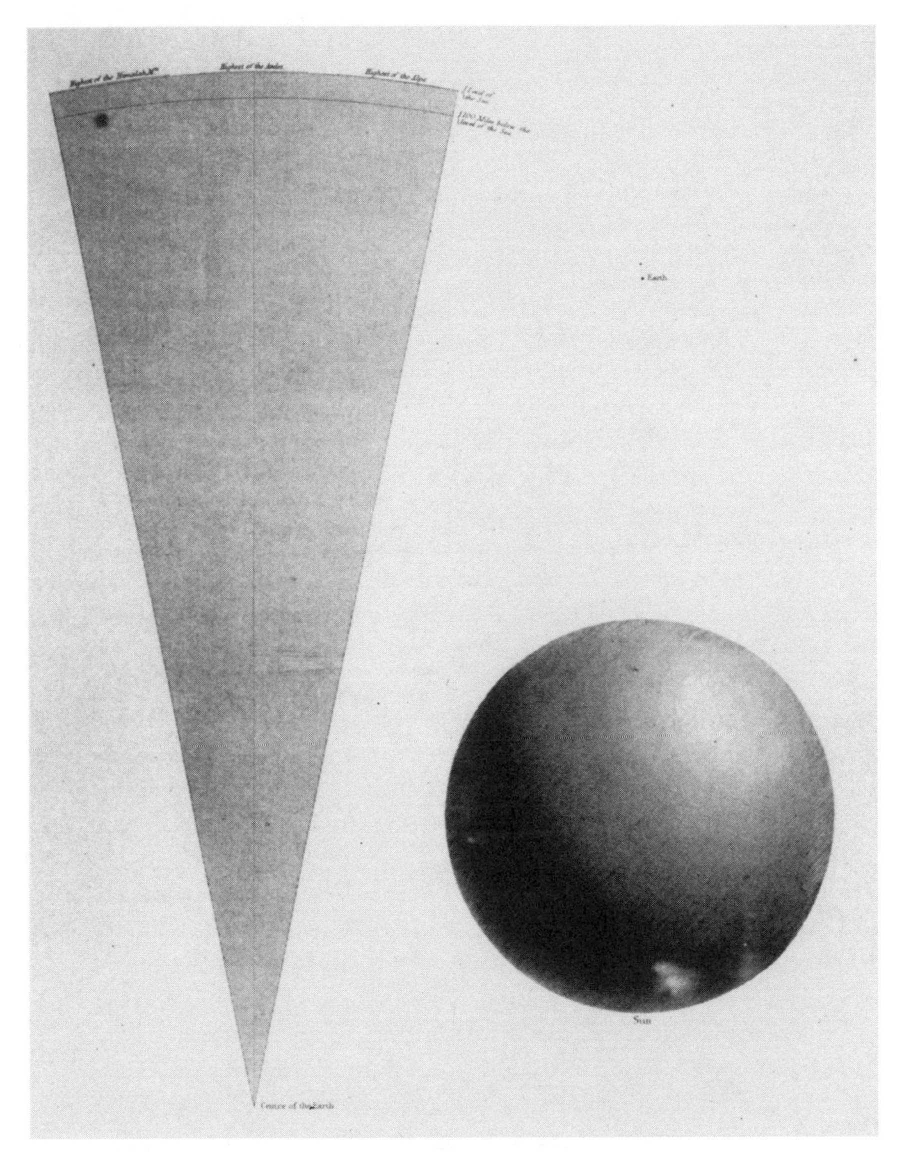

Figure 2.2. De la Beche's sketch of the proportion that the elevation of the highest mountains bears to the radius of the earth. Courtesy of the Library of Congress.

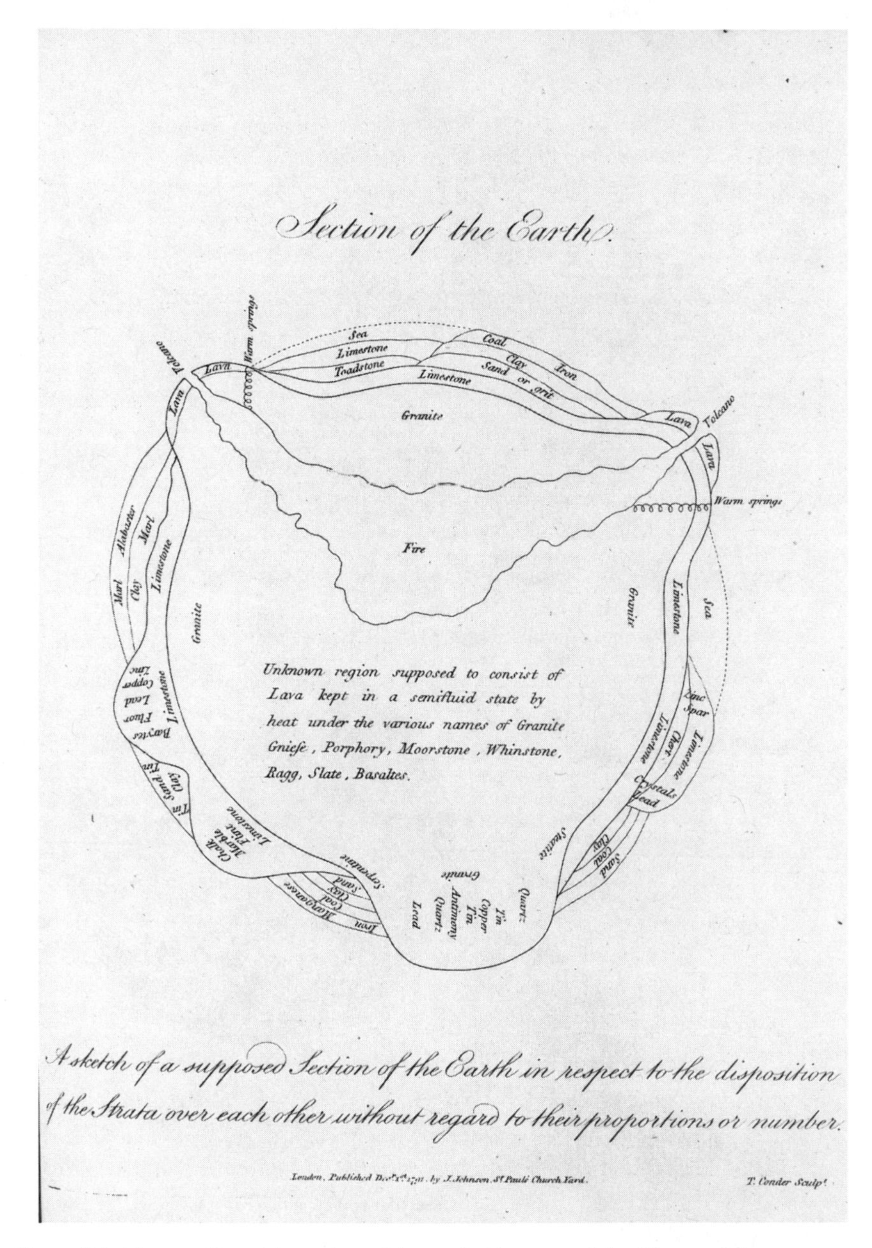

Figure 2.3. Erasmus Darwin's section of the earth. Courtesy of the Library of Congress.

picted freely communicating with each other (Figure 2.3). Erasmus Darwin's vision is an eighteenth-century one. Geologists grew sharper, but more circumspect, as their science grew.

The crust, then, was the agreed-on subject matter for geology, marking it off from other fields. As a young science the process of drawing lines, of making boundaries, was an important one. For example, in the introduction to their influential work *Outlines of the Geology of England and Wales,* Conybeare and Phillips advised readers that a geologist need not be a fully informed mineralogist. They also suggested that mineralogy was to be subordinate to geology, and for the purpose of geology, the mineralogist could be viewed as a "mere" mineralogist. Their instruction, intended as practical advice, also illuminates what was then an ongoing process of boundary formation between geology and mineralogy. They wrote that

> the number of mineral masses forming rocks of usual occurrence is so small, and the composition of those so simple, that a very limited knowledge of these sciences is sufficient for all introductory purposes, as far as the general outlines of Geology are concerned. Siliceous, argillaceous, and calcareous masses (substances with which every one is familiar under the common names of sand, clay, and limestone) constitute probably nine-tenths of these materials, and the compound rocks forming the remaining tenth consist principally of only four minerals, quartz, felspar, mica, and hornblende. These great masses contain . . . all the other substances included in the mineral kingdom; . . . *this knowledge which is the ultimate object of the mere mineralogist, is to the geologist only a subordinate acquisition.* . . . On this principle nearly one-half of the mineral species may be safely neglected in beginning a course of geology, nor is a knowledge of more than 100 species essential as a preliminary acquisition.[22]

Conybeare and Phillips's message is directed toward the initiate: the way into the field of geology was not through mastery of mineralogy. Spending years learning every mineral was not required. Memorizing Jameson's various manuals was not called for. Mineralogy had its place. Indeed, Phillips was himself the author of a text on mineralogy that traveled with Darwin aboard the *Beagle.*[23] But mineralogy was not the core of the subject; it was "subordinate" to the new subject. A hierarchy of knowledge was implicit. The beginning student was to master knowledge of the most common masses, for which commonplace recognition of the differences among sand, clay, and limestone provided a good start, and gain only such further knowledge as was necessary to proceed with the subject at hand—geology.

This process of boundary formation was a common occurrence in modern science, as new disciplines formed. For Darwin, the message with regard to mineralogy was not to devalue what mineralogical knowledge he had, but rather to concentrate on more purely geological questions. An expert mineralogist, not Darwin, would analyze the geological specimens he brought home from the voyage.

But what of the internal content of the science? Here the focus on the crust of the earth brought with it a key consideration for the content. To wit, was the crust uniform in construction around the globe? John Playfair, thought so. In his *Illustrations of the Huttonian Theory of the Earth,* which went with Darwin on the *Beagle,* he wrote,

> If it be said, that only a small part of the earth's surface has yet been surveyed . . . it may be answered, that the earth is constructed with such a degree of uniformity, that a tract of no very large extent may afford instances of all the leading facts that we can ever observe in the mineral kingdom. The variety of geological appearances which a traveller meets with, is not at all in proportion to the extent of country he traverses; if he takes in a portion of land sufficient to include primitive and secondary strata, together with mountains, rivers, and plains, and unstratified bodies in veins and in masses, though it be not a very large part of the earth's surface, he may find examples of all the most important facts in the history of fossils.[24]

By "fossils" Playfair included all kinds of matter dug up; the restricted modern meaning was not his. However, Playfair's position was clear: the crust of the earth was relatively uniform. If so, the geologist would have no need of extensive travel, let alone circumnavigation.

But James Hutton's view, as expounded by Playfair, was not in the majority—not in his lifetime, and certainly not in 1831. The majority opinion held that the earth's surface was more richly textured. It was not the case that "a tract of no very large extent" afforded sufficient compass for the geologist. Indeed, the reverse was true; global inspection was required. Possibly the influence of Humboldt had told. Even Werner's understanding of formations was viewed as too narrow. Greenough, writing "On Formations," commented,

> Virgil's peasant fondly imagined his native village a miniature of imperial Rome. With a corresponding love of generalization, Werner imagined Saxony and Bohemia a miniature of the world. He set down, more or less correctly, the order in which different rocks had arranged themselves in the districts with which he was acquainted, and concluded, that

> such must be the order of their arrangement in every other district. His theory was useful as a standard of reference, an incitement to inquiry, a clue to observation; but, unfortunately for Werner, his pupils viewed it in a different light.[25]

Complementing Greenough's sentiment, colloquially, Darwin himself was remembered by a Cambridge classmate to have said during their years at university, "It strikes *me* that all our knowledge about the structure of our Earth is very much like what an old hen w^{d} know of the hundred-acre field in a corner of which she is scratching."[26]

Geology had become focused on the connection between the universal and the particular, fixed locally and temporally. The focus was historical and stemmed from work done in the state mining bureaucracies and training schools of late-eighteenth-century Germany. Rachel Laudan, developing this line of thought, has argued that the foundation of geology lay in seventeenth-century mineralogy, which itself had two branches: chemical mineralogy, drawn from the ancient four-element theory and developed in modern times by (among others) Johannes van Helmont (1579–1644) and Paracelsus (ca. 1493–1541), and mechanical mineralogy, which emphasized the shape and arrangement of particles and was developed by Robert Hooke (1635–1702), René Descartes (1596–1650), Robert Boyle (1627–1691), and Nicolaus Steno (1638–1686). Mechanically minded mineralogists were also responsible for the first cosmogonic notions, particularly Steno with his understanding of the superposition of strata indicating relative age, and Descartes with his model of mountain building based on the fissuring and fall of portions of the earth's crust. The chemists eventually developed their own cosmogony, beginning with Johann Becher (1635–1682) and ending with Georg Stahl (ca. 1660–1734), and these cosmogonies "were predicated on the deposition of the rocks of the earth's crust from water."[27] The culminating figure in this tradition was Werner, who taught at the Bergakademie at Freiberg from 1775 to 1817. Laudan, following the lead of Alexander M. Ospovat and others, regards Werner as the chief originator of historical geology. Her argument turns on Werner's invention of the concept of geological formation:

> During the course of his career, Werner transformed the Becher-Stahl tradition from which he had taken so much. He made the time of formation of rocks, not their mineralogy, their most important character. Well aware that he was flouting the precepts of taxonomy [which dealt with natural kinds], he named bodies of rock formed in the same period 'formations,' and made these historical entities—formations—more important than chemical ones.[28]

Thus, Laudan minimizes Werner's Neptunism, ultimately discredited, in favor of emphasizing the historical, and ultimately successful, aspects of his work.

Ironically, in a book on Darwin, putting in a few good words on behalf of Werner is just, for Darwin himself did not. Darwin did not highly regard his own training at the hands of Werner's student Jameson. Nor did Darwin's elder colleague Lyell present Werner's work as a whole in a positive light.[29] Yet Darwin did encounter the Wernerian tradition in a way immensely useful to him during the voyage. This was through Humboldt, who had been a student of Werner's. Humboldt was part of what Laudan has termed the Wernerian "radiation."[30] As many others who were Wernerians, Humboldt abandoned the master's view that basalt was a precipitate. Rather, he adopted the Huttonian view of its igneous origin. However, Humboldt kept Werner's focus on formations. Moreover, with his fantastic energies, and his competency in French and English, as well as German, Humboldt read the current European literature and cited it. The key work by Humboldt that Darwin had with him aboard ship was *A Geognostical Essay on the Superposition of Rocks, in Both Hemispheres.* (Darwin owned the French second edition, but cited the English translation of the first edition during the voyage.[31])

Humboldt opened his work with a definition, and with a deft bow to the (unnamed) Huttonian igneous theory of the origin of basalt. He wrote that

> the word *formation* designates an assemblage of mineral masses so intimately connected, that it is supposed they were formed at the same epoch, and that they present, in the most distant parts of the earth, the same general relations, both of composition, and of situation with respect to each other. Thus the *formation* of obsidian and of basalt is attributed to subterranean fires; and it is also said that the *formation* of transition clay-slate contains Lydian stone, chiastolite, ampelite, and alternating beds of black limestone, and of porphyry.[32]

(Humboldt's view that basalt was of the same epoch worldwide would not stand.) Following his lengthy introduction, where he cited the major English geologists, including Smith, and did battle on points with Greenough, Humboldt turned to list formations. Significantly, he followed the standard groupings that had grown up from the mid-eighteenth century: primitive, transition, secondary, tertiary, volcanic. (By way of comparison, Sedgwick's lecture outline for 1821 followed an identical listing, under the title of "Division of formations," except for adding "Alluvial" as a category—though presumably not a temporal one—between "Tertiary" and "Volcanic."[33]) Most of the material Humboldt presented refers to rocks, though there is some discussion of the utility of fossils for identifying formations. Particularly useful to Darwin was the fact that Humboldt went back and forth between the New World and the Old. Although their routes in South America differed (Humboldt more to the north, Darwin to the south), Darwin used Humboldt's book as a ready reference on South American formations.

There are no graphics in Humboldt's book though he did discuss how his table of formations, given at the end of the book, might be arranged as a series of parallelograms, a device he used as early as 1804. (An expanded English-language edition of *The Animal Kingdom* by Georges Cuvier [1769–1832] did translate Humboldt's list into graphical form.)[34] Interestingly, Humboldt also suggested a system of representing formations using Greek letters (for example: α = granite, β = gneiss, γ = mica-slate, δ = clay-slate) that would allow one to describe a sequence of formations in something like sentence form.[35]

On the whole, though, geologists found graphical representations more compelling. De la Beche was one such author. He expressed his attitude best in the following comment: "Sections and views are, or ought to be, miniature representations of nature; and to them we look, perhaps, more than to memoirs, for a right understanding of an author's labours."[36] Darwin had two of De la Beche's well-illustrated works aboard ship: *A Selection of Geological Memoirs Contained in the Annales des Mines, Together with a Synoptical Table of Equivalent Formations, and M. Brongiart's Table of the Classification of Mixed Rocks* (1824), and the longer but more succinctly titled *Geological Manual* (1831).[37] For a traveling geologist, not yet well read in either the Continental or British schools, these books were godsends. We will discuss them in turn.

De la Beche provided, as did Humboldt, an international setting for geology. European geologists were often isolated from each other by language and by the geological differences among their homelands. Their difficulties could be overcome by travel and translation. De la Beche did both. He traveled extensively on the Continent as a young geologist and acquired a command of foreign languages.[38] De la Beche's publication of 1824 put within the easy reach of English-speaking geologists the important work being done by French geologists writing for the *Annales des Mines.* As the Belgian geologist J. J. d'Omalius d'Halloy put it in a paper originally delivered in 1813,

> [I]f Saxony, a country excavated to vast depths, in consequence of its mineral riches, has offered to the genius of Werner, an opportunity of establishing the first good system of geology; the environs of Paris, containing such abundant remains of living creatures, have given birth to *true philosophical geology,* that which drawing its conclusions from the

> knowledge of the organised bodies entombed in the bosom of the earth, can alone afford the certain means of comparison between distant formations, and will one day perhaps throw some light on the various catastrophes that have changed the surface of the globe, as it has already indicated the nature of the liquids in which some of these phenomena have occurred.[39]

A "true philosophical geology" was, in effect, to be more rigorously historical than Werner's. In the same volume, Brongniart made a similar point when writing eight years later about the chalk formation:

> My object is to shew the organic remains contained in it [the chalk] offer characters sufficient for recognising it in situations very distant from each other, when those [characters] drawn its consistence, stratification, colour, &c. have disappeared, and when its superposition is either obscure, uncertain, or difficult to be recognised.[40]

Brongniart thus argued for the primacy of fossil over mineral markers in the identification of formations. To this he added strong views concerning the history of life on the planet, namely, that the "distinct succession of generations of organic bodies" shows that "this crust has been formed in successive depositions, at different periods."[41]

In addition to bringing such important articles as Brongniart's to English-speaking geologists, De la Beche also provided an important service in attempting to affect a deeper translation. He devised what he called "A Synoptical Table of Equivalent Formations" as drawn by English, French, and German geologists.[42] De la Beche was constructing a geological column based not on one country only, or on the opinions of its geologists alone, but a column of wider applicability drawn from the researches of numerous individuals in different regions. The table is too large to reproduce easily. Yet its schema can be identified from its first lines:

ENGLISH	FRENCH	GERMAN
ALLUVIUM &. DILUVIUM.	TERRAIN DE TRANSPORT & d'ALLUVION (B). *Terrain de Transport* (Daubuisson.)	AUFGESCHWEMMTE–GEBIRGE

As De la Beche explained, the "B" stood for the opinion of Brongniart; the use of italics indicated that the term was synonymous with those immediately preceding them.

That De la Beche would group geologists by nationality suggests, to some extent, separate groupings. More importantly for Darwin, as of the 1820s there was a recognizably "English" school of geology into which he had been inducted by his training with Henslow and Sedgwick.[43] The single most authoritative text for that school was Conybeare and Phillips's *Outlines.* It was from this book's reworking of the successive list of formations that De la Beche took his characterization of the English school.[44] In their presentation Conybeare and Phillips attempted to substitute new language for terminology that was mid-eighteenth century in date, as, for example, "Supermedial order" for "Secondary class."[45]

But their substitutions, and those of others, failed; the old language held. Another mark of the English school at this period was the use of the term "diluvium," invented by Buckland, to refer to gravel and superficial debris apart from that "occasioned by causes still in operation," that is, apart from alluvial deposits, also known as "alluvium."[46] The term remained largely a British one, though Constant Prévost (1787–1856) took note of it in the *Dictionnaire classique d'histoire naturelle,* a set of which was part of the *Beagle* library.[47] (Cuvier also recorded Buckland's use of the term in the 1825 edition of his *Discours.*[48]) The term passed through the geological vocabulary in fewer than twenty years. Finally, Conybeare and Phillips intended their book to be convenient to the field geologist: "A small type and a thin paper have been preferred for the advantage of the traveller."[49]

Another book of the English school that accompanied Darwin, and to which he frequently referred, was Greenough's collection of eight essays entitled *A Critical Examination of the First Principles of Geology* (1819). The tone and organization of Greenough's book are quite unlike those of any other in Darwin's library. "Critical" was the key word, and even where Greenough was eventually shown wrong (as on the question of the diluvium), the directness of his questioning, salted by peevishness, remains fresh. Who could not be curious about, even charmed by, an author who opens his first essay, "On Stratification," thus: "Stratum is a word so familiar to our ears that it requires some degree of manliness to acknowledge ourselves ignorant of its meaning."[50] (The imperious tone of Greenough's remarks perhaps derives from a downward glance toward his working-class rivals and predecessors, the mineral surveyors.)[51] With similar insouciance, Greenough moved through the great questions of geology in his subsequent chapters: "On the Figure of the Earth," "On the Inequalities which existed on the Surface of the Earth previously to diluvian Action, and on the Causes of these Inequalities," "On Formations," "On the Order of Succession in Rocks, "On the Properties of Rocks, as connected with their respective Ages," "On the History of Strata, as deduced from their Fossil Contents," and "On Mineral Veins." His critical examination had greater force because of his central position in English geology. He was the first president of the Geological Society of London, and its president at the time the book was published. Furthermore, he was the co-compiler of *A Geological Map of England & Wales* (1820), whose publication was supported by the society. This map supplanted Smith's map of 1815.[52] Greenough's map was, in fact, the society's map.[53]

With regard to English geologists, it is worth noting what books did *not* accompany Darwin on the *Beagle*. Of Buckland he had relatively little. Thus, for example, Darwin does not seem to have had a copy of Buckland's *Reliquiæ Diluvianæ*.[54] However, Darwin did have on board Buckland's article on fossil quadrupeds that appeared—significantly—as an appendix to an account of an earlier voyage.[55]

There were other books that went into Darwin's cabin in 1831. Sedgwick had suggested he bring along "a book on Geology"—meaning a general work. Sedgwick seems to have had in mind Robert Bakewell's *Introduction to Geology* and Jean-François d'Aubuisson's *Traité de géognosie*.[56] Darwin took the second suggestion but not the first. D'Aubuisson had studied with Werner, but had become convinced of the igneous origin of basalt. *Traité de géognosie* was the "first competent treatment of general geology published in France."[57] It covered formations, the figure of the earth, causal agents affecting the surface of the globe, volcanoes, and various geological theories.

Important geological information was contained in reference works on the *Beagle:* J. B. Bory de Saint-Vincent's *Dictionnaire classique d'histoire* and an English-language edition based on, but not restricted to, Cuvier's *Animal Kingdom*. Other geological sources aboard the *Beagle* are listed in the bibliography.

Given Darwin's later interests, two additional groupings of books aboard ship call for notice: the material on coral reefs, and that on volcanoes and earthquakes. As early as 1962, D. R. Stoddart combed Darwin's early essay on coral reefs, written during the voyage, for references identifying Darwin's sources.[58] Most were from accounts of voyages and expeditions, but both De la Beche's *Geological Manual* (1831) and volume 2 of Lyell's *Principles of Geology* (1832) had sections on coral reefs.[59] Since, in outline, Darwin sketched his coral reef theory during the voyage, it was to these works that he looked for authoritative facts on reefs.

Darwin also carried with him three works on earthquakes and volcanoes: the classic article by John Michell (ca. 1724–1793), "Conjectures concerning the Cause . . . of Earthquakes"; Daubeny's *Description of Active and Extinct Volcanos*"; and Scrope's *Considerations on Volcanos*.[60] The subjects were important to him since he would be traveling to a region where volcanic eruptions and earthquakes might occur. Daubeny's work, recommended to Darwin by Sedgwick, was based on lectures given at Oxford. It was encyclopedic in its approach, listing volcanoes worldwide, and written from within the English school of geology. It favored a chemical explanation for volcanic action.

Scrope's book did not stem from the "English" school of geology. It avoided diluvial language, utilizing the term only in passing, as in a brief reference to the diluvium of Switzerland.[61] It distinguished "geognosy," which included mineralogy, from "geology," which included the study of earth processes. Its scope was also worldwide. Of particular interest is the map that exhibited what Scrope believed to be the parallelism of the principal trains of volcanic vents and the great mountain ranges of the globe. Points of eruption were colored in red; lines of elevation, in gray (Plate 2). Scrope's map of volcanic patterns presages Darwin's coral reef map of 1842. Both aim to represent large-scale patterns of movement. Both are also global in their scope, although neither was original in that sense. In England the tradition of global thematic mapping dates back to Edmund Halley (1656–1742), "who illustrated a number of his own scientific theories by cartographic means."[62] While such mapping was unusual for geologists, Scrope and, later, Darwin were in effect drawing on a long-standing tradition.

We now come to Lyell's *Principles of Geology*. Discussion of it was saved for last for two reasons. First, Lyell's influence on Darwin is already well known, and has eclipsed other influences, which therefore deserve the compensating prominence of prior treatment. Second, by 1831 only volume 1 of Lyell's *Principles* had been published. Thus, in examining the effect of reading Lyell on Darwin, we have to take into account that Darwin read the three volumes separately over the course of several years. He carried volume 1 with him as he boarded the *Beagle* in 1831; it was inscribed "From Capt FitzRoy." Volume 2 was published in January 1832; Darwin inscribed it "M: Video. Novem^{r}. 1832" when he received it. Volume 3, published in May 1833, probably reached Darwin while he was at the Falkland Islands in March 1834. This volume is not inscribed.[63]

The word "read" is important. Lyell's first volume had come to Darwin well recommended by its treatment in Herschel's *Preliminary Discourse*. It is possible that Lyell's and Darwin's paths may have crossed in May 1831, when Lyell visited Cambridge, but Darwin's primary encounter with Lyell was through his book.[64] Even later, though Darwin and Lyell had frequent, face-to-face contact after Darwin returned to England, Darwin chose to portray Lyell as author and himself as reader. Twice in his autobiography he tells the story of the impact on him of reading the *Principles*. First:

> I had brought with me the first volume of Lyell's *Principles of Geology*, which I studied attentively; and this book was of the highest service to

> me in many ways. The very place which I examined, namely St. Jago in the Cape Verde islands, showed me clearly the wonderful superiority of Lyell's manner of treating geology, *compared with that of any other author, whose works I had with me* or ever afterwards read.

And, then, again:

> When I was starting on the voyage of the *Beagle,* the sagacious Henslow, who, like all other geologists believed at that time in successive cataclysms, advised me to get and study the first volume of the *Principles,* which had then just been published, but on no account to accept the views therein advocated. How differently would any one now speak of the *Principles*! I am proud to remember that the first place, namely St. Jago, in the Cape Verde Archipelago, which I geologised convinced me of the *infinite superiority of Lyell's views over those advocated in any other work known to me.*[65]

The two italicized portions of these quotations are germane. Darwin valued Lyell's work over any other book on geology in the *Beagle*'s library.

The primacy of Lyell's *Principles* held on a number of levels. In content, Darwin shaped his logic to Lyell's to such an extent that he sometimes characterized his books as coming "half out of Lyell's brains [*sic*]."[66] This involvement extended over the many editions of the *Principles,* Lyell serving in effect as Darwin's alter ego. Lyell also elevated for Darwin notions drawn from authorship, books, writing, and reading. For example, Lyell frequently employed the metaphor of God as the "Author of Nature," a usage that Darwin would employ in his notes from the voyage.[67] Lyell also drew on language and books as a source for metaphor in geology.[68]

On a practical level, Darwin learned much from Lyell about authorship and books. He saw from Lyell's example that it was possible to write a book that spanned a scientific and a popular audience. This was something that Lyell had set out to do very consciously, largely for economic reasons. As Darwin would do after him, Lyell was seeking to supplement the regular income he received from family wealth with income from writing. With the *Principles* he succeeded in his aim. At one point, fearing competition from De la Beche, Lyell considered altering the *Principles* to keep his market share. His publisher John Murray advised against it. As Lyell wrote to his fiancée, quoting Murray,

> There are very few authors, or have ever been, who c^{d} write profound science & make a book readable.—& depend on it, if you try to abridge & condense more & still be *popular* you may fail.[69]

Many years later John Murray, as his publisher, would also guide Charles Darwin to the right balance of substance and accessibility in the *Origin of Species.*

What was it that so attracted Darwin to the *Principles*? While a full answer to that question will emerge over the course of this study, we can begin to address it now. Part of the appeal of the first volume of the *Principles* was the generous way in which it cast the subject matter of geology. Lyell's breadth is apparent from the first sentence of the book, which reads,

> Geology is the science which investigates the successive changes that have taken place in the organic and inorganic kingdoms of nature; . . . [including] the influence which they have exerted in modifying the surface and external structure of our planet.[70]

Lyell's definition of geology was not inconsistent with definitions offered by his contemporaries, for it included emphasis on the crust of the globe ("the surface and external features of our planet"). But he subordinated this interest to his concern with the "successive changes that have taken place in the organic and inorganic kingdoms of nature." He did not discuss the succession of formations found on the earth's crust until volume 3. Thus, it was not the conventional subject matter of geology that so attracted Darwin to Lyell on first reading.

Martin Rudwick has analyzed the structure of Lyell's overall argument in full detail. As he has shown, the three volumes hold together in their argument.[71] However, in looking at Darwin's appropriation of Lyellian geology, we have to consider the content of the three volumes of the *Principles* and the date that Darwin read them. For the formative first year of the voyage Darwin was working from volume 1 only. The dominant subject, discussed in chapters 10–26, pertained to "changes now in progress" in the inorganic kingdom of nature.[72] Such changes were divided into aqueous processes and igneous processes, including movements of the earth's crust. In the course of taking note of "changes now in progress," Lyell consulted and reported from a much wider variety of sources than was customary in geological writing. Among these were reports of voyages and of scientific exploration.[73] He used such reports to indicate the geological changes now in progress around the globe. To a traveler like Darwin, such literature was readily recognizable as useful, and of a genre toward which he might contribute.

In volume 1 Lyell considered two other general topics. First was his view of the history of geology. He began his work this way in order to assert his own identification with the Huttonian tradition and to criticize what he termed the view of the "cosmogonist."[74] He targeted those who sought to integrate biblical and geological history:

> A sketch of the progress of Geology is the history of a constant and violent struggle between new opinions and ancient doctrines, sanctioned by the implicit faith of many generations, and supposed to rest on scriptural authority.[75]

He then went on to praise Giovanni Quirini, "the first writer who ventured to maintain [that] the universality of the Noachian cataclysm ought not to be insisted upon."[76] But his real interest lay not with that seventeenth-century writer but with his contemporaries in English geology who traced the rough superficial gravels they termed "diluvium" to the biblical flood. These men, whom he left unnamed in volume 1, included the leading members of the English school of geology: Buckland, Conybeare, Sedgwick. Within Lyell's circle at the Geological Society of London, no one would have missed his intended targets. However, Darwin, having been brought up in geology by Sedgwick and Henslow, and not having been present in London for the oral arguments surrounding the reception of the *Principles*, does not seem to have responded instantly to Lyell's critique.[77]

The third key portion in volume 1 of the *Principles* was its middle, which included chapters 5–9. It was Lyell's most freshly added section: he composed it in January and February 1830 as he was correcting proofs for the volume.[78] Chapter 5 served as a bridge connecting the historical portion of the book to more direct discussion of Lyell's method. The following sentences capture Lyell's approach:

> The first and greatest difficulty . . . consists in our habitual unconsciousness that our position as observers is essentially unfavourable, when we endeavour to estimate the magnitude of the changes now in progress. In consequence of our inattention to this subject, we are liable to the greatest mistakes in contrasting the present with former states of the globe.[79]

Lyell was determined that geologists "be roused to exertion" in the search for discovering "operations now in progress."[80] In chapters 6–8 he presented a new theory of climatic change. Rather than relying on the presumption of a gradually cooling earth, Lyell accounted for changes in climate by positing the redistribution of land and sea on the globe. The cause of such redistribution was the presently visible process of the rise and fall of land. As a geologist he hoped to show that known changes in geographical features in the Northern Hemisphere (the only one then studied) corresponded to known fossil evidence for changes in climate.[81] In his presentation Lyell relied extensively on work done by Humboldt on isothermal lines.[82] His reliance on

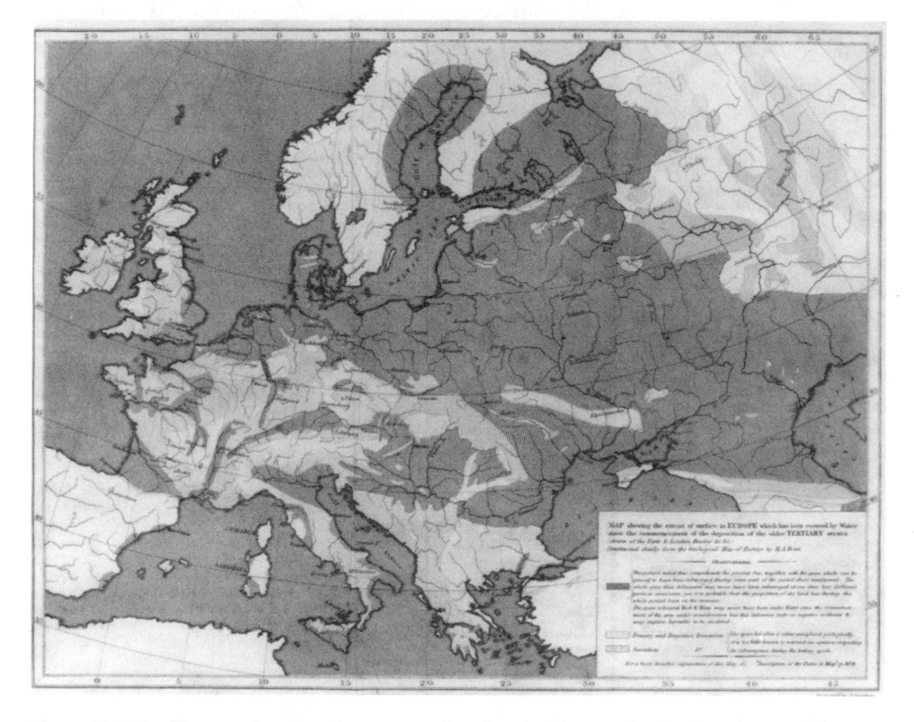

Figure 2.4. Lyell's map showing the extent of surface in Europe that has been covered by water since the commencement of the deposition of the older Tertiary strata. By permission of the Syndics of Cambridge University Library.

Humboldt would have ipso facto recommended the *Principles of Geology* to Darwin. Last, Lyell began his attack on the "geological proofs appealed to in support of the theory of the successive development of animal and vegetable life, and their progressive advancement to a more perfect state."[83]

The second volume of the *Principles* is best known for Lyell's attack on the theory of species transmutation originated by Jean Baptiste de Lamarck (1744–1829). However, this discussion occupies only the first portion of the book. The large middle of the book treats the geographical distribution of species, evidence for the recent origin of man, and arguments concerning the introduction and extinction of species. Also of importance in volume 2 is the final chapter on the formation of coral reefs. An important map accompanied this volume, "shewing the extent of surface in Europe which has been covered by Water since the commencement of the older Tertiary strata" (Figure 2.4). This map, drawn from the work of the geologist Ami Boué (1794–1881), complements all three volumes of the *Principles.* For volume 1 it makes

Lyell's climatic theory more easily comprehensible; for volume 2 it enhances Lyell's treatment of species migration; for volume 3 it suggests the physical conditions governing the formation of strata of various dates.

Overall, the entirety of volume 2 reflected an interest in "Changes in the Organic World now in progress," as the opening of the heading to the first chapter states. Darwin received this volume in November 1832. Thereafter he had in hand a logical presentation of what in Lyell's *Principles* can be called the species question. The directness and analytical clarity of Lyell's approach are apparent in his first inquiry: "whether species have a real and permanent existence in nature."[84] From November 1832 onward Darwin also had in hand Lyell's treatment of the question of the origin of coral reefs.

Unlike the first two volumes, volume 3 of the *Principles* closely resembled other works on geology that Darwin read. Its subject was the crust of the earth. Lyell's approach to the earth's crust was traditional in the sense of reviewing successive formations. However, he was original in his approach to the subject in two fundamental ways. First, he reinterpreted the classification of formations. As opposed to the Wernerians, he claimed that every rock type could be produced on the earth at any time. In inspiration this view was essentially cyclic or Huttonian, though Lyell's work reflected the progress in geology since Hutton's time. Lyell's approach to the study of formations is best displayed in the table that accompanies volume 3. Its argument is carried most succinctly in the heading to that table, which shows the "Relations of Alluvial, Aqueous, Volcanic, and Hypogene Formations of different ages." For example, in Lyell's view, rocks that his contemporaries commonly designated "primary" were not necessarily the oldest rocks on the planet.[85] He renamed them "hypogene," a term referring to rock type but with "no chronological import." For "stratified" primary rocks, which would be called "foliated" today, he proposed the term "metamorphic"; for unstratified hypogene rocks, "plutonic." Lyell stressed that his opinions did not necessarily require a reworking of the traditional understanding of the ordering of formations. Rather, it was a reinterpretation of their cause that he was after. As he concluded with regard to primary rocks,

> [I]n consequence of the great depths at which the plutonic rocks usually originate, and the manner in which they are associated with the older sedimentary strata of each district, it is rarely possible to determine with exactness their relative age. Yet there is reason to believe that the greater portion of the plutonic formations now visible are of higher antiquity than the oldest secondary strata. We shall also endeavour to point out, that this opinion is by no means inconsistent with the theory that *equal*

> quantities of granite may have been produced in succession, during *equal periods* of time, from the earliest to the most modern epochs.[86]

Both Lyell's belief in the uniformity of nature over time ("*equal* quantities . . . *equal periods*") and his desire to show how his own views could accommodate traditional conclusions are present in this passage. His assignment of a great age to the "plutonic formations now visible" conformed to received views.

Lyell's third volume also contained his division of the Tertiary into a subset of four shorter time periods, which he labeled Newer Pliocene, Older Pliocene, Miocene, and Eocene. What distinguished Lyell's approach from others was his statistical use of fossils as markers. As he put it, "[By 1828] . . . I had already conceived the idea of classing the different tertiary groups, by reference to the proportional number of recent species found fossil in each."[87] Lyell's concern was not with the "*characteristic* shells of the different tertiary formations . . . but the shells *common* to two or more periods, or common to some Tertiary period and to the *recent* epoch."[88]

Lyell's third volume also contained several lesser themes that would have been of particular interest to Darwin. Lyell described at length his own fieldwork in Italy, with reference to current issues, as, for example, his argument that no action from diluvial waves was visible on Etna.[89] He also addressed himself to current geological opinion. His remarks included an attack on the mountain-chain dating theory of Jean-Baptiste-Armand-Louis-Léonce Élie de Beaumont (1798–1874), a more directly critical treatment of Buckland's diluvial views, and reference to critics of his own first volume.[90] Among the latter, Lyell took issue with even "a friendly critic," who charged him "with endeavouring to establish the proposition, that 'the existing causes of change have operated with absolute uniformity from all eternity.'"[91] There were also references to very recent scientific meetings, which brought Darwin in touch with current affairs in London. Thus, Sedgwick's presidential address of 1831 to the Geological Society of London is quoted for its support of Élie de Beaumont's theory, while Conybeare's report of the same year to the British Association for the Advancement is quoted against the theory.[92]

As mentioned already, Darwin probably received volume 3 of Lyell's *Principles* in March 1834 while at the Falkland Islands. He "deferred" reading it until after his return from a three-week field trip (18 April–8 May 1834) up the Santa Cruz River.[93] By this time in the voyage, Darwin was working with assurance on his own. He could afford to read Lyell's new volume at leisure. Of the expedition up the Santa Cruz, Darwin wrote to Henslow about its importance for showing "a transverse section of the great Patagonian formation." He went on to link his own findings to the third volume of the *Principles:*

> I conjecture (an accurate examination of fossils may possibly determine the point) that the main bed is somewhere about the Meiocene [*sic*] period, (Using Mr Lyell's expression) I judge from what I have seen of the present shells of Patagonia.—[94]

Darwin was demonstrating to his mentor Henslow that he had absorbed Lyell's new terminology and was ready to begin using it.

Another striking instance of Darwin using Lyell's third volume occurred the next year while Darwin was exploring the geology of the Andes. Darwin wrote to Henslow in August 1835 of what he learned "reading Lyell":

> The geology of these Mountains pleased me in one respect; when reading Lyell, it had always struck me that if the crust of the world goes on changing in a Circle, there ought to be somewhere found formations which having the *age* of the great Europæan secondary beds, should possess the *structure* of Tertiary rocks, or those formed amidst Islands & in limited Basins. Now the alternations of Lava & coarse sediment, which form the upper part of the Andes, correspond exactly to what would accumulate under such circumstances.[95]

Lyell was training Darwin's eye.

Before leaving Lyell, we should note that Darwin did not use the terms "uniformitarian" and "catastrophist" during the voyage. These terms, like "right wing" and "left wing" in politics, have etched themselves indelibly into common vocabulary. William Whewell coined them in his review of the second volume of the *Principles.* He wrote,

> Have the changes which lead us from one geological state to another been, on long average, uniform in their intensity, or have they consisted of epochs of paroxysmal and catastrophic action, interposed between periods of comparative tranquility?
>
> These two opinions will probably for some time divide the geological world into two sects, which may perhaps be designated as the *Uniformitarians* and the *Catastrophists.*[96]

Later, after the voyage, Darwin did on occasion make use of the sectarian vocabulary. However, he generally preferred the term "gradual" to use in opposition to "catastrophist."

Also in the mail for Darwin at the Falklands in 1834 was the one-volume report, published in 1833, of the first two meetings (York in 1831, Oxford in 1832) of the British Association for the Advancement of Science. The report contained a number of interesting items. The two most important from the geological point of view were Buckland's account of the fossil remains of a *Megatherium* brought back to England by Parish, and Conybeare's report on the state of geological science, accompanied by a magnificent colored fold-out map illustrating a "Section across Europe from the North of Scotland to the Adriatic."[97] The report on mineralogy was given as a separate topic, supporting the notion that mineralogy and geology had diverged as disciplines.

With Darwin's receipt at the Falklands of the first report of the British Association, one can sense him responding to fresh geological news from home. He wrote to Henslow,

> I am now reading the Oxford Report.—the whole account of your proceedings is most glorious; you, remaining in England, cannot well imagine how excessively interesting I find the reports; . . . they cannot fail to have an excellent effect upon all those residing in distant colonies, & who have little opportunity of seeing the Periodicals.—[98]

Darwin also added a request for a report of the third meeting of the British Association at Cambridge in 1833. His shipboard library was expanding.

The Geological Society of London

In the first half of this chapter we looked at the field of geology from the point of view of the books that formed part of the *Beagle*'s library. However, books, by their nature, are static: they bear a date of publication, they reflect a moment in time. Science also requires longitudinal study. The question then becomes, what would be an appropriate vehicle for looking at the field of geology as it was presented to Darwin when he was most active? The relevant period runs from Darwin's early acquaintance with the subject through the mid-1840s, when he published his major geological work from the voyage.

In the 1830s geology was pursued under a variety of guises. Outside London local literary and philosophical societies played a role. In 1835 the founding of the Geological Survey offered paid employment, and hence an institutional nexus, for the pursuit of the science. Mining interests also fostered activities related to geology. Fossil collecting was a remunerative activity for a number of individuals. However, from the 1820s through the 1840s, for an English gentleman with a university degree living in London there was one outstanding institution devoted to the science: the Geological Society of London. It was within its fold that Darwin made his mark, and its activities offer the appropriate backdrop to understand his geology.

Convenient for historians, there exists a defined body of texts that facilitate study: the annual addresses of the presidents of the Geological Society of London.[99] These grand affairs were attended by invited men of rank, as well as by Fellows of the society. For the period under discussion, the addresses reflect a good portion of the science of geology as it was practiced in England and Wales, and to a considerable degree in Britain as a whole.[100] Simon Knell has argued persuasively that provincial geological science, based on a culture of collecting and museum display, was robust and creative during the period.[101] Yet, as an institutional embodiment for the more academic aspect of geological science, the London society dominated. Its meetings in Somerset House on the Strand, in central London, offered formal presentations of high order (Figure 2.5). Further, however gentlemanly, its proceedings, whether at the formal occasions or at its dining club—an elite within an elite—were never dull, at least in the golden decades from the 1820s through the 1840s. Eyewitness accounts testify to the frequently gladiatorial tone of the debates that followed the meetings, debates often balanced by conviviality over dinner. Evenings ran long, posing a difficulty for the salaried man. In 1848 Andrew Crombie Ramsay (1814–1891) wrote, "Got home from the Society by one in the morning. It is always too late there."[102] But still he came. The intellectual attraction outweighed the costs.

As a relatively young organization, founded in 1807, the Geological Society was established with the intention of furthering the discipline by collaborative research.[103] While the Royal Society of London, founded in the seventeenth century, included geology within its purview, and offered more prestige to its fellows, it was necessarily trained on a variety of subjects and was, by design, limited in membership. The British Association for the Advancement of Science, newly founded in 1830, provided another forum for geology. It was also current with the state of the discipline in the sense that by 1834 the British Association grouped geology with geography, in one section, rather than with mineralogy, which was instead grouped with chemistry.[104] Nevertheless, for all its ingenuity of design, the British Association was dependent on the Geological Society of London, rather than the other way around. A number of key members of the Geological Society also became key figures in the British Association, and indeed dominated meetings in its early years. And, most significantly, key members of the Geological Society used the summer meetings of the British Association for the rehearsal of disputes that would find their resolution in the autumn and winter meetings of the society.[105]

Figure 2.5. Somerset House. This image is reproduced by courtesy of the Geological Society of London.

By virtue of their office and their individual competencies and associations, the presidents of the society were in a position to speak authoritatively on behalf of the society and of the science. The presidents during the period under discussion were as follows:

1827–1829	William Henry Fitton (1780–1861), an old Huttonian and an expert in the stratigraphy of southern England
1829–1831	Sedgwick
1831–1833	Murchison, coworker with Sedgwick, and subsequently author of *The Silurian System* (1839)[106]
1833–1835	Greenough, a founder of the society, in his third presidency
1835–1837	Lyell
1837–1839	Whewell, from 1828 to 1832 professor of mineralogy at Cambridge
1839–1842	Buckland in his second presidency
1841–1843	Murchison in his second presidency

On this impressive list, one misses a few names of potential presidents—De la Beche for one—but a good percentage of the leading figures of the society for the period served.

On occasion there was some politicking for the position. Lyell was keen that Buckland not succeed him as president. There had been dissatisfaction on the part of Sedgwick and Murchison with Buckland's chairmanship of the geological section of the British Association at its meeting in August 1836.[107] Lyell and Buckland had also been at odds over the interpretation of superficial deposits, the "diluvium." Whewell—who had to be persuaded to take the job—was the preferred choice. Still, Buckland became president after Whewell.

Presidents were conversant with each other's work. Indeed, that work was often done in concert. Lyell, for example, studied under Buckland at Oxford and traveled with Murchison on the Continent. Murchison, in turn, traveled with Sedgwick. Sedgwick and Whewell, both Fellows of Trinity College, might combine more general scholarly travel as part of the "Cambridge network."[108] In 1824 they traveled to the Lake District together, and to Edinburgh in 1825.[109] Greenough, as mapmaker and keeper of that tradition within the society, worked with everyone else. For example, in 1816, with the intention of correlating formations of the continent with those in England, he traveled to Germany in the company of Buckland and Conybeare, visiting, among others, Werner and Johann Wolfgang von Goethe (1749–1832).[110] Associations in the upper ranks of the Geological Society were so close that Martin Rudwick, in his study of the origin of the idea of the Devonian system, selected ten geologists as key figures during the years 1834–1842. Half of these ten held the office of president of the Geological Society during the years we are considering.[111]

Both complementary and conflictual aspects to collaborative work by the society's members should be acknowledged. Murchison's energy and drive, so prominent in his extensive travel, were channeled to great effect after he had been tutored in the field by Sedgwick, who brought to bear not only his keen observational skills but also his university training in mathematics.[112] However, Sedgwick's mathematical skill did not rub off on Murchison. But Murchison's imperiousness and combative nature took their toll on his colleagues: on De la Beche who was a frequent target during this period, and, later, on Sedgwick, who suffered the sense, less silently as time went on, of having been shortchanged in his collaboration with Murchison. Still, Murchison and Sedgwick working together proved superior to either man working alone, and De la Beche's maps for the Geological Survey would not have had their long-term significance without the addition of Murchison's and Sedgwick's perspectives.

Institutional commitments and projects also mattered. Greenough's and De la Beche's commitment to map-making drove much of the society's agenda during these years. Similarly, the union of university and church during the period could on occasion make Sedgwick, Buckland, and Whewell distinct in their perspective from geologists who lacked their institutional loyalties. In addition, old philosophical alliances could come through: Fitton and Lyell were Huttonians, come what may. Formal education, or the lack of it, could also tell: university men, such as Whewell or Sedgwick, were better informed on a wide range of sciences than most of their non-university-educated peers in the Geological Society. Noticeably, Buckland, Lyell, and Whewell's work displayed the sophistication and embellishment of their classical training. And among the gentlemanly geologists of the society, there were distinctions of wealth and leisure. Lyell's independent income of £650 a year permitted him his career as a free-ranging author.[113] Buckland and Sedgwick held clerical appointments affording substantial income, £1000 and £600 per annum, respectively.[114] A *Punch* cartoon entitled "Geology of Society" ranged men from those in the "Primitive Formation" who dined at one o'clock and drank stout out of pewter, to those in the "Superior Class" who wore coronets.[115] The society's membership spanned a much narrower breadth, but some were clearly better placed to travel and write than others.

The officers and members of the Geological Society of London were so easily conversant with each other's work because theirs was a voluntary association born of mutual interest and of the habits of a common gentlemanly life. Their alternation between London society in the colder months and outside activities in the warmer months paralleled the life among gentlemanly circles generally. It is well known that the early mineral surveyors, such as John Farey (1766–1826) and Smith, were excluded from the society in its early years, not only by income but also by habits of life. Hugh Torrens has described the achievements of these men, and their perhaps surprisingly strong support from such an aristocratic and highly placed patron as Joseph Banks.[116] It was also difficult for a geologist to walk the line between receiving paid employment as a geologist, particularly of the society, and being a member of the society. This was the situation faced by Thomas Webster (1773–1844), a gifted stratigrapher who for a time served as both member of council of the Geological Society and a paid employee of the society. He left, stating that the society was a "bad lot."[117] The founding of the Geological Survey in 1835 eased the situation for combining paid employment and interest in the subject.

Darwin fit the gentlemanly profile of the society perfectly. Prior to his return to England he had written to Henslow, asking to be put forward for membership.[118] Henslow responded with a nominating letter to the society describing Darwin as "A.B. of Christ's College Cambridge, & resident in Shrewsbury, a gentleman well versed in geology & other branches of Natural History." Beneath Henslow's signature, to which he had added the assurance that his words were from "personal knowledge," were the signatures of Sedgwick, Robert Hutton (1784–1870), John Forbes Royle (1799–1858), William Clift (1775–1849), and Parish. (Henslow's alteration of "Member" to "Fellow" signaled a change in language dating to 1826 following the granting of the Royal Charter.)[119] The grouping of signatories on Darwin's application partly reflected his background—Henslow and Sedgwick from Cambridge, Parish as a government man with experience in South America—and partly some combination of convenience and positive predisposition, for Clift, Hutton, and Royle were (like Sedgwick) members of council in 1836.[120] Darwin had found a proper home (Figure 2.6).

A little over two years later, on 24 January 1839, he became a Fellow of the Royal Society of London. In the certificate of candidature for admission, he was described as a person "well acquainted with geology, botany, zoology."[121] The ranking of his specialties was significant. Among the sixteen signatories to his petition were such distinguished names in geology as Murchison, Whewell, and Herschel.

A digression is now in order in our discussion of the composition of the Geological Society, for women were excluded from membership in the Geological Society throughout the nineteenth century. (This was in contrast to the Zoological and Botanical Societies, where women were allowed as members from the beginning.)[122] Two notable early-nineteenth-century women with interests in geology were thus foreclosed from joining male geologists as

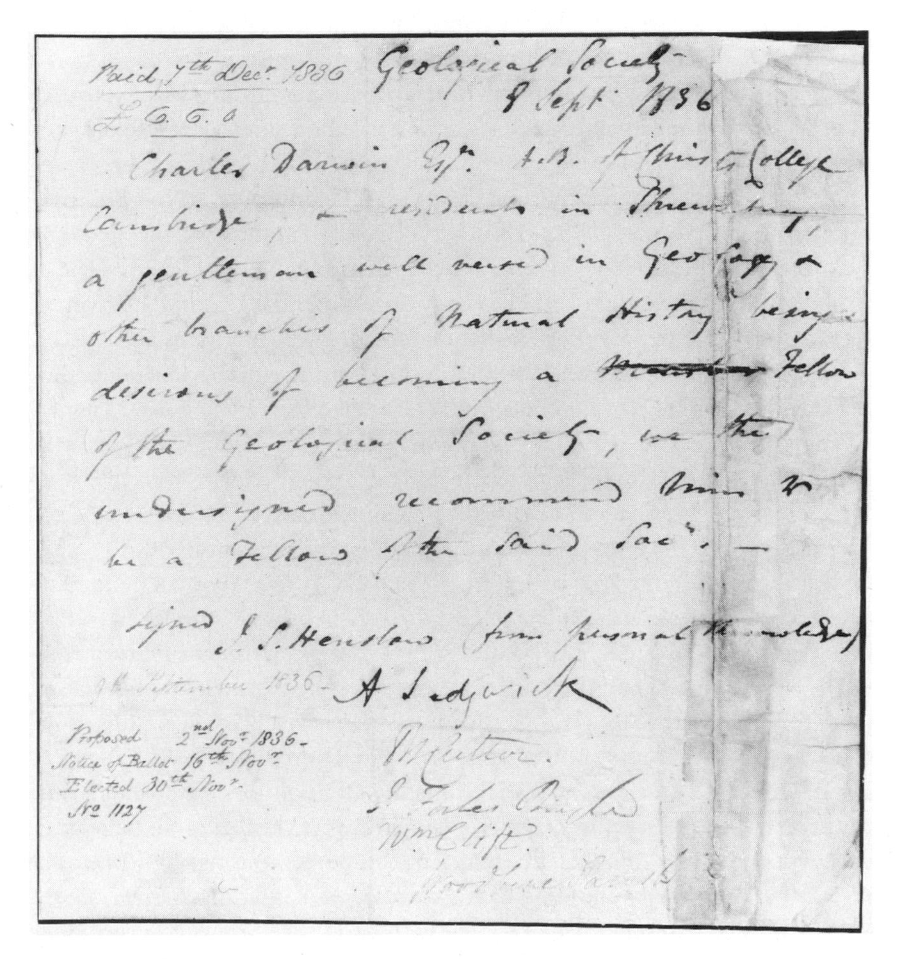

Paid, 7th Decr. 1836
£ 6. 6. 0

Geological Society
8 Sept. 1836

Charles Darwin Esqr. A.B. of Christs College Cambridge, & resident in Shrewsbury, a gentleman well versed in Geology & other branches of Natural History being desirous of becoming a ~~[illegible]~~ Fellow of the Geological Society, we the undersigned recommend him to be a Fellow of the said Socy. —

signed J. S. Henslow (from personal knowledge)
9th September 1836 – A Sedgwick
[illegible]
[illegible]
Wm Clift.
[illegible]

Proposed 2nd Novr. 1836 –
Notice of Ballot 16th Novr.
Elected 30th Novr.
No 1127

Figure 2.6. Darwin's admission certificate to the Geological Society of London. This image is reproduced by courtesy of the Geological Society of London.

colleagues in a formal association. Mary Anning (1799–1847), of Lyme Regis, the fossil collector, who was, in Murchison's words, the "indefatigable purveyor to the storehouses of our science," was honored by the society but was never a Fellow.[123] The other woman who was a prominent contributor to geological science was Maria Graham (1785–1842). Her observations on the effects of the 1822 earthquake in Chile were published in the society's *Transactions*.[124] Graham's publication was especially significant to Lyell and Darwin since it pertained to elevation. In his presidential address of 1834, Greenough challenged Graham's conclusions; she responded in a published rejoinder.[125]

A woman interested in geology might have had the opportunity to assist a male Fellow. Since geology in Britain was still pursued largely by unpaid practitioners, extra hands were often welcome. Buckland and Murchison were early beneficiaries of the labors of spouses who had independent interests in geology.[126] Darwin himself might well have taken a wife reared in a geological household, and one perhaps inclined to be helpful in its pursuit. In 1832 Lyell married Mary Horner (1808–1873), the eldest daughter of the geologist Leonard Horner (1785–1864) and his wife Anne Susan Horner (1789–1862). Among Mary's five sisters still at home was a potential wife for Darwin.[127] Certainly the "Misses Horner" had shown academic leanings, as is suggested by their joking in Maori with Darwin.[128] (On hearing of Darwin's impending marriage to someone outside the Lyell-Horner circle, Lyell's sister opined, "I suppose he will be buried in the country & lost to geology."[129]) Horner and Lyell, both of whom had interests in educational reform, were the leading Fellows advocating the admission of women to the society. Horner, as president in 1860–1862, pressed unsuccessfully for the cause.[130]

Darwin wed Emma Wedgwood on 29 January 1839. Her own cultural interests centered on music; she was a pianist sufficiently accomplished to have taken lessons from Frédéric Chopin (1810–1849).[131] In her inclination toward the arts rather than the sciences, Emma was like Darwin's other known love, Fanny Owen (1807–1887), a painter.[132] Perhaps Darwin sought someone his equal in maturity (both Fanny and Emma were slightly older than he), but of different tastes. In any case, Emma was of good fortune and well traveled; she had even seen the Temple of Jupiter Serapis at Pozzuoli, beloved of geologists. (Of the Temple, she noted its "immense columns" and that the "Circirone [*cicerone,* guide] told us the sea had been as high as halfway up the columns & the bases had been covered with lava."[133]) Emma had the added advantage of sharing Lunar Society grandparents with Charles, her first cousin. Perhaps more than any of the other prospective wives to Charles, she was thus prepared to tolerate the transmutationist speculations with which her fiancé was then occupied. She also balanced the marriage by not deferring to Darwin too readily. Thus, before their wedding day, she counseled him, "I hope you will manage to finish Glen Roy now & get shut of it."[134] But, a few days later, writing again to her "own dear old geologist," she allowed with equanimity that "I suppose you are going to the Geological today."[135] Unlike the women in the Lyell-Horner circle, Emma displayed no eagerness to join the gentlemen in their geology. She was at ease with the fact that in its institutional setting, geology remained a male preserve.

The presidential addresses were intended to be reviews of the progress of geology, chiefly in Britain, during the preceding year. The speakers varied in the manner in which they interpreted their mandate: Murchison, for exam-

Figure 2.7. De la Beche's sketch of a Geological Society meeting. This image is reproduced by courtesy of the Geological Society of London.

ple, included more discussion of the work of foreign geologists than did the others, whereas Whewell favored grand philosophical summary over discussion of individual papers. Yet all the authors understood that their remarks were to reflect the entire field. The Geological Society meetings were famous for the freewheeling discussions that took place following the formal reading of papers. Provision for these discussions was written into the society's bylaws, but publication of their content was prohibited (Figure 2.7).[136] However, one can gather from the annual presidential addresses which papers were discussed and attended to and which were not, as well as the tenor of debate. One can also gain a sense of the momentum of the society—where its leaders felt it had been and where it was going. Less personal than correspondence, yet more impressionistic and informal than papers, the presidential addresses allow us to listen in on the conversation of English geologists in the 1830s.

The most immediately striking feature of the conversation of English geologists in the 1830s was its optimism. These were boom times for geologists, and every president referred with obvious satisfaction to the society's prosperity and prospects for growth. Membership, excluding foreign and honorary members, climbed from approximately five hundred in 1830 to about eight hundred in 1840. But the growth in knowledge during this period is even more striking. It required twelve pages to print Fitton's address in 1828, but eighty-six pages for Murchison's in 1843. Even allowing for Murchison's expansive rhetoric, that was a dramatic increase. In addition, during the period, the science of geology was crowned with official recognition. On the recommendation of the Royal Society, the government sought the counsel of the Geological Society in such endeavors as establishing mapping conventions and standards of coloration for ordnance maps.[137] Such developments contributed to the public standing of the society and its authority, even though traditions for coloring maps developed but no fixed universal standard was adopted.[138]

On the evidence of the presidential addresses, geology was multipolar rather than unipolar in emphasis. Two ideas dominated the minds of British geologists during the 1830s. First was the notion that the primary work of geologists was to determine the true order of succession of strata. Second was the notion that geology should be a comprehensive science with regard to the earth. Put another way, geology was composed of a paradigm—stratigraphy—and a list, which included the remaining, and disparate, subjects that made geology, at least potentially, a comprehensive science of the earth. The meaning of "paradigm" employed here is taken from Kuhn when he wrote, "In learning a paradigm the scientist acquires theory, methods, and standards together, usually in an *inextricable mixture*."[139] This is a narrow construction of Kuhn's term, emphasizing its sense as an "exemplary problem solution" rather than in its broader sense as "disciplinary matrix."[140]

By using the term "paradigm" I do not mean to advocate Kuhn's model for identifying stages in scientific revolutions, though interesting analyses have been done for natural history and for geology on that score.[141] I also do not mean to identify any one individual with the paradigm. Indeed, part of the attraction of Kuhn's notion of paradigm is that it is focused on the field rather than on the individual. Even someone like Sedgwick, who was one of those working at the center of the stratigraphical paradigm, did not own it, though by honoring William Smith, who was not a member, the society in a sense was casting for the role of paradigmatic figure.[142] I also do not wish to see the term "paradigm" as being opposed to the historical and philosophical treatment of individual concepts within the sciences.[143] The widespread use of the word, both academically and popularly, testifies in part to its elasticity and ease of combination with a range of analytical ideas. For historians, it also has the advantage over such alternative expressions as "research programme" in that, from the point of view of common language, "paradigm" places greater emphasis on prior accomplishment.[144] With qualifications, then, paradigm is the right word for the case.

The paradigmatic emphasis on strata was demonstrated concretely in the manner in which the society arranged its collections. As a voluntary association, it depended on donations to maintain its collection. Its museum was to be for members and their guests only, useful to beginners and accomplished geologists, and specimens were to be available for loan.[145] In 1813, working under what John Thackray termed its "academic" faction, the council of the Geological Society voted to have the British collections arranged on stratigraphical principles, rather than geographically. Thackray noted,

> By 1819 the Museum contained about 16,000 registered specimens arranged in five separate collections: simple minerals and rocks arranged systematically; British rocks and fossils, the English mostly arranged stratigraphically and the Scottish and Irish by county; foreign rocks and fossils arranged by country; volcanic productions, and organic remains arranged systematically, together with some recent shells.[146]

In 1819 only the English portion of the collection could begin to receive stratigraphic treatment, but the principle of the need for such treatment was firmly established. As an aside, one may note that in his own house in London, Smith had set up his collection of fossils on sloping shelves to represent the different strata. In 1808 members of the Geological Society had inspected this collection, and no doubt observed its arrangement.[147]

If the paradigm of the Geological Society was mapping strata and, by the 1830s, elucidating the history of the formation of the earth's strata, what I have termed its "list" of interests included other questions that geologists wanted answered and all other topics that appeared to them geological. To some extent, the division between "paradigm" and "list" corresponds to the division Whewell drew in his presidential addresses between "descriptive geology" and "geological dynamics."[148] Certainly the desire to recognize division within the field stemmed from a common perception of its plenitude. Whewell's preference for causal issues affected his naming of the division. As measured by level of activity and achievement, what Whewell termed the "descriptive" side of geology was the more important of the two; hence, I prefer to label it "paradigmatic." Moreover, not all nonstratigraphical topics fitted under the rubric of "dynamics." Paleontology did not fall naturally into either of Whewell's two categories, though he tried at first to place it under "dynamics." Substituting the dichotomy of "paradigm/list" for "descriptive/dynamic" allows one to capture the fullness of the field while acknowledging the central nature, and dominance, of the paradigm. Contrary to Whewell's brilliant but prescriptive analysis, geology was not equally divided between "descriptive" and "dynamic" subject matter. In practice, geology was dominated by the stratigraphical enterprise. Other topics were subordinate. Stated more positively, in retrospect a clear sense of hierarchy of subject matter preserved harmony and cohesion within the field.

Several features of the presidential addresses illustrate the dominance of strata determination in the minds of English geologists. Most obviously, the notion of strata was the central principle around which the presidents structured their presentations. This was true for Fitton, who used De la Beche's table of British strata to organize his material.[149] It was also true, with variations, for all of the other presidential authors. What Whewell called "descriptive" geology, Buckland called "positive" geology. The referent for each was identical: the classification of facts based, as Whewell said, on the belief that the "key of all our geological knowledge of our country" is the "fixed order of strata, characterized mainly by their organic fossils."[150] What is more, discussions on stratigraphy took pride of place: whatever else presidents chose to discuss in their addresses, they first discussed contributions to knowledge concerning the order and character of strata. Another sign of the preeminence of interest in strata is the manner in which geologists assessed the lasting worth of their own and their colleagues' achievements. Murchison (1842) spoke to the point (and for his own ambitions): "The perpetuity of a name affixed to any group of rocks through his original research, is the highest distinction to which any working geologist can aspire."[151] In this vein, Greenough chose to repeat without correction or qualification the grounds on which the council of the Royal Society had awarded a medal to Lyell as author of the *Principles of Geology.* The council, declining to "express any opinion on the controverted positions contained in Mr. Lyell's work," credited the "comprehensive view" Lyell had taken of the subject, his emphasis on "existing causes," his description of "many tertiary deposits," and his "new mode" of investigating Tertiary deposits by "determining the relative proportions of extinct and still existing species."[152] Of the four grounds given by the Royal Society, two had to do with Lyell's work on Tertiary deposits. A modern reader of the *Principles,* innocent of the concerns of Lyell's contemporaries, might well be surprised to see it so much described as one on Tertiary geology. The contents of volume 2 are not mentioned at all directly. Possibly its content, largely devoted to Lyell's antitransmutationism, lay too far down the "list" of geology's concerns as a discipline, though the majority of geologists would have endorsed his conclusions. (On this point it is interesting how differently Darwin was reading the *Principles* at about the same time: all of Lyell's subject matters, including his gradualist opinions, were of interest. Here one sees a solitary reader moving with the author sympathetically point by point, while a corporate group is reading more selectively.)

One may well wonder how Lyell could have played so prominent a role in the society's affairs if his major treatise was given such qualified treatment. On this point Thackray showed that Lyell took particular care to attend to the society's goals. Of the forty-five papers Lyell read to the society, most added "some depth and supporting detail to the argument set out in the *Principles of Geology,* but their overall effect . . . [was] quite small."[153] In short, Thackray's argument suggests that Lyell saved his major theoretical argument for his book, for "the deduction of theories from facts was certainly acceptable [to the society], but anything that smacked of the all-inclusive theories of the previous century was deeply suspect."[154]

Whatever bending it required to make a work like Lyell's *Principles* fit the mold, the preoccupation of these geologists of the 1830s with strata paid handsome dividends, for their achievements rival those of any group of similar size working over a comparable period in any other scientific field. As one might expect, achievement came in stages. At the beginning of the decade the task was to fill in the knowledge of strata at home, notably in England and Wales. With Secondary strata reasonably well worked out by a previous generation of workers, including Smith, and Tertiary strata less present physically than on the Continent, an important group within the society—Sedgwick, Murchison, and De la Beche—set out to study the oldest fossil-bearing strata in Britain.

While the utility of fossils for identifying strata was well established (though constantly reemphasized in presidential addresses), the utility of lithology in identifying strata was also noted. Here is Fitton taking a judicious view:

> The whole series indeed, of the phænomena developed by recent examination in Scotland and the north of England, gives rise to the most interesting speculations on the questions of geological identity, and *of the relative value in geology of mineralogical and zoological characters,*—which has been so ably treated by Brongniart and other continental writers:—questions, which it is necessary to keep continually in view, and that acquire fresh interest and importance in proportion as we extend our researches to the remoter districts of the world.[155]

Further, at the beginning of the period, fossils were attended to primarily for their value as markers of strata, rather than for their purely biological significance. As is well known, the search for, and in, the oldest English and Welsh strata was successful, and from 1830 onward the presidential addresses proclaimed the work of Sedgwick, Murchison, and De la Beche, among others, in identifying what came to be called the Cambrian, Silurian, and Devonian formations. (The Silurian and Devonian were named, permanently, by the end of the decade; the Cambrian as a category had a longer and more fraught history.[156])

The second stage of achievement came as society geologists extended the scope of their results to Europe. By the middle of the decade, English geologists were finding what they termed "geological equivalents" of their own strata on the Continent. (The earlier work of Webster had been exemplary in showing the way.)[157] In 1838 Whewell spoke of "Home" geology, meaning "Northern European." He wrote,

> If we begin with geological facts, our attention is first drawn to that district on the earth's surface within which the facts have been subjected to a satisfactory comparison and classification, which may be considered, in a general way, as including England, France, Italy, Germany, and Scandinavia. The language which the rocks of these various countries speak has been, in a great measure, reduced to the same geological alphabet.[158]

A year later Whewell was pleased to report that the classification applicable to Northern Europe might be extended beyond the Alps:

> In the survey of the progress of our labours which I offered to your notice last year, I stated, that in proceeding beyond the Alps, and I might have added the Pyrenees, we no longer find that multiplied series of strata, so remarkably continuous and similar, when their identity is properly traced, with which we have been familiar in our home circuit. Yet the investigations of Mr. [William John] Hamilton [1805–1867] and Mr. [Hugh] Strickland [1811–1853] appear to show, that we may recognise, even in Asia Minor, the great formations, occupying the lowest and highest positions of the series, which are well marked by fossils, namely the Silurian and Tertiary formations; and also an intermediate formation corresponding in general with the Secondary rocks of the north, but not as yet reduced to any parallelism with them in the order of its members.[159]

This extension of the terminology of "Home" geology prompted Whewell to crow: "As if Nature wished to imitate our geological maps, she has placed in the corner of Europe our island, containing an *Index Series* of European formations in full detail."[160]

The third and final stage of achievement in stratigraphy came, then, as geologists from the society extended their results worldwide. By the 1840s the strata of the British Isles was seen, at least by the British, to be an index series—or as a paradigm—of universal applicability. As Murchison exclaimed with deserved satisfaction,

> The chief aim of this Society has been to gather sound data for classification; and, following out this principle, I have endeavoured to show, how the order of succession established in our own isles, is now extended eastwards to the confines of Asia, and westwards to the backwoods of America. From such researches, and by contributions from our widely spread colonies, we have at last reached nearly all the great terms of general comparison.[161]

To return to Whewell's metaphor, the rocks of the world had been reduced to the same geological alphabet.

As indicated by Murchison's and Whewell's statements, the articulation of the paradigm was proceeding rapidly. By the early 1840s, the paradigm of classifying and mapping strata worldwide had been shown effective. What the presidential addresses for the 1830s and 1840s reveal is the progress of that project year by year. Overall, they support the view of the "classifactory focus characteristic of the first half of the nineteenth century" in geology.[162]

As that project became realized, a secondary project, which had been subordinated to the first, could more freely emerge. This was paleontology. While the elements of that subject had been developing from the 1790s onward, and reached strong expression in England in the 1820s in the work of Buckland, Conybeare, and Gideon Mantell (1790–1852), among others, as a subject paleontology had not yet reached center stage in the discussions of the society, though it was of prime importance in France. There was a lag of several years between the paleontological insight expressed by the leaders in the field and its comprehension by the society as a whole, as represented in the format of the presidential addresses.[163]

The presidential addresses to the society, from Fitton's in 1828 to Murchison's in 1843, manifest the emergence of paleontology as a powerful subdiscipline. Fitton in 1828 did not treat the subject. In 1829 he treated "zoology" and "botany" under separate headings and merely provided a list of fossils found in British strata. In 1830 Sedgwick praised Cuvier and Brongiart. He also commented on the continuity of the fossil record, despite the occasional "lost page in the history of the world." But in the same year Sedgwick also rejected any connection of fossil continuity to transmutation. In 1831 he associated transmutation with the Huttonian hypothesis. (Just prior to his address of that year, Sedgwick, in the course of announcing Smith as intended recipient the first Wollaston Medal, also praised Smith's use of fossils in the identification of strata.) In 1832 Murchison discussed fossil botany and conchology. In 1833, on the death of Cuvier, Murchison credited him for his work on "lost types of creation." In his address, he also included separate sections devoted to fossil zoology and fossil plants. Greenough in 1834 devoted only a single paragraph to "Organic Remains" and, in the next year, 1835, used the phrase again as he discussed the work of De la Beche. In 1836 Lyell used the category "fossil zoology" and expressed special interest in "tracing the remains of the vertebrate animals through geological formations of every age." In 1837 he used the heading "Organic Remains" and applauded Buckland's paleontological work. In 1838 Whewell presented material on "Fossil Zoology" and celebrated Owen, who had been awarded the Wollaston Medal that year. In 1839 Whewell devoted a separate heading, with extended treatment, to "Palæontology." In 1840 Buckland did likewise, and in 1841 he did so again, but with multiple subheadings. In 1842 Murchison did not have a separate section labeled for the subject, but he did cast his presentation of strata with an eye to it: "palæozoic" geology. In 1843 he had a section specifically named "Palæontology."[164]

By 1843, then, it was clear to all in the society that the use of fossils for purely geological purposes to identify strata differed from the use of fossils for more zoological and botanical purposes.[165] Equally clear was that while the successive formations were understood on the basis of fossil evidence, the same evidence could potentially be used to construct a history of life on earth. Murchison had a particularly clear sense of this point. In his presidential address of 1843, he presented a "capsule" history of life. The passage is a long one but worth quoting in full, as it reflects the excitement of the subject matter as well as Murchison's Victorian rhetoric:

> Besides ascertaining where the great masses of combustible matter lie, we can now affirm, that during the earliest period of life, conditions prevailed, indicating a prevalence over enormous spaces—if not almost universally—of the same climate, involving a very wide diffusion of similar inhabitants of the ocean. We have learned, that in the earliest of these stages of animal life, no vestige of the vertebrata has yet been found, whilst in the succeeding epochs of the Palæozoic age singular fishes appear, which, in proportion to their antiquity, are more removed from all modern analogies. In each of these early and long-continued periods, the shells preserving on the whole a community of character, differ from each other in each division—and in that later formation, where a very few only of the same types are visible, they are linked on to a new class of beings, the first created of those Saurians, whose existence is pro-

> longed throughout the whole Secondary period; whilst we have this year seen reason to admit that even birds (some of them of gigantic size) may have been the cotemporaries of the first great lizards. With the close of the Palæozoic æra we have also observed a gradual change in the plants of the older lands, and that the rank and tropical vegetation of the Carboniferous epoch is succeeded by a peculiar flora. In the next, or Triassic period, we have another flora, whilst new forms of fishes and mollusks indicate an approach to that period when the seas were tenanted by Belemnites and Ammonites, marking so broadly these secondary deposits with which British geologists have long been familiar, and which, commencing with the Lias, terminate with the Chalk. And, lastly, from the dawn of existing races, we ascend through successive deposits gradually becoming more analogous to those of the present day, until at length we reach the bottoms of oceans so recently desiccated, that their shelly remains are undistinguishable from those now associated with Man, the last created in this long chain of animal life in which scarcely a link is wanting!—all bespeaking a perfection and grandeur of design, in contemplating which we are lost in admiration of creative power.[166]

As the sweep of this passage suggests, British geologists, as organized in the Geological Society of London, had expanded their insights beyond stratigraphy. There are aspects in the above passage to which some geologists would have objected. Lyell, for example, would not have shared Murchison's belief in universal or near-universal climatic changes.[167] But all geologists would have shared a belief in the use of fossils to reconstruct the past. This new goal came sharply into focus within the society only when its members had achieved their original goal, at least in outline, of determining the true succession of British strata. At this point, the work done by such members of the society as Buckland and Mantell on vertebrate fossils gained even greater meaning as it converged with knowledge of the succession of strata. By 1840, Buckland, with only a small amount of special pleading, could describe paleontology as "this most essential and perhaps most generally interesting branch of our subject."[168] Further, De la Beche, in 1845, as he wrote out "Instructions for the Local Directors of the Geological Surveys of Great Britain & Ireland," could refer, routinely, to "the Palæontologist" as a recognizable occupational title.[169]

But stratigraphy, with its quickly maturing partner paleontology, did not exhaust the interests of members of the Geological Society in the 1830s. The earth is more than its strata, and members of the society during the period insisted on theirs being a comprehensive science. This ideal had two dimensions. First, geologists wanted their science to continue to have interlocking ties with other sciences. They regarded their science as rather like a house with many empty or half-filled rooms. The earth was, after all, a planet, and geologists must reserve room in their science for the insights of physical astronomy. A similar argument was made for chemistry. While most geologists might prefer fieldwork to mathematical computation or laboratory analysis, some aspects of geology required these techniques.[170] Moreover, as geologists were pleased to point out, they had much to offer related fields: certain kinds of measurements to astronomers, rock specimens to chemists (one recalls here the work of Humphry Davy [1778–1829] and Jöns Jacob Berzelius [1779–1848]), and, ever increasingly, fossils to zoologists and botanists.

The second way in which geology was intended to be a comprehensive discipline was as a science of causes, or, as Whewell put it, of "dynamics." As is well known, debates over geological causation were heated during the 1830s and centered on the work of Lyell. While Lyell's impact on geology has usually been discussed in terms of his contributions to the long-running uniformitarian-catastrophist debate, the presidential addresses suggest that Lyell's work had, in addition, a more immediate and consensus-producing impact on the collective mind of the Geological Society. With publication of volume 1 of Lyell's *Principles* in 1830, three successive presidents announced their commitment to his claim that the Noachian flood be discounted as an agent for explaining the origin of what were then termed "diluvial" deposits. For two of these presidents, Sedgwick and Greenough, these announcements required them to abjure previously held views.

In addition to historical causes, two other classes of causes interested English geologists in the 1830s. The first referred to what Whewell termed "ulterior" causes. Chief among them were those pertaining to the interior of the earth. Despite the nebular hypothesis and the fact that a majority of geologists believed the interior of the earth to be molten, this class of questions had thus far received little attention in the presidential addresses, being, in Whewell's words, "obscure, but questions which we must ask and answer in order to entitle ourselves to look with any hope towards geological theory."[171] Whewell then went on to quote the work of William Hopkins (1793–1866), Herschel, and Charles Babbage (1791–1871)—men not active in the classification of rocks but in the sorts of mathematical questions pertaining to the earth that also interested Whewell.

The second class of causes, which Whewell termed "proximate," were more amenable to the investigation of geologists working with existing methods.[172] The proximate causes receiving the greatest attention in the 1830s were the opposing forces of elevation and subsidence. ("Elevation" refers to the raising of the earth's crust; "subsidence," to its lowering.) While Lyell in the *Principles* was chiefly responsible for focusing the attention of geologists on the action of these forces, he was not alone in invoking them. Murchison said in his 1832 address that "all inquirers agree in this fundamental opinion, that the earth's surface has been mainly brought into its present condition by numerous changes of relative level between the land and the sea."[173] In this passage, Murchison also credited "Mr. Lyell for his energetic attempt to elucidate the modes of action by which, in the ordinary course of nature, such revolutions may have been effected." In addition, Greenough said in his 1834 address, even as he prepared to attack Lyell's views, "Among the subjects which have for some years past engaged the thoughts of geologists none perhaps has excited so general and intense an interest as the Theory of Elevation."[174] Confirming Greenough's comment, Lyell's presidential addresses of 1836 and 1837 and Whewell's addresses of 1838 and 1839 devoted considerable attention to the subject of elevation and subsidence. Darwin's work from the *Beagle* voyage touching on elevation received great play in Whewell's 1838 address. Whewell said,

> Guided by the principles which he learned from my distinguished predecessor in this chair [Lyell], Mr. Darwin has presented this subject under an aspect which cannot but have the most powerful influence on the speculations concerning the history of our globe, to which you, gentlemen, may hereafter be led.[175]

Clearly Darwin's work had found a home in a portion of the society's agenda.

The 1840s saw a change. In his presidential address of 1841 Buckland introduced a new contender in the class of proximate causes: the glacial hypothesis, introduced to the society in 1840 by Louis Agassiz (1807–1873).[176] Soon the explanatory possibilities of glaciers attracted the close attention of British geologists. Moreover, since Agassiz's glacial hypothesis explained the origin of superficial deposits of detritus (the former "diluvial deposits") as well or better than Lyell's drifting iceberg hypothesis (itself based on the forces of elevation and subsidence), the victories of Agassiz's hypothesis, even if partial, drew attention away from elevation and subsidence as proximate causes. The effect on Darwin's geology of the new glacial theory coming out of Switzerland will be discussed in chapter 8.

Another aspect of geological dynamics in the period under discussion is the question of species. As Whewell pointed out in his address of 1838, the laws of change regarding species fall under the province of geology by virtue of bearing on the "history of the globe."[177] What is equally interesting is that Whewell made this point in order to discuss some of Darwin's work arising from the *Beagle* voyage.

Geology's "list" was not precisely equivalent to Whewell's term "dynamics," which, indeed, was a rhetorical organizing principle that he alone employed. Other subjects that were discussed, at least by some of the presidents, included igneous rocks (as a general heading), mineral veins, faults, craters of elevation, and mountain building. However, there was no general heading for petrology or for structural geology.

Now, to conclude: geology as practiced within the Geological Society of London during the 1830s and 1840s was multipolar in emphasis. One might say that it was composed of a paradigm and a list. The term "paradigm" carries with it the sense of dominance and centrality. In the case of British geology in the 1830s and 1840s, the paradigm involved the activities of mapping, classifying, and ordering that were central to the collaborative work of the Geological Society. "Paradigm" also implies success and accomplishment, which were strikingly evident in the society's endeavors of the period.

The term "list" suggests the disparate nature of other inquiries. The items on the list included the topics of elevation and subsidence, of volcanoes, and of the figure of the earth; the species question; and paleontology. With the advantage of hindsight, one might say that while paleontology was logically implicit in the study of the earth's crust, and the collection, naming, and description of fossils was proceeding apace, its emergence as a fully distinct subdiscipline in Britain awaited an established succession of strata. Cuvier's vertebrate paleontology was established earlier in France. When paleontology fully emerged, as in Whewell's address of 1839 and Buckland's presidential addresses of 1840 and 1841, it was treated under its own heading. Until then, the study of fossil remains "might at first sign seem to form a part of zoology rather than of geology," as Whewell said, when he set about to argue "how essential a part of our science the zoology of extinct animals is."[178] Later, as the Geological Survey of Great Britain got underway, De la Beche as its director would lead its workers to "detailed reconstruction of ancient environments."[179] This project would draw on both stratigraphy and paleontology.

Darwin's Place

Where, then, does young Darwin fit into this picture? Two anecdotes suggest that he was central to the field socially but slightly away from the center intellectually. As indicated by his certificate of admission to the society, Darwin had been proposed as a Fellow at its first meeting of the season on 2 November 1836. A few days later, on 11 November, Greenough wrote to De la Beche that

> Darwin dined at the Club and is about to publish a Volume of incredibilia on the effects of Plutonism in the countries he has traversed; he is an intelligent agreeable man.[180]

Greenough's offhand remark bears scrutiny for, in a brief compass, it suggests Darwin's social compatibility with the society's gentlemen—"he is an intelligent agreeable man"—while indicating the grounds of Greenough's suspicion. Greenough's slighting reference to Darwin's "incredibilia" suggests a lack of confidence on his part in the merits of the "Plutonic" enterprise. (By "Plutonic" Greenough would have included the intellectual descendant of Hutton, namely, Lyell.) In Darwin, Greenough recognized a geologist different from himself. He confided as much to De la Beche, who, like Greenough was involved with map-making and similarly skeptical of Lyellian theorizing.[181] In the late 1830s, Greenough was at work revising his map, the second edition of which the society published in 1840, and it was this enterprise that would have seemed paramount in importance to him, much more so than any plutonic "incredibilia."[182]

The second piece of anecdotal evidence reflecting on Darwin's place in the field is also from a letter. Dated 1846, ten years after the first letter, and written by Leopold von Buch (1774–1853) to Murchison, it goes to the same point:

> A map is always a decisive criterion of they who aspire to the rank of geologists[.] [E]very one who has not compiled a map, wants the necessary talent of combination. The spirited Darwin, with all his remarkable vivacity of mind, is for me no Geologist, only an able history maker of what nature as he believes, has done, and what never she did. . . . This man could never make a tolerable geological *map*.[183]

The trio of Greenough, De la Beche, and Murchison, who were involved in commenting on Darwin's position in the field, were themselves operating at its center, its paradigm. In contrast, Darwin was working more from geology's list than from its core paradigm, which, as von Buch correctly observed, involved map-making.

Some of this difference can be traced to Darwin's temperament, which inclined him toward high theory, and some, perhaps, to a familial alliance with Huttonianism. But most important was his role as global traveler. De la Beche and Murchison could return time after time, as often as was needed, to the same British or Continental site. Greenough, as the emblematic mapmaker, could incorporate their British results in the revisions for his map. In contrast, the young Darwin was always on the move. He was not in a position to contribute toward establishing the "*Index Series*" of formations, except, primarily, as a provider of specimens and observations from distant places—in effect a first-class colonial agent whose work could show its full value only after the initial work at home was suitably advanced.

In actuality, of course, von Buch was not quite fair in his assessment of Darwin as a mapmaker. From early on Darwin sketched the geological formations present in the areas he explored in South America. Of course, since these were unpublished maps, von Buch would have had no knowledge of them. But, in his disparaging comments, von Buch ignored Darwin's brilliantly synthetic published maps of coral reef distribution, which, to borrow von Buch's phrase, did indeed demonstrate a "necessary talent of combination" on Darwin's part. However, von Buch was absolutely on course in his criticism in the sense that Darwin's maps of coral reef distribution were not typical geological maps. They did not mark off stratigraphical boundaries. Nor had Darwin actually visited most of the islands he colored on his map; this would have been an anathema to von Buch. Further, Darwin was not directly engaged in establishing what would come to be called the "geological column," and this is what piqued some of his geological peers. What Darwin wrote in geology was consistent with and played off the main work of the Geological Society. But for the period of the 1830s and 1840s, its focus was on questions drawn more from geology's list than from its paradigm. Darwin's creativity flourished in the interstices of the discipline.

As we look at Darwin's entrance into the society, we should note how strongly he conformed to many of its best hopes, but in a way that embraced the society's multipolar interests. His fieldwork, as collector and observer, was exceptional, and its value was quickly recognized. The *Megatherium* bones he had sent back from South America were exhibited in 1833 at the Cambridge meeting of the British Association for the Advancement of Science.[184] In addition, Lyell summarized at length the contents of some his letters to Henslow, originally printed as a pamphlet by the Cambridge Philosophical Society, in his presidential address of 1836. He found Darwin's observations on the structure of the Andes, on fossil bones, on evidences for the recent rise of land in Chile, and on a petrified forest congenial to his own themes as a geologist.[185] The next year, with Darwin now home, Lyell again cited Darwin's South American work, this time with regard to his observation of evidence of recent elevation of the coast of Chile.[186]

Whewell, the next president of the society, was equally well disposed to Darwin's geology. A member of the powerful Cambridge network, Whewell was himself working outside the society's paradigm. Indeed, at one point Whewell referred to his rather "imperfect acquaintance with your science"—assigning to himself the status of outsider, though one available "to look at the subject in its largest aspect."[187] Like Darwin, Whewell was interested in the dynamical aspects of geology. If Lyell was a man of the paradigm and the list, Whewell was a man of the list solely. In Whewell's 1838 address, he praised Darwin for his coral reef paper, for his discussion of climate in the

Journal of Researches, for his fossil collection, and for his paper on earthworms.[188] In his 1839 presidential address Whewell again referred to Darwin's work on elevation.[189]

The concrete manifestation of Darwin's high status in the society was his speedy inclusion into the inner circle of its officers. Whewell pressed the office of secretary on Darwin as early as March 1837. Darwin had only just moved from Cambridge to London, and was not yet even in his own house, when he had to respond. He used the circumstance of his unfinished *Beagle* journal manuscript to beg off: "I really cannot accept the office.—I have to write the third volume of Capt. FitzRoy's account of our expedition." He also pleaded his "ignorance of English geology," a subject key to the society's disciplinary paradigm.[190] A few months later, in October 1837, Whewell came back to him. Darwin's *Beagle* manuscript was now complete. But Darwin wrote to Henslow "in loco parentis" to reiterate his reasons for not accepting the position of secretary:

> 1st. My entire ignorance of English geology, a knowledge of which would be almost necessary in order to shorten many of the papers before reading them, before the Society, or rather to know what parts to skip—Again my ignorance of all languages; & not knowing how to pronounce even a *single* word of French,—a language so perpetually quoted. It would be disgraceful to the Society to have a Secretary who could not read French. 2d. The loss of time. [Here Darwin detailed his work in progress on his geology and on the "government work"—meaning the *Zoology.*] My last objection, is that I doubt how far my health will stand, the confinement of what I have to do without any additional work.[191]

In the end, Whewell's appeal succeeded. Darwin was elected one of the two secretaries at its annual meeting of the society on 16 February 1838. He served for three years, until the annual meeting held on 19 February 1841.

The duties for the secretaries were substantial. The foreign secretary handled foreign correspondence, but the two domestic secretaries were in charge of the following:

> 1.—It shall be the duty of the Secretaries:
> 1°. To conduct the correspondence of the Society and Council
> 2°. To attend the general meetings of the Fellows, and the meetings of Council; to take minutes of the proceedings of such meetings
> 3°. At the ordinary meetings of the Fellows, to announce the presents

made to the Society since their last meeting; to read the certificates of candidates for admission into the Society; and the original papers communicated to the Society, or the letters addressed to it

4°. To make abstracts of the papers read at the ordinary general meetings, to be inserted in the minutes

5°. To see that all such minutes and abstracts of the proceedings . . . are entered by the Clerk in the several minute books

6°. To edit the Transactions of the Society; and to superintend the printing, and the making of the Index of each volume

7°. To give directions for placing in the cabinets all newly-presented specimens.

2.—The Secretaries . . . shall . . . divide among themselves the performance of the duties above enumerated and shall communicate to the . . . Council . . . which of those duties they have each undertaken to perform.

3.—The Secretaries . . . shall be Members of all Committees appointed by the Council.[192]

Since it was their usual task to read aloud papers at meetings, secretaries were prominent individuals in the society. Moreover, they were responsible for editing papers for publication in the *Proceedings*. This power included shaping papers to meet the council's standards with regard to expression, and apparently sometimes, its views.[193] Thus, the position of secretary was key to the society's functioning.

As he was considering the appointment, Darwin estimated that it would "*at least* cost me three days (& often more) in the fortnight" to do the society's business.[194] In his term of office, Darwin inevitably became deeply embedded in the daily workings of the society, and thus with its definition of the field of geology.

Another sign that Darwin had become embedded in the society's work is suggested by referee reports on papers being considered for publication in the *Transactions* of the society once they had appeared in the *Proceedings*. Fortunately, 231 reports for the years 1829–1842 are held in the society's archives, where they were catalogued in 1985. (The only other period for which referee's reports survive is 1818–1825.)[195] All three of Darwin's papers for which reports are available were reviewed positively for inclusion in the *Transactions*. Buckland reviewed one paper ("On the formation of mould"); Sedgwick, two others ("Volcanic phenomena" and "Raised beaches in Chile").[196] It is interesting that Sedgwick, rather than Lyell, reviewed the paper on volcanic activity. However, as a senior member of the society, he might be presumed to have standing to review publications from the whole range of the society's subjects.

Buckland reviewed "Mr Darwyns" paper (the spelling has a Welsh resonance) on the origin of mould, a subject somewhat out of Darwin's usual concentration on crustal motion, but of natural interest to Buckland as a paleontologist. As Thackray pointed out,

> In his report on Darwin's paper on the work of earthworms, Buckland suggests that a passage ascribing the origin of chalk to the activity of coral-eating molluscs should be deleted. The sentence appears in the *Proceedings* (Darwin, 1838), which is the record of Darwin's spoken paper, but omitted from the more formal *Transactions* (Darwin, 1840).[197]

Whatever his suggestion for deletion of "disputable" matter, better taken up in a separate communication, Buckland was full of praise for Darwin's main argument, referring to it as establishing a "new Geological Power."[198] Sedgwick's comment on Darwin's paper on the evidence of recent elevation on the coast of Chile was a perfunctory positive, but in his review of Darwin's paper on volcanic phenomena, he made an interesting observation, relevant to the society's attitude toward discussion of theory:

> To the President of the Geological Society.—
> I have read Mr Darwin's paper and think that it ought to be printed. If possible, the early, or historical part, should be made shorter. The concluding or theoretical part is not all clearly brought out, & might be reconsidered by the author with some advantage: Not with any view of altering his theoretical <views> opinion (for he only is responsible for them) but for the purpose of making them more definite & unequivocal—The main facts on which the paper hinges & the immediate deductions deduced from them appear incontrovertible.[199]

Sedgwick thus underlined Darwin's right to engage in theoretical discussion, as an individual author—"for he only is responsible." He also noted that the "main facts" of the paper and the "immediate deductions" were "incontrovertible"—that is, not subject to differences of opinion—whereas matters of "theoretical opinion" would be open to dispute.

This emphasis on characterizing a statement as an opinion, assignable to an individual and thereby not underwritten by the society, was a characteristic move during the period. Thus, during the debate over the new glacial theory in 1840–1841, a phrase in a presentation by Buckland that glaciers were the "only explicable" solution to a problem was altered in the printed *Proceedings* to "He is of the opinion."[200] Such an editorial practice indicated that the issue of theoretical inference was still in play within the society, or, at least, still a matter for consideration within the council, which controlled publication. In one referee report, Fitton criticized a title including the phrase "raised beaches" as assuming an inference before it was established, and even Lyell, in another report, referred to one point raised in a positively reviewed submission as "probably too theoretical for our journal."[201]

When Darwin himself turned referee, he was assigned five papers in the period 1837–1840.[202] Four of the five had to do with elevation in some way, a subject with which he had authority through his firsthand experiences in South America. The fifth paper pertained to observations undertaken during oceanic travel, again a subject with which Darwin was associated. This pattern of assignment would appear appropriate to a highly regarded but still junior member of the society. In his two "yes" and three "no" votes, Darwin drew on his own geological opinions. For example, the paper that drew his greatest enthusiasm was that on elevation in Denmark, for "the movements of the ground in that quarter of Europe have long been a subject of high interest, and therefore all additional information is valuable."[203] In 1838, on a "no" vote Darwin rejected a paper on limestones in Devon as having been adequately represented in the *Proceedings,* but then also remarked,

> A paper of this nature, if it be not accompanied by detailed sections, & if great care be not *evident* in guarding against sources of error,—which is not here shown, for instance in the absence of all allusion to the possibility that the present flat surfaces of the limestone may have been caused by denudation, instead of being the effect of a coral-reef <having> reached the surface of the water.—a paper not thus accompanied, your referee thinks is not worth of a place in the Transactions.—[204]

Underlying the reference to coral reefs was Darwin's own theory regarding their growth, to be discussed subsequently. Based on his own study, Darwin was alert to a greater variety of logical possibilities for the explanation of the limestone than was the author of the paper. For the moment, however, it is sufficient to note Darwin in the act of reviewing papers for publication in the *Transactions* of the society. This involved a quotidian processing and evaluating of claims to knowledge. In his work as reviewer for the society, and as one of its officers, Darwin demonstrated his full engagement with the field of geology as represented and defined by the Geological Society of London.

CHAPTER 3

SPECIMENS

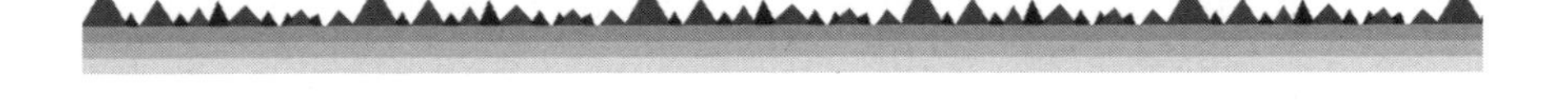

Whewell introduced me to Mr Darwin, with whom I had some talk;
he seems to be a universal collector.
CHARLES BUNBURY TO EDWARD BUNBURY, 13 MARCH 1837

Darwin as Collector

Natural history is centered on objects, but for the collector, particularly when storage and transport are difficult, the question is always which objects are worthy of attention. In natural history there is a creative tension between the intrinsic value of an individual object—whether its beauty or market price—and its value to the scholar in relation to a field of inquiry. Specimens of minerals and metallic ores, for example, are often both beautiful and valuable. Yet the worth of an individual object was not usually Darwin's goal; he was a scholar, not an aesthete or a prospector. For Darwin the geological specimen had value in relation to its location. In the most tangible form possible, it represented the rocks, or the fossils, of an identifiable stratum or surface feature. There were, of course, specimens too valuable on their own to pass up, even if their geological provenance were not known. One such category comprised fossil bones of extinct animals. Darwin did sometimes purchase individual specimens without knowing their exact provenance, but ordinarily he was insistent on knowing location. Indeed, he feared that others would not be so careful. At midvoyage, having heard that some of his fossil vertebrates were at the Royal College of Surgeons, he wrote home,

> I am in great fear lest Mr Clift [at the College] should remove the numbers or markers attached *to any of the Specimens.* Ask Erasmus [Darwin] to call on Mr Clift & state how anxious I am on this point. All the interest which I individually feel about these fossils, is their connection with the geology of the Pampas, & this entirely rests on the safety of the numbers.—[1]

Specimens were only part of the whole picture. The geology of a location had to be worked out, at least approximately, before specimens could be collected. Sometimes this involved studying a cliff face even while the ship was still at sea, as at the Cape Verde Islands, the *Beagle*'s first stop. At other times the sequence of beds was worked out inland. Ideally, specimens would then be collected from the entire known sequence of strata. Occasionally Darwin even recorded specimen numbers as part of the scheme of the strata he sketched (Figure 3.1). (Strata, and specimens, that were actually colored "like those pretty sections which geologists make of the inside of the earth" were rare; Darwin said his first view of such coloration was in the Uspallata range of South America's Cordillera.)[2] Other times no simple slablike diagram of strata was possible (one thinks here of William Smith's handsome and illustrative rule-drawn sections), and extended study produced more complicated pictures. One such example was of a formation found largely in the eastern plains of South America that interested Darwin because of its extent and because fossil mammal bones were often found embedded within it. Darwin referred to the formation by its local Spanish name "Tosca," of which he said, "This name probably originates in the term 'tierra *tosca*' or 'coarse earth'."[3] Darwin sketched a portion of it in cross section at Punta Alta, a place where he found an important cache of fossil bones (Figure 3.2).

The example of the tosca formation also suggests another point regarding Darwin's routine as a collector: his method of recording specimens. Among the specimens at the Sedgwick Museum are several pieces of tosca that had been wrapped in a folded piece of brown paper labeled "Pampean Rocks Tosca" (Figure 3.3). Another example of a specimen that has survived with Darwin's original wrapping intact is specimen 1079, a sample of a "common variety" of quartz from the Falkland Islands (Figure 3.4). These specimens suggest Darwin's possible technique. After gathering specimens in the field, he would wrap each in paper, inscribed with a brief description

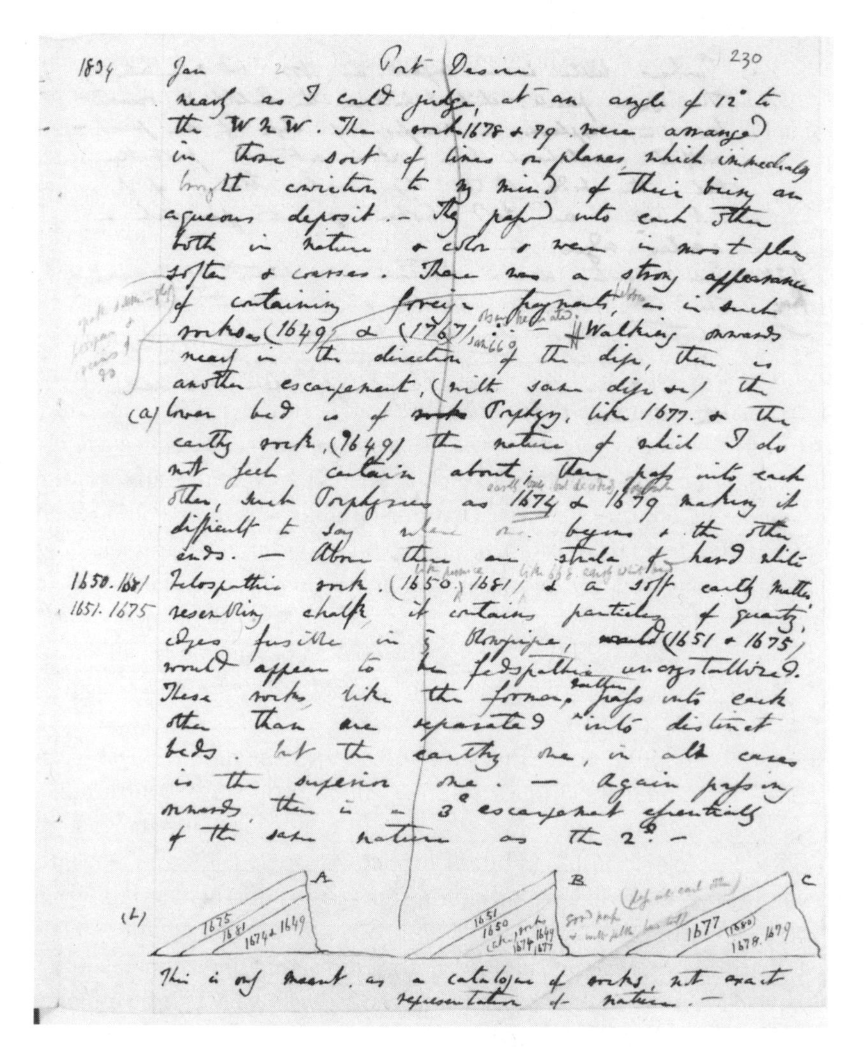
1834 Jan 2 Port Desire 230

nearly as I could judge, at an angle of 12° to
the W by N W. The rocks 1678 & 79 were arranged
in those sort of lines or planes, which immediately
brought conviction to my mind of their being an
aqueous deposit. — They passed into each other
both in nature & color & were in most places
softer & coarser. There was a strong appearance
of containing foreign fragments, as in such
rocks as (1649) & (1767) — Walking onwards
nearly in the direction of the dip, there is
another escarpement, (with same dip &c) the
(a) lower bed is of rock Porphyry, like 1677. & the
earthy rock, (1649) the nature of which I do
not feel certain about; these pass into each
other, such Porphyries as 1674 & 1679 making it
difficult to say where one begins & the other
ends. — Above these are strata of hard white
1650. 1681 Felspathic rock. (1650, 1681) & a soft earthy matter
1651. 1675 resembling chalk it contains particles of quartz,
edges fusible in the Blowpipe, (1651 & 1675)
would appear to be felspathic uncrystallized.
These rocks, like the former, rather pass into each
other than are separated into distinct
beds. but the earthy one, in all cases
is the superior one. — Again passing
onwards, there is a 3d escarpement essentially
of the same nature as the 2d. —

This is only meant as a catalogue of rocks, not exact
representation of nature. —

Figure 3.1. Darwin's recording of specimen numbers. From DAR 33. By permission of the Syndics of Cambridge University Library.

and location, place it in a carrying bag, and return to his base camp or to the ship. There he would number his specimens, affix labels to them, and record the numbers in his specimen notebook. Darwin's four geological specimen notebooks record 3913 items, most collected by him, over the course of the five-year voyage.[4] (It is possible that Darwin carried the specimen notebooks with him, but they are in such good condition, with the writing even and neat, that it is more likely they were used only where a desk of some sort was

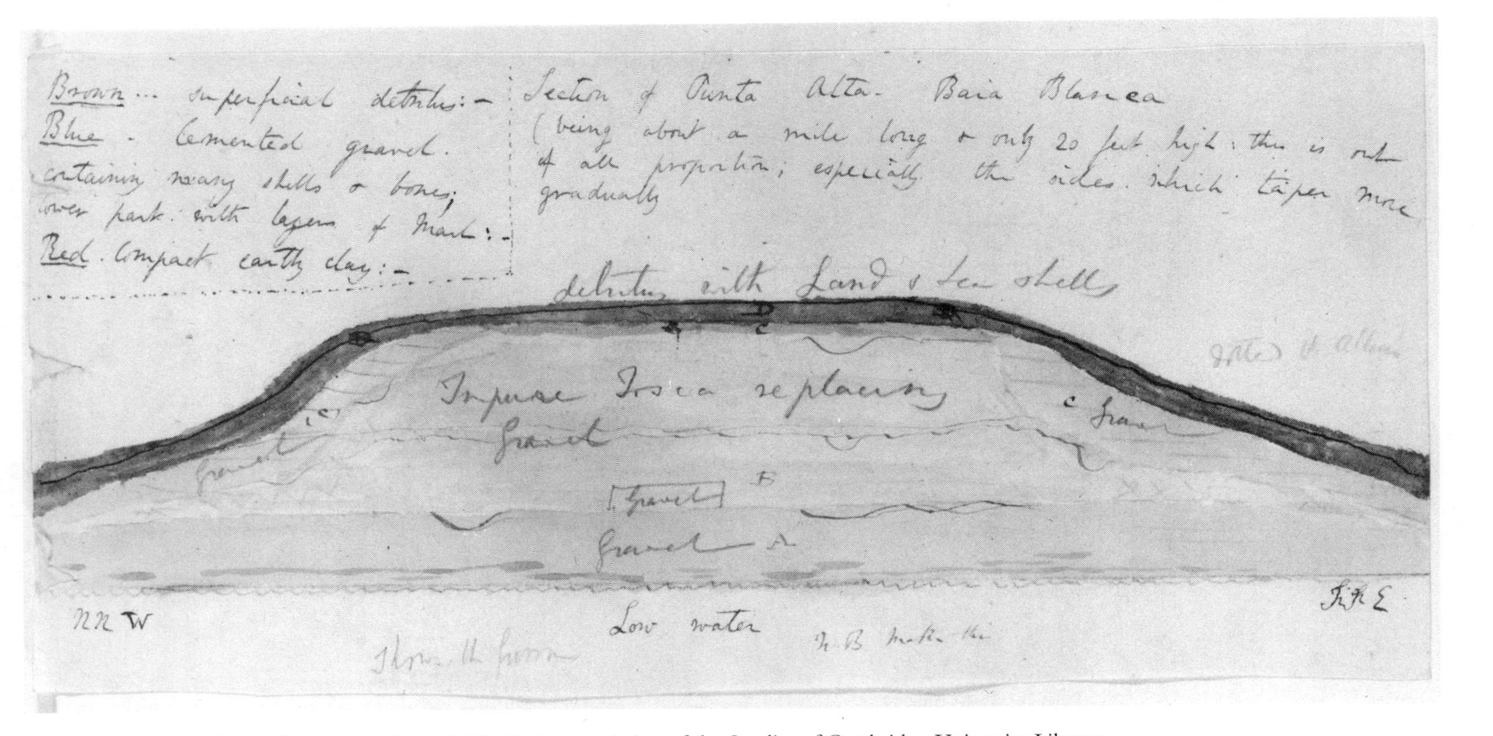

Figure 3.2. Cross section at Punta Alta. From DAR 44. By permission of the Syndics of Cambridge University Library.

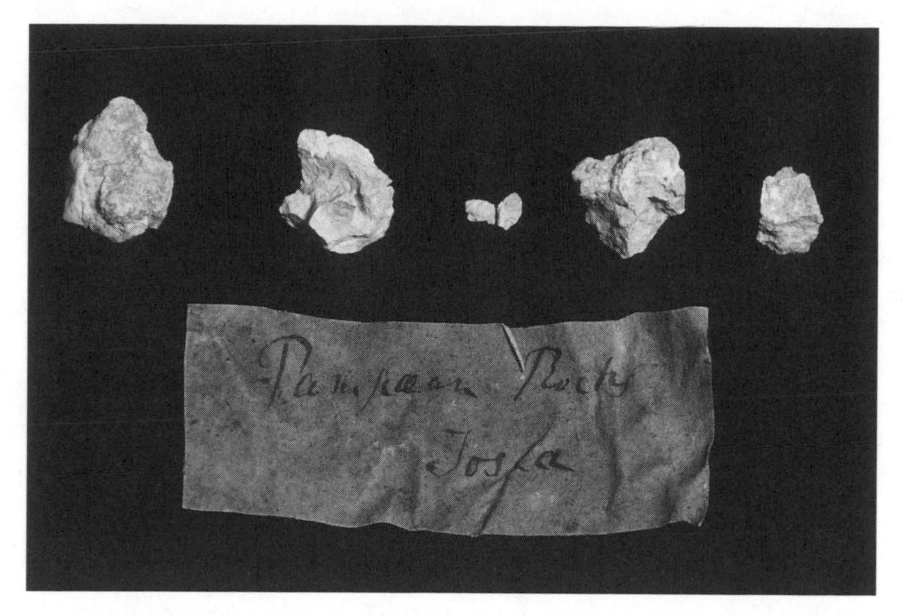

Figure 3.3. Specimen of tosca from the Pampas. By permission of the Sedgwick Museum, University of Cambridge.

available.) He described his general cataloguing technique for specimens as follows:

> Put a number on every specimen, every fragment of a specimen; and during the very same minute let it be entered in the catalogue, so that if hereafter its locality be doubted, the collector may say in good truth, "Every specimen of mine was ticketed on the spot." Any thing which is folded up in paper, or put into a separate box, ought to have a number on the outside (with the exception perhaps of geological specimens), but more *especially* a duplicate number on the inside attached to the specimen itself. A series of small numbers should be printed from 0 to 5000; a stop must be added to those numbers which can be read upside down (as 699. or 86.). It is likewise convenient to have the different thousands printed on differently coloured paper, so that when unpacking, a single glance tells the approximate number.[5]

Thus, as indicated in the inside of the front cover of the second specimen notebook, red stood for numbers 1000 to 1999, green for numbers 2000 to 2999, and yellow for numbers 3000 to 3999. Ordinary white paper would have been

Figure 3.4. Specimen of quartz from the Falkland Islands. By permission of the Sedgwick Museum, University of Cambridge.

used for numbers 1 to 999. The majority of Darwin's specimens now held at the Sedgwick Museum bear light-green labels, which were placed there some time after the arrival of the collection in 1897. As Dr. Graham Chinner at the Sedgwick Museum has pointed out, little corners of the appropriate colors from Darwin's original labels can be seen peeping out from under the later labels on a number of specimens, and Darwin's original labels are clearly visible on a few specimens. For example, some packets of sand, carefully wrapped, have the numbers 545, 546, and 547 written on their red labels. Darwin's specimen notebook records the sand as collected in Uruguay, and presumably is that referred to in the *Geology of South America* as the "loose sand . . . heaped into low, straight long lines of dunes, like those left by the sea at the head of many bays," a fact that served Darwin in his chapter "On the Elevation of the Eastern Coast of South America"[6] (Plate 3).

Let us turn next to consider what was entailed in the gathering of specimens. Since geological interests were paramount, specimens would normally be collected from their site of emplacement, rather than from fallen rock or loose deposits. Specimens representing strata were desired. Fitton, writing for the traveling geologist, in a publication known to be on board the *Beagle*, emphasized the value of the common rock:

> It seldom happens that large masses, even of the same kind of rock, are uniform throughout any considerable space; so that the general character is collected, by geologists who examine rocks in their native places, from the average of an extensive surface:—a collection ought therefore to furnish specimens of the most characteristic varieties;—*and the most splendid specimens are, in general, not the most instructive.*[7]

A traveling geologist should not try to fill a cabinet with "splendid specimens," but rather boxes and drawers with "characteristic varieties." In written instructions of a similar kind, Buckland suggested that the size of the specimen should be "that of a common flat piece of Windsor soap, taking not the outside bit, but the second slice that is struck by the hammer."[8] In his set of instructions for collections, Brongniart—who as an engineer perhaps valued precise expression over a familiar image—wrote simply, "The specimens ought to be almost square—about three inches or more on a side, and one and a half thick."[9] Here one sees the subordination of the aesthetic, and the mineralogical, to the geological. Darwin's collection is a case in point.

To obtain such samples required forceful removal, and hence that identifying and emblematic instrument of the geologist, the geological hammer. One owned by Darwin, and used on the *Beagle* voyage, is presently on display at Down House (Figure 3.5). In presenting a case for using a hammer with an ellipsoidal head to break primary and trap rocks "characterized by an uncommon degree of toughness or tenacity," the geologist John MacCulloch (1773–1835) argued for a weight of three pounds as a convenient general size.[10] Sedgwick's hammer kept at the Cambridge Museum named in his honor, weighs two pounds three and a half ounces, according to Mr. Rod Long, the curator. The weight of Darwin's hammer is not recorded but, from its appearance, is probably in the range of the ordinary hammers. As to variety, an advertisement from the firm of R. & G. Knight in London, proffered no less than seven hammers for sale, the first two of the form recommended by MacCulloch, the third resembling a miner's pick and "useful for detaching a fossil or other object by removing the surrounding matrix," the fifth and sixth trimming hammers for "reducing specimens of minerals and rocks suitable for the cabinet" (a practice seemingly not followed by Darwin), and the seventh an eight-inch pocket hammer that could be fitted with a leather case. This last was also advertised as forming part of the "Portable Mineralogical Chest" for travelers.[11] Presuming that he had more than one hammer with him, Darwin likely would have had one adapted to cutting away matrix from fossils, in addition to one for breaking rocks. One such hammer was later termed a "platypus" for the resemblance of its head to that of the animal.[12] In his advice to traveling geologists written for the Admiralty manual, Darwin recommended "a heavy hammer, with its

Figure 3.5. Darwin's geological hammer. © English Heritage Photo Library.

two ends wedge-formed and truncated; a light hammer for trimming specimens; some chisels and a pickaxe for fossils."[13]

Identifying the Specimens

What, then, of the specimens? Darwin's are typically in the recommended range of three inches square and three quarters of an inch, or less, in thickness. However, he determined the shape of the specimen primarily by the rock type, and how it broke; he did not attempt to trim his specimens to neat blocks. Many of them come in pieces; the splitting occurred as the rock was removed. Often the first-listed specimen, and the specimens listed as "do" (ditto), can be fitted together. They were originally part of the same rock unit, though the separate pieces have their own specimen numbers.

Simply collecting representative specimens involved judgment concerning differences in rock type. He identified some very broad categories of rocks, as, for example, sandstone, by visual and physical inspection. However, whether in the field or at a desk in his camp or on the ship, Darwin applied various tests to his specimens in the process of describing them. Many of these tests were well known to practicing mineralogists, including those who made their living from trades associated with mining.

Darwin's own knowledge of mineralogy came out of a more academic tradition, first begun as a schoolboy in the chemistry laboratory at home in Shrewsbury and continued in his attendance at Hope's lectures in chemistry and Jameson's in mineralogy at Edinburgh University. At Cambridge Darwin had the regular advice of Henslow, who had formerly taught mineralogy at the university, and perhaps the occasional assistance of others such as Sedgwick and Whewell. Although Darwin's introduction to field geology had come relatively late, his acquaintance with mineralogy had been relatively early. Even at school he was reported by one friend to have spent "some time, most evenings, with a blow-pipe at the gas-light in our bed-room."[14] This would account for the self-confidence with which he examined his specimens during the voyage. Moreover, some of the instruments from his home laboratory remained with him for life. In his last years he continued to use the "battered old chemical balance from his Shrewsbury days."[15]

Darwin's use of chemical methods in assessing geological specimens was not idiosyncratic. Sally Newcomb and Jan Golinski have shown that British experimentalists, working in laboratory settings, and in a tradition that emphasized the participation of relatively unskilled laborers, contributed significantly to the development of geology in the period from 1780 to 1820.[16] Brian Dolan has added a twist to this story. He argued for an association between the experimentally oriented mineralogy conducted at the University of Cambridge under Edward Daniel Clarke (1769–1822) and some of his successors, and the British amateur and commercial tradition supported by London merchants like John Mawe (1764–1829), who sold mineral specimens, along with tools and simple guides for their analysis.

As Dolan pointed out, Cambridge pedagogy was consistent with the portable laboratory work promoted by London shops operated by practical mineralogists. As a point of contrast, he suggested the Royal Institution, where electrochemistry, under Humphry Davy, was premised on large equipment operated by full-time expert practitioners.[17] The supportive link between academic and commercially oriented chemistry was, of course, quite familiar and congenial to the Darwin-Wedgwood families, and in Charles's generation it was brother Erasmus, five years older than Charles, who seized on it, not as a lifelong occupation but as the passion of his youth.

Charles's chemical education from his brother lasted at least three years, from 1822 to 1825, as well as some period of time before that. He spoke of it in his autobiography, emphasizing his brother's initiative as well as his own growing interest:

> Towards the close of my school life, my brother worked hard at chemistry and made a fair laboratory with proper apparatus in the tool-house in the garden, and I was allowed to aid him as a servant in most of his experiments. He made all the gases and many compounds, and I read with care several books on chemistry, such as Henry and Parkes' *Chemical Catechism*. The subject interested me greatly and we often used to go on working till rather late at night. This was the best part of my education at school, for it showed me practically the meaning of experimental science. The fact that we worked at chemistry somehow got known at school, and as it was an unprecedented fact, I was nicknamed "Gas."[18]

The particulars of the brothers' practice of chemistry have become known more completely with the publication of a series of letters written by Erasmus, when he was away at university, to Charles, then still at home in Shrewsbury.[19]

Taken in combination with Charles Darwin's autobiography, Erasmus Darwin's letters tell us several things about the practice of chemistry at that "snug" home laboratory.[20] First, the laboratory was an expanding enterprise, as Erasmus's sketches suggest. Their father, Robert Waring Darwin, funded the laboratory, as he did their other endeavors. The boys referred to the funds as their "cow."[21] Second, growing sophistication was evident from the equipment Erasmus purchased, such as ever more complicated blowpipes, as well as a small goniometer for measuring the interfacial angles of minerals. Third, chemistry and mineralogy ran together as Erasmus purchased specimens for his or his brother's home cabinet from Cambridge shops, such as that run by Clarke's former assistant. Erasmus wrote,

> I have bought 2 or 3 little stones from him; 2 specimens of uranite which is a very scarce stone, & some leaf copper, & a very odd looking thing like a petrefaction [*sic*] called a brain stone from its similarity to ye brain.[22]

Fourth, some of what Erasmus was learning of chemistry from lectures at Cambridge was being reported home (such as Henslow's having performed a test for arsenic using the blowpipe, which yielded the characteristic garlic-like odor). Fifth, Erasmus directed his brother's reading in chemistry, which included an impressive list.[23] Sixth, and finally, in 1825 Erasmus was beginning to ask original questions, as of the magnetic properties of a sediment found in tea, or a report, which he doubted, regarding the affinity of oxygen for saliva. On the latter question Erasmus requested Charles to test the claim in a series of analytical steps he outlined. Clearly Charles had a better laboratory than what Erasmus had access to at Cambridge, where instruction was by demonstration. As Erasmus wrote to Charles in January 1825, "I want you to try an experiment for me, & you remember you promised to do any I asked you & moreover to do them *immediately*."[24] Thus, Erasmus's letters suggest that Darwin had had a fair beginning at chemistry before he left Shrewsbury for Edinburgh in October 1825.

To Erasmus's credit, he continued to maintain a home laboratory in London, after his university days were over. While often cast as pursuing a life of "literary leisure," his sister Susan Darwin (1803–1866) found him in 1832 "quite buried alive in a little Lab he has set up in his Lodgings which makes him quite forget times & seasons"—a lab which four years later required "13 Cab loads of Glass bottles &c" to transport."[25] Perhaps Erasmus could be described as pursuing a life of laboratory leisure, as much as a life of literary leisure. In any case, after the *Beagle* voyage his younger brother Charles looked to him for advice on chemical questions. Herschel had reported on the crystallization of barium sulfate. Charles reminded himself to "ask. Erasmus. whether electricity would affect this.—"[26]

In observing and describing specimens, geologists considered the evidence provided by all the senses to be important. Sight could determine such characteristics as transparency, luster, and color. But touch, taste, and a sense of smell had their role as well, as Phillips's introduction to mineralogy indicated. On "odour," for example, Phillips referred to "some argillaceous minerals having a distinctive odour when breathed upon."[27]

The evidence provided by the senses was enhanced by equipment, modest in size but effective. To magnify the specimen, a hand lens was used. Thin sections, which permitted the microscopic examination of rocks by means of transmitted light, were not yet in use. (As early as 1833 Henry Witham [1779–1844] described the technique of making thin sections of fossil plants, as derived from the work of William Nicol [1768–1851], but the technique only became commonplace in the 1850s following the work of Henry Clifton Sorby [1826–1908].[28]) Traditional tools could be used to test for hardness. Listed in ascending sequence they would be a thumbnail, a copper penny, a piece of glass, and, finally, a steel knife. A small hand magnet was helpful on its own for identifying what would now be termed "ferromagnetic" material (for example, loadstone). A magnet in conjunction with the blowpipe was employed to test for reducible iron.

A goniometer was used to measure angles of crystal faces, these measurements then being compared to those provided in reference manuals. The first goniometer invented, now referred to as the "contact goniometer," gave rough measurements; the reflective goniometer, invented by Thomas Wollaston (1766–1828), produced very accurate ones. In 1822, while in Cambridge Erasmus ordered the first kind for the home chemistry laboratory "so that we shall be able to seperate [*sic*] the different crystalls [*sic*] in your cab[inet]."[29] Charles also took a second kind of goniometer with him on the voyage, possibly aware of its praise by Herschel.[30] The reflective goniometer was often the province of the experienced mineralogist. Darwin seems to have begun using it gradually over the voyage. From Montevideo in November 1833, he wrote to Henslow, "Can you send me out any book, which with instructions from yourself will enable me to use my reflecting Goniometer."[31] Henslow replied in July,

> I have not your letter bye [*sic*] me to answer your questions formally but I remember you enquire about a Goniometer—I would not advise you to bother yourself with one—It is an instrument of no use in the field, & of importance only in the hands of an experienced mineralogist in his *closet*. Phillips's book must be quite as much as you *need* for the detection of the few ingredients which form rocks—Any that you can't make out you must describe conditionally & we will set you to rights 10 years hence when you return—[32]

Indeed, as Henslow suggested, the reflecting goniometer was difficult to use. The face of the crystal to be measured had to be smooth, and the rays of light had to strike it in parallel. On the early machines, the chosen source of light was distant, the cross members of a window frame, for example.[33] Given

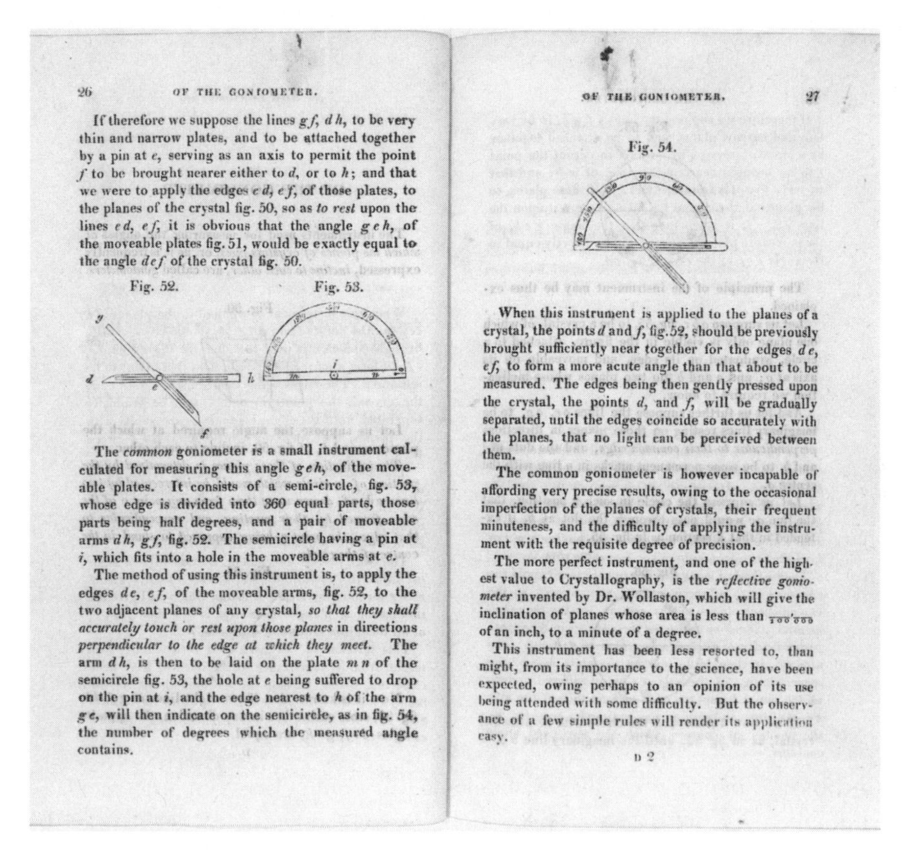

26 OF THE GONIOMETER.

If therefore we suppose the lines *g f*, *d h*, to be very thin and narrow plates, and to be attached together by a pin at *e*, serving as an axis to permit the point *f* to be brought nearer either to *d*, or to *h*; and that we were to apply the edges *e d*, *e f*, of those plates, to the planes of the crystal fig. 50, so as *to rest* upon the lines *e d*, *e f*, it is obvious that the angle *g e h*, of the moveable plates fig. 51, would be exactly equal to the angle *d e f* of the crystal fig. 50.

Fig. 52. Fig. 53.

The *common* goniometer is a small instrument calculated for measuring this angle *g e h*, of the moveable plates. It consists of a semi-circle, fig. 53, whose edge is divided into 360 equal parts, those parts being half degrees, and a pair of moveable arms *d h*, *g f*, fig. 52. The semicircle having a pin at *i*, which fits into a hole in the moveable arms at *e*.

The method of using this instrument is, to apply the edges *d e*, *e f*, of the moveable arms, fig. 52, to the two adjacent planes of any crystal, *so that they shall accurately touch or rest upon those planes* in directions *perpendicular to the edge at which they meet.* The arm *d h*, is then to be laid on the plate *m n* of the semicircle fig. 53, the hole at *e* being suffered to drop on the pin at *i*, and the edge nearest to *h* of the arm *g e*, will then indicate on the semicircle, as in fig. 54, the number of degrees which the measured angle contains.

OF THE GONIOMETER. 27

Fig. 54.

When this instrument is applied to the planes of a crystal, the points *d* and *f*, fig. 52, should be previously brought sufficiently near together for the edges *d e*, *e f*, to form a more acute angle than that about to be measured. The edges being then gently pressed upon the crystal, the points *d*, and *f*, will be gradually separated, until the edges coincide so accurately with the planes, that no light can be perceived between them.

The common goniometer is however incapable of affording very precise results, owing to the occasional imperfection of the planes of crystals, their frequent minuteness, and the difficulty of applying the instrument with the requisite degree of precision.

The more perfect instrument, and one of the highest value to Crystallography, is the *reflective goniometer* invented by Dr. Wollaston, which will give the inclination of planes whose area is less than $\frac{1}{100,000}$ of an inch, to a minute of a degree.

This instrument has been less resorted to, than might, from its importance to the science, have been expected, owing perhaps to an opinion of its use being attended with some difficulty. But the observance of a few simple rules will render its application easy.

D 2

Figure 3.6. Diagram of goniometer. By permission of the Syndics of Cambridge University Library.

the difficulties, it is not surprising that Darwin struggled with his reflecting goniometer.

Henslow's letter caught another aspect of Darwin's work, as defined by those who had arranged his travel: his findings were to be conditional. There were experienced men at home ready to set him right on his return. Still, Darwin did record angular measurements of specimens. With a tone of pride, he noted of a specimen found at Bahia, Brazil, in the first year of the voyage, "Hornblende determined by myself with goniometer"[34] (Figures 3.6 and 3.7).

Another way to identify a specimen was to put it in direct action with acids. The acid-bottle, containing muriatic (that is, hydrochloric) acid, was used routinely. Hydrochloric acid could ascertain whether carbonate was present in the rock, such as limestone and marble, or dolomite.[35] (A knife will scratch

Figure 3.7. Darwin's specimen notebook, a specimen, and a goniometer. By permission of the Sedgwick Museum, University of Cambridge.

carbonates, hence, the utility of acid in distinguishing between limestone and dolomite.) In his specimen notebooks Darwin referred to "acids" in the plural, probably meaning nitric and sulfuric acids as well as muriatic. One list of useful articles for the examination of minerals included "Bottles with ground Stoppers and Caps, to prevent the Fumes from escaping" and the bottles engraved "N," "S," and "M" for the three acids.[36] If there were fumes, presumably the bottles would have held concentrated acids.

Another important tool of analysis was the blowpipe. As used by the traveling geologist, it was remarkable for its simplicity and direct connection to the human operator. John Griffin remarked (1802–1877), "SIMPLE *blowpipes* are those through which air is blown through the mouth."[37] While styles of construction differed, these blowpipes required human breath. J. J. Berzelius (1779–1848) remarked that the blowpipe was brought to chemistry from the arts:

> Jewellers and other workers in metal on a small scale, avail themselves of the use of the blowpipe, to direct the flame of a lamp on pieces of metal supported on charcoal, so as to fuse the solder by which they are to be united.

> This instrument was long employed in the arts, before any one conceived the idea of applying it to chemical experiments, performed in the dry way, as it is called.[38]

From its original use, the blowpipe had become a welcome tool for practicing mineralogists. For example, Mawe, an earlier traveler to South America than Darwin, counseled his readers in an introductory work that gold might be distinguished from pyrite as follows:

> If a particle be placed on charcoal, and acted upon by the flame of the blow-pipe, if it be Gold it will melt, and retain its yellow colour; while the Pyrites will generally decrepitate, and burn with a faint blue flame, emitting the odor of sulphur, and be reduced to a dark colored [*sic*] ball of scoria, which will be attracted by the magnet; these effects prove that the Pyrites is [*sic*] a combination of Sulphur and Iron.[39]

The blowpipe was thus useful to the prospector and the miner. For the mineralogist, "this instrument is absolutely necessary, as his only resource for immediately ascertaining if the inferences he draws from external characters, such as form, colour, hardness, &c. be legitimate."[40] At the time the device served numerous purposes. There was also an expansion of its capacity when it was harnessed to the combustible powers of hydrogen and oxygen gases. Clarke at Cambridge took the lead in this high-temperature research in the 1810s, although it was not directly helpful to a traveling geologist such as Darwin (Figure 3.8).[41]

From the evidence of the specimen notebooks, Darwin seems to have made frequent use of the blowpipe to test the appearance of a substance after heating. This would have been consistent with the texts of Phillips and of Jameson.[42] Thus, for specimen 613, collected in June 1832 at Rio de Janeiro, Darwin recorded in his first notebook, "Quartz," "containing much iron" and, on the left side of the page, "Heated in blowpipe & pounded. particles strongly attracted by the magnet."

Chemical examination was also possible with the blowpipe. For example, early in the voyage, at Bahia in Brazil, Darwin noted that a thin bright coating of rock that overlay the granitic formation stained borax a pale yellow under the blowpipe.[43] The procedures for such analysis were well defined. Berzelius's chart, reproduced in the *Dictionary of Chemistry* (1823) by Andrew Ure (1778–1857), would have been a standard guide. One column of the chart listed the material ("the earths and metallic oxides") to be assayed. Another column of the chart listed the appearance of the material to be tested when heated with borax as a flux under an oxidizing flame (the external part

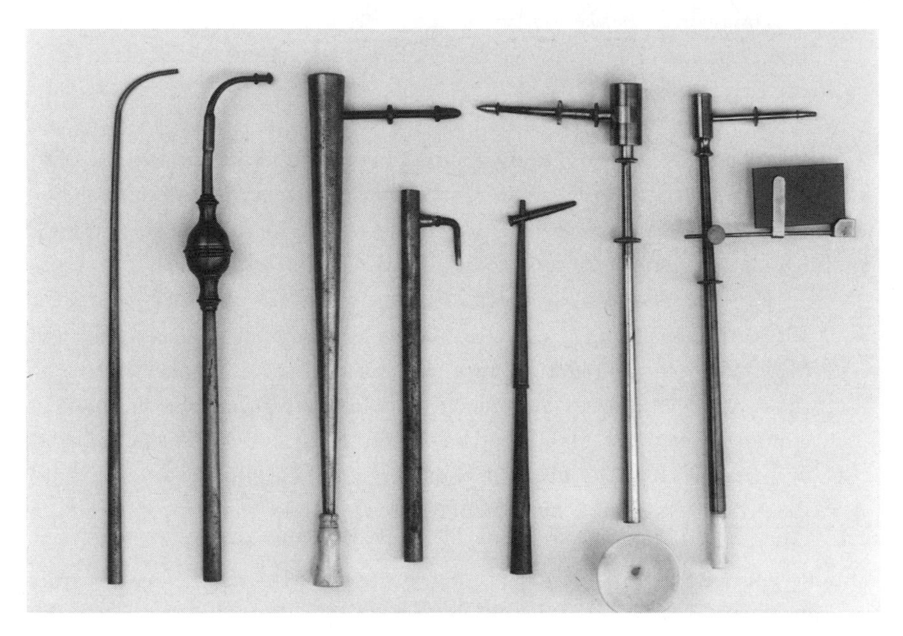

Figure 3.8. Blowpipes. By permission of Ulrich Burchard and Fotostudio Neuhofer Deggendorf.

of the flame) or a reducing flame (its internal part).[44] In this scheme, Darwin's results were not determinative, since a number of oxides yielded a yellow color under various analytical circumstances.[45] However, only selected specimens were subjected to blowpipe analysis. Moreover, it is doubtful that all the reagents described by such a thorough manual as Berzelius's would have been on the *Beagle,* or indeed even the manual itself. Interestingly, even a specimen (number 3895 as listed in the specimen notebook) that Henslow seems to have looked at after the voyage was referred back to Phillips's *Mineralogy,* which Darwin had in his possession early in his career. That book proved to be adequate for the tasks facing him as a traveling geologist. It would have complemented the relatively small number of reagents and pieces of equipment (for example, blowpipe, charcoal, borax) he would have carried with him.

One item that could have been useful to Darwin in rock identification is some sort of reference collection of specimens. His and Erasmus's boyhood collection might have served as the foundation for such a collection. However, there is no indication that he had such a collection on board ship, and the constraints of space aboard the vessel would argue against it.

Housing the Specimens

At intervals after Darwin collected his geological specimens, he sent them to Cambridge, where Henslow tended and stored them. After Darwin's arrival back in England, he returned to Cambridge for a few months in early 1837 to unpack his collection. He rented a row house on Fitzwilliam Street, near the present-day site of the Fitzwilliam Museum, to provide the space required.[46] There was no other obvious receiving point for the specimens. Their disposition was similarly yet to be decided. Before the end of the voyage, even his sister Catherine had broached the claims of the Shropshire and North Wales Natural History and Antiquities Society for "duplicates of curiosities, or specimens."[47] Certainly Darwin had been concerned with the question of disposition as early as 1831, even before the *Beagle* had left port. On the one side were the claims of central repositories in London, the capital and the obvious site for national reference collections. Because he had traveled on "a King's Ship," Darwin was keenly aware of the potential claims of such national institutions as the British Museum, while expressing concern over what he judged to be the then current efficiency of the institution.[48] The British Museum had included fourteen thousand geological specimens within its collections at its founding in 1753. However—as perhaps an indication of inadequate administration—a system of registration for geological specimens was not set up until 1837.[49]

On the other side were universities and other institutions. Owen's presence at the Royal College of Surgeons suggested an obvious home for the vertebrate fossils. Beyond that, choices were less obvious. In the notices to collectors cited above (footnotes 8 and 9), Brongniart and Buckland had advertised their own institutions as potential repositories, even including shipping instructions to entice potential donors to respond to their appeals promptly and directly. However, Darwin was British rather than French, and affiliated with Cambridge rather than Oxford. Darwin was aware that his alma mater, the University of Cambridge, with its attendant institutions such as the Cambridge Philosophical Society, was then attempting to build up its collections in anticipation of the development of museums devoted to natural history.

The university had some capacity to receive natural history specimens, including those in geology. It was limited, however, and in the late 1820s there was a campaign to expand lecture room facilities and to establish a natural history museum. In a notice dated 15 December 1827 Whewell and Henslow announced that the Cambridge Philosophical Society was collecting material for a museum of natural history.[50] In a printed statement of 9 December 1828 addressed to members of the university senate, Whewell made the case for museums, not omitting to evoke an emulatory or competitive spirit by citing the superior facilities at the University of Oxford:

> It is manifest, that with regard to the various sciences of Natural History, Museums easily accessible and well arranged are of paramount importance. And the continual progress of such sciences cannot be understood, except the Museums be capable of admitting the objects which illustrate this progress.
>
> Oxford has been supplied with casts of the most remarkable specimens of Osteology in the Paris collection, by the kindness of Cuvier and the liberality of the French Government; a donation which could not have been accepted here, for want of the room necessary for its reception.[51]

Eventually accommodation for geology and mineralogy did expand at Cambridge. Whewell's views prevailed. Provision was made for geological and mineralogy specimens on the ground floor of a new university library, though this space was not ready by 1836 when the *Beagle* returned.[52] However, when in 1897 the department at Cambridge received Darwin's geological specimens permanently, they were treated as a historical collection, to be kept distinct rather than integrated into the general collections.[53]

The disposition of Darwin's geological specimens after the voyage followed two general routes. Those that were of current interest to geologists captured the greatest curatorial attention and prominent placement. The fossils generally were most easily placed.[54] The great prizes of Darwin's collection—one may even call them the trophies—were the vertebrate fossils, which went quickly to the Royal College of Surgeons. On the other side were specimens of less interest to a national body or to any particular group of researchers. While the Geological Society of London did collect foreign specimens for its museum—after William Lonsdale (1794–1871) was relieved of the curatorship of the society's museum in 1836, owing to overwork—the museum's collections, particularly of foreign specimens, were not adequately and systematically tended until Leonard Horner (1784–1864) put them right in the early 1860s.[55] Presumably Darwin would not have seen in the society's museum a scholarly repository for his rock specimens in the years immediately following the voyage.

Thus, Darwin's geological collection was primarily of value to him, as a prospective author. The specimens were souvenirs, prompts to his memory, and tangible oases of stability that could be revisited, unlike their distant sites of origin that were forever lost, at least to Darwin.[56] This still left him with a dilemma. As he wrote to Henslow in his first letter after his return to England, "My chief puzzle is about the geological specimens, who will have the charity to help me in describing their mineralogical nature?"[57] The answer to that question was William Hallowes Miller (1801–1880), pioneer in crystallography and professor of mineralogy at the university from 1832 to 1880. As Henslow had done, Miller advised Darwin on "the determination of many minerals."[58] Miller did not publish a systematic treatment of Darwin's specimens.

The consequence of the differential treatment of Darwin's specimens was that those that made their way into general collections could more easily become part of an ongoing tradition of interpretation. In contrast, the specimens that remained in Darwin's hands, as with the majority of the geological specimens, were, in a sense, frozen in time, in the 1840s when his geological work was published. Specimens also tended to disappear from view as individual objects, since Darwin did not refer to them by their numbers in his published texts.

Using the Specimens to Interpret Geology

Considerably later, following Darwin's death and the arrival of his specimens at the Sedgwick Museum, geologists have had the opportunity to reexamine them. The first person to do this in a systematic way was the petrologist Alfred Harker (1859–1939) of the Department of Geology at the University of Cambridge. (The naming of the department appears to date from the opening of the Sedgwick Museum in 1904. Prior to that, what was to become the department was referred as the Woodwardian Museum.)[59] In 1907 Harker suggested reevaluating Darwin's specimens because

> they possess . . . a certain intrinsic value; inasmuch as an examination of these original specimens, with the advantages conferred by modern petrographical methods, may sometimes help towards a better understanding of the recorded observations.[60]

The particular specimens on which Harker concentrated were the rocks Darwin collected from the Cape Verde Islands. Harker noted that the specimen notebooks were written in "plain language." He also noted Darwin's use of "old-fashioned comprehensive names such as 'porphyry,' 'greenstone,' and 'basalt.'" Harker's points reinforce what has been said about the specimen notebooks thus far: that Darwin's terminology in them tends toward broad classificatory terms, as well as physical qualities including color, hardness, and texture. This was so partly because Darwin placed his more interpretive comments in a separate series of extended geological notes. Also, more refined terms for some igneous rocks, as, for example, "andesite" and "rhyolite," were not yet invented or in use.[61] An example is Darwin's frequent employment of the broad term "greenstone," which in John Challinor's dictionary is defined as "an old term for all those varieties of dark, greenish igneous rocks" that "mostly resolve themselves into diorites, dolerites (particularly), and somewhat altered basalts."[62] Interestingly, the rocks in Tasmania that Darwin described as "greenstone" have, on recent inspection, been described as dolerites, suggesting that, at least in this case, Darwin's identification was on target in the sense that it was consistent with current views.[63] This speaks well of his acumen as a field geologist. The broad character of Darwin's remarks in his specimen notebooks partly also reflected his status as a relative newcomer to the field of geology, one who worked while on the *Beagle* with the written literature available to him but without the opportunity for face-to-face consultation with experienced colleagues. On returning home he hoped to remedy that situation. He wrote to Henslow, "I am anxious to know, whether Prof. Sedgwick recommends any particular nomenclature for the rocks."[64]

The majority of the specimens are extant and housed at the Sedgwick Museum. By the time the specimens came to the museum, Darwin was already famous, and the collection itself has remained intact. New specimen labels were added, and a fair copy of Darwin's catalogue was made. Thin sections have been made from some specimens, an example of which is shown in Figure 3.9.

Periodically authors, most associated with the University of Cambridge, have returned to Darwin's specimens, either because of interest in the site associated with the specimen or because of interest in following Darwin's intellectual development. In 1933, twenty-six years after Harker first wrote on Darwin's geological specimens, Constance Richardson (1907–), formerly a student in the Cambridge department, published petrological analyses of Darwin's Galápagos specimens held at the Sedgwick Museum.[65] In 1947 Cecil Edgar Tilley (1894–1973), petrologist at Cambridge, examined the Darwin specimens from St. Paul's Rocks.[66] More recently, in 1991, Chinner of the Department of Earth Sciences at Cambridge assembled a large exhibit of Darwin's specimens, comprising a number of cases. This exhibit was on display in its complete form through 1998, and in reduced form as of 2003. It includes references to present-day interpretations of Darwin's specimens.

In 1993 David Norman, director of the Sedgwick Museum, mounted a second exhibit of Darwin's specimens, this time in connection with a conference in Santiago, Chile.[67] Finally, in 1996, Paul Pearson published an important study relating a number of Darwin's specimens to his views on the origin and diversity of igneous rocks.[68]

I have chosen three sets of specimens to illustrate their utility in interpreting Darwin's geological work. The first group of specimens is from St. Paul's Rocks; the second group, from the Galápagos Islands; and the third group,

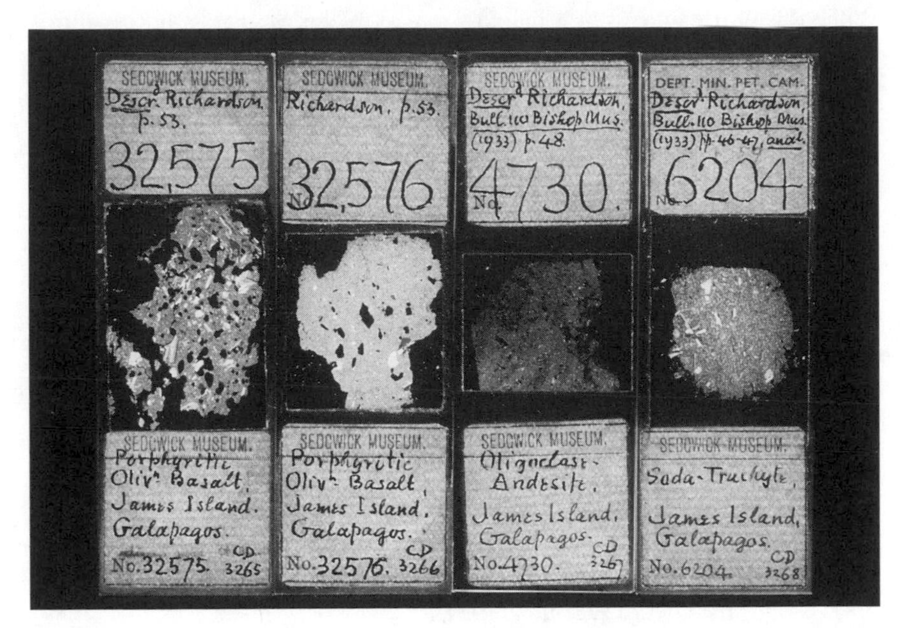

Figure 3.9. Thin sections of James Island specimens. By permission of the Sedgwick Museum, University of Cambridge.

from the Cape of Good Hope. At St. Paul's Rocks Darwin identified what the rocks were not and speculated as to what they might be. At the Galápagos he joined eminent geologists such as Scrope, Daubeny, Lyell, and Jean-Étienne Guettard (1715–1786) in gaining firsthand knowledge of a volcanic site. At the Cape of Good Hope he examined a site already known to geologists.

St. Paul's Rocks are uninhabited rocky islets in the Atlantic Ocean, lying just north of the equator (lat 0°23′N, long 29°23′W). This locale, at which the *Beagle* arrived on 16 February 1832, occupies a special place in the assessment of Darwin as a geologist. Although less than two months from England, Darwin immediately recognized the unusual character of the rocks lying beneath the white guano capping the islets. He initially identified them as serpentine and diallage, the former well known, the latter discussed by Jameson.[69] By the time he wrote his *Journal of Researches,* his view was slightly different, and more guarded: "Its mineralogical constitution is not simple, in some parts, the rock is of a cherty, in others of a felspathic nature; and in the latter case it contains thin veins of serpentine, mingled with calcareous matter."[70] The key point is that Darwin recognized that these were not volcanic rocks. In his notes on the island he then asked a question: "Is not this the first Island in the Atlantic which has been shown not to be of *Volcanic* origin?—"[71]

While this entry appears as a tail-end remark to his St. Paul's notes, and is clearly a later addition, it captures what became the thrust of his interpretation.

By the time Darwin wrote *Volcanic Islands*, published in 1844, his characterization of the rocks had become more guarded; they were "unlike any which I have met with."[72] He then proceeded to physical description, leaving behind mention of diallage but still including serpentine as a component. In the same book, Darwin placed his discussion of the distribution of volcanic islands in a worldwide context.[73] Noting the nonvolcanic origin of most mountains, he asked whether volcanic eruptions reached the surface through fissures formed during the first stages of the conversion of the ocean bed into land.[74]

In the words of Tilley, Darwin made the "first scientific landing" on these rocks.[75] Tilley noted that knowledge of the nature of the rock was

> due primarily to Renard (1882) and Washington (1930). The former gave a detailed account of them based on the collections of the Challenger Expedition (1873) and the latter examined the specimens brought back by the Quest Expedition (1921).[76]

Tilley emphasized that all subsequent researchers reiterated Darwin's point that the composition of the islands was unique among deep-water oceanic islands. Tilley himself restudied these rocks in the 1940s, noting that "the active interest now being taken in problems of sub-oceanic geology will lead to a systematic geophysical study of this part of the Mid-Atlantic Ridge."[77] There has thus been a continuing thread of interest among geologists in the special character of the rocks, from Darwin's visit in 1832 onward.

Of the twelve specimens Darwin collected at St. Paul's Rocks, Chinner placed eight on display in his exhibit. In his commentary Chinner added present-day classifications keyed to the theory of plate tectonics. He was thus taking the story from the point where Tilley had left it. Especially of interest were specimens 241 (for which a thin section was displayed) and specimens 242 to 245.

On technical grounds, the creation of thin sections reveals what had become a newly possible sense of the geological object. They allow boundaries between different components of a rock to be seen more clearly than is possible from inspection of the entire specimen and show properties of individual minerals. The rock, as a specimen, becomes more manageable. Further, thin sections lend themselves to accurate representation. Witham's early work was, in effect, an atlas of beautiful engravings of thin sections.[78] These engravings were based on drawings made using a camera lucida, a device with a prism fitted over the ocular piece of a microscope, coupled with a forty-five-degree mirror projecting a few inches to the right (for a right-handed person). By this means the observer can see the image in the microscope superimposed on a piece of paper placed below the mirror and can trace the outlines of the object being examined. When photography entered the technical repertoire, it also proved useful. Because thin sections are translucent, samples of specimens could be investigated in transmitted light, either directly or in photographic reproduction.

Thin sections of geological specimens are an excellent example of a development that spanned many scientific disciplines during the nineteenth century. Lorraine Daston and Peter Galison have described this development as the "full-fledged establishment of mechanical objectivity as the ideal of scientific representation."[79] During the nineteenth century, geologists went from specimens to thin sections of specimens, to mechanically supported drawings of thin sections, to photographs of thin sections—all stages affecting the geologist's understanding of objects and objectivity. Although thin sections came along too late for Darwin to make use of them during his active years as a geologist, he appreciated their impact on the field. In 1881, as he reviewed the "great steps in geology during the last fifty years," he noted, "The study under the microscope of rock-sections is another not inconsiderable step."[80]

In his exhibit text, Chinner identified the specimens as mylonized peridotites. Neither of these terms was part of Darwin's vocabulary. "Mylonite," a term from the 1880s, is

> a metamorphic rock produced by grinding and rolling out. It is characterized by a very close texture with banding, platy fracture, and, perhaps, new minerals. Hence, "mylonitization."[81]

The term "peridotite" was defined in the 1840s as an "olivine-rich basalt."[82] Chinner went on to pose the problem that these specimens were amphibole-rich peridotites, which would not be expected to survive the high temperatures associated with the melting processes along the Mid-Oceanic Ridge. Rather, in the words of his exhibit text, these rocks might be interpreted as material brought to the surface from a sub-Pangeal mantle, a remnant from the splitting of Africa and America one hundred million years ago.

Over the 160 years between Darwin's collection of the rocks on 16 February 1832 and the interpretations of their origin offered in the 1990s, we can see both continuities and discontinuities. The deepest continuities lie in the interest in oceanic islands, entertained by the successive scientific missions from the *Beagle* onward, and the overall interpretation of the islands as nonvolcanic in origin. The discontinuities include new systems of naming and analysis, the new technology offered by thin sectioning of specimens, and the new understanding of ocean basins from the 1940s onward, culminating in the current theory of plate tectonics.

Let us now turn to a moment late in the voyage—the *Beagle*'s visit to the Galápagos Islands in 1835. By this point in the voyage, Darwin had developed enormously in his theoretical reach as a geologist, but the collection and identification of specimens still played a fundamental role. As with the specimens from St. Paul's Rocks, specimens from the Galápagos Islands figured in work of unusual interest in subsequent geology. Chinner and Norman made this point in their exhibits, and Pearson in the study mentioned earlier. They all stressed the importance of the Galápagos experience for Darwin in his attempt to understand the origin of different lavas. What we will do here is to identify four specimens by number that have not been discussed in these earlier treatments, and to pose a problem that exists concerning the identity of one of them.

During the voyage Darwin speculated on numerous occasions the possible causes for the origin of various rocks. As in other matters, his inclination was to side with Lyell, Scrope, and the Huttonians—that is, to regard the various types as being generated continuously within and on the earth, rather than as being characteristic productions of successive periods of the earth's history. Thus, for example, he did not side with Humboldt, and others from the Wernerian tradition, who associated the production of trachyte with an earlier period in the earth's history than the current one.[83] Continuing with the vocabulary Darwin used in his specimen notebooks, two common families of volcanic rock are the basalts and the trachytes. In mineral composition, a distinguishing characteristic is the proportion of silica present. Rocks rich in silica are generally lighter in color. Basalt has less silica than trachyte and is generally darker.

By the time the *Beagle* reached the Galápagos Islands in 1835, Darwin was well prepared by his reading and prior experience to study them. He also knew his study would fill a gap in the systematic survey of volcanic localities done by Daubeny. Darwin's interpretation of the Galápagos lavas followed Scrope in the essential point: that the different kinds of lava could be produced from the same volcano, or at least from the same system of vents.[84] Darwin schematized the lavas he observed. He said of them that

> considering the Islands in the whole Archipelago, it may be remarked, that the Southern ones appear to be entirely composed of Basalt & Greystone. whilst the Northern division is more essentially Trachytic.—[85]

The "greystone" category derived from Scrope, who used the term to refer to an intermediate between basalt and trachyte.[86]

The portion of Darwin's work at the Galápagos Islands that has drawn the greatest attention from petrologists originated in observations at James Island. These observations, subsequently developed after considerable further reading and work over the course of the next nine years, assumed a prominent place in *Volcanic Islands,* published in 1844. Darwin's initial observation was made on 10 October 1835. His entry, made that same day, lists four significant specimens. It reads:

October 10th—

3265	Trachytic cellular Lava frequent large Crystals of glassy Feldspar.—commonest kinds low down
3266	Base (black grey) more abundant. (olivine?)
[3]267	do. [ditto] with scarcely any Crystals
[3]268	Compact Lava with many small Cryst glassy Feldspar Basis finely Crystalline.—both latter more central parts

The sequence of numbers probably represents the order of Darwin's collecting: he began climbing lower down on the slope, collecting (and labeling) number 3265 first, then proceeding upward, and inland, collecting specimen 3268 at the "more central parts." Some time later, when writing up his full geological notes for James Island, Darwin converted his on-the-spot observations into complete sentences. In the margin next to these sentences the specimen numbers are written at the appropriate points. The passage reads,

> The Trachytic Lavas in the lower parts of the Isd are very cellular & the imbedded Crystals of glassy Feldspar very large & abundant:—[*In margin:* "3265 3266 with Olivine? !] In the higher central part. the rock generally is more compact, the base blackish grey with scarcely any Crystals. & or they are abundant & small, the base itself being Crystalline.—[*In margin:* 3267 3268][87]

In describing a crater on James Island that appears to correspond to that described in Volcanic Islands as the key site, Darwin noted in the margin "3267:68 like."[88] Specimens 3265, 3266, 3267, and 3268 are housed at the Sedgwick Museum. They are illustrated in Plate 4.[89]

In 1933, without being aware of the precise significance of these specimens for Darwin's intellectual development, Richardson analyzed them with the aid of modern techniques. Her analysis supported Darwin's initial assessment in the sense that it indicated substantial differences among the specimens. Specimens 3265 and 3266 were treated under the heading "Basalts with Dominant Felspar Phenocrysts." ("Felspar" is a variant spelling of "feldspar.") Specimen 3265 was described as containing feldspar phenocrysts of basic labradorite; specimen 3266 "differs from the foregoing chiefly in containing only a few phenocrysts of the first generation and having none of augite, but a larger proportion of olivine."[90] Richardson described specimens 3267 and 3268 as nonbasaltic. She described specimen 3267 as an oligoclase andesite, being a "fine-grained, bluish-grey, occasionally slightly scoriaceous lava containing a few fragmentary phenocrysts of felspar which are probably labradorite."[91] Specimen 3268 was a "soda trachyte" described as "compact, greenish-grey, with a few small crystals of felspar visible to the naked eye. A thin section shows abundant phenocrysts of felspar, a few of augite, and also occasionally hornblende, olivine and magnetite set in a trachytic groundmass."[92] Some of the differences among the four rocks can be detected from a comparison of the thin sections[93] (see Figure 3.9).

An additional problem exists with specimen 3268. A survey of James Island by Alexander McBirney and Howel Williams, published in 1969, failed to produce evidence of a rock similar to it. They speculated that rock was not truly part of Darwin's collection:

> Special effort was made to find the soda trachyte reportedly collected by Darwin from James Island and carefully described by Richardson (1933), but we found nothing remotely resembling this unusual rock. Dr. S. O. Agrell furnished us with a thin section of Darwin's specimen, now in the Sedgwick Museum, Cambridge, and this is almost identical with the specimen meticulously described by Richardson. It is, however, quite unlike any rock we found in the entire archipelago; hence, until the presence of trachyte on James Island is confirmed, we cannot exclude the possibility that the specimen described by Richardson was erroneously included in Darwin's collection.[94]

However, McBirney and Williams did not have access to Darwin's contemporary notes from the voyage, where specimen 3268 is described in a way consistent with the physical specimen as it presently exists, and as Richardson analyzed it.

Darwin's long version of his notes, quoted earlier, referred to a "compact lava with many small Cryst glassy Feldspar." In his specimen notebook the specimen numbered 3268 is listed as "Compact <greenish grey> Lava with many small Cry[stals] of glassy Fels[par]." Richardson's description, cited earlier, corresponds to that in Darwin's original notes. Thus, the speculation of McBirney and Williams that the specimen may have been mislabeled must be held in abeyance. Possibly further investigation of the site may locate a specimen similar to number 3268 or, if not, provide a more definitive answer.

Gregory Estes, K. Thalia Grant, and Peter R. Grant have suggested that the specimen may come from the vicinity of the crater at Jaboncillos in the highlands of James Island.[95]

Darwin's judgment regarding these four specimens seems to have been the origin, at an observational level, of a sentence that opens chapter 6 of *Volcanic Islands.* Under the heading "*On the separation of the constituent minerals of lava, according to their specific gravities,*" Darwin remarked,

> One side of Fresh-water Bay, in James Island, is formed by the wreck of a large crater . . . of which the interior has been filled up by a pool of basalt, about 200 feet in thickness. This basalt is of a grey colour, and contains many crystals of glassy albite, which become much more numerous in the lower, scoriaceous part. This is contrary to what might have been expected, for if the crystals had been originally disseminated in equal numbers, the greater intumescence of this lower scoriaceous part would have made them appear fewer in number.[96]

From Darwin's initial listing of the specimens, his categorization of the lava changed from trachyte to basalt. (It was the "grey colour" that probably led Darwin to label specimens 3265 and 3266 as trachytes.) Further, rather than "glassy Feldspar" he referred to "glassy albite" (albite being a variety of feldspar). As viewed from the hand specimens, specimens 3265 and 3266 do not have huge feldspar crystals. In terms of distribution of crystals, what Darwin described was a comparatively subtle effect.

Thus, these specimens from James Island continue to provide a challenge. If they were drawn from the site that inspired Darwin's move to a new understanding of the causes of lava differentiation, they did not demonstrate the large phenocrysts of feldspar that are characteristic of some other specimens that Darwin collected on the Galápagos Islands.[97] However, it also may be useful to remember that the differences among the mockingbirds on the Galápagos Islands, which formed the observational basis for Darwin's notion of the transmutation of species, were also subtle.[98] Further, and despite the fact that Darwin referred to the particular geographical location of James Island, his ideas on the differentiation of lavas may have partly been derived from broader sources, as from the work of other authors.

In developing his ideas Darwin found Scrope's *Consideration on Volcanos* exceptionally useful. Scrope provided a highly speculative framework for the discussion of the fundamental issues in geology relating to the origin of rock, to the origin of continents, and even to the origin of "new tribes of organized beings, both vegetable and animals."[99] At the Galápagos Islands Darwin was interested in the types of lava he found, as well as their origin. In Scrope he had an author who had staked out a clear choice of positions: there was either "an original difference" in the nature of the lavas, or the variation was owing to "changes produced in this rock . . . during the process of elevation, which was probably accompanied in many instances by repeated intumescence and reconsolidation."[100] Scrope chose for himself the second position, as did Darwin. Scrope posited an original granitic source, from which he believed trachyte and basalt both derived. Thus, regarding the formation of trachyte, he wrote,

> It would not indeed be difficult to conceive the production of ordinary trachyte . . . from a granitic origin; for the process of intumescence . . . may easily be supposed to change the felspar crystals from compact to glassy; to dissolve the whole or the greater part of the quartz in the aqueous vehicle; forcing it to assume the crystalline form on consolidation.[101]

Scrope described a situation where heat and pressure might operate together to separate lavas into differing types:

> In like manner we may imagine the production of basalt to have been caused by the exposure, within the vent of a volcano, of an intumescent mass of granite to reconsolidation, effected by the augmentation of temperature, and consequent expansion, of its lower beds. In these parts the extreme heat may be supposed to volatilize the mica and other ferruginous minerals, while the intense pressure would separate them in a gaseous state from the felspar, thus leaving a felspathose lava with very little iron in one part of the chimney, and occasioning the crystallization of a highly ferruginous lava in another.[102]

Scrope also referred to the "elective aggregation between particles of the predominant ingredient in rocks."[103] Darwin would pick up and extend the themes of separation and aggregation.

Darwin acknowledged his debt to Scrope in *Volcanic Islands,* claiming as his own the particular facts he was presenting, as in regard to James Island, and suggesting that Scrope

> has overlooked one very necessary element, as it appears to me, in the phenomenon,—namely, the existence of either the lighter or heavier mineral, in globules or crystals.[104]

As he recalled to Lyell some years later, Darwin believed it was his focus on "globules or crystals" that distinguished his views from Scrope's: "Scrope did suggest . . . the separation of basalt & trachyte, but he does not appear to have thought about the crystals which I believe to be the Keystone of the phenomenon."[105]

As Davis A. Young has pointed out in an excellent summary treatment, Scrope's and Darwin's emphases varied largely because they did their fieldwork on different volcanic terranes. Scrope cut his teeth studying the gas-driven cataclysmic explosions at Mount Vesuvius. Darwin was intimately familiar only with the products of quiescent oceanic volcanoes. This prevented him from emphasizing the role of erupting gases in volcanic processes. Instead, he emphasized contrasts in mineral density. As Young also suggested, Scrope and Darwin were alike in looking for the source of variation of basalt and trachyte either on the surface or within the volcanic edifice.[106]

After the voyage, Darwin continued to speculate on the origin of rocks. The subject was treated in both the Red Notebook and in Notebook A.[107] He continued to consult other geologists to support and extend his ideas. His next primary source was Leopold von Buch's *Description physique del îles Canaries.*[108] From this work, Darwin received report of an instance at Tenerife similar to that from James Island. In both cases feldspar was more abundant in the lava in proportion to its depth in the flow.[109] (This is opposite of what would be expected because feldspar is lighter, that is, less dense, than any of the other materials in the lava.) In a key point Darwin also gained an idea from laboratory processes that might explain the settling of feldspar in lava. From von Buch he learned of the experiments of Étienne-Marie-Gilbert Drée (1760–1848) in which crystals of feldspar tended to sink to the bottom in various lavas in a crucible.[110] From a British Association report, Darwin learned of the industrial process, described by Hugh Lee Pattinson (1796–1858), whereby silver impurities were separated from lead by repeated rounds of fusion and crystallization.[111] The stirring of the liquid appeared to cause the formation of detached granules. This industrial example reinforced the notion that an analogous process might occur in nature. Even though Darwin had no precise analogue to stirring to explain the formation of crystals in nature, indirect evidence of the existence of such crystals ("Now we have plain evidence of crystals being embedded in many lavas") made his argument plausible, if not secure.[112]

By the time Darwin wrote *Volcanic Islands,* he was ready to present a well-worked-out treatment of lava differentiation. The core of his argument was as follows:

> In a body of liquefied volcanic rock, left for some time without any violent disturbance, we might expect . . . that if one of the constituent minerals became aggregated into crystals or granules . . . such crystals

> or granules would rise or sink, according to their specific gravity. Now we have plain evidence of crystals being embedded in many lavas, whilst the paste or basis has continued fluid. . . . Lavas are chiefly composed of three varieties of feldspar, varying in specific gravity from 2.4 to 2.74; of hornblende and augite, varying from 3.0 to 3.4; or olivine, varying from 3.3 to 3.4; and lastly, of oxides of iron, with specific gravities from 4.8 to 5.2. Hence crystals of feldspar, enveloped in a mass of liquefied, but not highly vesicular lava, would tend to rise to the upper parts; and crystals or granules of the other minerals, thus enveloped, would tend to sink.[113]

Harker later noted that the "actual diversity met with among igneous rocks" was "in the main attributable to process of differentiation," with Darwin's notion of the principal cause for that differentiation being the "*sinking of crystals* in a fluid magma."[114] Harker also drew the parallel between geology and species, suggesting that the "only practical alternative to magmatic differentiation . . . is the doctrine of countless special creations."[115]

Pearson properly credited Darwin for seeing the overall process of magmatic differentiation as a subtractive one. Using twentieth-century terminology, he noted that "the crystallization and removal of a mineral from a body of molten rock inevitably causes a chemical change in the remaining melt."[116] In this sense, Darwin's ideas were consistent with the key idea behind the currently accepted theory of "fractional crystallization" described in Norman Bowen's seminal work *The Evolution of the Igneous Rocks*.[117]

A final example regarding Darwin's interest in the origin of rocks is illustrated in the specimens he gathered at the Cape of Good Hope. Here the army surgeon Andrew Smith (1797–1872) guided him to a classic geological site comprised of a junction of granite and clay-slate at the undulating ground known locally as the "Lion's Head," the "Lion's Back," and the "Lion's Rump."[118] In the collection of the Sedgwick Museum are two specimens (numbered 3668 and 3669) from the rocks at what Darwin called the junction or the contact. Darwin described these specimens in his notebook as "imperfect Gneiss or Mica Slate, <near> at junction of Granite & Clay Slate. on Lions back."[119]

In his specimen notebook Darwin further described these rocks as follows:

> Layers very thin, scales of mica small: layers slightly contorted.—mica, black glittering feldspar granular yellowish. quartz in small quantities. very thin laminae.—but perfectly characterized. mingled with coarse granitic veins. if indeed they are veins & not parallel layers[.][120]

The clue to Darwin's longtime interest in these specimens lies in the last line. He was beginning to query the origin of the rocks. In his full geological notes, he compared the rocks at the contact, which included clay slate and granite, to the

> union of two <very> fluids of very different degrees of specific gravity, which although they may penetrate each other, for the time keep distinct. The forms of curvature are however different from what would happen in such a case.—By following the ridge of the hill from the Lion's rump to the Head. a beautiful contact of the Clay Slate & Granite may be seen (pointed out to me by D^r. Smith). The Clay Slate is here changed into a thinly laminated rock. composed of small brilliant scales of mica separated by layers of yellowish granular mineral, which I do not know. whether it is quartz. or feldspar; the laminae are undulating.—[121]

In his Notebook A from after the voyage Darwin referred to clay-slate as a "distinct formation" formed in deep water and to gneiss as metamorphosed clay-slate.[122]

In *Volcanic Islands* Darwin presented an extensive argument for the structure of the rocks at the Lion's Rump. In his view the clay-slate had been "violently arched" by a body of molten granite, which explained the distribution of granite and altered clay-slate.[123] In the *Geology of South America,* in the context of a general discussion of cleavage and foliation, he referred back to this site. Here he argued that

> where a mass of clay-slate, in approaching granite, gradually passes into gneiss, we clearly see that folia of distinct minerals can originate through the metamorphosis of a homogeneous fissile rock.[124]

He remarked that he was arguing against the views of contemporary geologists, including Sedgwick and Lyell, that the constituent parts of each layer were deposited separately as sediment before being metamorphosed. Darwin's view was that the apparent layering was not the product of sedimentary deposition but that "foliated schists indisputably are sometimes produced by the metamorphosis of fissile rocks."[125] Darwin related both cleavage and foliation to his emphasis on crustal motion, concluding that

> planes of cleavage and foliation are intimately connected with the planes of different tension, to which the area was long subjected, *after* the main

> fissures or axes of upheavement had been formed, but *before* the final consolidation of the mass and the total cessation of all molecular movement.[126]

Here again we see Darwin developing a strong theoretical position that evolved from his interpretation of specific sites and specimens. In chapter 9 we will show him making similarly strong judgments as he came to interpret fossil quadrupeds from South America.

CHAPTER 4

THE ROMANTIC THREAD

> If the labours of men of Science should ever create any material revolution, direct or indirect, in our condition, and in the impressions which we habitually receive, the Poet will sleep then no more than at present, but he will be ready to follow the steps of the Man of Science, not only in those general indirect effects, but he will be at his side, carrying sensation into the midst of the objects of the Science itself.
>
> WILLIAM WORDSWORTH, *LYRICAL BALLADS,* AS QUOTED IN *NOTEBOOKS*

Darwin's First Book

In assessing Darwin as a geologist, we must take into account his delight in his work. Part of that delight was, and is, commonly experienced. As one geologist put it, "A perfect day when one is in the field is one of the greatest things on earth."[1] But there was something more to Darwin's delight. He also valued, from a scientific point of view, recording his own subjective response to natural phenomena. In order to record his responses while on the voyage, he opened a journal or diary on 24 October 1831, well in advance of the *Beagle*'s actual departure. Darwin's diary, first published by his granddaughter Nora Barlow in 1933, conveys information lacking in the other written material from the voyage.[2] Containing almost daily entries, the diary was sent home at intervals as though it were letters, its tone suggesting immediacy and human presence. With the first installment sent home, Darwin added, "Remember . . . this, that it is written solely to make me remember this voyage, & that it is not a record of facts but of my thoughts.—"[3]

A journal was a convenient tool for a traveler interested in natural history. Each new day and each new place held interest. As Darwin wrote from his cottage at Botofogo Bay in Rio de Janeiro,

> In England any person fond of natural history enjoys in his walks a great advantage, by always having something to attract his attention; but in these fertile climates, teeming with life, the attractions are so numerous, that he is scarcely able to walk at all.[4]

For Darwin, the *Beagle*'s circumnavigation was, in a sense, a natural history walk round the world.

In 1837, having successfully avoided the possibility that his diary would be intermixed with FitzRoy's, as what he called a "joint stock concern," Darwin prepared to convert his manuscript into a book.[5] This he did from 13 March to the end of September.[6] His familiarity of the genre of travel narrative as a model clearly aided him. In 1832, as the voyage was beginning, John Maurice Herbert (1808–1882), a fellow student at Cambridge, reveled in the book Darwin would write:

> I have already begun to picture myself its appearance & the nature of its contents one of Murray's 4tos in Davidson's type? How will it be entitled? "Observations physical, political & moral, made during a voyage r^{d} the world in the years 1831—1835 by C. Darwin F.R.S., F.L.S. &c &c". You will of course stay its publication, till these hieroglyphical characters be affixed to your name[.]—[7]

Darwin had modestly warned him off ("By the way you rank my Nat: Hist: labours far too high: I am nothing more than a lion's provider"[8]), and continued keeping his journal. When it appeared, Darwin's volume was octavo rather than quarto in size, but with suitable honorary "hieroglyphical characters" attached. In volume 3 of the combined narrative from the voyage, the title page identified Darwin as secretary of the Geological Society; in the separately printed volume, his designation as "F.R.S." (Fellow of the Royal Society) was added.

At the end of the voyage, on the way to publication as volume 3 of the *Beagle* narrative, the manuscript lost its strictly chronological order, and absorbed new material, much of it scientific, in the form of what Darwin called "sketches."[9] Like Humboldt, but more succinctly, Darwin interspersed his personal reflections with summaries of his research. This intermingling led James Paradis to remark that Darwin's volume offered "two versions of South American landscape" that "reflect the aesthetic ideals of Romantic art and the system-building traditions of the geological and natural sciences."[10] Paradis's assessment is accurate. In effect, the *Journal of Researches* is two books. What unites them is a strong authorial presence, a dramatically expressive style, and a refined consistency in tone and feeling.

Darwin's *Journal of Researches* is the book of a young and passionate author. Both scientific and nonscientific statements appear in the book unmodified by the qualifications that come with age and experience. Large ideas are expressed grandly. One reads of the "never ceasing mutability of the crust of this our World," having just learned that "[w]e see the law almost established, that linear areas of great extent [on the earth's crust] undergo movements of an astonishing uniformity, and that the bands of elevation and subsidence alternate."[11] One also meets high rhetoric as Darwin takes up the species question for the first time. On this point he could not afford to be open, but his passion seeped through as he speculated on the reasons for the extinction of species:

> All that at present can be said with certainty, is that, as with the individual, so with the species, the hour of life has run its course, and is spent.[12]

Darwin favored dramatic expression—"never ceasing mutability," "astonishing uniformity," "the hour of life"—in his scientific sketches, as in his narration of events.

In the course of the voyage Darwin found his authorial voice. He did this partly with the help of his sisters at home, who were his chief correspondents. Caroline, Susan, and Catherine formed a rota. Once a month, one of the sisters took responsibility for writing their brother the news. In return, they responded to the "Diary," as sections of it arrived in Shrewsbury. They praised it when it was "exceedingly entertaining," when read "aloud to Papa in the evenings."[13] They also offered critiques. Caroline came to the point:

> I mean as to your style. I thought in the first part . . . that you had, probably from reading so much of Humboldt, got his phraseology, & occasionally made use of the kind of flowery french expression which he uses, instead of your own simple straightforward & far more agreeable style. I have no doubt you have without perceiving it got to embody your ideas in his poetical language & from his being a foreigner it does not sound unnatural in him[.]—[14]

The sum of Caroline's advice to her brother was: be yourself.

Eventually Darwin did separate himself from Humboldt in style; his words were fewer. Further, though he could have done so, he did not generally refer to classical literature. Instead he relied on his own experience. He knew some parts of Britain well and drew comparisons where apposite, as in comparing the black volcanic cones of the Galápagos Islands to Staffordshire iron-foundries, or in recollecting the small inns of North Wales when he found a comfortable inn in the Blue Mountains of Australia.[15]

His sisters, his first audience, assisted him in developing a pleasing style that would educate and entertain but not offend. Similarly, in the 1840s, his wife Emma would become his "model of the conventional Victorian reader."[16] Later, in the 1860s, when he was writing on the pollination of orchids he could assure his publisher, John Murray, that "[t]he subject of propagation is interesting to most people, & is treated in my paper so that any woman could read it."[17] Darwin could make such a remark so confidently because he had years before learned to treat intriguing but potentially explosive issues with tact and delicacy. Thus, he quite properly chose his daughter Henrietta Darwin (1843–1927) to edit the manuscript of his *The Descent of Man, and Selection in Relation to Sex* (1871).[18] Writing and reading were a family business.

Darwin's years after his arrival back in England in 1836 were relatively sedentary compared to his life during the voyage. "I am turned a complete scribbler,—" he remarked in 1838 to an old friend.[19] His daily direct experience of dramatically new scenery was over, but his career as an author was only beginning. In his first book, his narrative of the *Beagle* voyage, he forged a relationship not only with his scientific peers but also with a public beyond the reach of professional societies. With considerable success, he was able to convey to this broad audience his deep feeling toward nature. It is the geological aspects of nature that primarily concern us, but it is necessary to begin with geology as part of the whole.

Nature as Experience

Darwin's journal served him as a place to record his most profound feelings toward nature. Geology was one element of landscape, whose "chief embellishment" was vegetation.[20] As he suspected (Humboldt's descriptions had prepared him), tropical forests brought Darwin his greatest delight. The experience came early in the voyage, on 29 February 1832, as he walked in the forest at Bahia, also known as San Salvador:

> The day has past delightfully. Delight itself, however, is a weak term to express the feelings of a naturalist who, for the first time, has been wan-

> dering by himself in a Brazilian forest. Among the multitude of striking objects, the general luxuriance of the vegetation bears away the victory. The elegance of the grasses, the novelty of the parasitical plants, the beauty of the flowers, the glossy green of the foliage, all tend to this end.[21]

In his concluding remarks to his journal Darwin summarized his experience in some of the most lyrical language he was to use in his book:

> Among the scenes which are deeply impressed on my mind, none exceed in sublimity the primeval forests undefaced by the hand of man; whether those of Brazil, where the powers of Life are predominant, or those of Tierra del Fuego, where Death and Decay prevail. Both are temples filled with the varied productions of the God of Nature:—no one can stand in these solitudes unmoved, and not feel that there is more in man than the mere breath of his body.[22]

The phrase "temples filled with the varied productions of the God of Nature" in which no one can stand unmoved captures a high point in Darwin's evocation of a sense of sublimity. His vocabulary echoed the title of his grandfather's book *The Temple of Nature*.[23]

If Darwin's first temple was botanical, his second was geological. The mountains of Tierra del Fuego inspired in him deep feelings of awe. Tierra del Fuego was a difficult place for those aboard the *Beagle* to like. The climate was cold and damp; the skies often dark. Even the vegetation was inhospitable. Fierce winds brought the area's characteristic beech trees low to the ground, which, together with fallen branches, made walking through the woods difficult. In his notes on plants Darwin wrote, "The extreme dampness of the climate favours the course luxuriance of the vegetation: the woods are an entangled mass where the dead and the living strive for mastery."[24] With the dreary climate, it seemed unlikely that Darwin would find at Tierra del Fuego anything like the experience of sublimity that he had experienced in the tropics. However, curiously, he did. He phrased his experience in terms of opposites: life and death. He found beauty—or grandeur, his preferred expression—in both places. Of Tierra del Fuego he wrote,

> There was a degree of mysterious grandeur in mountain behind mountain, with the deep intervening valleys, all covered by one thick, dusky mass of forest. The atmosphere, likewise, in this climate (where gale succeeds gale, with rain, hail, and sleet), seems blacker than any where else. In the Strait of Magellan looking due south from Port Famine, the dis-

> tant channels between the mountains appear from their gloominess to lead beyond the confines of this world.[25]

The vocabulary in this passage is romantic in its sensibility—"mysterious," "grandeur," "gloominess"—as is the phraseology—"mountain behind mountain," "beyond the confines of this world." A photograph from Tierra del Fuego recalls the view (Plate 5).

Darwin was artful in constructing narrative that emphasized the dramatic elements in landscape. One example will suffice to make this point. In April 1835 he encountered a silicified forest in the course of his ride from Hornillos over the Uspallata range of the Cordillera.[26] With the understanding he already possessed, "It required little geological practice to interpret the marvellous story, which this scene at once unfolded."[27]

Darwin continued,

> I saw the spot where a cluster of fine trees had once waved their branches on the shores of the Atlantic, when that ocean . . . approached the base of the Andes. I saw that they had sprung from a volcanic soil which had been raised above the level of the sea, and that this dry land, with its upright trees, had subsequently been let down to the depths of the ocean. There it was covered by sedimentary matter, and this again by enormous streams of submarine lava. . . . [Again] the subterranean forces exerted their power, and I now beheld the bed of that sea forming a chain of mountains more than seven thousand feet in altitude.

Implicit in Darwin's story was his presence as observer: I saw, I beheld. His first-person account vivified the inference he drew of the rise and fall of land.

The emphasis on human perception was a thread in the Romantic vision of nature. In arguing the case for placing Darwin in this tradition, Paradis proposed Samuel Taylor Coleridge (1772–1834) as the exemplary Romantic author, most akin to Darwin in his view of landscape. To illustrate the romantic sense of landscape as centered in human perception, Paradis quoted Coleridge's poem "Lines Composed While Climbing Brockley Coomb" (1795): "Ah, what a luxury of landscape meets/My gaze!" Landscape here involved the "complex sense impression of some striking moment of experience in physical nature" that "could be recreated through memory in poetic and narrative description."[28]

Darwin's description of the intense feelings aroused in him by the Brazilian forest, or by the "mountains behind mountains" in Tierra del Fuego, fit the characterization drawn by Paradis. However, Darwin referred to reading Coleridge only after the voyage (along with Wordsworth, whose *Excursion* he read through twice), so that one must be cautious in speaking of any direct influence on him of the Romantic poets during the voyage.[29] More broadly one can connect Darwin's *Journal of Researches* with the Romantic movement by indicating their common ground in the literature of travel beginning with Cook. Alan Frost has made this argument connecting Cook's voyages with Coleridge, Wordsworth, and Robert Southey (1774–1843).[30] Darwin was heir both to the tradition of the Cook voyages and to the poetry partly inspired by them.

After the voyage, Darwin did read more widely in poetry. Showing similar interests, John Grant Malcolmson (1803–1844), a geologist, would write to Darwin in 1839 of travels to the Scottish Borders "in search of picturesque beauty and poetic associations."[31] Darwin would also have had his brother Erasmus, a confidant of Thomas Carlyle (1795–1881) and his wife Jane Welsh Carlyle (1801–1866), to guide his taste. One of Darwin's favored authors during the period immediately following the voyage was Wordsworth, whose reputation was undergoing a resurgence of critical regard in the 1830s.[32] Wordsworth's *Excursion* included numerous passages that heralded the spiritual power of nature. Such passages as the following would have resonated with the young geologist:

> Take courage, and withdraw yourself from ways
> That run not parallel to Nature's course
> Rise with the Lark! your Matins shall obtain
> Grace, be their composition what it may,
> If but with her's performed; climb once again,
> Climb every day, those ramparts; meet the breeze
> Upon their tops,—adventurous as a Bee
> That from your garden thither soars, to feed
> On new-blown heath; let yon commanding rock
> Be your frequented Watch-tower; roll the stone
> In thunder down the mountains.[33]

Drawing on his reading of the Romantic poets, Darwin cited Percy Bysshe Shelley (1792–1822) in the second (1845) edition of his *Journal of Researches*. There Darwin wrote, "One asked how many ages the plain [in Patagonia] had thus lasted, and how many more it was doomed thus to continue." He then quoted from Shelley's "Mont Blanc":

> None can reply—all seems eternal now.
> The wilderness has a mysterious tongue.
> Which teaches awful doubt.[34]

In 1838 Darwin cited Wordsworth in his theoretical notebooks. In an "Analysis of pleasures of scenery" Darwin commented on his own pleasure as a geologist. His note—partly quoted in chapter 1—reads:

> I a geologist have illdefined notion of land covered with ocean, former animals, slow force cracking surface &c truly poetical. (V. Wordsworth about science being sufficiently habitual to become poetical)[35]

To see what Wordsworth had in mind, consider the following passage from the *Lyrical Ballads* (1802):

> Poetry is the first and last of all knowledge—it is as immortal as the heart of man. If the labours of men of Science should ever create any material revolution, direct or indirect, in our condition, and in the impressions which we habitually receive, the Poet will sleep then no more than at present, but he will be ready to follow the steps of the Man of Science, not only in those general indirect effects, but he will be at his side, carrying sensation into the midst of the objects of the Science itself. The remotest discoveries of the Chemist, the Botanist, or the Mineralogist, will be as proper objects of the Poet's art as any upon which it can be employed.[36]

Wordsworth's phrase "impressions which we habitually receive" would seem to have sparked Darwin's interest in 1838, for he was then considering the origin of human imagination from an evolutionary perspective. However, in taking note of a romantic or experiential strand in the *Journal of Researches,* the key phrase is "carrying sensation into the midst of the objects of the Science itself." Darwin's *Journal of Researches* fused personal experience—"sensation"—with scientific subject matter. It carried sensation into the realm of science.

Among his shipmates Darwin was not alone in his attitude toward nature. Conrad Martens (1801–1878), the landscape artist who accompanied the *Beagle* for about a year in 1833–1834, was a man of comparable sensibilities. Several of his works appeared in small-scale, black-and-white reproduction in the first two volumes of the FitzRoy's *Narrative* where their beauty and fidelity to nature were not easily appreciated. Happily, Martens's work is now better known.[37] Like Darwin, Martens was attentive to accurate representation. His pencil drawings in his sketchbooks were made from nature, and he kept them throughout his life as reference tools for future paintings. He favored actual scenes, rather than imaginary ones, depicting them with attention to meteorological, botanical, and geological detail. Martens educated himself to this end. For example, one of his sketchbooks from the voyage records extracts he made from Lyell's *Principles of Geology*.[38] But Martens's goal was not scientific illustration. As he stated in a lecture of 1856, "The art of landscape painting lies not in imitating individual objects, but in imitating an effect which nature has produced."[39] This effect was necessarily on a human observer.

As Darwin stood in a tradition of travel writing, so Martens was heir to the British tradition of watercolor painting, having been trained by the painter and drawing master Copley Fielding (1787–1855). In addition, Martens greatly admired the work of Joseph Mallord William Turner (1775–1851), whose free treatment of light and color represents one of the high points in British art.[40] In an 1856 lecture, Martens directed attention to some of Turner's work "[where] will be found *breadth*, *grandeur*, and a total absence of petty details." ("Grandeur" was for Martens, as for Darwin, a term of high praise.) Martens also compared the activity of the artist and the writer: "knowing *what to do* in painting may be compared to knowing what *to say* in writing."[41] In Martens's view, the activity of the painter and the writer were comparable. Indeed, both Martens and Darwin viewed their work as complementary to each other. As a prosaic sign of that affinity, toward the end of the voyage, when Darwin reconnected with Martens in Sydney, he purchased two of his paintings.[42] One was of a scene from Tierra del Fuego, the other of a scene from the river Santa Cruz.

Geology as Spectacle

In his *Journal of Researches* Darwin described his pleasure in beholding the geological aspects of nature. For his readers he knew this would be a stretch of imagination. From an artistic perspective, he acknowledged, "Group masses of naked rock even in the wildest forms, they may for a time afford a sublime spectacle, but they will soon grow monotonous."[43]

Yet, of the sites he most remembered from the voyage, roughly half were geological. The most memorable scenery was that of the primeval forests of Brazil and Tierra del Fuego, of the vast plains of Patagonia, and of lofty mountain views from the Cordillera. Also on the list of "most remarkable spectacles" Darwin beheld

> may be ranked the stars of the southern hemisphere—the water-spout—the glacier leading its blue stream of ice in a bold precipice overhanging the sea—a lagoon island raised by the coral-forming polypi—an active volcano—and the overwhelming effects of a violent earthquake.

Figure 4.1. Osorno. Photograph by the author.

In the next sentence Darwin went on to note, explaining his taste, "The three latter phenomena, perhaps, possess for me a peculiar interest, from their intimate connexion with the geological structure of the world."[44]

The lagoon island was Keeling, in the Indian Ocean, visited in April 1836; a sketch of Darwin's theory of reef formation appears in the *Journal of Researches*.[45] The volcano was Osorno in Chile. While aboard ship, Darwin observed the volcano in eruption on the night of 19 January 1835 (Figure 4.1). He wrote of the experience:

> At midnight the sentry observed something like a large star; from which state the bright spot gradually increased in size till about three o'clock, when a very magnificent spectacle was presented. By the aid of a glass, dark objects, in constant succession, were seen, in the midst of a great red glare of light, to be thrown upwards and to fall down again. . . . By morning the volcano had resumed its tranquility.[46]

Just a month later, on 20 February, Darwin experienced an active earth once again:

> The day has been memorable in the annals of Valdivia, for the most severe earthquake experienced by the oldest inhabitant. I happened to be on shore, and was lying down in the wood to rest myself. It came on suddenly, and lasted two minutes; but the time appeared much longer. The rocking of the ground was most sensible. . . . There was no difficulty in standing upright, but the motion made me almost giddy.[47]

In the published version of his journal Darwin expanded his treatment of the earthquake to include a summary of his scientific work on the subject to that date, which entailed his conclusion that earthquakes were associated with a rise in the land.[48] To Darwin both aspects were valid, the record of his subjective experience and the record of his reasoned inferences. In the *Journal of Researches,* then, alongside summaries of scientific observations, one can speak of a poetics of geology—sensation carried into the midst of the objects of science itself. In complementary fashion, Martens, who was part of the company on the trip up the Santa Cruz, produced a series of fine sketches and watercolors from it.[49]

CHAPTER 5

A PROSPECTIVE AUTHOR

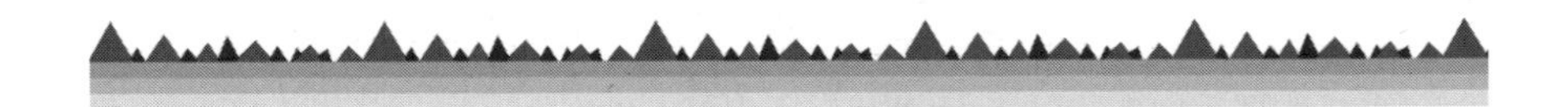

The Appeal of Authorship

On occasion Charles Darwin can seem like our scientific contemporary, for the subjects he engaged in remain engaging today, but in his role as author he belongs to the past. It is not customary today for scientists to write book after book, as Darwin did, or for these books to serve as the primary vehicle of scientific communication. For Darwin, however, the book was central. He wrote at least eighteen, depending on what one counts; in his *Autobiography* he entitled the section describing his most important work "An account how several books arose,"[1] and in his personal "Journal," begun in August 1838 after he had come to a mature sense of himself, he organized entries around his books. A characteristic entry is that for 1846: "Oct 1st. Finished last proof of my Geolog. Observ. on S. America; This volume, including Paper in Geolog. Journal on the Falkland Islands took me 18 & 1/2 months."[2] Further, almost always he had a book under way; when one was complete, the next was begun.[3] He called them the milestones to his life.[4]

Darwin's habits as a writer were established early in life and were nourished in a family that had already produced one major writer, Erasmus Darwin. Even before the voyage Charles Darwin wrote abundant vigorous prose, as witness his high-spirited and humorous account of a zoological walk taken while he was a student at Edinburgh.[5] His prose did not have the easy fluidity demonstrated by his contemporaries Alfred Russel Wallace (1823–1913) and

Thomas Henry Huxley (1825–1895); his son described his spoken sentence as "a system of parenthesis within parenthesis."[6] Nor did striking passages come easily. The "entangled bank" metaphor of the final paragraph of the *Origin of Species* developed over many years from an initial observation of the entangled beech trees at Tierra del Fuego.[7] But the copious notes from the *Beagle* voyage reflect his active intelligence and formed the basis for his early publications. Darwin himself recommended extensive note-taking along with section-drawing as a method for observing in geology, arguing, quoting Francis Bacon (1561–1626), that "[r]eading maketh a full man, conference a ready man, and *writing an exact man*."[8]

Geology figures prominently in Darwin's career as an author because it was the first subject on which he planned to write a book. At the first stop of the *Beagle*, São Tiago in the Cape Verde Islands, Darwin worked out some of the geology, and, as he recalled in later life,

> It then first dawned on me that I might perhaps write a book on the geology of the various countries visited, and this made me thrill with delight. That was a memorable hour to me, and how distinctly I can call to mind the low cliff of lava beneath which I rested, with the sun glaring hot, a few strange desert plants growing near, and with living corals in the tidal pools at my feet.[9]

The question "Why a book?" rather than articles may be answered only by surmise. Perhaps the book-length format seemed more commensurate with the scale of the voyage. Perhaps other authors inspired him. Certainly there were exemplary books before him—volume 1 of Lyell's *Principles of Geology* published the previous year, Leopold von Buch's treatment on Norway and Lapland, and Humboldt's massive *Personal Narrative*, which was heavily geological.[10] Whatever the attraction of the format, the rigor of Darwin's organizational scheme for his notes during the voyage is evidence consistent with his claim of having harbored an early ambition for authorship.

Parenthetically, the world did not reward Darwin's authorship with great financial benefit, at least in the early years. His net return on what turned out to be his first published book, the journal of the voyage, was a debit of £21 10s., spent on presentation copies.[11] Publishing his works on geology increased his loss. He had hoped his university would assist him in this endeavor, and in May 1837 he asked Henslow whether he had the opportunity of sounding out "any of the great Cambridge Dons"—presumably including Sedgwick and Whewell—on the matter.[12] On this occasion the Cambridge network failed to produce the desired result. Eventually, the three parts of his geology writings were published commercially by Smith, Elder and Company. Because the £1000 treasury grant for the publication of results from the voyage was exhausted in 1843 with the last part of the *Zoology,* Darwin and his publisher advanced sums to complete the geological work. Darwin's own contribution appears to have been on the order of £200 to £300.[13] It was thus as a financially burdened author that Darwin would remark while writing *Volcanic Islands,* "It is up-hill work, writing books, which cost money in publishing & which are not read even by geologists."[14] In 1846, as he was writing the third part of his geology treatise, he joked to his old Cambridge friend John Maurice Herbert that the "only object in writing a Book is a proof of earnestness" for geology "is at present very oral."[15] To his publisher he wrote, "[Y]et seeing that hitherto, only 216 copies of Coral Reef & 143 of Volcanic isl^{d} have been sold, I really doubt whether it be worth printing off more than 350 copies" of the new book, *Geological Observations on South America.*[16]

Eventual profit or not, Darwin's notes on the geology of the voyage were crafted with the goal of publication in mind. "Research" and "writing" ran together. His geological manuscripts from the voyage fall into five categories, with the emphasis running toward writing and analysis in the last three. They are as follows:

(1) The field notebooks.[17]
(2) The four geological specimen notebooks, presently on deposit at Cambridge University Library.
(3) The geological notes proper, generally organized according to locality, and running to 1383 pages, as opposed to 368 pages for the zoological notes.[18] The physical appearance of these notes suggests Darwin's method of composition. The notes are written on one side of the page only, with specimen numbers keyed into the notes and usually appearing in the left margin. Versos are left empty for footnotes, which are plentiful and give the manuscript a dense and almost prematurely scholarly appearance. Throughout the geological notes proper is evidence of later annotation, some dating from the voyage as Darwin revised and enlarged his own earlier work, and some dating from after the voyage as Darwin reworked his material for publication.
(4) Several synthetic essays, written toward the end of the voyage, including among them "Coral Islands," "Recapitulation and concluding remarks" on the geology of South America, and a set of 36 folio pages on cleavage.[19]
(5) Two notebooks, "Santiago Book" and "R.N." or the "Red Notebook," which Darwin used partly to prepare for publication.[20]

With the aid of these manuscripts one can reconstruct Darwin's overall vision of geology during the voyage. While I cannot be comprehensive, I will pick up Darwin's writing at several stages during the voyage: first, the beginning of the voyage, and then at several subsequent points, as his observations on individual sites yielded to essays on more general themes.

Establishing Methods in the Cape Verde Islands

For Darwin the *Beagle* voyage began with a major disappointment. The ship's first stop was to have been the Canary Islands, the place to which Humboldt had devoted much of his first volume, and which was the intended destination of the private trip planned by Darwin and some of his Cambridge friends before the *Beagle* offer was tendered. Owing to the presence of cholera in England, the *Beagle* was prohibited from docking in the Canaries on arrival. Unwilling to wait out a twelve-day quarantine, FitzRoy proceeded immediately to the Cape Verde Islands farther south. Of these islands Daubeny, who had been richly descriptive on the volcanic geology of the Canaries, thanks to information provided by von Buch, could say only that their structure was "too imperfectly known to detain us long."[21] Darwin was on his own.

Thus, Darwin for his first venture began his geologizing at Quail Island (presently Ilheu de Santa Maria), a speck of land about a mile in circumference lying in the harbor of Porto Praya, St. Jago (São Tiago), in the Cape Verde Islands. Figure 5.1 presents a painting of St. Jago by Conrad Martens, not while he was with the *Beagle* but sometime later, the Cape Verde Islands being a convenient port. Figure 5.2 shows a plan of Porta Praya.

Darwin called Quail Island his keystone to be used in unraveling the structure of the main island of São Tiago. His interpretation of both Quail Island and São Tiago is that they are volcanic islands that were submerged for some period beneath the sea, where they collected marine beds and then another layer of melted volcanic material. While on Quail Island he wrote that "the whole mass was then raised since which or at the time there has been a partial sinking.—I judge of this from the appearance of distortion. & indeed. the distant line of coast seen to the East [of Quail Island looking toward the mainland] which is considerably higher bears me out."[22] In conjunction with his interpretation Darwin provided a section drawing and a listing of beds from the lowest to the highest (Figure 5.3).[23]

Darwin also collected specimens and systematically listed them in his specimen notebook (Figure 5.4). Specimens from Quail Island are numbered from 12 through 81. (There is no record of any specimens numbered 1–11.) Specimens from Quail Island are presently stored in the drawer pictured in Figure 5.5. (Some boxes contain more than one numbered specimen, and not all specimens are extant.) The specimens are housed within the Department of Earth Sciences at the University of Cambridge.

Figure 5.1. Watercolor sketch of one of the Cape Verde Islands by Martens. Reproduced by courtesy of Richard Darwin Keynes.

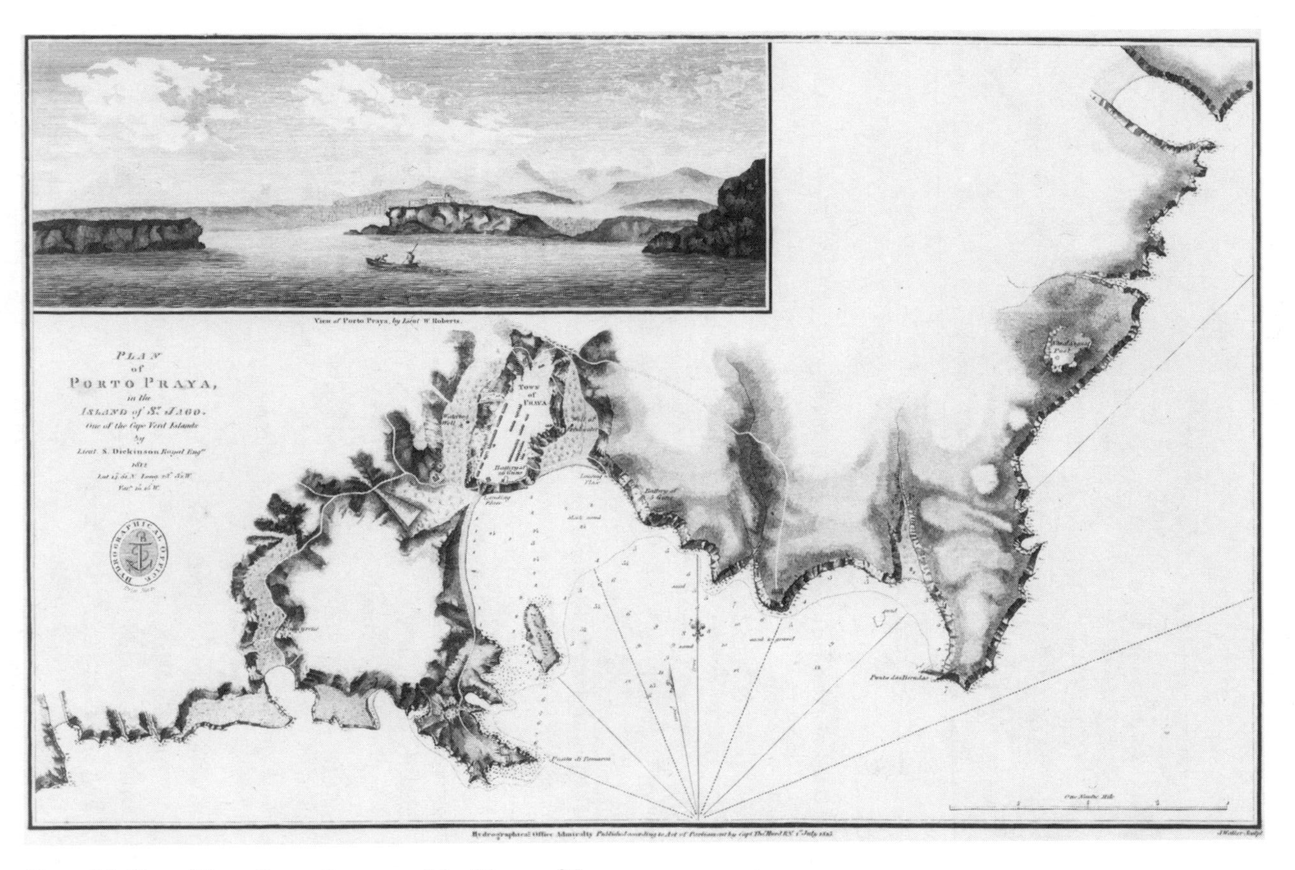

Figure 5.2. Plan of Porto Praya. Courtesy of the Library of Congress.

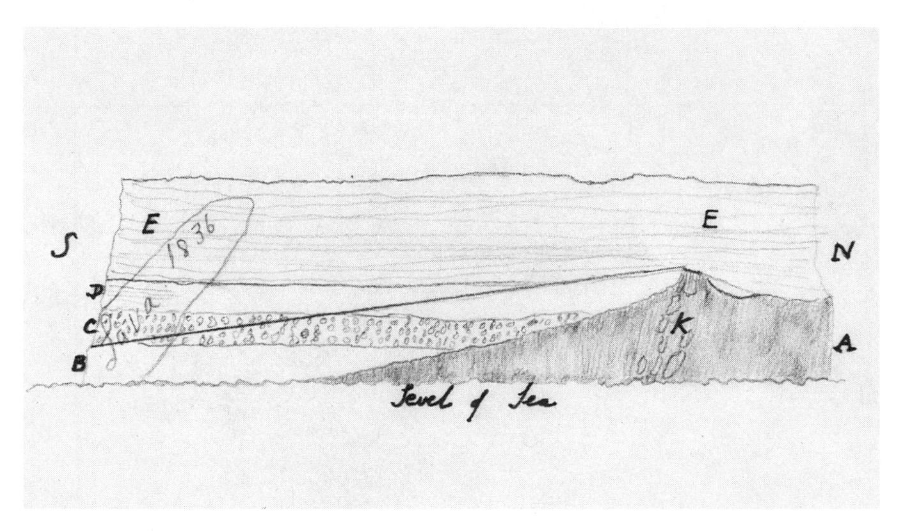

Figure 5.3. Section drawing of Quail Island, traced from DAR 32.1:16A. By permission of the Syndics of Cambridge University Library.

Darwin's description of bed "A" is as follows:

> The lowest rocks all contain included crystals[.] the most common is (1)Augite & Olivine (?) there are masses of (2) amygaloid.[*sic*] containing in cavities minute white crystals: Many of these rocks have undergone great (3) decompos[ition], in parts becoming either a yellow.red. or green clay. including the unaltered crystals of augite: these are all mingled & interlaced with great confusion: in other parts there are (4) dykes of a harder rock. which stand out from <<better>> resisting. the weather. the <<rock>> assumes a <<transverse>> columnar form: & from the cracks of such higher parts being filled with sand. I should think they must be of cotemporaneous origin.—

(Darwin's superscript numerals are keyed to notes regarding specimen numbers: (1) to specimens 52–58 and 64–69, (2) to specimens 48–52, (3) to specimens 58–59 and 68–69, and (4) to specimen 62, "cotemporaneous dyke.") A drawer containing specimens from this lowest or "A" bed is shown in Figure 5.6. The label "Limburgite lava" attached to the specimens reflects the work of Harker, who published his findings on the specimens in 1907. Harker worked from a specimen numbered 4704, which was a "thin slice" derived from specimen 48, as noted in the fair copy of Darwin's catalogue at the Sedgwick Museum.[24]

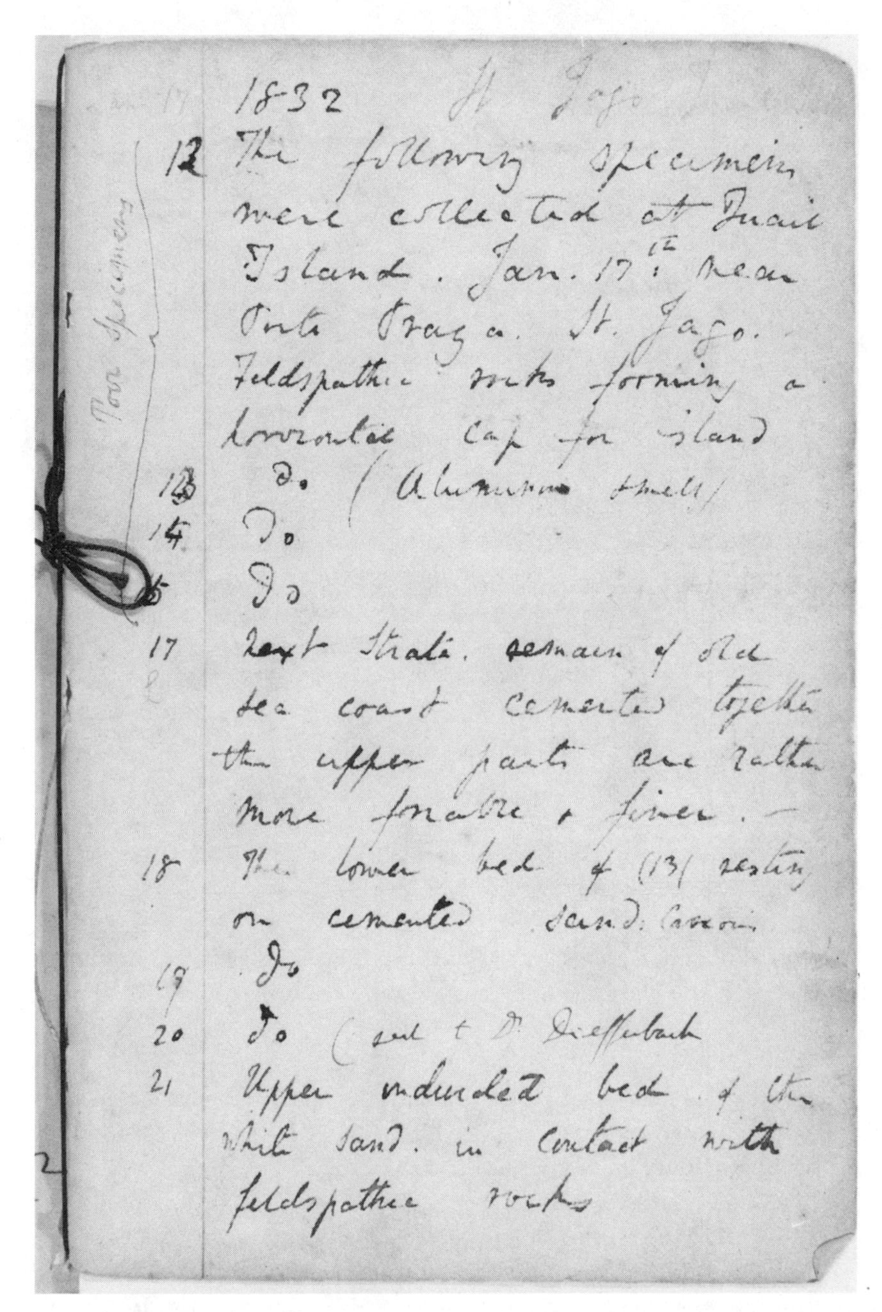

1832 St Jago

12 The following specimens were collected at Quail Island. Jan. 17th near Porto Praya. St. Jago.
Feldspathic rocks forming a horizontal cap for island

13 Do (Aluminous smell)

14 Do

15 Do

17 Next Strata. remain of old sea coast cemented together the upper parts are rather more friable & finer. —

18 The lower bed of (13) resting on cemented sand: Calcareous

19 Do

20 Do (sent to Dr Dieffenbach

21 Upper undulated bed of the white sand. in contact with feldspathic rocks

Poor Specimens

Figure 5.4. Page from Darwin's geological specimen notebook. By permission of the Syndics of Cambridge University Library.

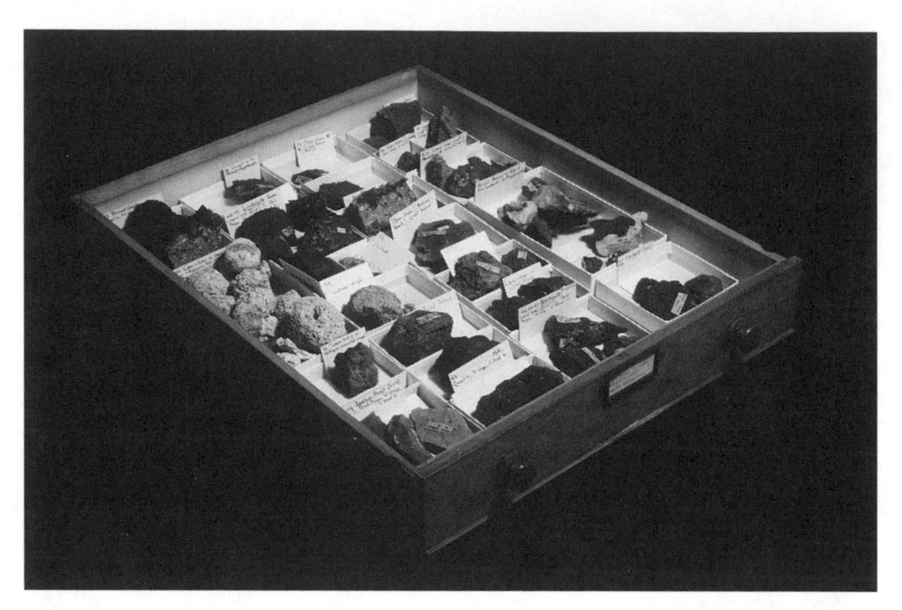

Figure 5.5. Drawer of Quail Island specimens. By permission of the Sedgwick Museum, University of Cambridge.

Darwin described bed "B" as follows:

> The lowest [(5)] bed is of more <<soily>> indurated character than the other & contains a remarkable number of Turritellae. & some few other shells. Patellae. &c &c

Darwin's superscript "5" refers to specimens 77–81 and 37–44. In the published version the bed is described as a "soft, brown earthy tuff."[25] His description of bed "C" follows:

> Several alternations. of coarse gravel. large rounded stones. & fine sand.—this contains a great variety of shells.—Cardium. Volutae a [(1)] rugged Ostreae in great profusion & adhering to the rocks on which they lived: patellae in vast numbers.—This is best seen on Eastern side of Island. & here 12 or 18 feet thick. on the West it is only a foot: on this side the gravel is cemented into a very hard conglomerate but as this process is now going on in the <<coast of>> island. perhaps it may <<be>> the work of the present day.—

Figure 5.6. Three subsets of Quail Island specimens. By permission of the Sedgwick Museum, University of Cambridge.

Darwin's superscript "1" refers to specimens 37–46, which include all the shells mentioned in the text. Darwin described bed "D" in the following way:

> This is a bed of very various characters[.] generally it is white <<or grey>> sand. (2) almost entirely formed of minute shells. & corallines.—The lower part of this is very remarkable. by being. almost composed of (3) concretionary masses of a *white* mineral: frequently being pisiform & having nucleus of some other rock <illegible> this & inferior beds. often have the appearance of a wall built with mortar: the upper (4) parts of this & in near contact with superior Feldspathic <<rock>> become very hard: it is then white spotted with yellow.—At the South of island. the sand is cemented into a fine (5) sandstone.—In these beds. besides numerous small shells. I found. crabs & perfect bones of a Echini.—

The superscript numerals refer to specimens: (1) 37–46, (2) 17–21, 27–34, (3) 25–26, (4) 21–24, and (5) 74. For specimens listed under superscript "1" Darwin noted, "The organic remains in these beds are certainly of a Tropical type. & as far as my knowledge goes. are the same as those of present day.—

<I should think this old coast one of no great duration?—>" For specimens listed under superscript "3" Darwin wrote, "Curious white concretions often becoming pisiform. lower bed of white sand." These are specimens 25 and 26, which are pictured in the middle box in Figure 5.6. Darwin also recorded the tests he performed on specimens 25 and 26: "Effervesces readily with Mur: Acid. gives precipitate with Oxalate of Ammonia.—Under Blowpipe becomes slowly caustic. & with [heat.?] Cobalt remains of a Violet colour.—Carbonate of Magnesia. (?) Carb of Lime."

Harker reexamined specimens 25 and 26 and noted that "the white balls, mostly from one to two inches in diameter, built up by 'Nulliporæ,' are interesting as the analogue, on a giant scale, of a certain type of oolite."[26] Harker's judgment regarding the identity of the white balls echoed Darwin's published view.[27]

Bed "E" Darwin described as follows:

> Above these beds as far as they extend & upon the Crystalline rocks (A) there is a horizontal cap of a hard [(a)] Feldspathic rock. which has the appearance of Basalt: in some parts it has the appearance of plains of stratification. & <<in>> others of 5 sided columns. occasionally curved.—About a foot of the lower part is [(b)] carious & harsh.—The [(c)] upper is slightly so. & in parts is divided into large concen<tionary>tric balls.—3 feet in diameter: giving a tesselated appearance to the rock.—[(1)] This formation varies from 6 to 20 feet in thickness.—[28]

"Feldspathic" was underlined in pencil. Darwin's alphabetical footnotes refer to the following specimens: (a) 12–15 and 71, (b) 18–20, and (c) 72–73. In his specimen notebook Darwin described numbers 12–15 as "Poor Specimens." (Specimens 12–14 are pictured to the right in Figure 5.6.) In *Volcanic Islands* Darwin referred these specimens to the "basaltic lava, superincumbent on the calcareous deposit." The lava is "of a pale gray colour, fusing into a black enamel; its fracture is rather earthy and concretionary; it contains olivine in small grains."[29] Harker's identification of specimen 13, made from a "thin slice" of it numbered 4703, is as a variety of basalt termed an "analcime-basalt."[30]

Darwin then concluded his description of the sequence of rocks with a remark on structure:

> The general height of Island is about 40 feet. but the beds in it often vary their relative height. in parts the grey sand is 30 feet above waters edge. in others close to it:[(2)] this is chiefly owing to dislocation of which there is a very marked one at (K)[(3)].—

"Dislocation" was underscored twice in pencil. Footnote 2 reads, "The Eastern side of island is much more regular"; footnote 3 reads, "By this fault a small portion of grey sand on extremity of shore has been carried down."

In reviewing Darwin's notes on Quail Island, one can identify elements of his practice that remained constant throughout the voyage, and reflected well both on his early training and on the seriousness with which he approached his investigation. At the level of observation, he sought to establish the sequence of rocks in a given locale, as he did at Quail Island, where the sequence was established for the entire island with rock samples collected for each distinct layer and where observations were summarized diagrammatically as well as in prose. Darwin's focus on stratigraphy, or more loosely, bedding, was consistent with the practice of colleagues at home, and of Sedgwick and Henslow, who had overseen his most recent field studies. Additionally, British geologists—especially Sedgwick—were interested in structural features, and Darwin's section drawing and notes on Quail Island show him observing this tradition, recording line of dip (the rule-drawn diagonal line on the section drawing) and a line of dislocation ("K").[31] (In notes on other locales he also recorded the strike of strata.)

Darwin was also working within established practice in other respects. Characterizing rock type presumed the application of traditional mineralogical categories, and Darwin's notes on Quail Island show him applying these categories and also using the blowpipe and muriatic acid for chemical analysis. The presence of fossil shells is attended to, and joined with speculation regarding the relative age of the deposits: see Darwin's entry cited earlier for bed "D." In later notes Darwin's interest in organic remains grew. For instance, in March 1833 at East Falkland Island he would note with respect to some shells that "it will be . . . preeminently interesting to compare these fossils, with those of a similar epoch in Europe: to compare how far the actual species agree & the comparative number of individuals of each sort[.]—"[32] Finally, in his geological work, Darwin remained within the boundaries of the common practice of the major British geologists of the 1830s. Thus, he did not follow the lead of Humboldt, from whom he took much in other respects, in pursuing the sorts of magnetic measurements that Humboldt favored and that were in fact gathered assiduously and successfully by FitzRoy and other officers during the voyage.[33] (Indeed, the *Beagle* delayed its departure from São Tiago as the "Captain is so much engaged with experiments on Magnetism.")[34] Despite his lack of direct involvement with the taking of magnetic measurements or their interpretation, Darwin did on occasion speculate as part of what John Cawood has termed the "cosmical tradition" in magnetic researches.[35] This he did in spite of his relative lack of mathematical sophistication, as compared to a number of his Cambridge colleagues and as compared to Humboldt.

On the level of interpretation, Darwin's geological notes, including those on Quail Island, are remarkably rich, fuller in theoretical material than the field notes of Sedgwick or Murchison.[36] Darwin was aware of his propensity to theorize and, in writing to intimates, lightly mocked his "geological castles in the air" and his "hypotheses" that "if they were put into action but for one day, the world would come to an end.—"[37] Much of his theorizing takes the form of responses to the current geological literature. But much also stands alone as Darwin speculated on the origin and process of the formation of the strata and landforms he viewed. His notes on Quail Island, and on São Tiago, contain examples of theorizing of both sorts. Whether in the form of debate with the writings of other geologists, or in direct comment on the immediate situation, Darwin's theoretically oriented notes are strongly recursive: first thoughts are corrected and adjusted as the voyage progresses, thus creating within the entire group of notes from the voyage a pattern of saved and discarded elements. On the revisions side, for example, Darwin ended his notes on Quail Island with the comment "I have drawn my pen through those parts which appear absurd.—"[38] Similarly, in January 1833, after "rereading this paper" (his earlier notes on Quail Island and São Tiago), Darwin wrote out a whole list of new comments and queries, which included the remark that he did not know "whether Quail Island was formerly an island," which had not seemed problematical earlier.[39]

On matters of theory two issues stand out in Darwin's notes on Quail Island and São Tiago: the first elevation, the second diluvium. In both cases Darwin was encountering questions that were still unsettled among English geologists. As such, Darwin might refer to the questions obliquely, for their significance did not require elaboration. Thus, once Darwin had satisfied himself that elevation had occurred at Quail Island and São Tiago, he took up the question of its rate of occurrence by alluding to then current interpretations of the appearance of the Temple of Serapis at Puzzuoli near Naples. Of interest to geologists at this site were three marble pillars that were marked at their middle reaches by the boring of marine organisms. Regarding the interpretation of this monument, and by inference, the nature of elevation at São Tiago, Darwin was challenging Daubeny and siding with Lyell. Of the Temple of Serapis, Daubeny had written in 1826,

> Had such been the case [that the ground on which the Temple stands has undergone a depression of approximately 30 feet and subsequently an elevation of nearly that much], it is probable that not a single pillar of the temple would now retain its erect posture to attest the reality of these convulsions.[40]

Since the pillars had remained standing, Daubeny dismissed the change in level of the land as improbable, and the rise and fall of the sea equally so (as requiring a worldwide alteration), and concluded that some purely local cause, such as the damming up of water around the temple, must be adduced to explain its present condition. Four years later in the first volume of the *Principles of Geology,* Lyell took up the question of the Temple of Serapis where Daubeny had left it. Both Daubeny and Lyell argued from the circumstance that the pillars now standing above the sea bore the markings of the perforations of marine organisms, thus indicating that the temple had at one time been submersed beneath the sea. Lyell implicitly accepted Daubeny's rejection of changed oceanic levels to account for the phenomenon. But Lyell rejected Daubeny's argument that elevation and subsidence of the land could not be involved. Lyell asserted that the pillars could have sunk and risen again without having fallen:

> That buildings should have been submerged, and afterwards upheaved, without being entirely reduced to a heap of ruins, will appear no anomaly, when we recollect that in the year 1819, when the delta of the Indus sank down, the houses within the fort of Sindree subsided beneath the waves without being overthrown.[41]

Lyell's argument was thus not only in favor of elevation and subsidence as geological forces (against those possessed of "an extreme reluctance to admit that the land rather than the sea is subject alternately to rise and fall"),[42] but also in favor of the belief that such forces could act so slowly and evenly that superficial features of the landscape were not disturbed.

At São Tiago Darwin rejected Daubeny's interpretation of the Temple of Serapis by arguing that "at St Jago in some places a town might have been raised without injuring a house":

> (a) D^r.Daubeny when mentioning the present state of the temple of Serapis. doubts the possibility of a surface of country being raised without cracking buildings on it.—I feel sure at St Jago in some places a town might have been raised without injuring a house.—[43]

By implication Darwin was also aligning himself in this remark with Lyell's gradualist understanding of the operation of the forces of elevation and subsidence. Hence, Darwin's comment concerning the Temple of Serapis, spontaneous and offhanded though it seems, corroborates his remark a few months later to Henslow regarding São Tiago that "there are some facts on a large scale of upraised coast . . . that would interest M^r. Lyell," as well as his recollection in his *Autobiography,* written much later in life, that "I am proud to remember that the first place, namely St. Jago, in the Cape Verde Archipelago, which I geologised, convinced me of the infinite superiority of Lyell's views over those advocated in any other work known to me."[44] As Paul Pearson and Christopher Nicholas have recently shown, however, Darwin's recollection of the suddenness of his conviction of the superiority of Lyell's views while at the Cape Verde Islands is an overstatement; his Lyellianism was achieved over time.[45]

To Darwin in 1832 the image of the Temple of Serapis would have appeared fresh and powerful, an image not yet worn into cliché. Within the context of Lyell's first volume, published by then, the image bracketed the book. It both served as a frontispiece and as the subject of a ten-page discussion that appeared in a prominent position near the close of the book, where Lyell was moving toward his summary argument (Figure 5.7).[46]

But the image of the Temple of Serapis was not Lyell's alone, though it subsequently has become identified with his name.[47] The image (or, rather, the written description of it) reached back into the history of geology, for Playfair had discussed the fall and subsequent rise of the Temple in his *Illustrations of the Huttonian Theory of the Earth* (1802), a work Darwin consulted frequently during the voyage.[48] Playfair treated the Temple of Serapis in his chapter "Changes in the apparent Level of the Sea." Here he argued for viewing such phenomena as the changed levels of the temple as indications that the land had risen rather than the sea fallen. In his view, adducing motion in the land rather than the sea was the simpler and therefore the preferable interpretation:

> To make the sea subside 30 feet all round the coast of Great Britain, it is necessary to displace a body of water 30 feet deep over the whole surface of the ocean. The quantity of matter to be moved in that way is incomparably greater than if the land itself were to be elevated. . . . Besides, the sea cannot change its level, without a proportional change in the solid bottom on which it rests.[49]

As part of his general argument Playfair discussed the case of the Temple of Serapis. The description of the temple he took from Scipione Breislak (1748–1826), who had ascribed these changes "to the motion of the sea itself"; Playfair countered by crediting "oscillations" in the level of the land.[50] With some differences, this was Lyell's explanation also, and while, curiously, he did not cite Playfair (he did cite Breislak), he was arguing from an established point of view.[51] Thus, by invoking a Lyellian image of the Temple of Serapis to account for appearances at the Cape Verde Islands, Darwin was establishing his

T. Bradley Sc.
Present state of the Temple of Serapis at Puzzuoli.
London, Published by John Murray, Albemarle St. June 1830.

Figure 5.7. Temple of Serapis. Courtesy of the Library of Congress.

tie to a tradition of interpretation. Although the image did not capture all that Darwin would eventually assign to the role of elevation, it was a significant point at which to begin.[52]

The second theoretical issue Darwin faced at the Cape Verde Islands concerned diluvium. Darwin referred to a small covering of loose rock on the western side of Quail Island as being "part of the long disputed Diluvium."[53] The "dispute" in question was over the origin of superficial deposits, particularly loose rocks and gravel. In the early 1820s the majority of members of what Sedgwick, at points, termed the "English school of geologists" supported the notion that superficial deposits could be divided into two kinds, "diluvial" deposits characterized by "great irregular masses of sand, loam, and coarse gravel, containing through its mass rounded blocks sometimes of enormous magnitude," and "alluvial" deposits of "comminuted gravel, silt, loam, and other materials." The former deposits were presumed to have accumulated as a result of "some great irregular inundation"; the latter, by the "propelling force of . . . rivers" or "successive partial inundations."[54] What gave emotional charge to the issue was that the "great irregular inundation" was identified with the "waters of a general deluge" of the Noachian flood.[55] The most publicly recognized spokesman for this interpretation of superficial deposits was Buckland in his *Reliquiæ Diluvianæ* of 1823. However, the two men who determined Darwin's course of geological study at Cambridge, Sedgwick and Henslow, were also closely associated with the diluvial interpretation, Sedgwick in the papers of 1825 just cited and Henslow in an 1823 paper.[56]

In passing, we should also note that Erasmus Darwin's work, though richly cosmological, did not integrate the Noachian flood into its images of the earth's history. Further, in the next generation, the geologist James Hall (1761–1832), whose work sprang from an interest in Huttonian geology compatible with Erasmus Darwin's own, published a diluvial theory accounting for the origin of superficial gravel and scored boulders that had no connection to the biblical deluge. This article, a masterpiece of scientific exposition, was known to a wide audience, being cited in one of the standard works of English geology, Greenough's *Critical Examination of the First Principles of Geology* (1819).[57] Thus, the tradition of interpretation represented in his own family, largely Huttonian, was consistent with contemporary Continental views regarding the history of the earth.[58]

At no point in his geological notes or correspondence from the voyage did Darwin connect diluvium with the Noachian flood. When Darwin used the term "diluvium," he was using it in a restricted manner as referring to a distinct formation whose physical characteristics had been described by his own guides in the field at Cambridge. However, the question of origin of the formation, part of the "long dispute," was still open, and Darwin's geological notes do suggest great interest on his part in superficial deposits, accompanied by a dramatic change in their interpretation. The terminal point of this change in view occurred by the end of the voyage, when he regarded most superficial deposits as products of marine deposition rather than of any sort of inundation. As he became more impressed with Lyell's views in the course of the voyage, he became correspondingly less inclined to invoke overland rushes of water—debacles—to account for "diluvial" formations. He remarked late in the voyage, "N.B. in general discussion introduce diluvium generally submarine."[59] In keeping with this shift in interpretation, he moved away from uncritical use of the term "diluvium" over the course of the voyage, drawing a distinction between the formation itself and what it was "called." His change of mind and his self-conscious distancing from what he perceived to be the standard view are explicit in his notes.[60]

For the sake of completeness in considering Darwin's application of the category "diluvium," we must also refer to the category "alluvium," for they were paired terms. In the 1830s, the term "alluvium" proved elastic, particularly in Lyell's hands. His views on the diluvium/alluvium question are scattered in the *Principles of Geology,* but the passages most useful for comparison to Darwin's views late in the voyage occur in volume 3, chapter 11. Of course, Lyell did not use the term "diluvium." However, he broadened the term "alluvium" significantly so that it could be associated with a range of depositional agents, including floods.[61] He thus subverted the distinction between alluvial and diluvial deposits. In addition, in a logically doubtful move (for he had previously defined "alluvium" as transported matter deposited on lands not permanently submerged), he claimed that certain superficial deposits he classed as alluvium (including what would formerly have been called "diluvium") had been deposited while the land was still submerged under the sea:

> Many of the most widely distributed of the British alluviums may we think be referred to the action of the sea previous to the elevation of the land; and for this reason we never expect to be able to trace all the pebbles to their parent rocks.[62]

Darwin increasingly took this view as the voyage progressed. When toward the close of the voyage he referred various deposits of "Diluvium" to the agency of marine deposition, he had behind him Lyell's authority.[63]

Yet even with the difference between Buckland's diluvialism and Lyell's notion of gradual elevation and marine erosion, there was a deeper continuity, for both credited the power of water as the primary agency in molding the land. Indeed, Herries Davies has characterized Lyell's understanding of the agency of marine erosion as "really nothing more than a uniformitarian version of the diluvial theory of landscape development."[64]

The dispute over diluvium was not over. During the middle decades of the nineteenth century, the issues it had raised, and temporarily joined, continued to be of the highest interest to a wide audience for a variety of reasons. In chapter 6 I discuss the general cultural dimension of the debate over "diluvium." In chapters 7 and 8 I touch on the more purely geological aspects of the debate over the interpretation of superficial deposits.

"Reflection on reading my Geological notes"

As the voyage progressed, the bulk of Darwin's geological notes increased as he systematically documented each site he visited. If we study his itinerary, and multiply in our mind the bulk of notes he took on tiny Quail Island by the number of sites visited, we can sense the large scale of his endeavors. Each site posed new questions, and it is instructive for present-day investigators, often possessing local knowledge of their own, to use Darwin's original notes to restudy these sites. The work of Paul N. Pearson and Christopher J. Nicholas at the Cape Verde Islands, Peter Lucas and Michael Roberts in Wales, of Greg Estes in the Galápagos Islands, of J. W. and J. M. Nicholas in Australia, and of Max Banks in Tasmania has already been cited. In addition, Patrick Armstrong's exploration of a number of sites from Darwin's 1831–1836 work should be noted.[65] Historically attuned geological tourism has brought researchers into contact with sites visited by Darwin. For example, as part of a 1994 meeting of the International Subcommission on Jurassic Stratigraphy, a field excursion was organized to revisit localities in the Cordillera described by Darwin.[66]

The value of such revisiting may be as great as that attending the Galápagos Islands, where preservation of a landscape has been ensured, or it may be simply satisfaction on the part of the researcher who finds himself or herself recognizing in the present a scene described in the past.[67] Certainly retracing Darwin's footsteps through his many stops is a worthwhile enterprise. Although my own study is, on the whole, broader in its intent, its goals are complementary to such reconstructions.

Occasionally—but increasingly as the voyage went on—Darwin began to work in a larger thematic framework, less driven by the particular site at hand. The format was that of the essay. The first of the essays I will treat was entitled "Reflection on reading my Geological notes." It was not organized for rhetorical effect. Of its ten folios, the first five and a half are diverse in subject matter and exploratory in tone. Halfway down folio 6 Darwin's tone changes to one of assertion. From this point on to the end of folio 10, Darwin's logic is sharp and directed.

In the essay Darwin developed, briefly, an imaginary narrative—in a sense, a theory—of the geological formation of South America that included a descriptive framework for the history of life on the continent. Elsewhere I have provided a transcription and analysis of the entire document, but here I will only sketch out the contents of its second half.[68] If we take Darwin at his word, that this essay reflects his notes from his site visits up to that point, we will see where all of these individual enterprises had led him.

A truncated presentation, in Darwin's own words, will provide the best summary of his thought:

> Looking at this whole part of Eastern side of S America. we must considers [*sic*] it as one grand formation.—In the Northern parts it seems to repose on the Crystalline rocks, some of which in their lines of cleavages & elevation. & mineralogical nature are allied to the Transition formations of Falkland Is^{ld}. & Tierra del Fuego.— . . .
>
> But with this exception the hiatus (as compared to Europe) between the Crystalline & Tertiary beds: . . . *is very remarkable.*—We shall presently . . . run over the proofs of repeated elevations:
>
> May we conjecture that these . . . began with greater$^{(x)}$ strides, that rocks from seas too deep for life . . . were rapidly elevated & that immediately when within a proper depth. life commenced. . . . The elevations *rapidly* continued; land was produced. on which great quadrupeds lived: the former inhabitants of the sea perished. . . . the present ones appeared.— . . .
>
> The study of this Geology is very instructive from the consideration of the greatness in extent. & perfect horizontally, <of the> & number of the Elevations: We have nothing here like anticlinal tilting on each side the strata into highly inclined position: it [is] rather a swelling of the Globe, on the largest & most regular manner. . . .
>
> It becomes a problem. how much the Andes owes its height. to Volcanic matter pouring out?.—how much to horizontal strata tilted up.? how much to these horizontal elevations of the surface of continents?— . . .
>
> It is impossible not to be struck with the vast scale on which geological facts take place in S. America[.]— . . . With the exception of the Granite. this vastness (is in comparison to other countries) more apparent than real.—In Europe. We have the chalk from Ireland to Poland &c &c &c.—The real difference. consists perhaps. in which I have alluded to a greater rapidity in the elevation. . . . As Patagonia has risen

> from the waters in so late a period, it may be interesting to consider whence came its organized being [*sic*].—I have conjectured the absence of trees in the fertile Pampas & rich valleys of B. Oriental. to be owing to no Creation having taken place subsequently to the formation of the superior Tosca bed.[69]

The flow of this passage suggests Darwin's interest in the geological agency of elevation. It also suggests his interest in correlating the history of life with the history of geological formations. In his "Reflection" of 1834 Darwin's treatment of the history of life is hardly transmutationist, but it is sequential. Passages from these notes on the history of life will be referred to again in chapter 9.

"Elevation of Patagonia"

The next text I would like to treat is a synthetic essay of the sort Darwin inserted at several points into his ongoing run of geological notes. The essay is entitled "Elevation of Patagonia" and dates from mid-1834, some time after the expedition up the Santa Cruz River, which ran from 18 April to 8 May, but before the *Beagle* had sailed far up the west coast of the continent.[70] The essay was also written after Darwin had in hand volume 3 of Lyell's *Principles of Geology*, which figures prominently in Darwin's discussion. Another work to which Darwin made reference in the essay was Whewell's "paper on tides and currents," which he reminded himself to consult.[71] "Elevation of Patagonia" is an important essay for several reasons. It was written roughly halfway into the voyage, at the conclusion of the *Beagle*'s work on the eastern coast of South America. It thus represents the state of Darwin's opinion on the action of elevatory forces at an obvious and natural juncture.

In content the essay suggests Darwin's developing interest in continental elevation. In tone it reflects its midway position, being both assertive to the point of challenging Lyell's authority and tentative and open-ended in its treatment of the relation of elevation of the plains to that of the Andes and of retaining the lowering of sea level as at least a logically possible, though unlikely, explanation for the phenomena of the plains.[72] Another interesting feature of the essay is that it shows Darwin working closely with officers of the *Beagle* in the activities of measuring the height and breadth of the plains of Patagonia, and of estimating the slope of the adjacent ocean floor. For his part Darwin carried with him on the voyage a set of aneroid barometers, the reason for which he had explained to FitzRoy while preparing for the voyage: "Several great guns in the Scientific World have told me some points in geology to ascertain which entirely depend on their relative height."[73] On FitzRoy's part, the charting of the coastline of Patagonia was at the heart of his mission as commander and surveyor, for as the hydrographer Beaufort had noted in his official memorandum of instructions, "To the southward of the Rio de la Plata, the real work of the survey will begin."[74]

While Darwin's concern was primarily with the interior and FitzRoy's with the coastline, their interests overlapped because FitzRoy was required as part of his survey to supply the "perpendicular height of all remarkable hills and headlands," and Darwin took as his point of departure the height of the escarpments running along the coast.[75] In addition, Darwin depended entirely on measurements taken by the ship's officers and men for estimating the slope of the sea bottom. The culminating event of FitzRoy and Darwin's collaboration was the expedition up the Santa Cruz River, which provided Darwin with what he termed his "best opportunity" of observing facts regarding the elevation of Patagonia.[76] Finally, from the point of view of literary continuity, the essay "Elevation of Patagonia" is of interest because it served as the foundation for Darwin's later treatments of the same subject in the *Journal of Researches* and the *Geological Observations on South America*. These published treatments will be touched on briefly following discussion of the unpublished manuscript.

In "Elevation of Patagonia" Darwin's primary contention was that the southern plains had risen by successive elevations propelled by forces acting over a large area. His evidence was the succession of plains whose height had been measured by himself and by officers of the ship. As part of his essay Darwin drew a composite figure representing the height of these plains, those drawn connected having "been seen close together, generally forming Escarpements" (Figure 5.8).[77] He did not label his sketch, but from his text their probable identity can be known. For the purposes of the sketch, he provided round figures for several measurements. Reading from left to right the figures signify 40- to 60-foot plains near the River Chubut (40-foot plains were also measured at Buenos Aires, 50-foot plains at Santa Fé on the Paraná River); 100-foot cliffs at St. Joseph (Golfo San José), New Bay (Nuevo Gulf) (later Darwin referred to the coastline of La Plata as elevated to 100 feet)[78]; plains at Port Desire/St. George's Bay (Puerto Deseado/Golfo San Jorge) of 60 (100 feet in text), 250, 350, and 580 feet; plains at the Santa Cruz River and inland of 350, 710, and 840 feet; plains at Bird Island/St. Julian (Isla del Pajaro [48°44′ S], San Julián) of 580, 950, 350, and 100 feet; and an estimated plain of 1200 feet at St. George's Bay. The breadth as well as the height of these plains interested Darwin, and he reckoned the 580-foot plain to extend over a distance of more than 200 miles, the 350-foot plain over a distance of more than 550 miles, the 250-foot plain even farther, and the

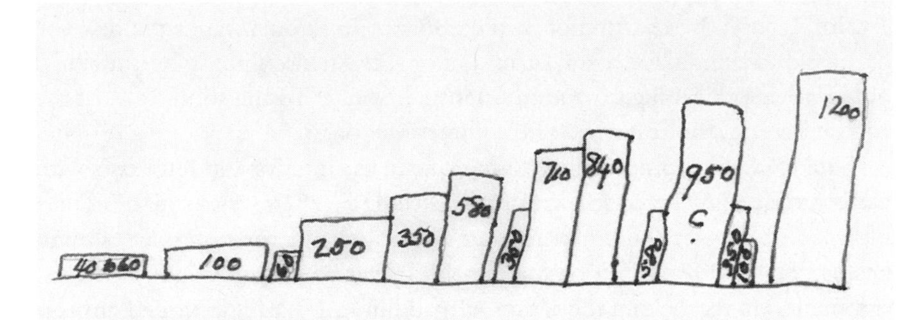

Figure 5.8. Darwin's sketch of the height of the Patagonian plains. This figure has been traced from the original, which is in ink drawn over a partial sketch in pencil. From DAR 34.1:57. By permission of the Syndics of Cambridge University Library.

50-foot plain over a 300-mile distance as measured in a north-south line in the Pampas.[79]

In addition to the height and breadth of the plains, Darwin was interested in the slope of the bottom of the sea adjacent to the Patagonian coastline, for if the slope of the sea bed were approximately that of the adjacent plains, one might infer that the force elevating the land was acting concentrically with the globe. On the other hand, if the slope on the land were far steeper than that beneath the sea, one might posit an elevatory force acting along the line of the Andes. Unfortunately, Darwin found only "scanty" data on the slope of the sea bottom.[80] His most complete measurements came from the ocean floor extending out from the Santa Cruz River. Here the slope of the sea bottom was considerably less than that on the land. However, Darwin distrusted these data since he believed the presence of the Falkland Islands might have made the bottom shallower than it would have been otherwise. Thus, for the actual estimated slope (54 fathoms over a distance of 137 miles), he substituted a figure for the slope he believed more likely to have existed in the ancient sea (80 to 100 fathoms over the same distance). Working with a comparable estimate for the original slope of the land from the Andes to the head of the Santa Cruz River (100 fathoms over a distance of 160 miles), he concluded that if there was a tilt from the Andes, it was very small, "only 100 or 200 ft in a hundred miles."[81] However problematic his data and conjectural his premises, Darwin concluded in favor of elevations "concentric with form of globe (or certainly nearly so)."[82]

What then was the nature of these elevations? The key word in Darwin's interpretation is "successive"; that is, he believed the elevations to have been identifiable and distinct rather than smoothly continuous. He formed his view by close scrutiny of the relation of two plains at Santa Cruz. The 355-foot cliffs to the south of the anchorage, though appearing level to the eye, actually rose over a distance of six miles inland to a height of 463 feet, where the plain formed the base of a 710-foot escarpment, most of the rise coming in the last half mile. Since Darwin could find no line of a former beach within this six-mile expanse, he thought it probable that the 108-foot rise represented a discrete event: "I do not mean it is necessary that it rose in five minutes or a day. *but* in so short a time that no beach—no coast-line could be formed."[83] Reviewing all his data, Darwin then concluded, "I think we clearly prove 7 or 8 successive elevations."[84] (He did not, however, equate the existence of distinct plains with discrete elevations. Thus, the 350-foot plain was formed by steps, with evidence for some successive elevations lost, a point on which he cited Lyell.[85]) Darwin then drew the larger conclusion:

> Considering how little of the coast was examined with geological views, I think it is quite astonishing the agreement in height <of> in the *different* series of plains widely apart. I feel quite convinced that the whole <part> <plains> <<of the modern formations>> of S. America from above C. Horn to near B. Blanca, <<a distance of nearly 1200 miles>> <rose> was <raised> formed into dry [land] and <elevated> <<uplifted>> to its present height by a succession of elevations which acted over the whole of this space with nearly an equal force.[86]

From the vast extent of this elevation Darwin drew an important conclusion regarding the source of the elevatory force. He asserted that the elevation of the plains required the "<<gradual>> expansion of [the?] central mass [of the earth].—acting by intervals on the outer crust," and he explicitly contrasted his own view with that of Lyell, who had posited elevatory forces operating at relatively small distances beneath the surface of the crust.[87]

In "Elevation of Patagonia" Darwin also discussed the date of elevation of the plains, not in numerical terms, but in the relative sense indicated by Sedgwick: "As the historians of the natural world, we can describe the order of the events which are past; . . . but we define not the length of time during which they were elaborated."[88] In dating the elevation of the plains, Darwin relied on animal remains. His single most important fact was the prevalence of marine shells retaining their original color and found up to a height of 400 feet over a distance of nearly 800 miles from St. Joseph (Golfo San José) in the north to St. Sebastian (Cabo San Sebastián) in the south. He believed the retention of color in the shells spoke to a recent date for the elevation of the plains. (He also identified a number of shells but did not attempt at this time to use the information in dating.) The similarity in height of a number of the plains also convinced him that their rise had been "cotemporaneous."[89] He believed this rise extended beyond the Patagonia plains to include the Pampas to the north and, very probably, the Andes to the west. He tied the date in the rise of the Pampas to that of the Patagonian plains using not shells, in which the Pampas were deficient, but what he believed was the similarity between certain unspecified fossil bones found there and others found along a low, coastal—and hence recent—plain in Patagonia. The Andes too, he suspected, were part of the "cotemporaneous" rise, though he made this point tentatively: "If on some future day I shall be able to prove that the West coast has been elevated <to a . . . > within the same period.—it will almost render it certain that the whole <S. part of> continent has < . . . > been elevated.—"[90] Darwin concluded his treatment of the elevation of Patagonia by placing himself on the Lyellian end of the spectrum of contemporary geological opinion regarding the history of the earth:

> As some authors have supposed the elevations of continents took place when the agents of change were in a state of greater activity, it appears to me a fact of high interest that <such> so large a part of < . . . > extremity of S. America has been uplifted in a period during which recent shells exposed to atmospherical change have retained their color [*sic*] and animal Nature.[91]

Darwin then added a thought jarring to the present-day reader. Expanding on a conjecture by FitzRoy, with whom he was then in frequent conversation, Darwin speculated that the recent date of the elevation of South America might account for the distribution of centers of ancient populations on the continent.[92]

En route to publication Darwin's essay underwent revision. In neither the *Journal of Researches* nor the *Geological Observations on South America* was mention made of the interior of the earth, or of Lyell's and Darwin's divergent views on the subject. Nor was the dating of the elevation of the continent correlated with human history. In both publications very much more was made of the gravel deposits of Patagonia and their interpretation. Yet, overall, the treatment of the plains of Patagonia between manuscript and publications was continuous. In the *Journal of Researches* Darwin presented an "imaginary section of the plains near the coast" rather than discussing actual plains; however, the elevations referred to in the diagram (580, 350, 250, and 100 feet) were clearly drawn from the more familiar plains (Figure 5.9).[93]

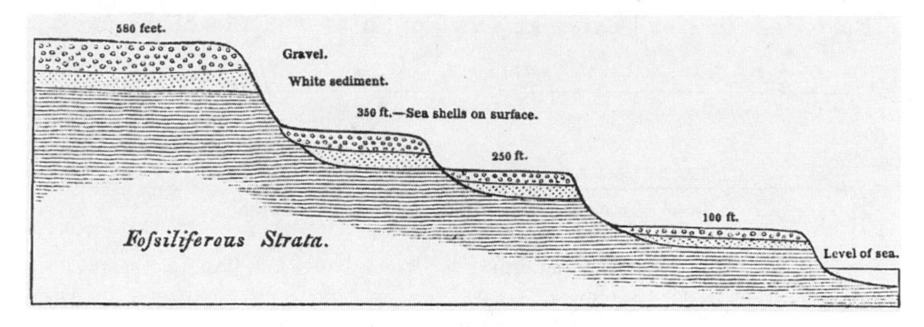

Figure 5.9. Section drawing of the Patagonian plains. From *JR*:202. By permission of the Syndics of Cambridge University Library.

There was, however, one striking departure between the manuscript and the *Journal of Researches*. It pertained to the manner in which elevation had occurred. In the *Journal* he wrote,

> At first I could only understand the grand covering of gravel, by the supposition of some epoch of extreme violence, and the successive lines of cliff, by as many great elevations. . . . Guided by the "Principles of Geology," I came to another, and I hope more satisfactory conclusion.[94]

He then proceeded to argue for elevations "at a perfectly equable rate," with the cliffs being formed during periods of "repose in the elevations." He continued, "Accordingly as the repose was long, so would be the quantity of land consumed, and the consequent height of such cliffs."[95] Since Darwin's earlier view of elevations as discrete and successive was also formed in consultation with the *Principles of Geology*, one may speculate that his citation of the book in this revised context might have been only rhetorical, or have reflected a rereading of the work or, possibly, contact with its author.

In the *Geological Observations on South America*, discussion of the elevation of the Patagonian plains occurs in the first chapter. Placement of the subject reflects Darwin's decision to organize the work according to the age of deposits, with the most recent treated first. It also reflects his continuing confidence in the strength of his argument. As he wrote to Lyell in September 1844 after having written the first sixty pages of the book, "The two first chapters, I think will be pretty good, on the elevation & great gravel terraces & plains of Patagonia & Chile & Peru.—"[96] As one might expect, Darwin's treatment of the plains of Patagonia is smoother and more complete in this later version. The shells have now been described and dated by Alcide

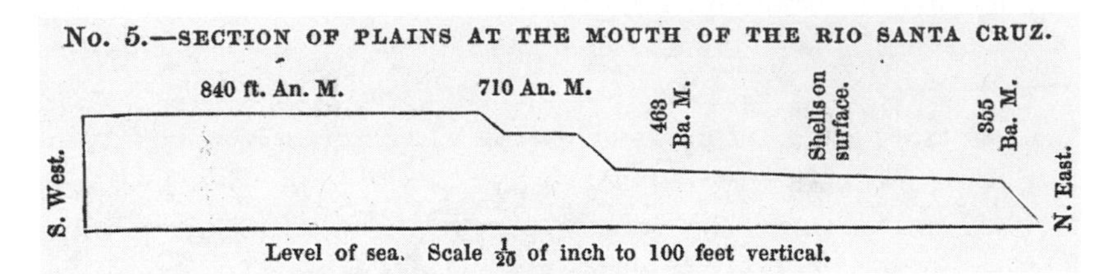

Figure 5.10. Section of plains at the mouth of the Santa Cruz River. "An. M." stands for angular or trigonometrical measurement, "Ba. M." for barometrical measurement. From GSA.8. By permission of the Syndics of Cambridge University Library.

Dessalines d'Orbigny (1802–1857), with additional work by George Brettingham Sowerby (1788–1854).[97] Presentation of the plains is polished. Five are figured, including the section of plains at the mouth of the Santa Cruz River shown in Figure 5.10.[98] Evidence for the similarity in height of plains across wide distances is presented.[99] The subject of the Patagonian gravel, whose origin was briefly considered in "Elevation of Patagonia," has now become of major interest.[100] On a key question Darwin has retained the view, expressed in the *Journal of Researches,* that the rate of elevation is gradual and that cliffs are formed during periods of repose. In *Geological Observations on South America,* however, Darwin is more tentative: "in Patagonia the movement may have been by considerable starts, but much more probably slow and quiet."[101] Interestingly, in his book Darwin did not forget the 108-foot rise at Santa Cruz to which he had devoted attention in "Elevation of Patagonia" ("I particularly looked out for ridges in crossing this plain, . . . but I could not see any traces of such"),[102] even while he left behind his initial interpretation of that rise as representing a single elevatory movement.

"Recapitulation and concluding remarks"

The last group of manuscripts I should like to consider date from the last fifteen months of the voyage. They all point toward publication and are critical to understanding Darwin's posture as a prospective author. These manuscripts are complex, and their dates cannot be firmly established in every case since Darwin was then in the business of rewriting. I will not explicate the full contents of the manuscripts here. Even so, I will discuss their importance in the scheme of Darwin's geological work.

The first of these texts derives from Darwin's last months in South America and is entitled "Recapitulation and concluding remarks." The title refers to the opening sentence of the text: "Before finally leaving the shores of South America I will recapitulate those conclusions and facts which appear to me to be most worthy of attention."[103] "Recapitulation" contains literary indications that suggest Darwin wrote it with future readers in mind, as well as for his own benefit. It is polished in appearance, Darwin having had his servant copy it in a fair hand; at one point in the text, space was left for citations from "various Authorities" supporting Darwin's view of elevation of the Americas; and the text closes with a rhetorical peroration worthy of Humboldt: "in laying the foundation of South America, Nature chose a simple plan—but upon that basis she has raised a magnificent structure!"[104] In short, this "geological memorandum"—Darwin's term—was moving toward publication.[105]

In "Recapitulation" Darwin summarized his geological work in South America, citing with approval previous work done by Humboldt and acknowledging his own Huttonian heritage. In the first half of the essay he reviewed the principal divisions of the formations of the sections of the Cordillera he had visited.[106] A hand-drawn and colored map of the southern portion of the continent is also included with the essay.[107] Its generality suggests a similarly summary and schematic approach (Plate 6).

In the second half of the essay Darwin took up the history of the development of the Cordillera and, with that, the subject of continental elevation. In 1834 in "Elevation of Patagonia" he had argued for the elevation of the eastern plains of the continent and had expressed interest in studying the Andes in relation to that elevation. In the later "Recapitulation" he broadened his view of elevation to include the Andes: "in the Cordillera the whole of the present elevation above the sea is owing to a gradual and horizontal upheaval."[108] (In context, "gradual" is to be read as "slow, prolonged, and perhaps scarcely sensible" rather than as occurring at a constant rate over time; "horizontal" is used in contradistinction to "angular.")[109] In favor of this view he pointed to the equability of the basins situated in the summits of the Andes, such as Titicaca, to the great Mexican platform, and especially to the equability of the Patagonian plains.[110] Thus, he came at the Andes from the point of view of the plains. However, his work in the Andes did strengthen his general argument for elevation as the explanatory agent. Formerly he had allowed that the subsidence of the sea might have produced the effects he had seen in Patagonia.[111] By 1835–1836, with his view of mountain building set firmly in the context of continental elevation, he could write in the "Recapitulation,"

> The Andes and their accumulated detritus, together with this one great bed of crystalline rocks compose South America.—The whole has been raised from beneath the ocean into dry land by the action of one connected force.[112]

The key phrase here is "one connected force." In points he was later to expand, he argued that earthquakes and elevatory movements "are so intimately connected, that if one is part of a regular series the other probably will obey similar laws."[113] He adopted a similar view of volcanic activity.[114] Darwin described the force behind volcanic activity as a "gradual movement of the internal fluid mass" of the earth's interior. This view was consistent with that taken in the essay on Patagonia, though not yet quite the thin-crust model of the earth Darwin was later to adopt.[115]

Interestingly, in "Recapitulation" Darwin did not make much of the elevatory effect of the Concepción earthquake of 20 February 1835, which he was later to do, presumably since the elevation observed did not fit the model for elevatory action that he then held. The earthquake had produced higher elevation on the offshore island of Santa Maria than on the mainland. At the time of writing "Recapitulation," Darwin was positing the notion of elevation occurring as a curved enlargement of the earth's crust. If the elevatory movement were centered on the continent, this ought to have produced greater elevation inland than offshore.[116]

"Coral Islands"

The next text of interest is the essay "Coral Islands." Darwin's original text is dated 1835 and presumably would have been written in the interval between 26 November 1835, when the ship departed Tahiti, and 21 December, when the ship arrived at New Zealand. In a sense, the subject of coral reefs was Darwin's only direct assignment as a geological author. Beaufort's memorandum of instructions for the voyage was unusually detailed regarding the inspection of coral reefs.[117] Extending this point, advance notice of the voyage in the *Athenaeum* characterized proposed work on reefs as "the most interesting part of the *Beagle*'s survey" affording "many points for investigation of a scientific nature beyond the mere occupation of the surveyor."[118] As the only man with geological training aboard ship, the scientific aspects of the subject would be expected to devolve on Darwin. Still, Darwin does not seem to have taken it in hand until 1835, when the Pacific itinerary was close upon him. Nonetheless, he arrived at the question at a propitious moment. During the preceding decade the question of the origin of coral reefs had been transformed into something like a set piece following the determination of a key parameter regarding the growth of reefs. The French naturalists Jean René Constant Quoy (1790–1869) and Joseph Paul Gaimard (1796–1858) had observed that organisms forming coral reefs operate at relatively shallow depths beneath the surface of the water.[119] This startling new finding gained currency quickly. It formed the basis of assessments of reef formation made by De la Beche, Frederick William Beechey (1790–1869), and Lyell.[120] As Beechey put it,

> The general opinion now is, that they have their foundations upon submarine mountains, or upon extinguished volcanoes, which are not more than four or five hundred feet immersed in the Ocean; and that their shape depends upon the of the base whence they spring.[121]

This was the legacy of interpretation Darwin inherited from previous authors. However, when he took up the problem, he also brought with him an imagination shaped by his own recent experiences in South America. He held before his mind's eye a vision of a rising continent shaped by the sea during its elevation.

In his *Autobiography* Darwin claimed that he had thought out his theory of the origin of coral reefs while still in South America; his statement is confirmed by jottings in his "Santiago Book."[122] In notes that appear to date from 1835, before his September departure from the west coast of South America, Darwin wrote,

> As in Pacific a Corall bed, forming as land sunk. would abound with. those genera which live near the surface. (mixed with those of deep water) & what would more easily be told the Lamelliform: Corall forming, Coralls.—I should conceive in Pacific. wear & tear of Reefs must form strata of mixed. broken sorts & perfect deep-water shells (& Milleporæ).—
>
> Parts of reefs themselves would remain admidst these deposits, & filled up with infiltrated calcareous matter.—Does such appearance correspond to any of the great Calcareous formations of Europe.—
>
> Is there a *large* proportion of those Coralls which only live near surface.—If so, we may suppose the land sinking.[123]

The key idea here is in the opening phrase: a coral bed forming as the land beneath it sank. There is also reference, with qualification, to the recent observation of Quoy and Gaimard regarding the relatively shallow depth beneath the surface of the water at which reef-building corals live. Thus, in the first paragraph quoted, Darwin referred to "those genera which live near the surface." In the last paragraph quoted, Darwin framed his qualification as a question: "Is there a *large* proportion of those Coralls which only live near surface[?]" However, Darwin was prepared to presume the affirmative. Thus, he concluded that "if so, we may suppose the land sinking." Thinking as a geologist, he also asked himself whether any of the "great formations of Europe" might have been formed in this way.

Darwin's next comments on reef formation, hitherto unremarked, occur in his notes for the Galápagos Islands. The *Beagle* stopped in the islands from 15 September to 20 October 1835, and the bulk of the notes in question appear to have been written between the date of departure from the Galápagos and 15 November, the date of arrival at Tahiti. In these notes Darwin tested the received notion of reef formation against his observations at the Galápagos. Interestingly, his first points were made against himself. He noted that sixteen craters of "sandstone" he observed at the Galápagos were like lagoon islands (atolls) in being raised more on one side than the other. He did not regard the parallel between these craters and atolls as exact but remarked, "I am so much the more bound to point out their coincidence, as I am no believer in the theory of Lagoon Islds. being [illegible] on the circular ridges of submarine craters.—"[124] He then drew attention to another circumstance at the Galápagos favoring the received explanation for the origin of coral atolls:

> There is another circumstance connected with this subject, which is of some interest. Five of the great Volcanic mounds of Albemarle & Narborough Is^{lds}. which are surmounted by Craters having a diameter of between two and three miles, appear to the eye to be of an equal elevation. Three have been measured by angular observations: two in Albemarle Isd. are respectively 3720 & 3730 ft. & that of Narborough Isd. 3720 ft: high.—Inspecting the chart, one is tempted to exclaim; on such foundations ready placed at an equal height, the Lithophytes. might soon raise to the surface, their circular ridges of Coral rock.—[125]

Clearly the communality in height of the three craters encouraged speculation along the lines of the received opinion, against which Darwin had already positioned himself. The next topic regarding reef formation that Darwin took up in his Galápagos notes was neutral as between the received theory and his own. It pertained to the absence of coral reefs at the Galápagos Islands. He considered the possibility that the quantity of calcareous matter was deficient in the region, but his primary suggestion, for which he credited FitzRoy, was that the sea surrounding the islands was too cold for reef-building corals, "a tribe of Animals, which seem . . . only to flourish where the heat is intense.—"[126] To this remark he attached a long entry, part of which was added after the *Beagle* visited Tahiti, in which he recorded sea temperatures from the "Weather Journal" kept aboard ship.[127]

There are no comments on coral reefs in the regular series of notes on Tahiti. Darwin's rich yield from Tahiti, in regard to his theory of reefs, was rather deposited in "Coral Islands." In the essay Darwin recorded his new solution to the problem that had still vexed him at the Galápagos Islands: the origin of coral atolls. This solution linked the presence of volcanic islands encircled by reefs with the formation of atolls. In a well-phrased description of a moment of discovery, Darwin wrote of his reasoning when viewing the coral-framed island of Ei Meo (Moorea) from the heights of Tahiti:

> I was forcibly struck with this opinion.—The mountains abruptly rise out of a glassy lake, which is separated on all sides, by a narrow defined line of breakers, from the open sea.—Remove the central group of mountains, & there remains a Lagoon Is^{d}.—[128]

Darwin imagined the mountains being removed by subsidence of the ocean floor, the reef maintaining itself by upward growth. He also pointed out that his new understanding of the origin of atolls solved the difficulty posed by Beechey that the size of some atolls exceeds that of any known craters.[129]

"Coral Islands" also provides evidence of Darwin's systematic reading of the literature on coral reefs, which was largely available on board the ship.[130] In the essay Darwin attempted to integrate previous work with his own insights. Finally, he came full circle by returning to the vision that had brought him to his new interpretation of coral reef formation: elevation in South America balanced by subsidence in the Pacific Ocean basin. In turn, the two essays—"Recapitulation and concluding remarks" and "Coral Islands"—represent the yin and yang of a greater unity. The closing paragraph of "Coral Islands" draws attention to this duality:

> Before finally concluding this subject, I may remark that the general horizontal uplifting which I have proved has & *is now* raising upwards the greater part of S. America & as it would appear likewise of N. America, would of necessity be compensated by an equal subsidence in some other part of the world.—Does not the great extent of the Northern & Southern Pacifick include this corresponding Area?—[131]

This balancing of uplift and subsidence was the key to his understanding of the formation of coral reefs.

"Cleavage"

Another essay from the latter part of the voyage was entitled "Cleavage."[132] In it Darwin summarized his work on rock cleavage in South America. As he wrote by way of introduction, “In my geological account of each district I have described all the facts respecting the [illegible] cleavage & stratification of the rocks & likewise the more obvious conclusions which might be drawn in each particular case.—I will now take a general review of the whole phenomenon as seen in S. America.—”[133] As the essay “Recapitulation” with its emphasis on elevation reflected Darwin’s tie to the tradition of Hutton and Lyell, that entitled “Cleavage” reflected his indebtedness to his study of structural geology under Henslow and Sedgwick. James Secord has shown that Sedgwick was particularly interested in distinguishing bedding, or stratification, from slaty cleavage.[134] It was therefore consistent with his earlier training that Darwin would open his essay on “Cleavage” by defining the terms “stratification,” “cleavage,” and “fissures”:

> By the term *Stratification* I mean those planes of division (resulting from <some> changes of circumstances,) which occur in matter, that has been deposited beneath <the sea> <<water>> & which <particles> <<matter>> was <successively> arranged <<in an undisturbed manner>> by the attraction of Gravity.—
>
> By *cleavage,* a fissile structure.—The laminae not being necessarily parallel to the plane of Stratification.—
>
> By *Fissures,* planes which are seen in laminated Stratified & non-Stratified rocks:—These when they occur in the two first cases are recognized as of a different nature.[135]

Darwin followed these distinctions in the essay by considering the instances of rock cleavage, as opposed to stratification, that he had recorded as present in South America, supplementing his own observations with those of others.

The second source Darwin drew on for his “Cleavage” essay was the discussion of “loxodromism” in Humboldt’s *Personal Narrative.* Humboldt had written that

> there exists in no hemisphere a general and absolute uniformity of direction, but that in regions of very considerable extent, sometimes on several thousand square leagues, we observe that the direction, and still more rarely the inclination, has been determined by a system of particular forces. We discover at great distances, a parallelism (loxodromism), a direction, of which the type is manifest amidst partial perturbations, and which often remains the same in primitive and transition soils.[136]

Darwin used Humboldt’s notion of loxodromism to organize his discussion of cleavage. The appeal of the concept of loxodromism lay in its recognition of the operation of large-scale forces affecting rocks over regions of great geographical extent. Humboldt had recorded the presence of loxodromism in the equatorial areas of South America he had visited. Darwin extended Humboldt's work southward. Integrating some of his own work with that of Humboldt, he wrote, "Justly to appreciate the importance of this subject, it is necessary to bear in mind, that the Area of which we are speaking, extends from 12° North. nearly to 35° South Latitude; a space of 2800 miles.— Conclusions drawn from so vast a territory become at once applicable to the Geology of the Whole Globe.—"[137]

Darwin speculated about forces that might have produced identical cleavages in rocks extending over large areas. He believed them similar to forces producing metamorphism and elevation, though his comments were suggestive in tone rather than absolute. He included among the possible forces those of chemical attraction and of electricity ("Do currents of Electricity flow . . . for long periods in certain directions, deep within the earth?").[138] In the end, however, he left the question open while asserting an "intimate connection" among lines of metamorphic action, lines of elevation, and cleavage.[139] Much of the material developed in "Cleavage" eventually found its way into a chapter of the *Geological Observations on South America*.[140]

Notebooks

Throughout the voyage Darwin engaged in speculation of a theoretical nature. In the essays discussed thus far, his theoretical insights had matured into material suitable for formal presentation: complete sentences, paragraphs, sequential development of an argument. However, there was considerable distance to be covered between insight and argument. Toward the end of the voyage Darwin began to use notebooks to store some of his speculations. This process became one of the hallmarks of his work in the years immediately following the voyage, but it began as the voyage drew to a close.[141]

In the last year of the voyage, when the *Beagle* was more often than not at sea, Darwin also used some of his time to prepare himself for reentry into the English scene. On 9 July 1836 he wrote to Henslow asking him to activate Sedgwick's offer of 1831 to propose him as a member of the Geological Society.[142] Darwin also used this period to prepare his collections for dispersal to specialists, to write the synthetic essays of the sort discussed earlier, and to consider how he might best present his geological work. Particularly in the Santiago Book and in the Red Notebook, Darwin recorded what were in effect instructions to himself as to how to proceed as an author. Thus, in the Santiago Book he reminded himself, "Before concluding the cleavage paper.

consult the VI Vol. of [Humboldt's] Pers. Narra[.]" This he clearly did, for he quoted extensively from Humboldt's treatment of loxodromism toward the conclusion of the "Cleavage" essay.[143]

Authorial issues might involve questions of placement. Thus, he remarked in the Red Notebook, "Introduce part of the above [on the earth's interior and crust] in Patagonian paper; & part in grand discussion."[144] Or the instruction might involve a question of presentation:

> In discussion on Porph. Breccia. I should state to gain confidence, that it was sometime before I fully comprehended origin[.][145]

Or Darwin might remark on what he was *not* going to cover, as in this intriguing aside:

> In the History of S. America we cannot dive into the causes of the losses of the <<species of>> Mastodons. which ranged from Equatorial plains to S. Patagonia. To the Megatherium.—To the Horse. = One might fancy that it was so arranged from the forseight [*sic*] of the works of man[.][146]

In his directions to himself Darwin also addressed questions of theme, the common elements he wanted to emphasize throughout his geological writing. On the most general level he mentioned two themes most frequently: first, the contrast between South American and European geology and, second, the symmetry or equilibrium of the globe based on the vertical motion of its crust.

Darwin believed the primary contrast between South American and European geology to be the relative simplicity of the former as compared to the latter. He spoke of "troubled England" and its complicated geology versus that of Patagonia.[147] The complicated nature of English geology, and indeed of European geology generally, resulted from the more complicated pattern of movement to which the strata had been subjected:

> In Europe proofs of many oscillations of level, which in the nature of strata & Organic remains does not appear to have taken place in the Cordillera of South America.[148]

Since it appeared that "the forces have acted with far more regularity in S. America," he advised himself to bear that in mind while writing.[149] Yet he did not intend for his results to be of interest only for South America, and his orientation toward global geology emerged strongly in his reflections from late in the voyage. He advised himself,

> In a preface, it might be well to urge, geologists to compare whole history of Europe, with America; I might add I have drawn all my illustrations from America, purposely to show what facts can be supported from that part of the globe: & when we see conclusions substantiated over S. America & Europe. we may believe them applicable to the world.—[150]

In considering global geology Darwin's point of departure was to emphasize the motion of the earth's crust. Thus, when writing on South America, he reminded himself,

> Amplify on importance of proving extent & recency: of upheavals. & <<manner>> over whole America.—Explaining generally Continental upheaval. so important in understanding valleys, diluvium, escarpements & successive lines of formations &c.—Showing that they are not mere Local effects, as so many Authors suppose.—[151]

The presence of the term "diluvium" should be noted. Darwin was pressing for his own marine interpretation of such deposits.

In an expansive passage he set out the plan for his book in the most general terms:

> Read geology of N. America. India.—remembering S. Africa. Australia..Oceanic Isles. Geology of whole world will turn out simple.—[152]

The "simple" geology of the whole world was based on a proper understanding of the "general movements of the earth," which were the "vertical movements" of elevation and subsidence. These movements were "great" ones producing "great continents"—"not mere patches as in Italy proved by Coral hypoth."[153] In short, Darwin intended to cast his book in the largest terms. At the end of his Red Notebook, he referred himself back to the long passage quoted just above, "NB. [nota bene] P. 73 General reflections on the geology of the world[.]"[154] His full presentation of the "geology of the world" will be treated in chapter 7.

As a part of discussing Darwin's notebook entries, notice should be taken of their rough and unfinished qualities at many points. Thus, for example, in a long passage from the Red Notebook, Darwin raised a number a questions pertaining to both elevation and subsidence and questions on the formation of rocks. Questions of causation and prediction remained: "In a subsiding area. we may believe the fluid matter instead of afflux (always slightly oscillating as that of a spring) moves away. Will geology ever succeed in showing a direct relation of a part of globe rising, when another falls.—" He followed that quickly by noting, "Volcanos must be considered as chemical retorts.—neglecting the first production of trachyte, look at Sulphur. salt.lime, are spread over whole surface; how comes it they do not flow out together? How are they eliminated.—"[155] These were very much questions of unfinished business. Rhetorically, as Alan Gross has pointed out, such passages record Darwin speaking to himself, asking questions, offering possibilities. In addition, Gross added that some of the Red Notebook's comments are so cryptic that "we reach an important limit beneath which rhetorical analysis cannot operate."[156]

Keeping to the theme of finished and unfinished business, we may note in the Santiago Book and the Red Notebook the prominence of loose ends. In the Santiago Book, Darwin noted, "Every conclusion is of consequence with respect to Erratic Blocks."[157] This sweeping statement suggested the openness of the question of erratic boulders. Other unfinished business, particularly prominent in the Red Notebook, pertained to the relation of geology and species. And, throughout the notebooks, there were questions of traditional geological import such as the origin of formations.

Becoming an Author

Darwin's intention remained secure after his arrival home. As he wrote to his cousin William Darwin Fox (1805–1880) on 6 November 1836, "So that about this day month [*sic*], I hope to set to work: tooth and nail at the Geology, which I shall publish by itself."[158] That he did not put his plan into full effect does not compromise the sincerity of his intent, though it does require a brief accounting for what became his actual pattern of publication.

Part of Darwin's plan for the presentation of his geology findings proceeded smoothly. The Geological Society of London, as he had hoped, provided an immediate audience, made the more congenial by the circumstance that Lyell was then its president. The society had been aware for some time of Darwin's existence (though misidentifying him as "F. Darwin" of "St. John's College" in the *Proceedings*), for at its meeting of 18 November 1835 Sedgwick had summarized some geological passages from his letters to Henslow.[159] But with Darwin's return to England his work came to center stage. On 4 January 1837 he presented evidence for the recent elevation in Chile.[160] On 3 May 1837 he described the deposits containing extinct mammals in the neighborhood of the Río de la Plata, a paper complementary to Owen's naming of the *Toxodon Platensis* at a meeting of the society on 19 April.[161] On 31 May 1837, with Lyell active in scheduling the presentation of the paper, Darwin argued his theory of coral reef formation.[162] At the first meeting after the summer recess, on 1 November 1837, Darwin presented a paper on the formation of "vegetable mould," which he believed was due to the digestive process of the common earthworm.[163] Ostensibly this paper, drawn from observations made in England, was distinct from researches from the voyage. Yet there was a connection between this paper and the coral theory, for, as Darwin's cousin Elizabeth Wedgwood (1793–1880) remarked, "[My father] desires me to tell you he is very much struck with your hypothesis of chalk being made by fishes—if fish made Chalk Hill I dont see why worms may not make a meadow."[164] The next spring, on 7 March 1838, Darwin presented what was in effect a continuation and enlargement of his first paper on elevation.[165]

If Darwin did not set to work "tooth and nail" on his geology manuscript soon after his return to England, the fault lay not in his involvement with the Geological Society but in his alternate writing project, converting his journal into a book.[166] On 28 May 1837 when he was better than two-thirds through his rewriting, he wrote to Henslow,

> I suspect I have begun at the wrong end, I ought to have published detailed Geology, & Zoology first; & then all general views might have come out in as perfect a form, as the subject permitted.—[167]

But he quelled his hesitations and finished the manuscript. In the autumn of 1837, with the journal off his desk, he applied himself to geology. He did this even though his intellectual life had become immensely, and richly, complicated by his private adoption of transmutationism.

From October–November 1837 until June 1838 Darwin wrote up the material ultimately to be treated in the second volume on the *Beagle*'s geology. Places treated included volcanic islands in the Atlantic, the Galápagos Islands, New Zealand, Australia, and the Cape of Good Hope. There was a draft in the hand of Syms Covington (1816?–1861) for this material, but it has not survived.[168] In October 1837 he wrote to Henslow that "by giving up society & not wasting an hour," his geology might take him a year and a half, but in January 1838 he was already canvassing Henslow with the thought of splitting the geology into parts.[169] Early in 1838 Darwin's publishers advertised a volume for later that year to be entitled *Geological Observations on Volcanic Islands and Coral Formations*.[170] While Darwin did not leave behind a justification for this division of subject matter, one may speculate that it was appealing to him since it would allow him to link elevation (volcanic islands) and subsidence (coral reefs) while leaving aside the large, and thus less tractable, mass of material on South America. In any case, he continued to refer to a forthcoming single volume on volcanic islands and coral reefs as late as September 1838.[171] On 5 October 1838 he began work on the coral reef text.[172] The first unambiguous indication that he had revised his plan for publication still further appears in a letter dated 24 October 1839, where he referred to his "hope in a couple of months to have a very thin volume 8vo on Coral Formations published."[173] Even this schedule proved too ambitious, however, for a few months later in February 1840 he had to write to Lyell that his coral manuscript was in a state of "such confusion" that he could not send it for Lyell's use.[174] While his timetable eluded him, Darwin did remain with the order of publication he had established in 1839, the volume on coral reefs being the first volume of the geology published.

Other circumstantial reasons for delay in publication of the geology manuscript are various and include Darwin's ill health and his work on the species question. However, in 1838 Darwin confronted a purely geological issue that forced him to reevaluate much of his earlier work and thus also contributed to the delay. This was the theory of ice ages put forward by Louis Agassiz. Its intersection with Darwin's work is represented briefly in his paper on the "parallel roads" of Glen Roy in Scotland, read before the Royal Society of London on 7 February 1839, and more fully in the important appendix to his journal that he referred to as his "Addenda on Theory of Erratic Blocks."[175] These developments will be described in chapter 8. However, simply as a factor that promoted delay in publication the impact of Agassiz's work can be noted. In any event Darwin persevered in his ambition to publish his geology findings during the voyage, and however altered, more complex, or chastened in tone the final volumes were from the "grand discussion" he once imagined, they were also refined from having passed through a critical fire.

CHAPTER 6

NEGOTIATING GENESIS AND GEOLOGY

We admire the power by which the human spirit has measured the movements of the globes, . . . ; genius and science have burst the limits of space, . . . Would there not also be some glory for man to know how to burst the limits of time, and . . . to recover the history of world, and the succession of events that preceded the birth of the human species?

GEORGES CUVIER, "PRELIMINARY DISCOURSE" TO THE *RESEARCHES ON THE FOSSIL BONES OF QUADRUPEDS* (1812)

Time! Time! Time! we must not impugn the Scripture Chronology, but we *must* interpret it in accordance with *whatever* shall appear on fair enquiry to the *truth* for there cannot be two truths.

JOHN HERSCHEL TO CHARLES LYELL, 20 FEBRUARY 1836

Diluvialism

Religious and geological opinions overlapped in mid-Victorian Britain. Two intertwined strands of religious influence are evident in Darwin's scientific notes from the voyage: a traditional understanding of divine providence, and the use of the terminology of diluvialism. Darwin's use of providentialist language—"Author of Nature," the "Creation," the "creation'—was general in tone. It did not presume the subordination of geological forces to the requirements of human existence. One does not find in Darwin the sentiment expressed by Hutton that volcanoes were instruments designed "to prevent the unnecessary elevation of land and the fatal effects of earthquakes." Indeed, Darwin set aside just this use of final causes:

> Earthquakes part of necessary process of terrestrial renovation & so is volcano a useful chemical instrument.—Yet neglecting these final causes.—
>
> What more awful scourges to mankind than the Volcano & Earthquake.—[1]

The key phrase here is "neglecting these final causes." The notion of design, whether expressed in the Aristotelian language of final causation, or in the religious language of providence, became a key issue when Darwin began to address the subject of transmutation.[2] However, on strictly geological issues, it does not seem to have troubled him. There was no escaping the second issue of diluvialism, however. It brought together purely geological questions of interpretation of landscape with broader questions of interpretation of the past.

As the field of geology emerged into autonomy in the late eighteenth and early nineteenth centuries, practitioners faced the task of integrating what was known of human history, much of it contained in sacred narrative, with what could be told from topography and the strata of the earth. Further, during the same period, and in the same countries where geology was establishing itself as a discipline, an enlightened skepticism cautioned against credulity in all areas of knowledge. On occasion skeptics challenged the opinions of geologists. The *philosophe* [François Marie Arouet] Voltaire (1694–1778) argued against the developmental view of the world: "Men have not been fish. . . . I cannot repeat too often that we are not gods who can create a universe with a word."[3] Other critics unjustly ridiculed the allegedly extraterrestrial origin of meteorites as "folk tales" and "fairy tales."[4] In the task of combining human history with geology, similar skepticism was expressed with regard to the use of ancient texts. Nonetheless, one strand of that skepticism called for subjecting such texts to scrutiny rather than disallowing them. On this basis, Moses was read as "neither prophet nor scientist, but *historian*."[5] Genesis and geology might share common ground.

To determine how geologists, and those linked to them, went about reconciling Genesis and geology, we will sample the opinion of several key figures. The chief persons to be treated include Cuvier, Jameson, Buckland, Sedgwick, Henslow, Lyell, FitzRoy, Darwin himself, John Phillips, Leonard Horner, and several contributors to *Essays and Reviews* (1860). All these authors wrote from a background of Protestant Christianity.

Cuvier set the terms of the debate. Jameson, Buckland, Sedgwick, and Henslow were initially in agreement with Cuvier; Lyell broke with him publicly to the greatest effect. FitzRoy is interesting as a counterpoint to the others. Darwin himself appears as a student, observer, and, eventually, partisan.

Georges Cuvier was the author of what one commentator has referred to as the "Cuvierian compromise," but what I would describe in a more favorable light as the "Cuvierian synthesis."[6] Cuvier found a way of combining what he termed "civil history"—much of it drawn from sacred narrative—with his profoundly original understanding of the history of the earth. His synthesis is contained in the "Discours préliminaire" (preliminary discourse) to his great multivolumed work, *Recherches sur les ossemens fossiles de quadrupèds* (1812). This "preliminary discourse" was republished separately in English in 1813 and in French in 1825, both under new titles, and went through numerous editions and printings throughout the nineteenth century.[7] The message of the work remained constant, however, and its appeal is easy to see, even from the perspective of the present day. First, Cuvier's approach to Genesis is resolutely historical. The text is treated as one of the "histories of nations" that are useful for the "real facts" they contain among various "interested fictions."[8] The validity of such texts as Genesis rested wholly on the truth of empirical claims, which could be tested by comparison with other ancient writings. Thus, Cuvier stressed that the Pentateuch was "received as authentic by the Samaritans as well as by the Jews" and that Egypt, whose history stood behind that of the Jews, was "universally allowed by all the nations of the west to have been the most anciently civilized kingdom on the borders of the Mediterranean."[9] Second, Cuvier judged that the text of Genesis was written by Moses about thirty-three hundred years ago, which was centuries after the flood, but immeasurably long after the "history of thousands of ages which preceded the existence of the race."[10] Although he was not explicit on the subject, Cuvier seemed to suggest that events treated in the Bible could be dated only when they were events in human history for which there were human witnesses. The Bible, therefore, could not be used as a source through which to establish the age of the earth. Thus did Cuvier circumvent a subject that had troubled other authors. Finally, when Cuvier did introduce the subject of the biblical flood, he did so only in the most general terms. The flood appeared as "an event of an universal catastrophe, occasioned by an irruption of the waters," with a date not "much farther back than five or six thousand years ago."[11]

On the geological side of the synthesis, Cuvier drew on the work that has formed his lasting scientific achievement, the evidence he provided for the extinction of species as a repeated occurrence in the history of life on earth.[12] His reconstructions of fossil quadrupeds provided the main text of the work to which the "preliminary discourse" was the introduction.[13] In this context the deluge described in ancient literatures appears as but one of a series of violent episodes with a force sufficient to produce profound geological changes including extinctions of species. Evidence for the occurrence of such violent episodes might include overturned strata, "heaps of *debris* and rounded pebbles," and, in the most recent case, the preservation of "carcases [*sic*] of some large quadrupeds which the ice had arrested."[14]

Cuvier's synthesis appealed to the geologist who might wish to emulate him in serving as "an antiquary of a new order."[15] Such a geologist would seek to retain a unified view of history that encompassed traditional accounts and yet, on the geological side, absorbed the destabilizing notion of species extinction. There were weaknesses to Cuvier's argument in the "preliminary discourse"—the connections between biblical description and ascribed geological causation were loosely drawn, a perspective was taken on world history that was slighting of non-Western testimony, indeed harsh, even for the time, as regards the judgment of "Negroes"—and Lamarck's transmutationism was pushed aside.[16] Nonetheless, the strengths of the synthesis were apparent to many.

We must next follow the appropriation of the Cuvierian synthesis by the British geologists Jameson and Buckland. Before doing so, however, it would be well to take note of work by Toby Appel on the subject of Cuvier's religious views. She argued that the relatively greater prominence given to religion in the writings of Cuvier's British followers than by Cuvier himself is chiefly due to the circumstances under which the scientists operated rather than to differences in belief. Appel pointed to the unsympathetic attitude toward lay interpretations of scripture in Catholic France and to the tradition of natural theology operating in Protestant Britain, with its emphasis on adaptation and contrivance that Cuvier shared. Appel also drew attention to the professionalism of French science and to the dominance in France of the physical sciences, in which area religious discussion was no longer welcome.[17] To her arguments against downplaying the importance of religion to Cuvier, it might be added that in later editions of the "preliminary discourse" Cuvier used the term "diluvium," which had been promoted by Buckland.[18] Of course, judgment of interior religious conviction, that is, of belief, is always problematic, as Rudwick has properly emphasized.[19] Judging by his public persona in assessing Cuvier's affiliations, one is left placing emphasis on the company Cuvier kept, that is, on Buckland, and his own continuing official support, as a public official, of his fellow Protestants in France.

In considering Darwin and some of his British contemporaries, it is helpful to divide the group in two: Jameson, Buckland, Lyell, Sedgwick, and Henslow on the one hand; FitzRoy and Darwin on the other. Members of the first group were older, all born in the eighteenth century, and their reactions to the issues addressed by Cuvier's "preliminary discourse" were fresher and more direct. FitzRoy and Darwin, born in the first decade of the nineteenth century, came on the scene as authors after the wave of reaction marked by publication in 1830 of volume 1 of Lyell's *Principles of Geology*. Furthermore, FitzRoy and Darwin are separable from the others through their common experience during the 1831–1836 voyage of H.M.S. *Beagle*, summarized in their joint publication of 1839. In the discussion of the views of both groups, we must emphasize that the issue summarized by the term "Cuvierian synthesis" was a complicated one. More people were involved than those mentioned here. Issues overlapped or came in chains: the deluge and its alleged geological traces, the antiquity of the human species, the fixity of species, and the rate of geological change. Then too, even for the prominent figures, the reasons for shifting opinions are not fully known, though recent examination of lecture notes for several of them has proved illuminating. Yet, despite complexities, some salient points suggest how the issue stood for Darwin and those around him by the end of the 1830s.

In 1813, Cuvier's "preliminary discourse" was introduced to the English-speaking world in an edition for which the preface and notes were supplied by Jameson, holder of the Regius chair in natural history at the University of Edinburgh since 1803. In the opening sentence of his preface Jameson demonstrated that he had not understood, or preferred to ignore, Cuvier's argument. Rather than following Cuvier in suggesting how the occurrence of the biblical flood might be inferred from the combined testimony of records of ancient peoples, Jameson appealed to revelation:

> Although the Mosaic account of the creation of the world is an inspired writing, and consequently rests on evidence totally independent of human observation and experience, still it is interesting . . . to know that it coincides with the various phenomena observable in the mineral kingdom.[20]

Although this sentence was dropped after the second edition, damage had already been done, for some of Cuvier's British readers continued to be ignorant of his subtle and scholarly method.[21] In any event, Jameson's five editions of Cuvier's "preliminary discourse" gave the work wide currency and provided Jameson the opportunity to expand his notes, which are particularly valuable to the historian for the record they provide of his changing opinions. Joan Eyles, Leroy Page, and James Secord have noted this fact, and in a fine piece of detective work Secord has shown how very far Jameson's opinions on species had changed by the date of the last edition.[22]

Still within the decade of the 1810s, Buckland, reader in mineralogy (1813) and geology (1818) at Oxford, entered into a more significant engagement with Cuvier than had Jameson. Buckland's reputation in the history of geology had suffered from an obscurantist taint that inevitably settled on an author, later dean of Westminster (1845), who, in 1823, titled one of his books *Reliquiæ Diluvianæ*—"Remains of the Deluge." But Susan Faye Cannon reoriented our view of Buckland by measuring his opinions against contemporary views within the Anglican church. The evangelical John Sumner, later archbishop of Canterbury (1848), was his foil. In Cannon's view,

> Buckland's insistence on the actual evidence of a deluge was partly an answer to Sumner's insistence that the Mosaic records were much more reliable than geological evidence. . . . For years one of Buckland's roles was to keep room clear for an independent evaluation of scientific evidence within the Anglican community, in spite of increasing pressures from Evangelicalism and, later, from Tractarianism.[23]

Further, in a major study, Nicolaas Rupke has reassessed Buckland's career against the background of university politics. In this light, Buckland's bow toward historical knowledge, including biblical texts, is seen as a means to find a place for the fledgling science of geology in the Oxford curriculum, which was weighted toward the classics.[24] In any case, Cuvier and Buckland found in each other kindred spirits, Buckland pursuing Cuvier's paleontological leads and serving as his host in England in 1818.

The one feature in their relationship to which I should like to draw attention is that Buckland contributed to both sides of the Cuvierian synthesis. On the geological side Buckland's efforts are well known. In regard to the biblical flood, for example, he listed what he believed were its signs: rough gravels, valleys of denudation, and particular kinds of organic remains.[25] Indeed, his enumeration of signs was so empirical and his interpretation so sharp that it invited reaction. Buckland spoke straightforwardly in favor of a "transient deluge, affecting universally, simultaneously, and at no very distant period, the entire surface of our planet."[26] On the historical side, Buckland's contributions have not been seen as clearly. This is so partly because he did not follow Cuvier directly in arguing for the historical significance of the flood on the grounds of a comparison of ancient literatures. When speaking of texts, he turned rather more comfortably to Bacon's contrast between the "Book of God's Word" and the "Book of God's Works."[27] Yet, in his practice over a lifetime, Buckland inclined toward Cuvier's intent to seek documentation for ancient human society and to integrate that history into geological understanding. Indeed, the very name Buckland chose for diluvian gravel—diluvium (flood, deluge, or inundation)—signaled in its Latin origin the union between historical and geological knowledge that he favored.[28] Also, Buckland hoped that one day human remains would be found in the diluvium.[29] There were other signs of Buckland's proclivities as a historian. In 1842 he suggested that the Regius chair of modern history be used to teach ethnography, and in 1844 he contributed a paper to the first meeting of the British Archaeological Association.[30] In sum then, even when in 1832 a beleaguered Buckland fell silent on the question of equating the biblical flood with the last geological upheaval, and then in 1836 retracted his previous position equating the two, he remained true to the balance of Cuvier's interests and to his program.[31]

Darwin and the Diluvium

Darwin enters this story in the 1820s, first at Edinburgh (1825–1827), where he attended the lectures of Jameson, and then at Cambridge (1828–1831), where he received training from Sedgwick and Henslow, both of whom were initially in accord with Buckland. As Rupke has argued, there was a split between Edinburgh and the English universities on issues in geology, including that surrounding the diluvium, and it is therefore of interest that Darwin's own movement traced the lines of the dispute.

On 25 November 1826, Jameson signed the preface to his fifth and last edition of Cuvier's "preliminary discourse." That same month Darwin signed the matriculation book for his second year at Edinburgh. One of the classes for which he enrolled was Jameson's in natural history, which included geology. In notes to his fifth edition of Cuvier's book, and presumably also in lectures, Jameson described how opinion on the geological deluge was split:

> We have been frequently requested to give the two views, in regard to the universal deluge, namely, that which maintains that it is proved by an appeal to the phenomena of the mineral kingdom; the other, which affirms that that great event has left no traces of its existence on the surface or in the interior of the earth. M. Cuvier's Essay, and Professor Buckland's Reliquiae, are the best authorities for the first opinion; while numerous writers have advocated the second.[32]

While Jameson now clearly ranged himself with the latter group, it is interesting that he chose to present the opinion as divided. When faced with a choice, he equivocated. Similarly, in Secord's interpretation, Jameson softened his shift on the concomitant issue of transmutation by publishing his views anonymously.[33] Jameson's own state of mind may well reflect his milieu, however, for on 26 December 1826 Cuvier was proposed but blackballed for honorary membership in the Plinian Society, a university group devoted to natural history, of which Darwin was a member.[34]

In Edinburgh, the man from whom Jameson drew most considerably in his new views was John Fleming (1785–1857), a minister in the Church of Scotland (1806) and later professor of natural philosophy at King's College, Aberdeen (1834). Fleming argued brilliantly against Buckland, and drew a rebuke from him for the immoderacy of his tone, but what is striking in much of Fleming's argument is his own biblical literalism.[35] He has a very different attitude toward the biblical text than does Cuvier. Cuvier had attempted to conciliate ancient traditions and had read loosely in order to do so. Fleming went back to a close reading of the biblical text, all the while professing against a union of geology and revelation as "indiscreet."[36] Thus, Fleming could argue against the Cuvier-Buckland notion of a violent flood on narrow textual grounds. In a passage later alluded to by Lyell, Fleming wrote,

> But if the supposed impetuous torrent excavated valleys, and transported masses of rocks to a distance from their original repositories, then must the soil have been swept from off the earth to the destruction of the vegetable tribes. Moses does not record such an occurrence. On the contrary, in his history of the dove and the olive-leaf plucked off, he furnishes a proof that the flood was not so violent in its motions as to disturb the soil, nor to overturn the trees which it supported.[37]

This same attitude on Fleming's part allowed him to choose a traditional and low number—six thousand years—in referring to the dispersion of the human race over the earth's surface.[38] Thus, although Fleming's argument was effective against Buckland, he did not supply a substitute for Cuvier's synthesis between geological and human history.

In the later years of the decade of the 1820s, resistance increased to the Buckland-Cuvier formulation on the flood. In his book on the geology of central France, Scrope did not adopt Buckland's term "diluvium" in his description of surface deposits either in his text or in the beautifully drawn views and sections accompanying the volume. He also argued against the diluvial hypothesis as an explanation for the excavation of the valleys in the area he studied. On the historical side, his primary contribution was to argue for "almost unlimited drafts upon antiquity" in geological reasoning.[39] The force of Scrope's work was magnified because of its adoption by Lyell, who used it in constructing his arguments for present causes and against a universal flood. These arguments formed the cornerstone of the *Principles of Geology.* In a broad sense Lyell, like Cuvier and Buckland, was interested in forming a synthesis between human history and geology. But where Cuvier and Buckland sought to anchor their synthesis in a discrete event, Lyell's synthesis was methodological. To Lyell, just as the historian had to consider past society in understanding the present, so did the geologist have to consider former landscapes in order to explain present landscapes. As he wrote, "It is easy to imagine the general law by which the present course of Nature is governed, viz., that in each period, the earth's surface and its inhabitants should be influenced by their former existence."[40] More narrowly, as Gillispie wrote in a judgment that still stands, the *Principles of Geology* "administered the *coup de grâce* to the deluge."[41]

In addition to specific arguments, Lyell was important to the "Genesis and geology" debate for his ability to accommodate to change. On the scientific side he began as a student of Buckland's at Oxford, met Cuvier in Paris, and in his early years defended their position on the flood.[42] Within a few years he had developed his own opposing views. Further, over the course of a long life, he had occasion to alter his views on the antiquity of man and the transmutation of species. Numerous scholars have addressed the issue of Lyell's intellectual accommodation.[43] What is perhaps less well known is the extent of Lyell's accommodation in daily life. Raised an Anglican, Lyell joined the large number of men and women in Victorian Britain who sought in Unitarianism a middle ground of belief and practice. He eventually became part of the Unitarian congregation in London that assembled in Little Portland Street.[44] A member of the congregation recalled that "Sir Charles's pew at Portland-street was in full view from that in the gallery in which the Manchester College students sat, and my distinct recollection is that during my own undergraduate years (1868–1872) Sir Charles was regular in his attendance."[45] Thus, while Lyell was a modernizer by virtue of his campaign in the *Principles of Geology* against Mosaic geology and by virtue of his opposition to Anglican hegemony in such important areas of national life as education, he remained a religious man.

Unlike Lyell, Sedgwick and Henslow remained within the communion of their birth. Indeed, their Anglican ties bound them together as much as their science, as is evident from the tender affection shown by Sedgwick to Henslow in his final illness.[46] From their combined geological fieldwork on the Isle of Wight in 1819 through their activities on behalf of science at the university in the decades of the 1820s and 1830s, their interests remained mutual and sympathetic. The sequence of Sedgwick's positions on the flood is known: his positive view of Buckland's treatment of the flood in 1825, together with his adoption of a distinction between diluvial and alluvial deposits; his backing away from Buckland's position in 1826–1827; and his utter abandonment of a geological role for the biblical flood in 1830–1831 following publication of the first volume of Lyell's *Principles of Geology*.[47] Henslow's views on the flood are less well known than Sedgwick's, but are of interest because of the greater intimacy between Henslow and Darwin than between Sedgwick and Darwin.

Like Sedgwick, Henslow had responded to Buckland's diluvialism with a corroborative contribution. In 1823 Henslow put forward a hypothesis of a nonmiraculous cause for the deluge, one that would employ the ordinary means of nature and also coincide with the account given in Genesis. He suggested that the nucleus of a comet composed of aqueous vapor might have fallen to earth, descending in the form of rain to create a flood. Water gained in this process might then have been partly absorbed by the solid portion of the earth; this last point Henslow regarded as key in his argument.[48] So far as is known, Henslow did not comment again in print regarding the scientific aspect of Buckland's diluvialism; presumably he followed Sedgwick and other leading English geologists in altering his views on diluvium in the late 1820s and early 1830s.

Yet even if Henslow did not leave behind a record of the entire range of his opinion, he did make numerous remarks over the course of his career suggesting how one ought to read sacred texts in the light of scientific evidence. These remarks are important as indicating the direction from which Henslow approached the subject of the historicity of the flood. They are also important for suggesting the point of view Darwin would have heard expressed at Cambridge from the teaching officer of the university with whom he had the greatest contact and with whom he had considered reading divinity.[49] In his *Autobiography*, Darwin emphasized, alongside Henslow's sheer goodness, his orthodoxy: "He was deeply religious, and so orthodox, that he told me one day, he should be grieved if a single word of the Thirty-nine Articles [of the Church of England] were altered."[50] If examined more closely, however, Henslow's views as expressed at the time were not as set as Darwin's anecdote suggests, and indeed were in flux on points regarding biblical interpretation.

Henslow's interest in scriptural interpretation was intense in the late 1820s. In the preface to the published version of a sermon on the "First and Second Resurrection," preached in 1829, he argued for the study of prophecy:

> In critical and philological enquiries the Bible must be studied like any other book, but that we may comprehend what is spiritual, we should remember that the Prophets, among whom are the Apostles, had all of them the same long vista of futurity before their eyes, and in looking down it, each was suffered to catch certain hasty and partial glimpses of events to come, to be seen only through the distorting medium of types and visions, as the Holy Ghost thought fit to represent them.[51]

This quotation illustrates both the seriousness of Henslow's attention to scripture, including prophecy, and his provision for differing modes of interpretation. The next relevant text, scientific rather than religious in character, is Henslow's *Descriptive and Physiological Botany* of 1836. After speculating on the botanical history of the earth, Henslow closed his book with the following statement:

> The commentator who wishes us to pay attention to his interpretations of the sacred text, must not proceed upon the supposition that there has been any thing written in the Bible for our learning, which can possibly be at variance with the clear and undeniable conclusions deducible from other and independent sources. If the letter does not announce a particular fact *revealed* in the works of the creation, a true believer will immediately infer that the letter (though it have the authority of inspiration) was not intended to teach that fact. When the philologist has ably interpreted the letter, the aid of the natural historian may still be needed before the divine can safely pronounce upon the exact scope and meaning of the instruction which it was intended to convey.[52]

Despite some obscurity (the question of what would count as a revealed fact), Henslow's statement suggests a criterion by which to judge the propriety of interpretations of scripture: where the literal meaning of a text contradicts a known scientific fact, one may presume that the text was not intended to be read for that meaning. Gauging the intent of the author of the text thus becomes paramount, and one may at least speculate that Henslow's elaboration of this view at a prominent point in this book reflected, among other things, recent rethinking of diluvial theory within his Cambridge circle.

What we find in Henslow, then, is a willingness to take the Bible very far at points with regard to its literal meaning but an insistence that truth is one, and that therefore literal readings yield where scientific findings call them into question. Such a view would accommodate either an integration of geological history with the Noachian flood or a disengagement of the two. Thus, with characteristic caution, Henslow could have altered his position on Buckland's diluvial theory without altering his views on scriptural interpretation. Presumably Darwin would have known something of Henslow's approach, as in the latter half of Darwin's time at Cambridge, he "took long walks with him on most days."[53] In his autobiographical recollections of his thoughts on religion while a student and during the voyage, Darwin cast himself as seeking to be convinced of the simple truth or falsity of the scriptures, but one may posit that at the time Henslow's more nuanced view was available to him as well.[54]

Once Darwin had left the care of Sedgwick and Henslow, he passed into the domain of FitzRoy, with whom he lived on intimate terms for much of the five years from December 1831 to October 1836. What is known of FitzRoy's views is primarily from his published account. There FitzRoy figures in three capacities: as a representative of the Crown who on occasion might negotiate government business; as a surveying officer responsible for work set out by the hydrographic office of the navy; and, more privately, as a man committed to promoting Christian missions. FitzRoy's activities to promote Christianity among the Fuegian Indians are well known. In addition, FitzRoy interested himself throughout the voyage in missions—commenting, for example, on denominational differences among missionaries—and on efforts to translate the Bible into foreign languages. Protestant in sentiment, and passionately anti-Roman, FitzRoy approached scripture from the point of view of individual interpretation. It was from this position that he addressed the questions that concern us in the chapter of his book entitled, "A very few Remarks with reference to the Deluge."

As with others discussed in this chapter, FitzRoy's views were in transition. However, whereas the others were moving away from linking human history and geological history according to the terms of Genesis, FitzRoy was moving to reassert the bond. As he described his change of views,

> Much of my own uneasiness was caused by reading works written by men of Voltaire's school; and by those of geologists who contradict, by implication, if not in plain terms, the authenticity of the Scriptures.
>
> While led away by sceptical ideas, and knowing extremely little of the Bible, one of my remarks to a friend [possibly Darwin], on crossing vast plains composed of rolled stones bedded in diluvial detritus some hundred feet in depth, was "this could never have been effected by a forty days' flood," . . . I was quite willing to disbelieve what I thought to be the Mosaic account . . . though knowing next to nothing of the record I doubted:—and I mention this particularly, because I have conversed with persons fond of geology, yet knowing no more of the Bible than I knew at that time.[55]

In his remarks on the deluge, FitzRoy then proceeded to read Genesis literally, even taking the "days" of creation to refer to twenty-four-hour days. Oddly, though, he combined biblical quotation with elements drawn from contemporary science, including that of Lyell and Darwin. This mixture of sources gives FitzRoy's intended synthesis an imbalance: on the one side, a fixed-point notion of human history drawn from a sincere but untutored reading of Genesis, and on the geological side, observations made honestly but with only very passing connection to the work of those operating within established traditions of geology. On some points FitzRoy's arguments are utterly unconvincing, as in his explaining away of the fossil record. On other points, where his own direct observations came more into play, he is at least plausible. Thus, for example, FitzRoy argued that the compressed appearance of the shelly deposits at Port St. Julian in Patagonia indicated that they had once been subjected to a great weight, which he interpreted as having been the biblical flood.[56]

How could FitzRoy have come to the conclusions he did? Even without knowing the full circumstances of his turn toward biblical literalism, one can point to two possible factors: his education was at the Royal Naval College, not at a university, and while in a general sense he was a man of science, his knowledge was in geography and meteorology rather than in geology. These circumstances help to explain the sharp differences between his views and those of the other British authors we have discussed. Unlike them, he did not have at hand the university-inculcated Baconian canon of a "two books" tradition of interpretation. As James R. Moore has argued correctly, the Baconian tradition was a political compromise.[57] Yet it was also a productive compromise, for it enabled adjudication of claims regarding knowledge, without eliminating either biblical studies or natural philosophy. These rules of jurisdiction FitzRoy did not learn or, if he learned them, did not observe. On the geological side, FitzRoy appears to have been ignorant, possible willfully so, of the vast progress in stratigraphy and paleontology. It is thus ironic that it was FitzRoy who presented Darwin with a copy of the first volume of Lyell's *Principles of Geology.* Later, after the voyage, FitzRoy would not have been made easy by the relaxed agreement between his former shipmate and Lyell on the subject of the flood. In 1839, as she was about to read the just-published narrative from the voyage, Darwin wrote to his sister, "You will be amused with FitzRoy's Deluge Chapter—Lyell, who was here to-day, has just read it, & says it beats all the other nonsense he has ever read on the subject.—"[58]

FitzRoy, who became a Fellow of the Royal Society in 1851, continued in the literalist direction in the decades after the voyage while remaining current with ongoing discussions in science. For example, in an 1858 letter he expressed the view that "the *constant* error of Ethnologists has been the *absurd* . . . idea that man *rose* from a savage state." This letter is consistent with ideas FitzRoy had expressed nineteen years earlier in his *Beagle* account under the heading "Remarks on the early migrations of the human race." However, in 1858 he buttressed his argument by citing the work of the German-born Oxford philologist Friedrich Max Müller (1823–1900). In the same letter he expressed hope that Murchison might affirm the notion of a great flood and

"Bible Science rightly understood."[59] Two years later, in 1860, FitzRoy again affirmed his views in a highly public setting where, during the debates over the *Origin* at the 1860 meeting of the British Association for the Advancement of Science, he held a Bible aloft, and, in the report of the *Athenaeum*, "regretted the publication of Mr. Darwin's book, and denied Prof. Huxley's statement, that it was a logical arrangement of facts."[60]

Sources for considering Darwin's views on diluvial geology are plentiful, but they must be read for what is absent as well as what is present. The most noticeable feature of Darwin's writing on the issue is that he used the term "diluvium" throughout his notes from the *Beagle* voyage of 1831–1836, but not, except with qualification, in his publications after returning home. His use of the term reflected what he heard when he was in residence at Cambridge, where both Sedgwick and Henslow had adopted Buckland's vocabulary. (There were those, like Scrope, who never used that vocabulary.) By the time Darwin had returned from the voyage, Buckland's diluvial geology was sufficiently discredited that even its vocabulary was no longer in vogue among the best-informed geologists in Britain. The second striking feature of Darwin's notes from the voyage is that he used the term "diluvium" only with reference to the formation of superficial gravels and boulders. There is frequent reference in his notes to the possible agencies of deposition for these gravels, but there is no mention of a biblical flood or, indeed, of any event described in an ancient text. This absence of reference is consistent with Sedgwick's warning, issued in 1831, against conflating the imperfectly understood phenomena of superficial gravels with the biblical flood:

> It was indeed a most unwarranted conclusion, when we assumed the contemporaneity of all the superficial gravel on the earth. We saw the clearest traces of diluvial action, and we had, in our sacred histories, the record of a general deluge. On this double testimony it was that we gave a unity to a vast succession of phaenomena, not one of which we perfectly comprehended, and under the name diluvium, classed them all together.[61]

Finally, the third feature of interest in Darwin's notes is that he shifted his explanation of the causes for diluvium during the course of the voyage. Initially, like his teachers, he traced superficial gravels to debacles rushing overland. By the end of the voyage he traced most superficial gravels to submarine deposition. As suggested in the previous chapter, in this he was following Lyell's lead.

After returning to England, Darwin also contributed to the other side of what I have termed the "Cuvierian synthesis." He did not do so in the manner of Cuvier and Buckland, by interpolating his reading of ancient texts with knowledge derived from geological evidence. Indeed, by this period his autobiography suggests that he no longer accepted the historical claims of the Genesis narrative.[62] (Darwin's change of views on religious matters was shaped by his own complex religious heritage—part Anglican, part Unitarian, and all touched by the legacy of his grandfather Erasmus.[63]) Nor did Darwin pursue the study of antiquities, which interested a number of his peers. But he did take up the subject of human origins in a series of private notebooks, opened following his adoption of the transmutationist hypothesis in 1837.[64] This exploration of human origins was thoroughly integrated with his geological views. In sum, for Darwin, by the close of the 1830s the task of integrating human and geological history was very much in process, even without the unifying event of the biblical flood.

In this light the Cuvierian synthesis, signified in Britain by the term "diluvium," seems not so much erroneous as premature. To supply all of the elements for a stable synthesis would require the efforts of many scholars—Darwin among them—over the course of succeeding decades. In the meantime, the presence of unstable, if recurring, solutions, such as FitzRoy's, served in part to underline the absence of consensus. A complicating feature for the debate was that some of the ancient history of human society was presented in sacred texts. Thus, much of the debate focused on questions of interpretation, that is, whether the ancient narratives ought to be read comparatively, or literally, or with consideration of the intent of the author. Who had the right to pronounce on such issues was also at stake. With the flood no longer under consideration, most British geologists wished to step away from the texts and interpretations that had figured in the debate. They wished to separate Genesis from geology; they were demonstrating a "desire for autonomy which has, from time to time, involved repudiating and suppressing links between . . . fields."[65] Thus, certain combinations would have seemed in poor taste to geologists in the 1840s, such as that suggested by the title Buckland planned to give the second volume of his book on the flood: *Reliquiae Diluviales et Glaciales*—"Diluvial and Glacial Remains."[66] Although Buckland had repudiated the biblical connection of the last flood and had integrated Agassiz's glacial key into his work, his use of diluvial language, recalling the scriptures, would have seemed as atavistic to his fellow geologists as his Latin title.

If there had not been a climate for the free expression of geological opinion in Britain in the 1820s and 1830s, and if such a forum as the Geological Society of London had not existed, the weaving together of biblical and geological history as occurred with the diluvium debate may well have retarded the development of the science. However, as it happened, the debate probably moved the science forward, for it brought questions of chronology front and center. Much has been made, often in homiletic fashion, of the difficulty for humans of comprehending what John McPhee has described as "deep time," and, indeed, in the period under discussion, the Huttonians particularly pressed this point.[67] But the geologists addressing the diluvium issue were working with what might be termed "shallow time." John Phillips expressed this notion in 1837, following the diluvial dispute, when he remarked that he had adopted "as the limit of least antiquity of the scale of stratified rocks, the traditionary [*sic*] age of the human race."[68] This was in the course of a discussion of "geological time," a phrase Phillips may well have been the first to use. Phillips wrote,

> The very first inquiry to be answered is what are the limits within which it is possible to determine the relative dates of geological phenomena? For if no scale of geological time be known, the problem of the history of the successive conditions of the globe becomes almost desperate.[69]

Phillips's answer to this "almost desperate" problem was that the "scale" of geological time would be provided by the scale of "the series of stratified rocks" in which the "*lowest were formed first, the uppermost last.*" Here the metaphor of "deep" time was key: in his book Phillips took care to list the thickness of various strata, which would be a key to identifying their relative ages, and noted that the "total length of the scale" amounts to "more than ten miles, and is seldom to be estimated at less than five."[70]

Measuring the depth of strata allowed Phillips to approach the question of the total time elapsed in the formation of the crust, though he did not offer a number. Thus, effectively, Phillips separated the question of the age of the earth from the age of its stratified crust. For Phillips human existence was part of the history of the stratified crust, at least in theory. Yet even the upper strata were so ancient that they antedated "the race of man now existing there," though not necessarily a "former race of men."[71] (The elasticity in Phillips's views on this last point may provoke surprise.) In any case, writing in 1837—after the decline of the diluvial theory, but before Agassiz's views were current—Phillips represents a good stopping point for a discussion for the Genesis and geology debate in the 1830s. Phillips regretted the sudden rise and equally sudden fall of the "diluvial" theory.[72] But he did not entirely abandon the term "diluvium" and, on a deeper level, respected the analyses that had been made using it.

The 1830s closed with the Geological Society of London, as represented by Lyell and Sedgwick, having effectively combined to disconnect historical geology from the Noachian flood. The particular geological issues it had engaged regarding surface formations remained to be solved, but without a scriptural import. A young author such as Darwin could proceed without defending himself against the charge of irreligion if he were to avoid the subject of the deluge. Such a man as Sedgwick would be there to answer critics.[73] One set of negotiations had been concluded to the satisfaction of the majority of practicing geologists within the Geological Society. Beyond the issue of the biblical flood, opinion was hardly unanimous. Whewell had identified two emerging "sects": "Uniformitarians" and "Catastrophists" (religious vocabulary readily serving his purpose). To some extent these two opinions corresponded to religious differences: cyclically minded Uniformitarians pulling toward deism, historically minded Catastrophists toward Christianity. Whewell, firmly of the Church of England, honed in on that distinction as he wrote of Lyell: "he appears to forget that the geological series, long and mysterious as it is, has still a beginning."[74] Yet whatever their sectarian differences, both uniformitarian and catastrophist geologists were legitimate members of Whewell's "geological world."

The next decades brought with them a chain of issues on both sides of the geological/cultural divide. On the scientific side, the issue of transmutation came once again to the forefront with the "sensation" caused by the anonymous publication in 1844 of *Vestiges of the Natural History of Creation,* and fifteen years later, with Darwin's own *Origin.*[75] On this issue the Geological Society provided no cover for transmutationist-minded geologists. The leading lights of the society stood squarely against transmutation. That, after all, had been the thrust of Lyell's second volume of the *Principles.* Sedgwick was particularly strong on the subject and became the most prominent and effective of the anonymous author's scientific critics ("I think you could smash him and I wish you would" urged one colleague).[76] Inevitably, then, the controversy over transmutation moved beyond the learned societies into the larger public domain.[77]

On the religious and cultural side, the question of the proper canons of biblical interpretation became of general interest with the rise of critical scholarship of a historicizing nature. A number of geologists interested themselves in this movement, as one parallel to their own. Thus, for example, Leonard Horner in his 1861 presidential address to the Geological Society of London advocated removing the traditional date of 4004 B.C. for the beginning of the world from the standard English version of the Bible, pointing out that it had been a later addition to the text, based on seventeenth-century scholarship, and conflicted with the "truths of geology." On receipt of a copy of Horner's address, Darwin replied to him, "How curious about the Bible! I declare I had fancied that the date was somehow in the Bible.—You are coming out in a new light as a Biblical critic!"[78]

More generally, some liberalizing Anglican scholars sought philosophical common ground for biblical criticism and for recent science. This move culminated in the publication of *Essays and Reviews* in 1860. The author who dealt with geology, under the title of "Mosaic Cosmogony," was Charles Wycliffe Goodwin (1817–1868), an alumnus of the University of Cambridge. Goodwin criticized prior attempts to conciliate scripture and geological findings, and he named geologists (Buckland, Hugh Miller [1802–1856]) in doing so. His own conclusion was that "theological geologists" had been disrespectful toward the intent of the Mosaic narrative, and a greater "respect for the narrative which has played so important a part in the culture of our race" was required. Certainly "no one contends that it can be used as a basis of astronomical or geological teaching."[79]

This was a line of thought appealing to the majority of British geologists in the 1850s, including Darwin and his immediate circle. Among the authors of *Essays and Reviews,* Darwin was a sympathetic figure. The essay by Baden Powell (1796–1860) singled out Darwin and the *Origin* for praise, calling it "a work which must soon bring about an entire revolution of opinion in favour of the grand principle of the self-evolving powers of nature."[80] As a further indication of how party lines were then being drawn, Frederick Temple (1821–1902), an *Essays and Reviews* author, preached a sermon at University Church in Oxford, on 1 July 1860 as the British Association met, very much under the shadow of the recent publication of Darwin's *Origin.* One person in attendance understood Temple's message to be that "he espoused Darwin's ideas fully."[81] Reciprocating the favor, a number of religiously liberal men of science, including Darwin, rallied behind the authors of *Essays and Reviews* when their views were threatened with censure by authorities within the Church of England.[82] By 1860, in England, negotiating Genesis and geology had resolved into questions of what kind of religion, what kind of science.

CHAPTER 7

TOWARD SIMPLICITY

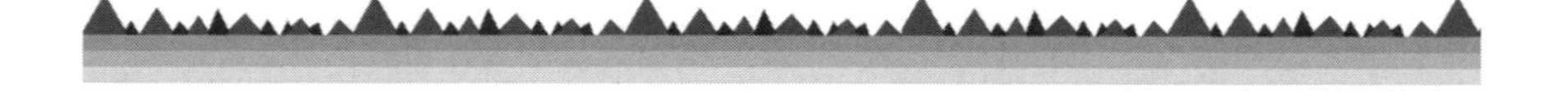

Earthquakes. times, nature of undulation. effects on buildings: <wave> cracks, Springs: Mineral. Springs: effects on neighbouring Volcanoes

Road to Valdivia; Englishmen there?.—Concepcion. country?.

rise in ground at same time as Valparaiso.?—
DARWIN, *BEAGLE* NOTEBOOK 1.11

Elevation and Subsidence

Darwin believed that elevation and subsidence were the primary movements determining the appearance of the earth's crust. Presently geology is governed by the theory of plate tectonics, which is founded on an understanding of the lateral motions of the earth's crust. It therefore requires some historical imagination for the reader to enter Darwin's understanding. The task is not made easier by the circumstance that Darwin did not publish a "theory of the earth" per se. The closest he came to enunciating such a theory in a title was in his early presentation of 31 May 1837 to the Geological Society: "On Certain Areas of Elevation and Subsidence in the Pacific and Indian oceans, as Deduced from the Study of Coral Formations." Here the motions of the land were brought to the fore: coral reefs figured as the object of instrumental rather than primary interest. Even this emphasis was lost, however, when Darwin brought his book to publication in 1842. Its title was *The Structure and Distribution of Coral Reefs.* Thus, it was perhaps no wonder that the eminent marine geologist Henry William Menard (1920–1986), in his foreword to the reprint of that work, noted in a tone of surprise, "At times, Darwin seems to be concerned with the origin of coral reefs only so far as it gives evidence for regional subsidence of the sea floor."[1] Indeed that was the case.

Historians have explained reasonably adequately why Darwin was not more direct in declaring his intentions publicly. First, theories of the earth were out of favor with early-nineteenth-century British geologists. Scrope, who used the phrase "a new theory of the earth" on his title page of his *Considerations on Volcanos* (1825), was rather the exception. Second, as Rudwick has argued, geologists in the Geological Society of London favored the "exemplars" format: theory should be embedded within the treatment of particular cases, rather than brought forward on its own.[2] Finally, while there was an acceptance of theoretical reasoning within the Geological Society, this did not translate into an invitation for a geologist to cast himself in the specialist role of theorist.[3] Darwin's speculations on a wide variety of geological topics, recorded primarily in his Red Notebook and in Notebook A, are of interest retrospectively. However, they would not have struck a contemporary geological audience as appropriate for formal publication if expressed in so free a form.

The Ideal of Simplicity

Darwin's classic statement of his ideal as a theorist occurs in his Red Notebook, written in 1836 on the last leg of the *Beagle* voyage. The entries are difficult to follow but worthy of attention for the insight they allow into the stream of Darwin's thoughts. Here is the passage in its entirety:

> Humboldts. fragmens.
>
> Read geology of N. America. India.—remembering S. Africa. Australia.. Oceanic Isles. Geology of whole world will turn out simple.—
>
> Fortunate for this science. that Europe was its birth place.—Some general reflections might be introduced on great size of ocean; especially Pacifick: insignificant islets—general movements of the earth;—Scarcity of Organic remains.—Unequal distribution of Volcanic action, Australia S. Africa—on one side. S. America on the other: The extreme fre-

> quency of soft materials being consolidated: one inclines to belief all strata of Europe formed near coast. Humboldts quotation of instability of ground at present. day.—applied by me geologically to vertical movements.[4]

"Geology of whole world will turn out simple" is the key statement in the rush of associated phrases recorded in the notebook entry. A statement farther along in the passage—"Humboldts quotation . . . applied by me geologically to vertical movements"—explains what is meant by that simplicity, for a few pages later, Darwin framed his thoughts as a question: "Will geology ever succeed in showing a direct relation of a part of globe rising, when another falls[?]—"

The notebook entries just quoted feature a rhetoric of simplicity. They also refer to passages from Humboldt. Some information may be useful on these points.

The ideal of simplicity as a value in judging scientific theories is an enduring one. Darwin might have encountered it in a number of settings, as, for example, in the closing paragraph of John Herschel's *Preliminary Discourse:*

> It is in this respect an advantageous view of science, which refers all its advances to the discovery of general laws, and to the inclusion of what is already known in generalizations of still higher orders; . . . Yet it must be recollected that . . . *every advance towards generality has at the same time been a step towards simplification.* It is only when we are wandering and lost in the mazes of particulars . . . that nature appears complicated:—the moment we contemplate it as it is . . . we never fail to recognise that *sublime simplicity* on which the mind rests satisfied that it has attained the truth.[5]

Herschel's declaration in favor of simplicity as a feature of the most profound science was read by Darwin while he was still in Cambridge. It could well have made a lasting impression.

With regard to Humboldt, the phrase "Humboldts. fragmens" holds the clue. It refers to Alexander von Humboldt's *Fragmens de géologie et de climatologie asiatiques,* published in two volumes in Paris in 1831. On 18 August 1832 Charles's brother Erasmus wrote that he had got the work "which I suppose was the one you want"; the flyleaf of the second volume is inscribed "Chas Darwin Monte Video Novem: 1832" so the month of arrival of the book is known.[6] In the concluding paragraph of his first essay on his coral reef hypothesis, written in 1835, Darwin referred to the book:

> Before finally concluding this subject, I may remark that the general horizontal uplifting which I have proved has & *is now* raising upwards the greater part of S. America & as it would appear likewise of N. America, would of necessity be compensated by an equal subsidence in some other part of the world. Does not the great extent of the Northern & Southern Pacifick include this corresponding Area?—Humboldt carrys a similar idea still further[.] In the Fragmens Asiatiques, P 95. he says . . .

A passage then followed in French, of which Darwin included a translation in his notes:

> Humboldt <Fragmens Asiatiques P 95> in a similar manner considers that the epoch of the sinking down of Western Asia coincides with the elevation of the <plateau> platforms, of Iran, of central Asia, of the Himalaya, of Kuen lun, of Thian chan, & of all the ancient systems of mountains, directed from East to West.—[7]

In the long passage from Darwin's Red Notebook quoted at the head of the present section, Darwin's reference to Humboldt was less direct, and unfortunately he gave no page number. Darwin's sentence reads, "Humboldts quotation of instability of ground at present. day.—applied by me geologically to vertical movements." In *Fragmens asiatiques* there is no passage that conforms precisely to Darwin's phraseology, but the following passage will give some idea of Humboldt's views:

> Volcanicity, that is to say, the influence which the interior of the planet exerts on its exterior envelope in the different stages of its cooling, on account of the unevenness of aggregation (of fluidity and of solidity), of the materials which compose it, this action from the inside to the outside (if I may express myself that way) is today very weakened, confined to a small number of points, intermittent, less often shifted and very simplified in its chemical effects. Volcanicity only produces rocks around small circular openings or in less well known longitudinal crevasses, only manifesting its power from great distances, dynamically shaking the crust of our planet in linear directions, or in places (circles of simultaneous oscillations) which remain the same over a great number of centuries.[8]

From such a passage, however tortured its prose, Darwin may have gained the idea of "Humboldts quotation of instability of ground." Yet since Humboldt also referred in this passage to the "different stages" of the cooling of the earth—a subject excluded or, at the least, bracketed in the Lyellian tradition—we must turn to address points bearing on that question.

The Bracketed Subject of Origins: The Nebular Hypothesis

In the 1830s Whewell, wordsmith to British natural philosophers, used the phrase "nebular hypothesis" to describe the cosmogonical theories of William Herschel (1738–1822) and Pierre Simon Laplace (1749–1827).[9] Herschel, the pioneer of observational cosmology, proposed that stars are formed by the condensation of the nebulous objects he had observed in the sky. The formation of the sun was thus part of a larger process. To explain the origin of the solar system, Laplace, whose work followed quickly after Herschel's, relied on a mechanical-thermal process by which the planets could have been formed from a hot cloud surrounding the sun. As Brush has emphasized, the phrase "nebular hypothesis" was used during the nineteenth century to embrace both Herschel's and Laplace's differing but related theories.

From the start both theories had ties to natural history. Herschel's approach to stellar astronomy relied on natural historical images. The discourse of taxonomy and classification tied his cosmology to natural history. As Simon Schaffer has suggested,

> "Order", for Herschel, was marked by the stability not of the system of the heavens as we now see it, but by the stability of a cyclical system of natural types which was maintained by the action of gravity through time.[10]

By use of the analogy of a life cycle, nebulae could be understood as in the process of evolving as stars, even though the process was too slow to observe the change in a single nebula. In the first edition of his theory, Laplace cited as a predecessor the natural historian Georges-Louis Leclerc, the Comte de Buffon (1707–1788), who had proposed a naturalistic cosmogony.[11] Laplace was also soon to draw on the newly certain knowledge developed by Cuvier regarding the extinction of species. In a later edition Laplace explicitly cited conclusions then being drawn by Cuvier with regard to the extinction of species, as illustrating "a tendency to change in things, which are apparently the most permanent in their nature."[12] Cuvier also linked himself to Laplace by dedicating his *Recherches sur les ossemens fossiles de quadrupèds* (1812) to him.[13] It is thus perhaps not surprising that the nebular hypothesis would eventually be drawn into the species question, most famously in the first chapter of the anonymously published *Vestiges of Creation* (1844).[14] Since Darwin was actively working both on evolutionary ideas and on large geological questions in the late 1830s, it is a matter of interest what relation his work bore to the nebular hypothesis.

Darwin seems to have approached the nebular hypothesis as a bracketed issue. That is, it was a subject not off-limits, but one that receded in favor of the other scientific questions that seemed more urgent. While Darwin did not discuss the subject in his published work in the 1830s and 1840s, it does appear in Notebook A. The key factor affecting Darwin's approach to the nebular hypothesis was the guarded attitude toward it expressed by Lyell and by John Herschel.

In treating Lyell's opposition to the nebular hypothesis, it would be straightforward to cite his choice of uniformity as a philosophical assumption. However, in fact, his opposition also had a heuristic basis. As he said in a lecture at the Royal Institution delivered on 2 May 1833,

> If instead of speculating on the works of mysterious agents during the original formation of the earth, geologists had been employed for the last half century or more in instituting a close comparison between effects as minute as these ripple marks on the sea with the internal structure and arrangement of the strata in mountain regions [the subject of Lyell's May lecture], we should . . . have arrived at the solution of nearly all the most difficult problems in the science.[15]

This heuristic impulse—to discourage one line of inquiry and to encourage another—was similarly present in the *Principles of Geology.* Thus, Lyell could counsel against the "hypothesis of gradual refrigeration" until scientists could establish "whether there be internal heat in the globe." While indicating his own preference against associating such heat with the "original formation of the planet," he argued more to postpone than omit consideration of the subject.[16] Another tack that he took in such discussions was to insist on "positive" or "geological" facts for consideration of such views.[17] By this he seems to have meant facts that were observable on the earth's crust.

Rather surprisingly, John Herschel did not champion his father's nebular hypothesis as an appropriate line of inquiry for geologists. That John was an astronomer of the first rank in his own right, as well as the son of William Herschel, added significant weight to his opinion. While not arguing for the "eternal duration" of the earth, the younger Herschel wrote that

> if we would speculate to any purpose on a former state of our globe . . . we must confine our view within limits far more restricted . . . than either the creation of the world or its assumption of its present figure.

To John Herschel, such were the "favorite [*sic*] speculations of a race of geologists now extinct." Rather, geologists ought to "aim at a careful and accurate examination of the records of [the globe's] former state, which they find indelibly impressed on the great features of its actual surface."[18]

The reference to the "actual surface" of the globe brings to mind John Herschel's ideal of *vera causa*, or "true cause":

> Some consider the whole globe as having gradually cooled from absolute fusion; some regard the immensely superior activity of former volcanoes, and consequent more copious communication of internal heat to the surface, in former ages, as the cause. Neither of these can be regarded as real causes in the sense here intended; for we do not *know* that the globe has so cooled from fusion, nor are we sure that such supposed greater activity of former than of present volcanoes really did exist.[19]

Following this passage Herschel went on to treat as a "cause on which a philosopher may consent to reason" Lyell's proposal for understanding climatic change in terms of the distribution of land and sea on the globe. The influence of Lyell and Herschel thus tended in the direction of discouraging the new generation of geologists from relying on the nebular hypothesis, or accepting too readily its related conclusion—the notion of a cooling earth.

Herschel addressed the question of the relations of astronomy to geology in a paper published in the 1835. In it he reviewed the astronomical causes—with emphasis on changes in the earth's orbit—that might influence geological phenomena, particularly in regard to climate. After citing Laplace and Jean Baptiste Joseph Fourier (1768–1830), he concluded that "the limits of the excentricity [*sic*] of the earth's orbit are really narrow" and that the "mean as well as extreme temperature of our climates would *not* be materially affected" by changes within those limits, and that "a period of 10,000 years would elapse without any perceptible change in the state of the data of the case we are considering." This left him free to assert that the causes for climatic change were to be sought elsewhere.[20]

Overall the effect of Herschel's argument was to support Lyell's position, and, indeed, he wrote to Lyell that

> I hope your example will be followed in other sciences, of trying what *can* be done by existing causes, in place of giving way to the indolent weakness of a priori dogmatism—and as the basis of all further procedure enquiring what existing causes really are doing.[21]

Given, then, that Herschel stood behind Lyell in his approach to geology, Darwin's relation to the nebular hypothesis was understandably muted. His guides to the subject of geology did not send him down that road. Darwin did not refer to Laplace in his notebooks. Among his few references to the nebular hypothesis in his notebooks, one to William Herschel appears in the context of his speculations concerning the figure of the earth. As his point of departure, Darwin probably used the description of William Herschel's views in the fifth (1837) edition of Lyell's *Principles of Geology:*

> It has long been a favourite conjecture, that the whole of our planet was originally in a state of igneous fusion, and that the central parts still retain a great portion of their primitive heat. Some have imagined, with the late Sir W. Herschel, that the elementary matter of the earth may have been first in a gaseous state, resembling those nebulæ which we behold in the heavens, and which are of dimensions so vast, that some of them would fill the orbits of the remotest planets of our system. It is conjecture that such aëriform matter (for in many cases the nebulous appearance cannot be referred to clusters of very distant stars), if concentrated, might form solid spheres; and others have imagined that the evolution of heat, attendant on condensations, might retain the materials of the new globes in a state of igneous fusion.[22]

In a notebook entry written in late 1838 or early 1839, Darwin observed,

> Assuming from Sir W. Herschel's views earth originally fluid, then cooling process must go from surface towards the interior,—who knows how far that may have penetrated. lower down the temperature may be kept up far higher from circulation of heated fluid or gases under pressure.—[23]

Here we find Darwin willing to entertain the premise of an originally fluid earth, cooling down from the surface. However, in a skeptical vein, he then went on to suggest the difficulty of understanding the nature of processes operating in the interior of the earth ("circulation of heated fluid or gases under pressure"). This was also a recurrent theme of Lyellian derivation: the interior and exterior of the earth were not to be casually analogized.

In Notebook A, Darwin signaled that his own geology would not wait for the resolution of issues regarding the earth's interior. On 24 February 1839, a short time after writing his note on William Herschel, Darwin recorded at the front of Notebook A that he had abstracted his notebook as far as it concerned "Geolog[ical] Observat[ions] on Volcanic Islands & Coral Formation[.]"[24] His geology would go forward.

Next we will address a question raised recently by Brush, as to whether it was the "Nebular Hypothesis rather than Darwinian natural selection" that provided an evolutionary worldview, or paradigm, for modern science.[25] While Brush's question is intriguing, two difficulties appear at once. First, the nebular hypothesis is not directly comparable to "Darwinism natural selection" in any point-for-point manner. The nebular hypothesis includes a variety of ideas put forward over many years. In contrast, Darwinian natural selection as articulated in the *Origin of Species* was, comparatively, a narrowly defined theory. Second, nebular hypotheses were constructed with recourse to the discourse and conclusions of natural history, as well as to that of astronomy. Both Buffon and Cuvier influenced Laplace. After Cuvier, the extinction of species was taken as a certainty, which argued for a treatment of time as irreversible. To follow on Brush's use of the Kuhnian vocabulary of paradigms and revolutions, one could then say that extinction was the anomaly that mobilized a revolution. (Lamarck's transmutationism, rejecting extinction, was one response to the idea of extinction.) The point to make here is simply that a geological conclusion—a belief in the extinction of species—infiltrated speculations about the heavens.

More positively, the nebular hypothesis provided a provisional cosmological backdrop for evolutionary views. Its tone, associated with the starry heavens, was also brighter, and hence more inherently appealing, than the extinction-derived tenets of paleontology. What the nebular hypothesis, in any of its varieties, could not do in the nineteenth century was to provide an agreed-on set of conclusions and thus a firm foundation for evolutionary theory.

Meteorites

Darwin was also interested in meteorites, the extraterrestrial nature of which Ernst Chladni (1756–1827) had established.[26] In one notebook entry from the late 1830s Darwin pursued the subject of meteorites with two thoughts in mind: the possible utility of meteorites for understanding the chemistry of the earth, and the question of the effect of the fall of meteorites on the earth's surface. On this latter point he queried,

> What must be the effect of all the meteoric stone which must have fallen on the whole globe since the Cambrian system. In Ures dictionary between 1768 & 1818 that is fifty years—90 <showers of> stones are

> recorded as falling; many of these were not single, but are described as many, (one even 3000) This ninety includes all actually counted.—The weight <<or size>> is given of 25 stones.—

Darwin then went on to estimate that 100 pounds of stone might fall in any given year, which over the course of fifty thousand years would yield 2500 tons of meteorites landing on the earth. He then asked, "If world increased a tenth; would the perturbation be serious? if so other cause besides thin vapour bringing planets to an end?"[27]

In his speculation regarding meteorites Darwin demonstrated an interest in their chemical and astronomical aspects. Possibly a meteorite could be an "old Planet"—a thought he traced to David Brewster (1781–1868). Possibly understanding meteorites might explain the order in which volcanic materials proceeded from the interior of the earth, as trachytes emerging before basalts. On the astronomical side, he wondered whether the weight of fallen meteorites would be sufficient in "bringing planets to an end," presumably by altering their orbits. He also noted, obliquely, that this might provide another cause besides "thin vapour" (a phrase associated with nebular hypotheses) in explaining the duration of a planet. In sum, for Darwin, the effect of the impact of meteorites on the earth was an open topic. Once again, however, it was bracketed for discussion in private notes rather than in publications.

The Figure of the Earth

In contrast to the nebular hypothesis, and to the subject of meteorites, the question of the figure of the earth captured a considerable amount of Darwin's attention. However, the figure of the earth, like the nebular hypothesis, though to a lesser extent, was a bracketed subject. The Huttonian tradition had gone through a Lyellian filter by the time it reached Charles Darwin in the 1830s. Whereas Erasmus Darwin had freely hypothesized concerning the interior of the earth, his grandson Charles Darwin was more cautious in speculating in print about the interior of the earth. (This Lyellian reticence sparked some complaint. Whewell took a swipe by asking, "What does Mr. Lyell intend to substitute for the Plutonic cookery of these elder assertors [Hutton and Hall] of the constancy of nature?[28]) While Darwin was hardly a rote follower of Lyell, he did prefer to reason from geological processes observable on the earth's surface to consideration of the earth's interior. The crust was where one began.

The central theme that runs through Darwin's comments in the Red Notebook and Notebook A is a belief that the earth's crust is thin. This claim is expanded in a number of entries, which are best summarized by the following quotations, listed in the order in which they appear in the notebooks. In these quotations, one sees Darwin circling a question, returning to points he had raised earlier.

> ⸮ If crust very thick would there be undulation? would it not be mere vibration? but walls & feeling shows undulation ∴ crust thin.—Concepcion earthquake [RN:154]
>
> From the lost & turned about position of strata, prooff [*sic*] thickness not very great; where piece turned over axis or hinge no doubt fluid.— [A:13]
>
> Does the isothermal subterranean line moves [*sic*] upward from effects of Elevation if not crust much thinner beneath ocean than above it
>
> 
>
> ———
>
> no because heat proceeds from great body of mass.—The last speculation becomes important with respect to thickness of crust broken up.——My view of Volcanos &c &c
>
> ———
>
> This view will bear much reflection on method of cooling—Very difficult subject. PP—
>
> ————————
>
> I think from dislocation taking place chiefly beneath water & volcanos. crust must be thinner <under water> but cause most difficult (better conductor) [A:77–78]
>
> In Glen Roy paper I show crust yield easily. & if easily must be thin: <beside mere fracture> [A:133]
>
> From strata being not only vertical, but turned over in many parts of the world. —argument strong in favour of thin crust theory.—[A:136]

In these comments Darwin was drawing on his own experience of sensing the undulations of the 1835 earthquake and on his direct observations of overturned strata. His use of the term "isothermal" suggests his indebtedness to Humboldt and how that train of thought led him to consider the question of the transmission of heat within the globe. The phrase "Glen Roy paper" refers to Darwin's publication of 1839 entitled "Observations on the Parallel Roads of Glen Roy, and of Other Parts of Lochaber in Scotland, with an Attempt to Prove that They Are of Marine Origin."[29]

The Glen Roy paper contained one of Darwin's strongest statements on the relation of the crust to the interior of the earth. He interpreted the so-called parallel roads or shelves at Glen Roy in Lochaber, Scotland, as former sea beaches, which he believed resulted from the slow intermittent elevation of the land. He used the appearance of the roads to draw conclusions regarding the exterior and interior structure of the earth. These conclusions are contained in the last section of the article, entitled "Speculations on the Action of the Elevatory Forces, and Conclusion."[30] In retrospect, it is surprising that Darwin would draw strong conclusions relating the crust of the earth to its interior from what seems today so local and superficial a feature of topography as the parallel roads of Glen Roy. (The roads are presently interpreted as strand lines of former lakes.)

It is by factoring in Darwin's South American work, written up in the "Connexion" article of 1838, that his approach to the question of elevation appears internally consistent. In both instances (South America and Glen Roy) Darwin believed he was observing the effects of elevatory forces operating beneath the earth's crust. The thinness of the earth's crust was premised on the belief that "if . . . the crust did not yield readily, partial elevations [as at Glen Roy] could not be so gradual as they are known to be, but they would assume the character of explosions."[31] The image Darwin suggested to relate the exterior and interior surface of the globe is that of ice riding over the tidal waves:

> We may almost venture to say, that as the packed ice on the Polar Sea, with its hummocks and wide floes, rises over the tidal wave, so did the earth's crust with its mountains and plains rise on the convex surfaces of molten rocks, under the influence of the great secular changes then in progress.

In his view a molten fluid nucleus of the earth, *acting in local areas,* caused the motion of the crust. He noted the identical curve of the shelves of Glen Roy to that of the ocean surface. He deduced from their curvature

> first, that the district of Lochaber formed only a small part of the area affected; secondly, a confirmation of the view, which I deduced from the phenomena observed in South America, that the motive power in such cases is a slight additional convexity slowly added to the fluid nucleus; and thirdly . . . that we thus obtain some measure of the degree of homogeneous fluidity of the subterranean matter beneath a large area,

namely, that its particles, when acted on by a disturbing force, arrange themselves in obedience to the law of gravity.

He believed a "slight additional convexity added to the fluid nucleus" caused the alteration in the earth's crust.

Thus, Darwin concluded in favor of a fluid interior to the earth. ("Fluid" rather than "liquid" was the preferred adjective.) His treatment of this question was framed by the particular time in which he took it up—the late 1830s—and by the authorities on whom he relied, including not only Lyell and Humboldt but also, prominently, John Herschel. The range of Darwin's views on the interior of the earth are most succinctly reflected by quoting selections from the Red Notebook and from Notebook A:

> As the rude symmetry of the globe shows powers have acted from great depths, so changes, acting in those lines. must now proceed from great depths.—important.— [RN:123]
>
> Metamorphic action: <most> coming so near surface most important [RN:156]
>
> On Lyell's idea of whole centre of earth same heat, then change in form of fluid centre would lift with it isothermal line, but if heat from centre, then crust of solid earth would be thicker.— [A:79]
>
> Consider profoundly all consequences of EXTREME FLUIDITY of earth.—study different forms of earth as shown by arc.—read Herschels astronomy with oscillations of level.—

> will point

> be the one which generally yields.—Will this not explain *littoral* mountains & volcanos.—Why on *one* coast? [A:113]

At no point in the notebooks did Darwin present a precise numerical estimate as to the thickness of the earth's crust, but there is in his 1838 article a hesitantly framed suggestion ("all the calculations . . . if they can be at all trusted") that "the earth's crust is not much more, and perhaps less, than twenty miles in thickness."[32]

In all of Darwin's notes and publications, the earth's interior was regarded as hot. A point on which there is discussion in the notebooks is the nature of the earth's crust as a conductor of heat. For example, he asked himself, "How comes it in volcanos that have gone on for thousands of years that surface [of earth] does not become hot?—this looks as if [surface] bad conductor[.]—"[33] He also consulted literature on experiments that might illumine the question. He cited one report of oil freezing under certain conditions and asked, "[W]ill it bear on central fluidity?"[34]

It is significant that Darwin's views on the figure of the earth were gained and articulated in the late 1830s. The image of the earth as a having a thin crust surrounding a hot and fluid interior was then dominant. The image was the starting point for the contraction theory proposed by Élie de Beaumont, which Darwin had cited without attribution (and only to dismiss) in his 1838 article.[35] The image was also held by the Huttonians. On its own it could be combined with the notion of a cooling earth, or not. The great strength of the image was that it accommodated geological observations of a mobile crust, as suggested by the occurrence of earthquakes, and of the accessibility of fluid matter to the exterior of the earth, as in volcanic eruptions. Its weakness was that so little was known about the interior of the earth.

In the 1830s Darwin attended to the views of two of his contemporaries on the subject of the figure of the earth. They were John Herschel and William Hopkins. Darwin met Herschel in Africa at the Cape of Good Hope in June 1836. As already noted, Herschel was one of the foremost astronomers of the day, then mapping the stars of the Southern Hemisphere, and also someone, like Darwin, who was reading Lyell closely. In February 1836, inspired by the fourth edition of *Principles of Geology,* Herschel had written to Lyell with "a bit of theory" regarding changes in the earth's surface. Darwin may have known of Herschel's ideas from personal communication at the Cape, from the later printed communication, or, most likely, from both.

The quoted portions below are taken from the second edition of Charles Babbage's *Ninth Bridgewater Treatise,* a work Darwin owned and read. Herschel wrote to Lyell,

> Has it ever occurred to you to speculate on the probable effect of the transfer of pressure from one part to another of the earth's surface by the degradation of existing and the formation of new continents—on the fluid or semi-fluid matter beneath the outer crust? . . .
>
> Granting an equilibrium of temperature and pressure within the globe, the isothermal strata near the centre will be spherical, but where they approach the surface will, by degrees, conform themselves to the configuration of the *solid* portion; that is, to the bottom of the sea and the surface of continents. . . .

> According to the general tenor of your book, we may conclude, that the greatest transfer of material to the bottom of the ocean, is produced at the coast line by the action of the sea; and that the quantity carried down by rivers from the surface of continents, is comparatively trifling. While, therefore, the greatest local accumulation of pressure is in the central area of deep seas, the *greatest local relief* takes place along the abraded coast lines. Here . . . should occur the chief volcanic vents.[36]

Herschel's depiction of the transfer of matter from the continents to the ocean floor was consistent with Lyellian cyclism. However, there was an originality to it that went beyond anything in Lyell's text. It was therefore not surprising that Darwin turned to it.

Darwin's discussion of Herschel's idea in Notebook A is extended and skeptical:

> Arguments against Herschel's view of cause of continental elevations (1) the alternation of linear bands of movement in Indian & Pacific Oceans.—(2^{d}—) does not explain first formation of continents, if globe be considered as condensed vapour.—inequalities are required to start with (& does not Hersche[l] theory imply tendency to equilibrium.) 3^{d}—. there are mountains in the moon, which though not very analogous (see Edinburgh. Phil. Journal <][CD]>, no great chains like Andes or Himalayas, but great circular mountains, yet so analogous, that as we see mountains formed (& mountains are effect of continental elevations) we may conclude that elevation is independent of spreading out matter by action of sea.—as no sea exists there.—[37]

Darwin's first objection pertained to his own understanding of patterns of elevation and subsidence gained from study of the distribution of coral reefs. His second objection turned the nebular hypothesis against John Herschel's ideas, and raised the question of how the inequalities on the surface of the earth originated if the earth was originally "condensed vapour." The third objection compared the topography of the moon to that of the earth, suggesting that "as no sea exists" on the moon, Herschel's mechanism for explaining elevation by the "spreading out" of matter on the sea floor must be wrong. Later in the notebook, Darwin also recorded that "Erasmus [Charles's brother] suggested to me that Herschel's theory offers no explanation of intermittent action of elevatory force[.]—"[38] This latter point was integral to Darwin's understanding of the elevation of continents. Overall, as regards Herschel's "bit of theory" for transferring matter from continents to the sea floor and then back again, in the form of new continents, Darwin was interested but did not become an advocate. What is perhaps more important here is the demonstration of the power of Lyell's program to win adherents who pressed the questions he had raised.

The final person to be discussed in this section is Hopkins, a prominent member of what Susan Cannon has termed the "Cambridge network" in early Victorian science. Hopkins's career developed differently from that of the typical Cambridge undergraduate, for he did not enter the university until 1822, when he was thirty years old. Ultimately his greatest contribution to science was as a private tutor of mathematics to, among others, William Thomson (as of 1892 Lord Kelvin) (1824–1907). Sedgwick introduced Hopkins to geology in about 1833. Like Darwin, Hopkins accompanied Sedgwick to the field in Wales.[39] Inscriptions on Darwin's copies of two of Hopkins's early papers suggest Cambridge ties and personal acquaintance: Darwin's copy of Hopkins's "Researches in Physical Geology" (1835) is inscribed "Professor Henslow from the Author," while his copy of the follow-up *Abstract of a Memoir on Physical Geology* (1836) is inscribed "C. Darwin—from the Author."[40]

Darwin turned to Hopkins for advice on geological matters where geometry and physical astronomy played a role. Thus, for example, Darwin wrote in Notebook A, "What a curious investigation it would be to compare, the time of the earthquake of Chile, with that of the passage of the moon.—" He made a later annotation: "Ask Hopkins.—"[41] We also know that he requested information from Hopkins on a complicated pattern of strike and dip of the foliated mica-schist he had observed at Cape Tres Montes, Chile.[42] However, the pattern and content of Darwin's publications on geology were affected by Hopkins's work in other interesting ways.

In 1839 Hopkins initiated a series of three papers on the subject of the figure of the earth.[43] This happened to be just after Darwin had argued for a thin-crust model of the earth in his 1838 and 1839 papers. Like Darwin's Glen Roy paper, Hopkins's papers were published in the *Philosophical Transactions of the Royal Society of London*, where they would be assured of widespread and respectful attention. Hopkins's papers articulated a mathematical treatment of the figure of the earth that would end the dominance of the thin-crust model in English geology. Under the general heading "Researches in Physical Geology," Hopkins assayed three topics: "On the Phenomena of Precession and Nutation, Assuming the Fluidity of the Interior of the Earth" (1839), "On Precession and Nutation, Assuming the Interior of the Earth to Be Fluid and Heterogeneous" (read 1839, published 1840), and "On the Thickness and Constitution of the Earth's Crust" (1842). His mathematical argument, replete with differential equations, assumed as a hypothesis the original fluidity of the earth as a body and its subsequent cooling and consolidation. It was thus consistent with some version of the nebular hypothesis, though Hopkins did not use that phrase.

Hopkins's chief interest was in examining the pressures exerted on the shell of the earth and its fluid interior mass by the sun and the moon and by the centrifugal force resulting from the earth's motion. He concluded that for the earth to maintain its shape, its solid crust must be of a certain minimum thickness:

> Upon the whole . . . we may venture to assert that the minimum thickness of the crust of the globe which can be deemed consistent with the observed amount of precession, cannot be less than one-fourth or one-fifth of the earth's radius.[44]

Since the earth has a radius on the order of 4000 miles, "one-fourth or one-fifth" of that amount would be 1000 or 800 miles. Either figure (1000 or 800 miles) obviously contrasts sharply with the estimate that Darwin employed of a crust that was on the order of 20 miles in thickness. Further, according to Hopkins, if the earth has a thick crust, geologists must think of the lava feeding volcanoes as "forming subterranean *lakes,* and not a subterranean *ocean.*"[45]

Obviously such conclusions were incompatible with Darwin's views as enunciated in his 1838 "Connexion" and 1839 "Glen Roy" articles. What is striking is that after 1839 one does not find in Darwin's geological writing a defense of the thin-crust model of the figure of the earth, nor indeed a discussion of the question of the figure of the earth. While there is no documentary evidence I know of to connect this to Hopkins's articles from *Philosophical Transactions,* the timing is significant. Of course, other factors may be relevant in Darwin's retreat from the issue, as, for example, the fate of his Glen Roy theory—to be discussed in chapter 8—but clearly Hopkins's approach to geology had changed the nature of the discussion. For his part, Hopkins did not deign to challenge Darwin in print.

Hopkins's impact on British geologists can be gleaned from his 1847 report to the British Association for the Advancement of Science.[46] Here Hopkins quoted from his earlier *Philosophical Transactions* articles, which dealt with abstract mathematical questions, and also treated the more geologically commonplace subjects of the operation of volcanoes and earthquakes. The 1847 article included positive treatment of Herschel's speculations on the effects of the weight of accumulating deposits in effecting crustal motion, and respect for Herschel's and Babbage's views on the accumulation of such deposits effecting isothermal lines beneath the earth's crust. Hopkins cast doubt on theories of a cooling contracting earth, such as that held by Élie de Beaumont.[47] The article ended with a section entitled "On the Observations required for the determination of the Centre of the Earthquake Vibrations, and on the Requisites of the Instruments to be employed." Hopkins's comment reflected a new era, then just beginning, in the systematic collection of earthquake data. This era was marked by the use of instruments in measuring the vibrations caused by earthquakes. The new age yielded a subfield in geology, "seismology," so named by its founder Robert Mallet (1810–1881).[48]

The late 1840s mark a convenient divide in examining Darwin's work in regard to a number of geological questions, including that of the figure of the earth. This was largely a matter of circumstance, for by the early part of the decade Darwin had transferred his primary interest to the subject of transmutation, and in 1846 had completed the last volume in his trilogy on the geology of the voyage of the *Beagle.* He was no longer a full-time geologist. Still, he remained in contact with a number of the persons and issues encountered in the 1830s and early 1840s. For example, on earthquakes Darwin and Mallet became frequent correspondents, and both published in the 1849 Admiralty *Manual of Scientific Enquiry.* In his article "Earthquakes" for the volume, Mallet described instruments useful for giving the "direction of transit of the earth-wave and its dimensions."[49]

Before leaving the subject of Darwin and Hopkins, it is necessary to note, by way of completing the story, that Hopkins was among the most perceptive of Darwin's critics following publication of the *Origin of Species.* Hopkins's review of that work was entitled "Physical Theories of the Phenomena of Life," which suggests the stance Hopkins was to take. In Hopkins words, "The great defect of this theory is the want of all positive proof," and he went on to note the "remarkable ductility" of such theories.[50] Darwin complained to a fellow natural historian, Asa Gray (1810–1888), that Hopkins's review was written "with not more of the arrogance of a mathematician, than might have been expected."[51]

Ultimately, a positive effect of Hopkins's objections to the *Origin* was to provoke Darwin to be more articulate concerning his own philosophy of science. As Darwin wrote to Gray after the dust had settled over Hopkins's review,

> I believe that Hopkins is so much opposed because his course of study has never led him to reflect much on such subjects as Geograph. Distribution, Classification, Homologies &c&c; so that he does not feel it a relief to have some kind of explanation.—[52]

Thus, Darwin answered Hopkins's attack on the *Origin* not by claiming that he had provided mathematical demonstration for his theory of the kind Hopkins wanted, but by suggesting a difference in goals and approaches. Darwin now claimed that he wanted to include in his explanation matters that Hopkins had ignored. As time went on, Darwin did develop better accounts in his correspondence and in his published work of his explanatory strategy. During the 1830s he had confided his thoughts on such subjects to his private notebooks; after 1859 he returned to such subjects in a more public forum.

On a more personal level, Darwin had the satisfaction of seeing his son George Howard Darwin continue work on geological themes relating to elevation and subsidence but with the training in mathematics that his father lacked. George Darwin, as an applied mathematician, stood in the tradition of Hopkins, but he worked on questions of geology and cosmogony that reflected his father's interests. The elements of this combined heritage are prominently displayed in an 1877 article drawing on his father's observations on subsidence in the Pacific Ocean. (As his father had done years earlier, George Darwin estimated areas by fitting bits of paper to a globe.) The title of the article indicates its joint parentage: "On the Influence of Geological Changes on the Earth's Axis of Rotation."[53] Thomson served as a referee for the article prior to its publication. Thomson, himself a student of Hopkins, in turn became a mentor to George Darwin. While George Darwin followed Thomson in arguing for a solid earth, he also continued to ask questions similar to those suggested by his father in the 1830s and 1840s.[54]

David S. Kushner provided us with the most complete account of what he aptly terms the "ironic triangle" of Charles Darwin, William Thomson, and George Howard Darwin. Kushner described George Darwin's assent to Thomson's line of argument regarding a solid earth, but an assent accompanied by a resistance to dogmatism and an acknowledgment that "we know far too little as yet to be sure that we may not have overlooked some important points." George was his father's son, as well as Thomson's protégé, and his dual loyalties are present in his approach to questions bearing on geology and evolution.[55] One can also imagine that his father counseled him on the reversals in geological opinion that he had witnessed in his lifetime.

The question of the figure of the earth had changed from the late 1830s, when a thin crust was the dominant view, but it had not been resolved. Brush has suggested an array of six possibilities considered by late-nineteenth- and early-twentieth-century geologists: "CL" (thin solid crust, liquid interior); "SL" (thin solid shell, a "shell" being defined as thinner than a "crust"; liquid interior); "SLS" (thin solid shell, liquid interior, solid core); "S" (solid); "CLS" (solid but with a thin liquid layer under crust); and "SLG" (solid shell, liquid interior, gaseous center).[56] As Brush has emphasized, despite the diversity of opinion, the dominant view at the end of the nineteenth century was of a solid earth. Presently, a century later, the figure of the earth is pictured quite differently, as layered (an outer crust, a solid mantle, a liquid outer core, and a solid inner core, or "CSLS") rather than as solid throughout.[57] What is noteworthy concerning Charles Darwin's career in geology with respect to the then current understandings of the figure of the earth is that he did his major work in geology just as the image of a thin-crusted earth was being challenged. His own geology was shadowed by this ambiguity, but rather than being paralyzed by it, in a characteristic move he artfully sidestepped difficulties that neither he nor his contemporaries could solve with the methods and instruments at hand.

A Role for the British Association for the Advancement of Science?

Now let us return to the 1830s and to the core question of elevation and subsidence. Since this decade also saw the establishment of the British Association for the Advancement of Science, one might well ask if the question of elevation and subsidence figured in its agenda. The answer is that it did, though not to great long-term effect. At the first meeting of the association in 1831, the Geological and Geographical Committee proposed that Robert Stevenson (1772–1850), engineer to the Northern Lighthouse Board, Edinburgh, prepare a report on the "waste and extension of the land on the East coast of Britain, and the question of the permanence of the relative level of the sea and land."[58] A subcommittee was appointed in 1833, and included, among others, Greenough, Sedgwick, and Whewell as well as Stevenson. A report, written by Whewell, was issued in 1838.[59] The report opened with discussion of the question of "how far the position of the earth's surface is permanent—and what ought to be understood by 'level of the sea'." It largely discussed the technical aspects of how such measurements of sea level should be made, with Whewell concluding for a mean sea level, rather than following the practice of "surveyors and naval men" who assumed the "surface of *low water* of spring tides to represent this level." Under Whewell's direction, and with the financial support of the British Association, land markers were set out at several points running from the Bristol Channel to the English Channel. (The report, however, noted that the Netherlands had already set up a zero-point marker in Amsterdam to enable such studies.)

While Whewell was the dominant figure on the 1830s committee, the presence of Greenough was significant. In his 1834 presidential address to the Geological Society, Greenough noted that in recent years no subject at hand "excited so general and intense an interest as the Theory of Elevation."[60] He then proceeded to attack Lyell's understanding of elevation as expressed in the *Principles of Geology*. Greenough was a thorough empiricist, and he questioned whether Lyell had proved that elevatory movements had in fact occurred to the extent that Lyell proposed. Greenough also applauded the then current efforts at the association to study changes in land level in Britain, and expressed a hope that other countries would undertake similar investigations.

In effect, the British Association report on changes in land level in Britain was a trial run for a large-scale sponsored research program that was never realized. The program had begun with the implicit encouragement of Herschel, who had outlined the virtues of such a project in his *Preliminary Discourse*. There he had imagined the "industry of observers scattered all over the world" who would "ascertain, by very precise and careful observations at proper stations on coasts," the actual direction in which changes of relative level are taking place between the existing continents and seas.[61] But no global series of observing stations was established, and the project languished. From the historical perspective, however, it is noteworthy that the subject of elevation did receive attention from the scientific establishment during the mid-nineteenth century.

Elevation: Coastlines, Mountains, and Continents

As the British Association program and Greenough's comments suggested, establishing benchmarks for measuring levels was best done at sea level. Darwin's first contribution to the debate over elevation within the Geological Society occurred as a result of the Lyell-Greenough clash. In volume 1 of the *Principles of Geology,* Lyell claimed that the coast of Chile had undergone recent elevation.[62] On the testimony of Maria Graham the earthquake was said to have raised the Chilean coastline three to four feet for a distance of one hundred miles.[63] Lyell accepted this estimate as fact and incorporated it into his general argument for the present operation of geological forces affecting the earth's crust.

In his presidential address to the Geological Society of 1834, Greenough vigorously attacked what he believed were unsupported claims being made for the diverse action of elevation in explaining a host of phenomena. Particularly at issue was the credibility of widespread elevation along the coast of Chile, said to have been caused by the earthquake of 1822.

As it happened, a second major earthquake occurred in Chile on 20 February 1835. This earthquake had authoritative witnesses, including FitzRoy and Darwin. Greenough had been openly suspicious of the credentials of Graham, a well-traveled Englishwoman. Possibly it was partly her gender that rendered her an unreliable witness in Greenough's eyes. In any case, Lyell was pleased to use FitzRoy's testimony in his presidential address of February

1836, and he was forceful in pointing out its application to his own general argument for the present power of geological forces. In a second presidential address to the society a year later, Lyell returned to the point, this time with the additional testimony of the *Beagle*'s naturalist. Lyell again rehearsed the dispute and its happy ending:

> In Mr. Darwin's paper you will find many other facts elucidating the rise of land at Valparaiso, and he has also treated of the general question of the elevation of the whole coast of the Pacific from Peru to Terra [*sic*] del Fuego. Beds of shells were traced by him at various heights above the sea, some a few yards, others 500 or even 1300 feet high, the shells being in a more advanced state of decomposition in proportion to their elevation.[64]

In his address Lyell went on to describe Darwin's extension of three of Lyell's own ideas. First, while Lyell (following Graham) had already associated earthquakes with the raising of the Chilean coastline, Darwin went further to suggest that elevation was occurring regardless of the occurrence of earthquakes. Second, Darwin expanded the geographical area under consideration. Lyell discussed elevation of the coastline of Chile from Lima southward. Darwin included the entire southern half of the South American continent. (Among Darwin's observations to which Lyell referred were the parallel terraces at Coquimbo in Chile.) The third issue pertained to the rate of change. In his February 1837 presidential address Lyell argued that modern upheavals of land proceeded by "insensible degrees" as well as by sudden starts. Darwin's paper to the Geological Society of 4 January 1837 had supported this contention. Robert Edward Alison (fl. 1830s), who had assisted Darwin in studying the region around Valparaiso, said in an 1835 letter to Darwin that "if a rising has taken place, it has been per gradus & not per saltum."[65]

Darwin's focus on such elevation as viewed on the South American coastlines remained continuous from the *Beagle* period through publication in 1846 of his *Geological Observations on South America*. He opened that work with two paired chapters. Chapter 1 was entitled "On the Elevation of the Eastern Coast of South America"; chapter 2, "On the Elevation of the Western Coast of South America." As he said on the first page of chapter 1, he was interested in the "elevation of the land within the recent period, and the modification of its surface through the action of the sea." By emphasizing the recent period, Darwin was demonstrating a Lyellian interest in existing causes.

One particularly important instance that Darwin treated in his 4 January 1837 presentation to the Geological Society and also in chapter 2 of *Geolog-*

ical Observations on South America (1846) pertained to Coquimbo in Chile. In Darwin's words, at Coquimbo there was "a narrow finger-like plain gently inclined towards the sea . . . extend[ing] for eleven miles along the coast, with arms stretching up between the coast-mountains, and likewise up the valley of Coquimbo."[66] At Coquimbo, along the sides of the plain, the naval captain and author Basil Hall had described five terraces, which he believed to have been the beaches of a former lake, formed "no matter how."[67] Rather than having been formed by a draining lake—an explanation, as Lyell remarked, that had been adopted for the terraces of Glen Roy—Lyell suggested that the terraces had been formed during the "successive rise of land" from the sea.[68] By the time he published his *Geological Observations* in 1846, Darwin was ready to stress the differences between the terraces of Coquimbo and those of Glen Roy. (The former were less strictly horizontal than the latter.) However, the overarching explanation for both remained the gradual, though intermittent, elevation of the sea from the land.

Darwin sometimes viewed scenery through the eyes of previous beholders. His imagination followed theirs. Such was the case at Coquimbo in Chile, which he visited on 19 May 1835, and of Glen Roy in Scotland, which he visited from 28 June to 5 July 1838. Aboard ship Darwin had Hall's account of Coquimbo, which drew on the lacustrine explanation for the parallel roads of Glen Roy. Hall had written of his ride up the valley of Coquimbo, where "the most remarkable thing we saw was several series of horizontal beds, along both sides of the valley, resembling the Parallel Roads of Glen Roy" examined by Thomas Dick Lauder (1784–1848).[69] Darwin also had on board Lyell's *Principles of Geology,* which cited Hall but suggested a marine explanation for Coquimbo. In journeying to Glen Roy, Darwin was in a sense completing the Lyellian reinterpretation of a landscape. A presentation of the two figures published by Darwin indicates the similar morphologies of the two sites (Figures 7.1 and 7.2).

Thus far we have discussed the elevatory action as it could be observed at or near the sea. We must now turn to discuss Darwin's treatment of the Andes. He made one excursion (from 14 to 27 August 1834) to the base of the Andes, and another (from 18 March to 10 April 1835) to the Andes proper. During the latter trip he and his companion, Mariano Gonzales (dates unknown), began in Santiago (Chile) and crossed the Cordillera via the Portillo Pass. They ended at Mendoza (Argentina) and returned to Santiago by way of the Uspallata Pass. (One author, well known to Darwin, had described the Portillo Pass as "the most direct from Mendoza to St. Jago [Santiago]" and as at the point "where the Cordillera divides itself into two chains." The same author had taken note of the "grand pass of Uspallata, in front of Mendoza" and had reckoned the Portillo Pass as being "thirty leagues south of Mendoza."[70])

Figure 7.1. Glen Roy. Sketch by Albert Way (1805–1874). By permission of the Syndics of Cambridge University Library.

Several factors affected Darwin's work in the Andes. Let us begin by listing the handicaps under which he worked. First, the interior of the continent was much less well known than the coastline. At Coquimbo Darwin was reviewing a site that Hall had described earlier. In the Andes Darwin was working with largely new material in the sense that no one had provided detailed geological section drawings or views of the sites he visited. Indeed, Dr. Victor Ramous at a recent geological congress suggested, "The geological understanding of the Central Andes started with the pioneering exploration of Charles Darwin in 1835."[71] (In the 1830s Darwin's peers in England were aware of the novelty of his effort. Without his knowledge, and before his return, extracts from his letters to Henslow, including long quotations regarding the Andes, were read before scientific gatherings at home.[72]) Second, it is inherently more difficult to measure the present action of elevatory forces within the interior of a country than on the shore, where the level of the sea provides a standard. Third, Darwin did not come to the experience with a theory of mountain building readily at hand. Lyell had not presented him with one, nor had he adopted one from other authors. In 1834 he noted to Henslow, half-jokingly, that "I have not one clear idea about cleavage, strati-

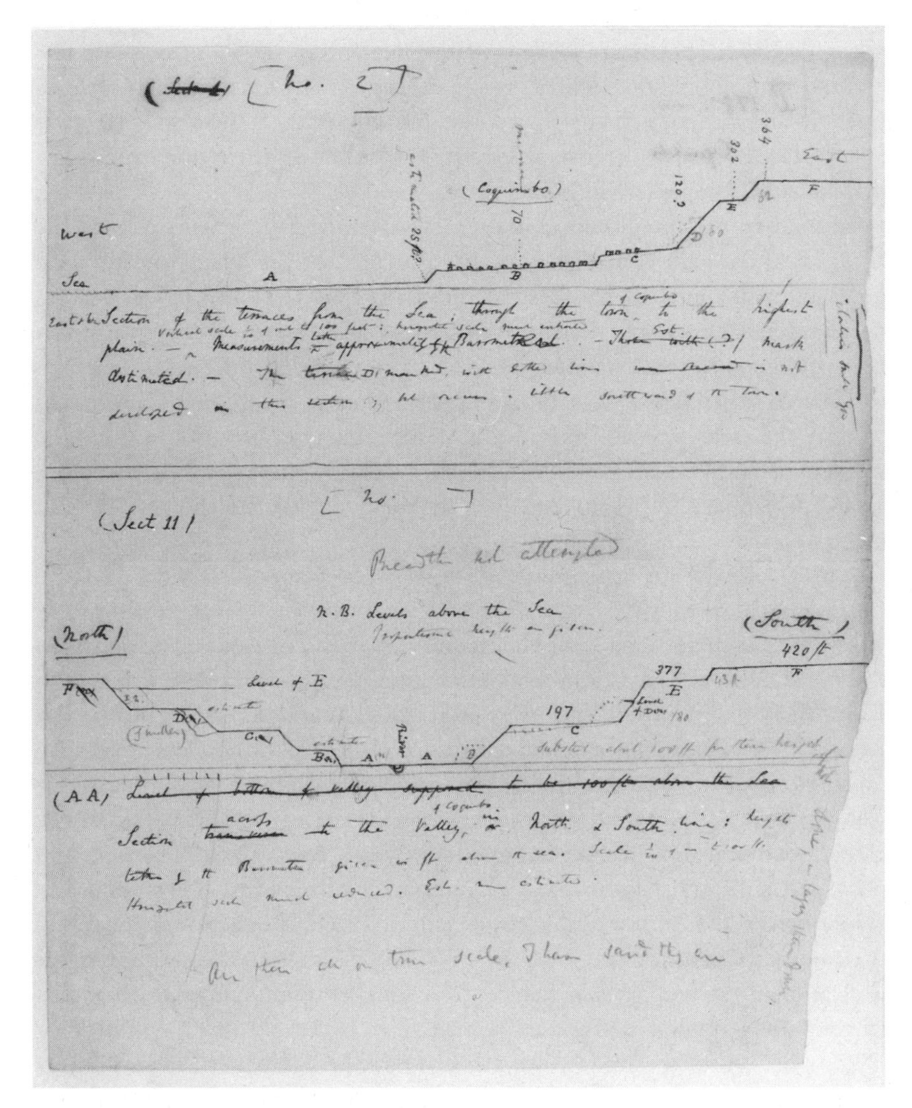

Figure 7.2. Coquimbo. From DAR 44.30. By permission of the Syndics of Cambridge University Library.

fication, lines of upheaval.—I shall persuade myself there are no such things as mountains."[73] Upon returning from crossing the Cordillera at two passes, he again wrote to Henslow that "no previously formed conjecture warped my judgement."[74] (This third "disadvantage" could, of course, be turned to advantage: Darwin could construct his own understanding.)

Let us now list Darwin's advantages. First, in the Andes Darwin was working with good exposures. Unlike in Patagonia and in the Pampas, there was little soil and vegetation. In his notes, dozens of small sketches of bedding sequences and structural features as strike and dip testify to how much he was able to see. Eventually he assembled his sketches to produce the magnificent plates that accompanied his book (Plates 7 and 8).

Second, by the circumstances of the *Beagle*'s itinerary, Darwin explored the Andes in the fourth year of the voyage. Bearing out the adage that the best geologist is the one who has seen the most rocks, by 1834–1835 Darwin was an experienced geologist, equal to the task of assessing the geological structure of even such grand landscapes as that presented by the Andes. Further, in what were strokes of luck for his appreciation of the geological forces at work in the area, his excursions in the Andes came after he had witnessed the volcano named Osorno (lat 41°7′ S, long 72°30′ W) in eruption (November 1834 and January 1835) and had been shaken by the Concepción earthquake of 20 February 1835.

Third, Darwin was not entirely without predecessors. While the geological literature on the area was sparse, he had the use of several then-recent works by British authors. Basil Hall has already been mentioned in regard to the "parallel roads" at Coquimbo. Two other authors were also important: John Miers (1789–1879) and Alexander Caldcleugh (d. 1858). Both men were part of what might be called the informal empire, or network, of British interests. (Darwin stayed with Caldcleugh in Santiago; he wrote that Caldcleugh "took an infinite degree of trouble for me.—"[75]) Miers and Caldcleugh had traveled in the Andes a few years previous to Darwin, and possessed substantial knowledge of mining, a subject integral to geology. While neither Miers nor Caldcleugh included detailed section drawings—their interests were the practical commercial ones associated with mining—it is clear from Darwin's notes that he drew on their commentary with regard to geography, mining, and some geology.

On portions of the Andes, Darwin also had the use of such older authors as Juan Ignacio Molina (1740–1829), whose work Darwin bought in Valparaiso in 1834; the contemporary French author Claude Gay (1800–1873); and, of course, Humboldt. In addition, on general geological topics related to volcanoes, Darwin drew particularly on Scrope and Daubeny, as well as Lyell. He also took with him, or bought while in South America, a number of maps such as that entitled "Plan of the route from St Jago de Chile to Mendoza describing the Pass over the Andes" published by W. Eaden, Charing Cross (London), and dated August 12, 1821.[76]

Darwin interpreted the Cordillera of the Andes at roughly three intervals: in 1835 during his excursions there; in 1838 when he published his highly theoretical article entitled "On the Connexion of Certain Volcanic Phenomena in South America; and on the Formation of Mountain Chains and Volcanos, as the Effect of the Same Power by which Continents Are Elevated"; and in 1846 when he published his book *Geological Observations on South America*. The relative proportion of theory to fact shifted over the course of these three intervals. During the voyage, theorizing was present but not dominant. Rather, fact gathering was of the most intense kind since Darwin was traveling over difficult terrain, collecting specimens, and sketching features of the geological landscape as he went. In 1838, theory was in the forefront. By 1846 Darwin consciously sought to reverse that trend. While he was planning the book, he wrote to Lyell that he was "determined to have very little theory & only short descriptions."[77]

During the voyage Darwin held what is probably best regarded as fundamentally an image, rather than a theory, of the elevation of the Andes. In his Diary he wrote, "At a remote Geological æra, I can show that this grand chain consisted of Volcanic Islands, covered with luxuriant forests."[78] Or, put slightly differently, the Cordillera "formerly consisted of a chain of Volcanoes from which enormous streams of Lava were poured forth at the bottom of the sea.—"[79] This was a theme that he had also explored, with variations, in his essay "Reflection on reading my Geological notes" written in 1834. There he mused about the rise of land from the sea ("Perhaps the first opening of the N. & S. crack in the crust of the globe. forming the Cordilleras") and suggested that during a "swelling of the Globe," the Andes as a "longitudinal crack" would probably occur during an elevation of such a longitudinal enlargement. He then queried himself as to how much the Andes owed its height to volcanic matter "pouring out" and how much to horizontal strata tilted up or how much to horizontal elevation of the surface of continents.[80]

As Darwin got into the field himself, he drew a more complicated picture of the formation of the Andes. Of course, from maps, such as that provided by Miers, he was aware that the Cordillera, in the area surrounding the Uspallata and Portillo Passes, is not a single chain but a double chain separated by a high plain.[81] Darwin attended to the already known division. But he also identified other "crevices" running from north to south between the Andes and the Pacific Ocean, and provided a small sketch to that effect (Figure 7.3). Significantly this "theoretical drawing" captures Darwin's conclusion that there had been numerous lines of elevation in the region. (See, for example, the seven axes of elevation identified in the "Section Up the Valley of Copiapo to the Base of the Main Cordillera" provided in his *Geological Observation*.[82]) Using the evidence of structure and, where possible, rock type, Darwin also attempted to date the upheaval of the various chains or ridges relative to each other. He was so convinced by his field evidence of the variety of dates

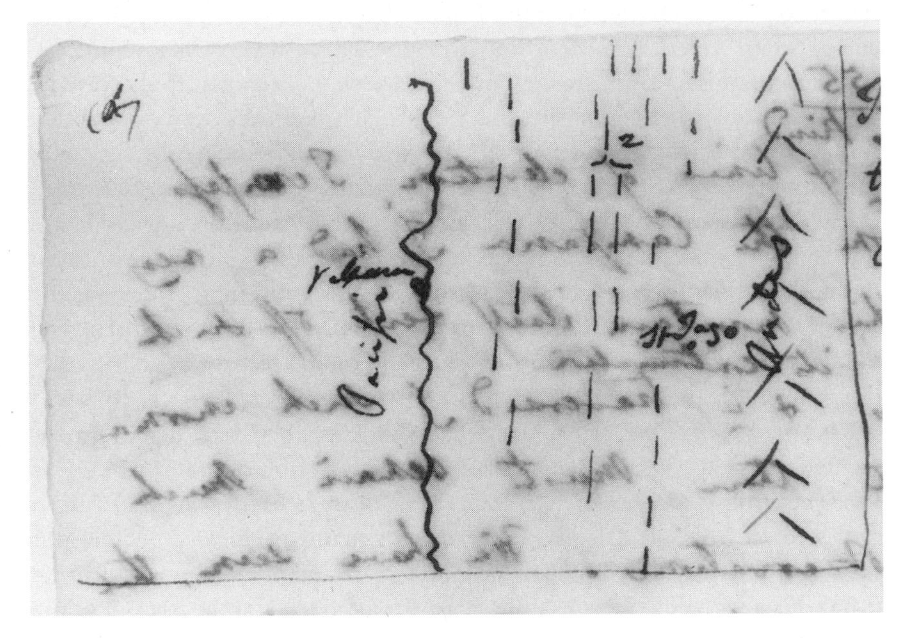

Figure 7.3. Theoretical drawing of crevices of upheaval in the country between the Andes and the Pacific. From DAR 36.1:438^{v}. By permission of the Syndics of Cambridge University Library.

of origin that he seems never to have seriously considered Élie de Beaumont's notion of the simultaneity of origin of parallel lines of mountains. When Darwin finally referred to Élie de Beaumont, in a remark written sometime after his initial commentary on South America, the reference was dismissive:

> The elevation, & formation of igneous & sedimentary causes, of the Portillo [eastern] range subsequently to the existence of the Peuquenes [western range] as dry land—. . . . This is contrary to M. Elie de Beaumonts [*sic*] theory: & what is more important shows that the Cordilleras, one of the best defined ranges on the surface of the globe, has not risen by one blow. but by accession after *considerable* intervals.—It may also be remembered that in the Portillo there are two distinct granites.—[83]

For Lyell, on the psychological level, Élie de Beaumont was a major adversary; for Darwin he was not.[84]

The phrase "risen by one blow" suggests the traditional argument of change *per gradus* or *per saltum*, gradually or suddenly—literally "by degree" or "by jump." Darwin remained a gradualist in his discussions on South America, and often noted evidence in its favor. For example, when writing about evidence for elevation at Valparaiso, he remarked that "shells from the higher parts are in a much more advanced stage of decomposition, than those lower down: which very certainly shows that the rise could not have been per saltum.—"[85] Yet the contrast is not always clear between gradualist and saltationist in Darwin's *Beagle*-period writing. (Darwin did not use the terms "uniformitarian" and "catastrophist" during the voyage.) Some of the analyses he made would appear to the present-day reader as catastrophist in their foreshortening of geological time. For example, Darwin speculated that the Andes had risen since humans inhabited the area.[86] It also must be kept in mind that Darwin's understanding of time-scale was subject to quick amendment. In April 1835 he wrote Henslow dating the formation of the eastern chain of the Andes to Tertiary times.[87] Four months later, in August, following an excursion from Valparaiso to Coquimbo and Copiapó, he wrote to Henslow that some of views had altered and that "I believe the upper mass of strata are not so very modern as I supposed.—"[88] In any case, during the voyage, dating was provisional; the subject was to be readdressed prior to publication, following the expert examination of Darwin's collection of fossil shells.

Let us now turn to the second occasion when Darwin expressed his ideas on the formation of the Andes. This was his "Connexion" article of 1838. While it had less effect on the subsequent course of geology as a discipline than did other portions of his geological work, it is the single most important article for understanding his intellectual position. The article began with observations on the earthquake that struck Concepción on 20 February 1835. It documented the relation of that earthquake to a measurable rise of land along the west coast of Chile. It then moved to documenting the simultaneous or temporally proximate eruptions of volcanoes across large areas. Next, Darwin connected this phenomenon to a presumed cause—the movement of molten material beneath the earth's crust, which he estimated as on the order of twenty miles in thickness. From these Darwin concluded,

> In a geological point of view, it is of the highest importance thus to find three great phenomena,—a submarine outburst, a period of renewed activity through many habitual vents, and a permanent elevation of the land,—forming parts of one action, and being the effects of one great cause, modified only by local circumstances.

He went on to suggest that "no theory of the cause of volcanos which is not applicable to continental elevations can be considered as well-grounded."[89]

Darwin then reproduced some tables from Humboldt, and one of his own, showing the historical periodicity of volcanic eruptions in the Americas from 1796 to 1835. On checking records Darwin noted that he had been particularly struck by the near-simultaneous eruption on 20 January 1835 of three volcanoes—Osorno, Aconcagua, and Cosequina—at distant points along the Cordillera. (Darwin listed Aconcagua as 480 miles north of Osorno, Cosequina as about 2700 miles north of Aconcagua.) He also supplied a map of the region around Concepción, illustrating the proximity of a chain of volcanoes to the east (Figure 7.4). He concluded that the "subterranean forces manifest their action beneath a large portion of the South American continent, in the same intermittant [*sic*] manner as . . . they do beneath isolated volcanos."[90]

Darwin then turned to consider earthquakes. He believed that they "are caused by the interjection of liquefied rocks between masses of strata," that they "generally affect elongated areas," and that when they are accompanied by an elevation of the land in mass "there is some additional cause of disturbance." He suggested that earthquakes and volcanic eruptions were alike in relieving subterranean force. He also asserted that the interjection of "a wedge-formed linear mass of rock" that formed the axis of most great mountain chains worked similarly to relieve subterranean pressure. The process of interjection of material "must have formed one step in a line of elevation." Thus, Darwin linked the occurrence of at least some earthquakes, the eruption of volcanoes, and the elevation of mountain chains. (He left room for earthquakes that occurred when the "fluid matter, on which I suppose the crust to rest, should gradually sink instead of rising.")

Darwin's last move in the article was to associate the rise of mountain chains with continental elevation. Darwin entitled this section of his paper "Theoretical Considerations on the Elevation of Mountain Chains." Ironically (given that Hopkins was soon to oppose a thin-crust model for the earth's structure), Darwin used Hopkins's work of 1835–1836 to support his argument. In his 1835 paper for the Cambridge Philosophical Society, Hopkins had demonstrated, as Darwin put it, "that the first effect of equably elevating a *longitudinal* portion of the crust of the earth, is to form fissures parallel to the longer axis." Further, once uplifted, strata would not usually settle back into their previous position but appear broken as in a mountain chain. Darwin reproduced figures from Hopkins's article to support his point (Figure 7.5).

Creatively, Darwin sought to apply Hopkins's conclusion to a vast scale, that of the Cordillera. He queried what it was that prevented the "very bowels of the earth from gushing out" upon such elevation. Darwin then traced the formation of mountains to a "succession of shocks similar to those of Con-

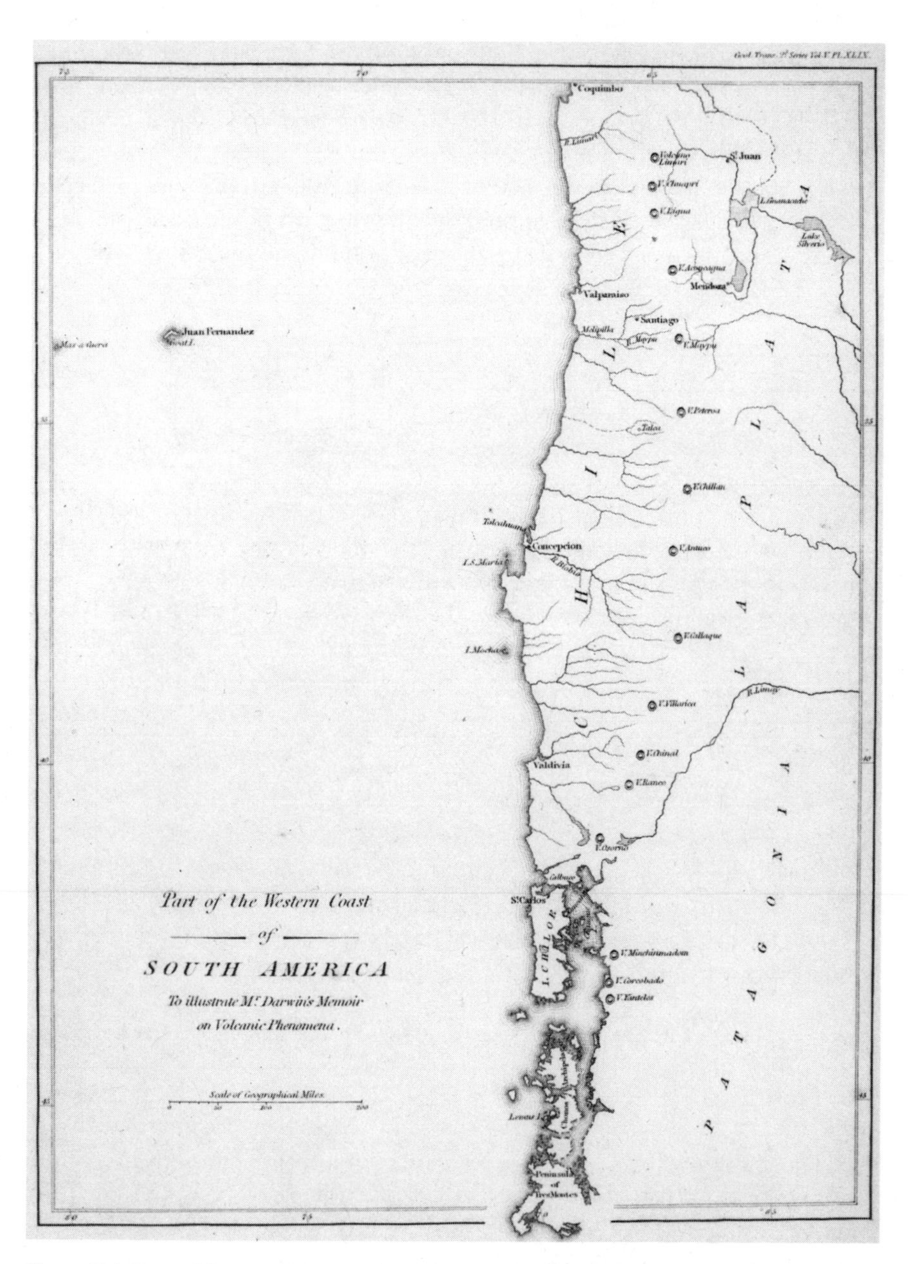

Figure 7.4. Part of the western coast of South America. By permission of the Syndics of Cambridge University Library.

Theoretical Considerations on the slow Elevation of Mountain-Chains.

The conclusion that mountain-chains are formed by a long succession of small movements, may, as it appears to me, be rendered also probable by simple theoretical reasoning. Mr. Hopkins has demonstrated, that the first effect of equably elevating a *longitudinal* portion of the crust of the earth, is to form fissures, parallel to the longer axis (with others transverse to them, which may here be neglected) of the kinds represented in the accompanying diagram (No. 1.), copied from that published in the Cambridge Philosophical

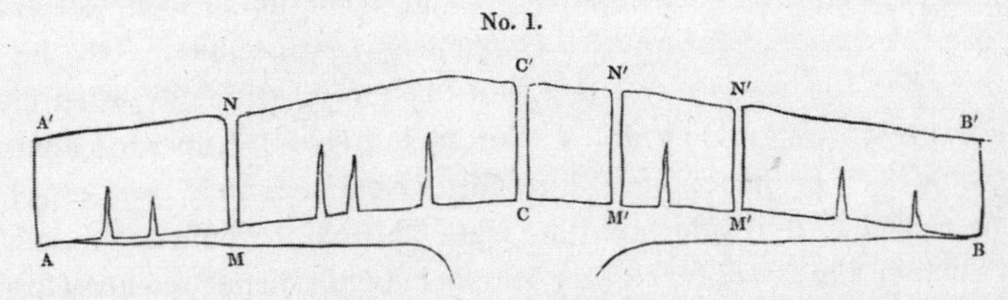

Transactions. But he further shows, that the square masses, now disjointed, will,—from the extreme improbability of the force uplifting them, when separate, equably, or from their settling down afterwards,—assume some such position as that given in Diagram No. 2. In the Cordillera, which may be taken

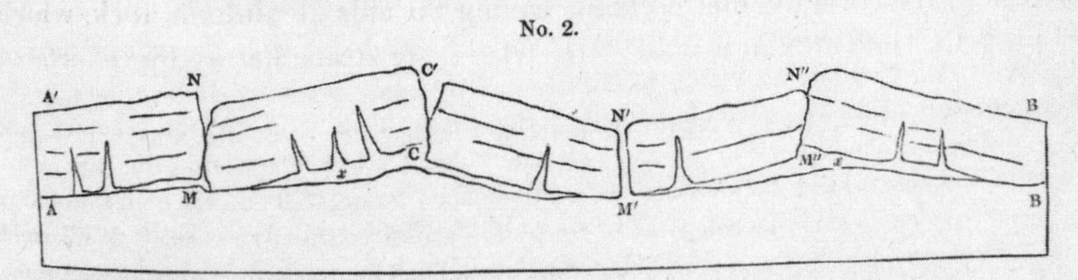

as a good example of the structure of a great mountain-chain, the strata in the central parts are inclined more commonly at an angle above 45°, than beneath it; and very often they are absolutely vertical. The axis of the lines of dislocation is formed of syenitic and porphyritic masses, which, from the

Figure 7.5. Hopkins's sketches. By permission of the Syndics of Cambridge University Library.

cepcion." He suggested that the "formation of a fissure through the whole thickness of the crust would be the effect of many efforts on the same line, and that during the intervals the rock first injected would become cooled." In Darwin's view, a mountain chain, with its axis of plutonic rock, is the effect of "an almost infinite series of small movements." His ringing conclusion was that "mountain-chains are the effects of continental elevations" and that their growth rate was nearly measured by the "growth of volcanos."[91]

In his conclusion to the article Darwin both rephrased his points and considered some then-current alternatives or objections. Using the map of South America, which illustrated the length of the "great chain" of the Andes, Darwin noted the "grandeur of one motive power, which, causing the elevation of the continent" produced "as secondary effects, mountain-chains and volcanos." ("One motive power" was a hypothetical entity referring to an uplifting power whose parameters Darwin was working to identify.) In the same concluding section, he argued for the extension of his conclusions to the "entire globe." And, finally, he also argued against the belief, then widespread, of the "secular shrinking of the earth's crust."

The belief was based on the premise that the inside of the earth contracts faster on cooling than does the solid crust, leaving space between them that would lead to the occasional collapse of the crust. Against this notion Darwin simply asked how the "slow *elevation,* not only of linear spaces, but of great continents" could be thereby explained. He also noted, without naming the author, Herschel's queries concerning the effect on the earth's fluid interior of deposited sedimentary beds. Similarly, he raised the question, originating from Herschel, as to the possible effect of the attraction of planetary bodies on a "sphere not solid throughout."[92]

In recent years historians of science, and especially Mott Greene, have provided the general background against which Darwin's work on crustal motion can be situated.[93] However, as Frank Rhodes has also shown, Darwin's 1838 paper had relatively little effect on the history of geology. It did appear in German translation in 1841 in the *Annalen der Physik und Chemie,* but otherwise the article might well have appeared to Darwin as stillborn. When presented to the Geological Society of London, it was received as part of the ongoing debate among geologists over the intensity of presently active geological forces, and their adequacy to explain geological phenomena. The comments of De la Beche and Phillips were to this end: De la Beche questioning whether the strata of the Alps, for example, would remain sufficiently flexible to be "tilted, convoluted, or overturned" by "gradual small shoves" and Phillips simply noting Darwin's allegiance to Lyell's views.[94] As Phillips and others of the society would have seen, Darwin's implicit critique of Élie de Beaumont's theory and his catastrophist view of the elevation of mountain chains was hardly novel; it derived from Lyell. (Rhodes noted that Darwin did not actually read Élie de Beaumont until December 1837.)[95] Darwin's reference to Herschel's views was too oblique to have merited attention.

What stands out in retrospect is Darwin's use of Hopkins. In him Darwin might have found an ally. However, Hopkins never supported Darwin's conclusions about the structure of the earth and, indeed, as we have shown, began in the 1830s to argue against the traditional model of a thin-crusted earth and for a model of a thick-crusted earth. Added to these difficulties was the fact that Darwin was working in South America, a remote locality for British geologists. In short, Darwin did not carry his contemporaries with him. Further, within the next few years, Hopkins was to declare his views on the earth's crust to considerable acclaim, Mallet was to begin work on the experimental study of earthquakes, and the Rogers brothers—William (1804–1882) and Henry (1808–1866)—were to offer a new interpretation of the origin of the Appalachian mountain chain that relied on lateral deformation as well as vertical uplift.[96] There was no dearth of other approaches.

In his date book Darwin noted for 27 July 1844 that he "Began S. America"; for 1 October 1846 he noted that he "Finished last proof of my Geolog. Obser. on S. America." He observed that the project engaged him for "18 & 1/2 months" and that was "now 10 years since my return to England."[97] It was also eight years since his theoretical article of 1838. Now, as he approached the book on South American geology, he was determined to subordinate the expression of theoretical views to the recording of empirical findings. In the book Darwin did not discuss either the interior structure of the earth or the question of the thickness of the earth's crust. Nor did he attempt to expand his insights on South America into a global theory. Similarly he did not address the then highly contested glacial hypothesis. The subjects of cleavage and foliation do figure in the book. Elevation is used routinely as a geological force; "axes" of elevation of various chains are identified. See, for example, Darwin's colored "Sketch—Section of the Valley of Copiapo to the Rise of the Main Cordillera" among the set of plates he included in the book.

Yet, on continental elevation Darwin was subdued while never abandoning the principal argument of his 1838 article. His arrangement of subject matter in the book suggested his intention: the notion of continental elevation was present but subordinate. Darwin opened the book with two key chapters devoted to proofs for elevation on the eastern and western coasts of South America, but it was not until the last pages of the last chapter that he turned to continental elevation. There he finally quoted his 1838 article, reiterating his argument:

> In South America, everything has taken place on a grand scale, and all geological phenomena are still in active operation. . . .
>
> The Cordillera from Tierra del Fugeo to Mexico, is penetrated by volcanic orifices, and those now in action are connected in great trains. The intimate relation between their recent eruptions, and the slow elevation of the continent in mass, appears to me highly important.

Darwin went on to suggest that the changes causing these effects cannot "have been too complicated."[98] The word "simple" went unused, but the meaning was there. Overall, Darwin made much of the repeated raisings and lowerings he believed the continent to have undergone. The final section in his book is devoted to this point.[99]

While Darwin was generally more circumspect with regard to grand theory in 1846 than he had been in 1838, he could be more precise on dating, and on good theoretical grounds. In the 1840s geological dating meant situating a formation at its proper position within the stratigraphical column (that is, to use vocabulary of the day, the "geological formations in the order of their superposition"). Such dating, or positioning, was done with the use of fossils. In the 1840s no numerical estimate of the length of geological time represented by the stratigraphical column was available. As a good sign of the international character of geology, it was the French naturalist d'Orbigny who worked on a portion of Darwin's collection from the Andes. He consented to study Darwin's fossils, though he disagreed with various remarks made in the *Journal of Researches*.

D'Orbigny identified the key fossils as from the Neocomian stage of the Cretaceous system.[100] Darwin also made use of work by von Buch, who had studied a South American specimen collected on another expedition. Von Buch had not used the term "Neocomian" in his description. Before going into print, Darwin also sought the help of a British zoologist and geologist, Edward Forbes (1815–1854), paleontologist to the Geological Survey. In correspondence with Darwin, Forbes supported d'Orbigny's general conclusion and interpreted it for Darwin:

> D'Orbigny & Von Buch no doubt mean the same thing. Exogyra or Gryphæa couloni is a characteristic Neocomian fossil. Von Buch's conclusion of the beds forming a passage from the oolites to the chalk is exactly what I believe to be true. Von Buch wrote before the "Neocomian" was investigated. The relative position would be as follows:

England	*Switzerld*	*S America*
Lower Green Sand		
	Neocomian	
Wealden		Your beds.
	oolites	
ool. beds		

> D'Orbigny probably looks on the Amer beds to be purely "Neocomian" since he holds the doctrine of definite divisions or formations everywhere.[101]

Eventually, Forbes was the person who described and provided figures of eleven Secondary fossils that appeared as an appendix to Darwin's *Geological Observations on South America*. With the backing of both Continental and

British geologists, Darwin then proceeded to use the label "cretaceo-oolitic" to describe his specimens. He suggested that "a great portion of the stratified deposits of the whole vast range of the South American Cordillera belongs to about the same geological epoch."[102]

Incidentally, in this assessment of the age of the deposits from the Cordillera, Darwin did not include the spectacular stand of fossil trees he found at Agua del Zorra (Agua de La Zorra). He believed them to be of the same age of the upper beds of the Patagonian Tertiary formations. As to the identity of the trees, Darwin's expert was Robert Brown (1773–1858), keeper of the botanical collections at the British Museum. Brown identified the fossils as coniferous, partaking "of the character of the Araucarian tribe (to which the common South Chilian pine belongs), but with some curious points of affinity with the yew."[103] Present-day geologists have revised the dating of the deposit to late Triassic. The fossilized trees have been removed from Agua de La Zorra to museums, but some fossil tree stumps are still visible, and there is a marker at the site commemorating Darwin's work.[104]

Also on dating we should mention that George Brettingham Sowerby described sixty species of Tertiary fossil shells from South America collected by Darwin. His son George Brettingham Sowerby Jr. (1812–1884) provided figures for the specimens. Both descriptions and figures appear as appendices to Darwin's book. The work of the Sowerbys, and of Forbes and d'Orbigny, was essential to Darwin since their expertise alone permitted the assignment of specimens to chronologically ordered formations. Experts, of course, do not always agree, and the surviving correspondence between Darwin and his experts suggests a process of judgments being made and then confirmed or altered, as fresh eyes looked at specimens or descriptions.[105] The social setting of their work is also of interest—Forbes representing the newly professionalized position of paleontologist to the Geological Survey, the Sowerbys the older tradition of conchology associated with independent collecting and publishing.

Subsidence: On the Evidence from Coral Reefs

Darwin's coral reef theory never stood alone. It connected to his understanding of volcanoes and of continents. As of 1838 it was Darwin's intention to publish a single volume to be entitled *Geological Observations on Volcanic Islands and Coral Formations.* Instead, this book was divided into two books—one on coral reefs, the other on volcanic islands. However, if one returns to Darwin's initial intention and reads the two volumes together, the connections stand out. The colors indicated by the legend on the map of the distribution of coral reefs reproduced in Plate 1 are telltale: blues for subsidence, reds for elevation. As Darwin noted, "The two distinct colours . . . mark two distinct types of structure."[106] More particularly on the reds, "Vermilion spots & streaks" stood for "active volcanoes"; "Red," for "Fringing Reefs."

Darwin's understanding of volcanoes in relation to reefs is suggested by his comment in *Volcanic Islands* (1844), where he wrote,

> A connection . . . between volcanic eruptions and contemporaneous elevations in mass has, I think, been shown to exist, in my work on *Coral Reefs,* both from the frequent presence of upraised organic remains, and from the structure of the accompanying reefs.[107]

Volcanic islands as areas of uplift were confirmed by his own experience, beginning at the Cape Verde Islands. It was also reinforced by the South American work.

"Fringing reefs" (or shore reefs) was Darwin's own category, described in chapter 3 of *Coral Reefs.* He seems to have first used the phrase in his 1837 presentation to the Geological Society. It described reefs skirting an island, or part of a continent, but without the interior deep-water channels that characterized barrier reefs. He believed this reflected a close relation in the horizontal extension of the reef with the probable slope beneath the sea of the adjoining land. The island of Mauritius, where Darwin worked from personal knowledge, was his type case. The *Beagle* had arrived at Mauritius on 29 April 1836, directly following its stop in the Keeling Islands on 1–12 April.

In associating volcanoes with elevation, Darwin also deployed an argument derived from months of study of charts, both during and after the voyage, that volcanoes were largely absent in the "great areas of subsidence tinted pale and dark blue," and, conversely, with the "coincidence of the principal volcanic chains with the parts coloured red, which indicates the presence of fringing reefs; and . . . the presence . . . of upraised organic remains of a modern age."[108]

The association of Darwin's notion of continental elevation with his coral reef theory was equally close. He believed areas of volcanic eruption were undergoing elevation. As he wrote in *Volcanic Islands,*

> This fact of the ocean-islands being so generally volcanic, is, also, interesting in relation to the nature of the mountain-chains on our continents, which are comparatively seldom volcanic; and yet we are led to suppose, that where our continents now stand, an ocean once extended. Do volcanic eruptions, we may ask, reach the surface more readily

> through fissures, formed during the first stages of the conversion of the bed of the ocean into a tract of land?[109]

Darwin's own answer to this question was, clearly, yes. Volcanic eruptions are the beginning of a process leading to the creation of land.

Volcanoes as precursors to continents or archipelagos represent the short story. But alongside that sketch was the more nuanced view that identified patterns of elevation and subsidence in the ocean basins of the world. Darwin's interest in identifying such bands was clear in his essay "Coral Islands" (1835). It was also the main point of his first public presentation on the subject of reefs, made to the Geological Society of London. The title of that paper was significant: "On Certain Areas of Elevation and Subsidence in the Pacific and Indian Oceans, as Deduced from the Study of Coral Formations." The rhetorical emphasis was on elevation and subsidence. The study of coral formations was valued preeminently as an avenue to the understanding of the history of the motion of the earth's crust. Barrier reefs and atolls were markers, or monuments, to the former presence of land. Fringing reefs indicated stability in land level, or a rise in land level.

By emphasizing elevation and subsidence in his approach to coral formations, Darwin was taking his study toward one goal and, in effect, moving away from another. An alternative emphasis, as William Montgomery has argued, was an interest in fossilized reef as a possible explanation for the source of calcareous formations.[110] In his earliest notes on coral reefs Darwin did pursue that question. In his Santiago Book he queried, "Does such appearance correspond to any of the great Calcareous formations of Europe[?]—" As time went on, however, relating fossilized reef to formations in the geological record did not place high on his agenda, though there is the occasional reference.[111] Rather, his own subsequent research was directed toward using coral reefs to understand crustal motion. Here he was following an interest also present in the Santiago Book, where he noted, "As in Pacific, a Corall bed. forming as land sunk." Or as he could also say in the Santiago Book, as a good Lyellian, "The Coral theory rests on the supposition of depressions being very slow & at small intervals" (Figure 7.6).

This brings us to the question of evidence. As Darwin liked to recall, and as his Santiago Book notes suggest, he had arrived at his theory deductively, before he had seen a coral reef. Only after 1835 did he develop the evidence that would corroborate his theory and refine it.

Darwin began his search for evidence at the Keeling Islands, which are composed of coral reef. Patrick Armstrong has argued (I think correctly) that Darwin may have influenced FitzRoy to make the call at these islands.[112] They were not an obligatory stop, nor, given the season at which the *Beagle* reached

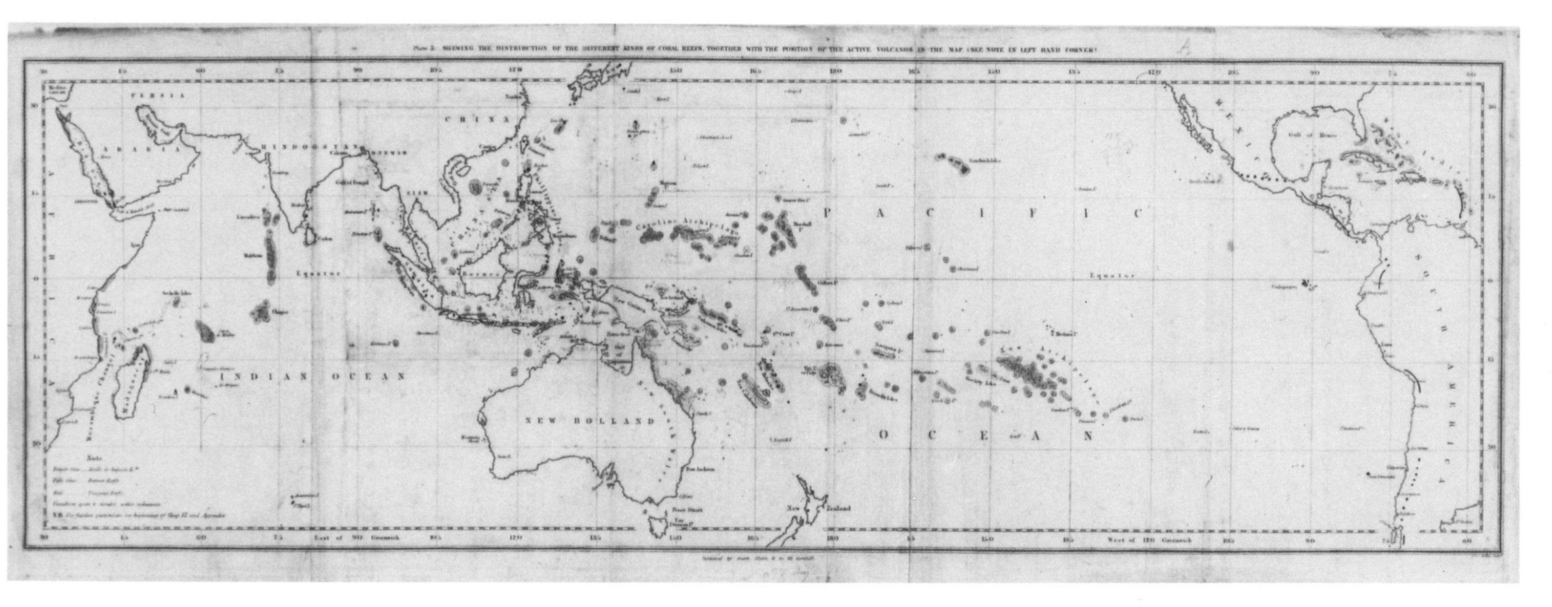

Plate 1. Darwin's map of the distribution of coral reefs. By permission of the Syndics of Cambridge University Library.

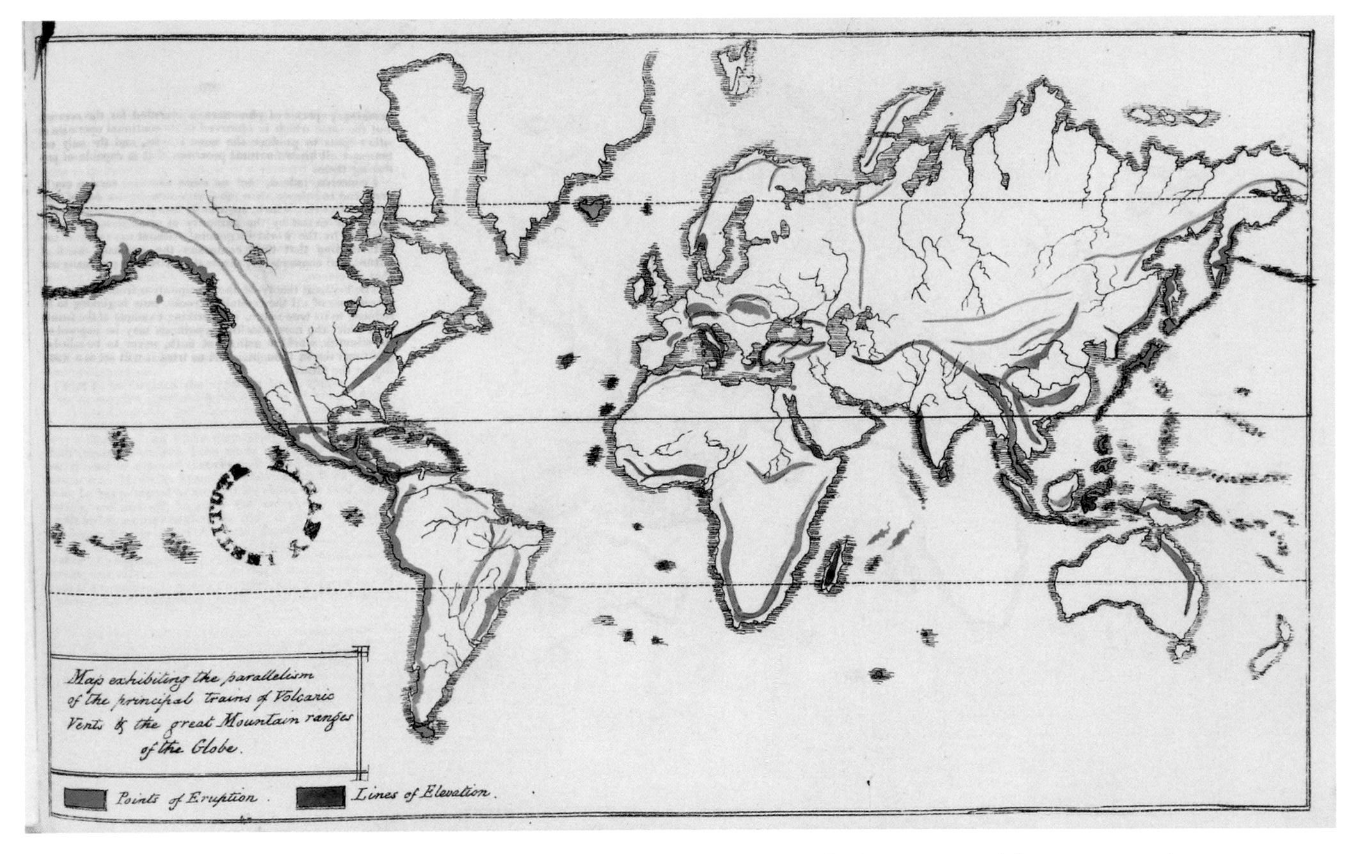

Plate 2. Scrope's map exhibiting the parallelism of the principal trains of volcanic vents and the great mountain ranges of the globe. Courtesy of the Library of Congress.

Plate 3. Various specimens in packets and sauce bottles. By permission of the Sedgwick Museum, University of Cambridge.

Plate 4. James Island specimens. By permission of the Sedgwick Museum, University of Cambridge.

Plate 5. Scenery in Tierra del Fuego. Photograph by the author.

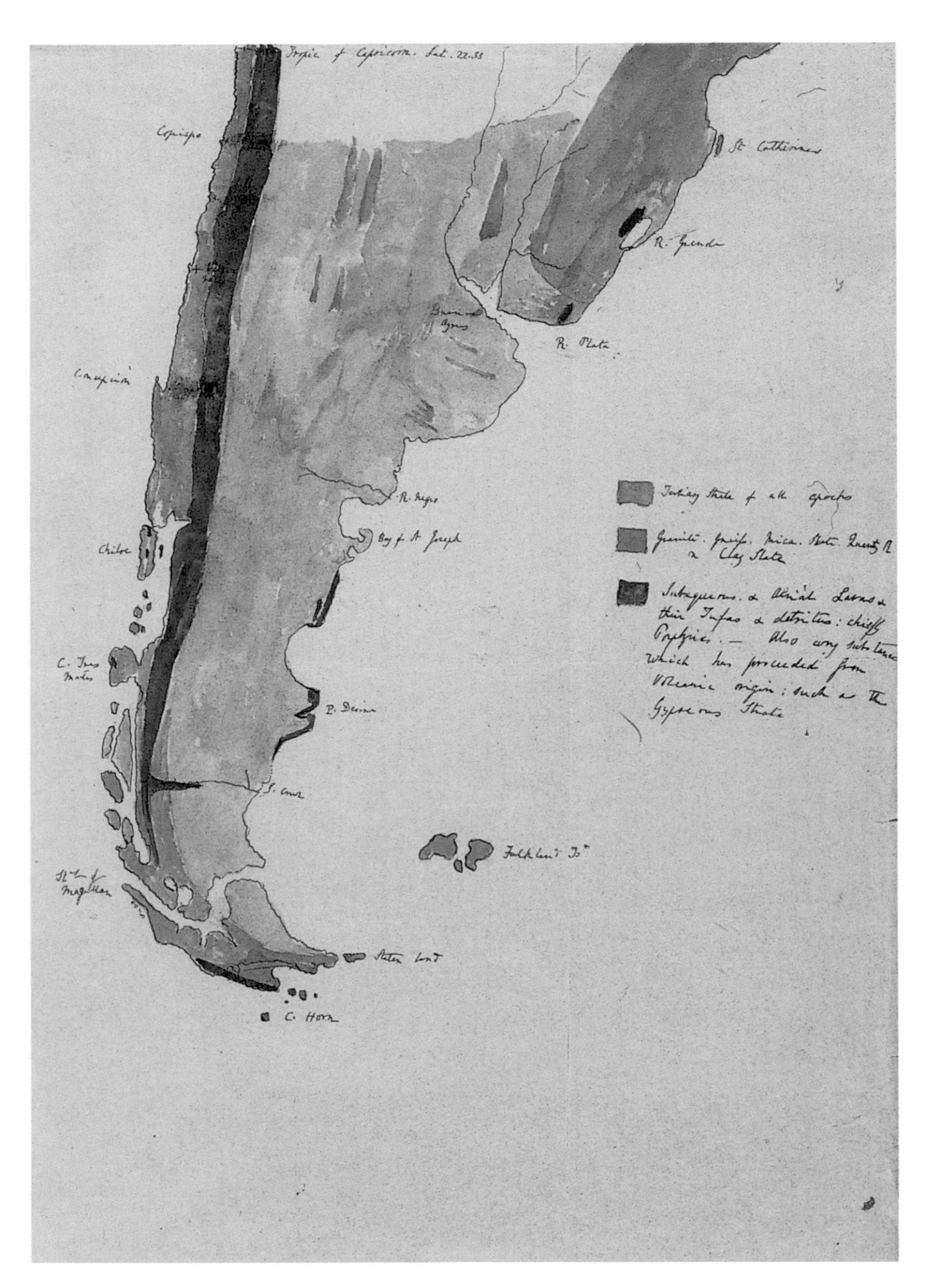

Plate 6. Darwin's hand-drawn and colored map of southern South America. The legend reads: for blue, "Tertiary Strata of all Epochs"; for pink, "Granite. Gneiss. Mica. Slate. Quartz R & Clay Slate"; and for purple, "Subaqueous & Aerial Lavas & thin Tufas & detritus: chiefly Porphyries.— Also any substance which has proceeded from Volcanic origin: such as the Gypseous Strata." From DAR 41. By permission of the Syndics of Cambridge University Library.

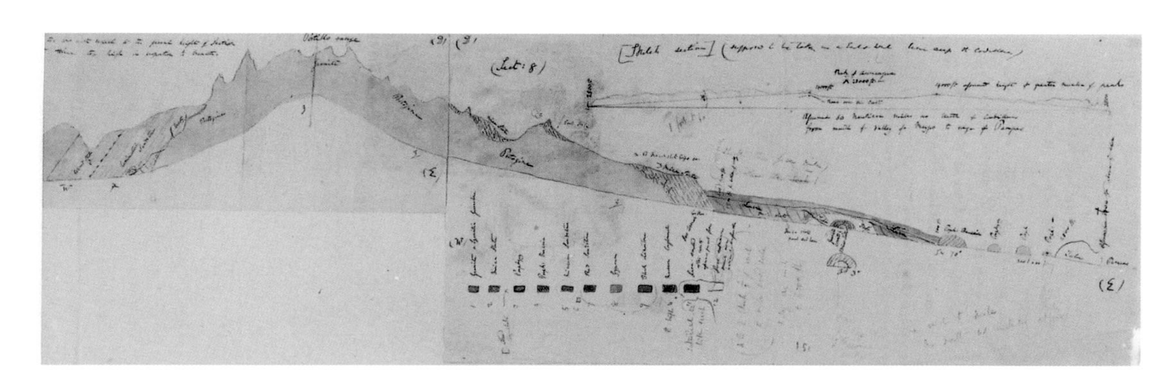

Plate 7. Cordillera sketch. From DAR 44. By permission of the Syndics of Cambridge University Library.

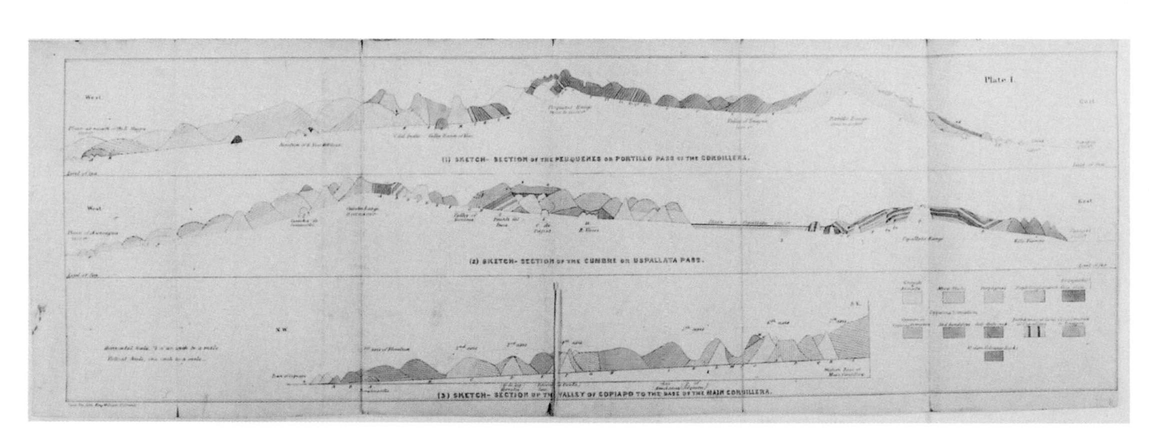

Plate 8. Published Cordillera sections. By permission of the Syndics of Cambridge University Library.

GREENS.

Nº	Names	Colours	Animal	Vegitable	Mineral
46	Celandine Green.		Phalæna. Margaritaria.	Back of Tussilage Leaves.	Beryl.
47	Mountain Green.		Phalæna Viridaria.	Thick leaved Cudweed, Silver leaved Almond.	Actynolite Beryl.
48	Leek Green.			Sea Kale. Leaves of Leeks in Winter.	Actynolite Prase.
49	Blackish Green.		Elytra of Meloe Violaceus.	Dark Streaks on Leaves of Cayenne Pepper.	Serpentine.
50	Verdigris Green.		Tail of small Long-tailed Green Parrot.		Copper Green.
51	Bluish Green.		Egg of Thrush.	Under Disk of Wild Rose Leaves.	Beryl.
52	Apple Green.		Under Side of Wings of Green Broom Moth.		Crysoprase.
53	Emerald Green.		Beauty Spot on Wing of Teal Drake.		Emerald.

Plate 9. Syme's color chart. By permission of the Syndics of Cambridge University Library.

Plate 10. Martens's painting of Mt. Sarmiento. By permission of Mark Smyth.

Plate 11. Cwm Idwal. Photograph © Dr. Kenneth Addison, St. Peter's College, University of Oxford.

Plate 12. Boulders at Cwm Idwal. Photograph by the author.

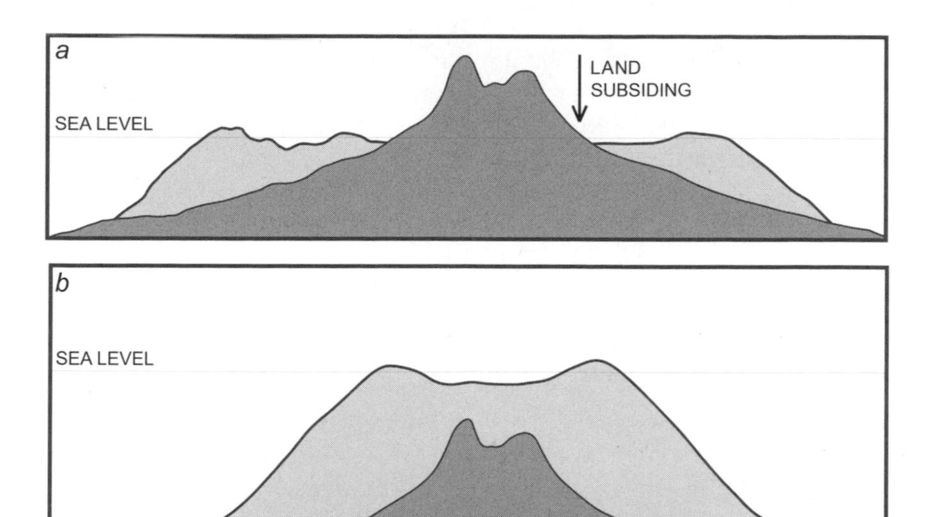

Figure 7.6. Darwin's theory of the formation of coral atolls. UMBC Cartographic Services.

Australia, were they on the recommended route. As late as February 1836 the visit was scheduled only as "possible." As the *Beagle* came up the west coast of Australia, one can envisage Darwin talking with FitzRoy on the virtues of converting that possible stop into a reality. FitzRoy himself noted in his *Narrative* that "many reasons" had induced him to select the islands for study.[113] Darwin's interest may well have been one of those reasons, strengthening the general interest in coral reefs held by the Hydrographic Office. (In one of the odder bits of documentary evidence regarding the *Beagle* voyage, a commentary by James Clunies Ross [dates unknown] of the Keeling Islands suggested that FitzRoy took Darwin along on the voyage on a "far too independent footing"; the preferred "savant" being "a humble toadyish follower—who would *do* the Natural History department.")[114] During the stop at the Keeling Islands both FitzRoy and Darwin were operating at top speed and in top form. The chart that FitzRoy made remained the standard one until the main atoll was resurveyed during World War II.[115]

Darwin's work at the Keeling Islands was fundamental for his published work on coral reefs. He referred to the islands at length in all of his subsequent publications touching on reefs. Further, the subject of the Keeling Islands was the starting point for his book on coral reefs. Because of the nature of the questions needing answers, Darwin was necessarily working more closely with the chart-makers during the Keeling stop than was usual. According to his theory, atolls, as those at Keeling (there was a larger and a smaller atoll), were the remains of islands that had gradually sunk below the surface of the sea. This was the insight he had gained at Tahiti, where he stood on its heights and gazed toward Eimeo (Moorea). At Keeling he hoped to test that insight against reality. It was the surveyors, with their boats and plumb lines, who were necessary to identify the structure of the reef beneath the surface of the ocean. Identifying this structure occupied his time. Unlike his geological notes on other points visited, his notes for Keeling are brief and dominated by his sketches of the reef.[116]

The key points at "the Keelings" (as FitzRoy referred to them) were to map their horizontal contours and to define their profile. As the reefs were largely under water, understanding their structure required, in Darwin's words, the "eye of reason" rather than the "eye of the body."[117] FitzRoy described his first view of the larger (southern) island from five miles out: "A long but broken line of cocoa-palm trees, and a heavy surf breaking upon a low white beach, nowhere rising many feet above the foaming water."[118] In his *Diary*, Darwin described the mode of exploration of the larger island. He employed two of his favorite words—"grandeur" and "simplicity"—in his description:

> When we [FitzRoy and I] arrived at the head of the lagoon we crossed the island & found a great surf breaking on the windward coast. I can hardly explain the cause, but there is to my mind a considerable degree of grandeur in the view of the outer shores of these Lagoon Islands. There is a simplicity in the barrier-like beach, the margin of green bushes & tall Cocoa nuts, the solid flat of Coral rock, strewed with occasional great fragments, & the line of furious breakers all rounding away towards either hand.[119]

Darwin, with the assistance of soundings made under the direction of FitzRoy, created a profile of the reef. Measurements were made at high and low tide. (At the Keelings FitzRoy was also working with a new tide-gauge of his own invention.) The fair copy of Darwin's hand-drawn profile is reproduced in Figure 7.7. The more abstract published version appears as Figure 7.8. This published version, which follows Darwin's hand-drawn profile, is preserved in his manuscripts.[120]

As the published text suggests, Darwin carefully noted that the reef above water level was dead. He also attended to the species of corals growing at different points on the reef, noting, for example, that at the outer margin of the reef, where the coral was alive, he found the genus *Porites,* "which forms great

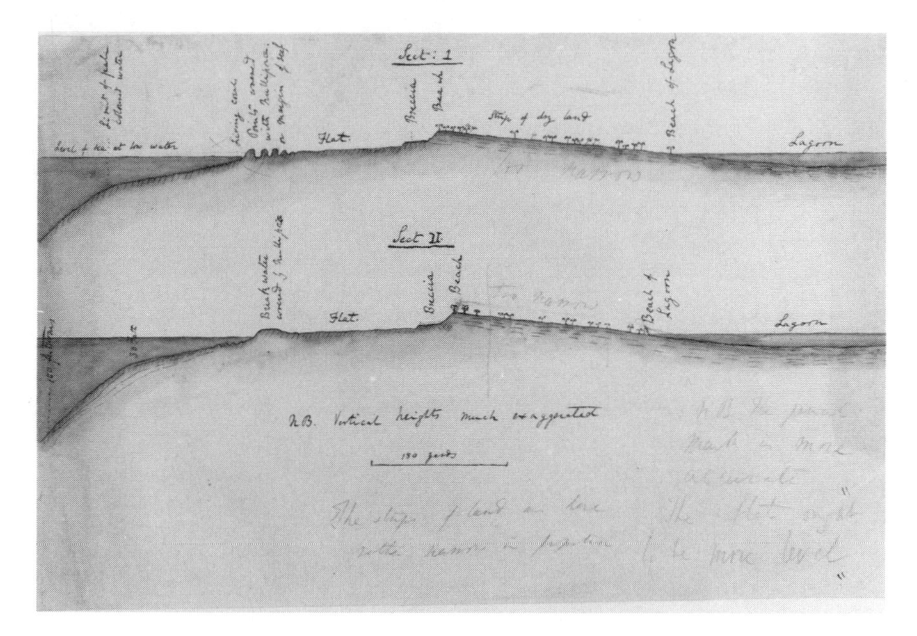

Figure 7.7. Hand-drawn sections of Keeling Island reef. From DAR 44:24. By permission of the Syndics of Cambridge University Library.

irregularly rounded masses . . . from four to eight feet broad."[121] He also carefully studied the tallow on the bell-shaped lead FitzRoy used for sounding the depth of the ocean floor. His observations confirmed Quoy and Gaimard's findings that coral-producing organisms did not grow far beneath the surface of the water: "Out of twenty-five soundings, taken at a greater depth than 20 fathoms, everyone shewed that the bottom was covered with sand."[122] Finally, FitzRoy provided him with a sufficiently deep sounding that he could at least speculate on the foundation on which the Keeling atolls had grown.

In what was an extremely deep measure for that time, FitzRoy showed that there was no bottom to be found at 1200 fathoms, measured on the southeast side of the larger island at a distance of 2200 yards from the line of breakers. Darwin concluded from FitzRoy's measure that "the submarine slope of this coral formation is steeper than that of any volcanic cone."[123] This comment spoke against Lyell's presumption that coral atolls grew directly on top of volcanic craters, which, in Lyell's view, determined their distinctive shape. Darwin related the doughnut shape of the atoll, with its interior lagoon, to the inability of the reef-forming organisms to thrive in still water. In other

CHAPTER I.

ATOLLS OR LAGOON-ISLANDS.

SECTION FIRST, KEELING ATOLL.

Corals on the outer margin—Zone of Nulliporæ—Exterior reef—Islets—Coral-conglomerate—Lagoon—Calcareous sediment—Scari and Holuthuriæ subsisting on corals—Changes in the condition of the reefs and islets—Probable subsidence of the atoll—Future state of the lagoon.

Keeling or Cocos atoll is situated in the Indian Ocean, in 12° 5′ S. and long. 90° 55′ E.: a reduced chart of it from the survey of Capt. FitzRoy and the Officers of H.M.S. Beagle, is given in Plate I. fig. 10. The greatest width of this atoll is nine miles and a half. Its structure is in most respects characteristic of the class to which it belongs, with the exception of the shallowness of the lagoon. The accompanying woodcut represents a vertical section, supposed to be drawn at low water from the outer coast across one of the low islets (one being taken of average dimensions) to within the lagoon.

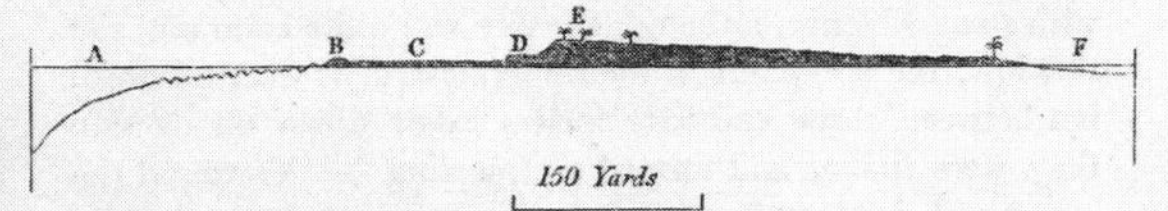

A—Level of the sea at low water: where the letter A is placed, the depth is 25 fathoms, and the distance rather more than 150 yards from the edge of the reef.

B—Outer edge of that flat part of the reef, which dries at low water: the edge either consists of a convex mound, as represented, or of rugged points, like those a little farther seaward, beneath the water.

C—A flat of coral-rock, covered at high water.

D—A low projecting ledge of brecciated coral-rock, washed by the waves at high water.

E—A slope of loose fragments, reached by the sea only during gales: the upper part, which is from six to twelve feet high, is clothed with vegetation. The surface of the islet gently slopes to the lagoon.

F—Level of the lagoon at low-water.

Figure 7.8. Published profile of Keeling atoll. From *CR*, 5. By permission of the Syndics of Cambridge University Library.

words, the shape of the atoll reflected the nature of the coral-producing organisms rather than the shape of the base of the platform on which the reef was built.

FitzRoy's plumbing the ocean with a line 7200 feet in length was sufficiently remarkable for its time to require comment. As Helen Rozwadowski has discussed, the earliest serious effort to plumb the deep ocean took place during the 1818 expedition of John Ross (1777–1856). Prior to that, explorers sounded with a 200-fathom line, which was long enough to reach what is today known as the continental shelf.[124] Once ships were in deep water, and were in no risk of running aground, depth was irrelevant. By performing deep sounding in "blue water," FitzRoy was in the vanguard of exploration of what would eventually become the science of oceanography. His work at the Keeling Islands permitted Darwin to carry forward his theory within newly expanded limits, though it would not be until the twentieth century that the ocean floor could be studied directly.

The question of the structure of the reef below the water's surface was of greatest interest to Darwin. While at the Keeling Islands, he speculated as to the answer in a sketch of a cross section of hypothetical reef. Stoddart reproduced this sketch, which he aptly termed an "adventurous resort to visual theorisation."[125] At the bottom level, just above 1000 fathoms (6000 feet), he wrote "Greenstone ?" (Greenstone was a then common field term for a dark greenish igneous rock.[126]) Above the notation "Greenstone ?" Darwin pictured an alternation of layers of coral and breccia. The sloping sides of the structure are marked as "sand." Thus, the geometry of the submerged reef that Darwin imagined had coral present at a depth approaching 1000 fathoms. This was a rough measure of the amount of subsidence he envisioned having occurred at the Keelings. Both FitzRoy's actual measure, and Darwin's imaginative sketch, were operating in the same range.

If subsidence of that amount had occurred at the Keeling Islands, what might have been its cause? The answer would have to be elevatory movements elsewhere. This is absolutely clear in Darwin's writing from 1835 onward. Thus, for the Keeling Islands, Darwin sought out some indications of compensatory elevatory movement. This he found in Sumatra, also in the Indian Ocean. In the words of his 1837 presentation to the Geological Society,

> He then proceeded to show that within the lagoon of Keeling Island, proofs of subsidence might be deduced from many falling trees and a ruined storehouse; these movements appearing to take place at the period of bad earthquakes, which likewise affect Sumatra, 600 miles distant. It was thence inferred as probable, that *as Sumatra rises* (of which proofs are well known to exist), *the other end of the lever sinks down; Keel-*

> *ing Island thus acting as an index of the movement of the bottom of the Indian Ocean.*[127]

In the same presentation Darwin deduced the generalization that "linear spaces of great extent are undergoing movements of an astonishing uniformity, and that the bands of elevation and subsidence alternate."

Directly following its stop at the Keeling Islands, the *Beagle* made its way to Mauritius, also in the Indian Ocean. This island, taken by the British from the French in 1810, had been well explored by French voyagers and naturalists. Drawing on reports by French authors, Darwin needed only a few days to satisfy himself that an island which showed signs of recent elevation would have a distinctive profile of reef. As he wrote in *Coral Reefs*, "A fringing reef, if elevated in a perfect condition above the level of the sea, ought to present the singular appearance of a broad dry moat within a low mound."[128] Darwin's classification was complete.

Darwin's Relationship to Lyell

In chapter 2 we discussed Darwin as a reader of the *Principles of Geology*. Here we go beyond the voyage to the period when Darwin and Lyell became colleagues, frequent correspondents, and friends. Theirs was a unique tie. Darwin, for example, never formed a similar relation with any other geologist, and Lyell expressed himself more fully to Darwin than to any other geologist.[129] They did this without Darwin ever losing a slight deference to Lyell as the elder man. Thus, though they were intimates, Darwin would not have been quite so forthcoming to Lyell as he was, say, with his brother Erasmus. Nor was he quite so at ease as he was with Joseph Hooker (1817–1911), a younger man with whom he formed a similarly deep, and parallel, relationship on botanical matters such as that he had with Lyell on geological matters. In a formal sense, in geology Lyell always remained the senior partner.

Following the return of the *Beagle* in 1836, Lyell initiated contact with Darwin, who noted, "Mr. Lyell, has entered in the *most* goodnatured manner, & almost without being asked, into all my plans."[130] Lyell quickly brought Darwin to center stage at the Geological Society. He apparently set the date for the reading of the coral reef paper, and was much involved with Darwin's reading of his first paper on elevation.[131] Lyell accepted and promulgated Darwin's new ideas of coral reefs, not only in later editions of the *Principles of Geology*, but also in lectures in the United States in the 1840s, where Darwin had had an early supporter in Dana.[132]

More generally, Lyell accepted various suggestions by Darwin over the years. Thus, for example, in 1853—fifteen years after his "Connexion" paper of 1838 was written—Darwin urged Lyell to incorporate a portion of its argument into a forthcoming edition of the *Principles*, which Lyell did.[133]

With a fine sense of reciprocity, Darwin promulgated Lyell's views in numerous publications. For example, following his reading of the *Principles* during the voyage, Darwin dropped his use of the term "diluvium." However, as a junior partner to what was a recognizable position in British geology, Darwin was not required to lead the charge against opponents. This was largely an advantage, for it saved him time and trouble. Nonetheless, in his correspondence with Lyell, Darwin's continuing engagement with theoretical issues is clear. What follows are three illustrations of this engagement: first, an instance involving gradualism; second, the "craters of elevation" controversy; and third, the question of the operation of elevation and subsidence. In the first case Darwin was applying Lyell, in the second case Darwin was seeking middle ground in a controversy, and in the third case Darwin was taking Lyellian tenets but pushing them to the point of transformation.

The French naturalist and geologist d'Orbigny, whose 1826–1833 route in South America had overlapped that of Darwin, was writing up and publishing his results during the same period.[134] A noted stratigrapher, d'Orbigny's interpretations were nonetheless catastrophist by comparison with Darwin's, who railed at him in a letter to Lyell:

> I am astounded & grieved over d'Orbigny's nonsense of sudden elevations. . . .
>
> Because at 12,000 ft he finds the same kind of clay with that of the Pampas he never doubts that it is contemporaneous with the Pampæan debacle, which accompanied the right Royal salute of every volcano in the Cordillera. What a pity these Frenchmen do not catch hold of a comet, & return to the good old geological dramas, of [Thomas] Burnett & [William] Whiston[.]—[135]

Yet when Darwin went into print against d'Orbigny, he would merely contrast their theories of the origin of the Pampean formation, one "that of a great debacle," the other "that the Pampean formation was slowly accumulated at the mouth of the former estuary of the Plata."[136] Lyell could be counted on to stand firm against geological debacles. Darwin left the larger battle to him.

The "crater of elevation" theory arose from Continental geologists. It was inspired by Humboldt, invented by von Buch, and championed by Élie de Beaumont. The theory was strongly opposed by Lyell in the *Principles of Geology*.[137] The theory held that molten lava could not solidify on steeply sided inclines. Therefore, craters surrounded by circular walls of steeply inclined lava were believed to have been produced by elevatory forces pushing up the volcanic material from beneath rather than by volcanic eruption. As von Buch put it,

> *Volcanos* are the constant chimneys, the canals uniting the interior of the earth with the atmosphere, which spread around themselves the phenomena of eruption from craters that are of small extent, and are only once in operation. *Craters of elevation,* on the contrary, are the remains of a great display of power from within, which can and actually has raised islands of several square miles in extent, to a considerable high.[138]

During the voyage, Darwin used the skeptical phrase "so called craters of elevation" when he wrote about the large crater at Mauritius.[139] However, in *Volcanic Islands,* he constructed a position that used his own (modified) version of the concept to account for the appearance not only of Mauritius but also of St. Helena and St. Jago, both of which possess large craters bounded by a ring of basaltic mountains. Darwin justified his adoption of the "craters of elevation" theory on the grounds that the average inclination of the slopes of the surrounding mountains was steeper than what could have been maintained by lava flowing down a sloping surface.[140] Darwin put his stamp on the concept by suggesting that the central portion of a "crater of elevation" had been formed by differential elevatory action, rather than by the arching of the surface of the ground as Élie de Beaumont would have had it. In his 1849 article for the Admiralty handbook, Darwin also suggested the systematic study and mapping of "craters of elevation," with particular attention to the inclination of the streams of lava, which Élie de Beaumont asserted did not consolidate well on a slope greater than two or three degrees.[141]

The "craters of elevation" controversy touched Darwin relatively little. His compromise on the subject was so tied to his own understanding of elevation that it did not bring him to a position identical to von Buch's. However, Darwin's compromise in *Volcanic Islands* does illustrate his willingness to challenge Lyell on certain points. Ultimately, through repeated field studies in the 1850s, Lyell disproved the "craters of elevation" theory in a careful field study of lava slopes, where he found evidence that lava had consolidated at the steep incline of thirty-five degrees at the Val del Bove.[142] Like other geologists, Darwin left behind the theory of craters of elevation.

The third illustration of Darwin's relationship to Lyell involves the theory of coral reefs. Here Darwin adopted Lyell's ideas but transformed them. Lyell treated coral reefs in volume 2 of the *Principles of Geology.* He suggested the Pacific Ocean as an area of subsidence. Coral islands were presumed to rest on submerged mountaintops; coral atolls, to rest on submerged volcanic craters. To Darwin, Lyell had taken insufficient notice of the vastness of the Pacific Ocean. Such a huge area must have undergone correspondingly vast subsidence, greater than Lyell imagined. Therefore, it was a "monstrous hypothesis" to posit the existence of barely submerged mountains and volcanoes simply to account for the occurrence of coral reefs in mid-ocean.[143] Better, Darwin thought, to imagine reefs forming on platforms that they themselves had built as the floor of the ocean subsided beneath them. This led to the conclusion that coral reefs might well be thousands of feet thick.

From our present vantage point, it is clear that Darwin's departure from Lyell on the issue of coral reefs was one of degree. In method, in the choice of problem, and in the elements addressed to the solution of the problem, Darwin demonstrated that he had read the *Principles of Geology* closely. His departure was not one involving the correction of error or the adoption of contrary modes of thought. The major change he made in Lyell's notions of elevation and subsidence was to work with them on a larger scale—continents rather than patches, and long-term changes in level rather than temporary oscillations. As Darwin put it in 1836 in a notebook jotting, "The great movements (not mere patches as in Italy proved by Coral hypoth. agree with great continents)."[144] "Patches as in Italy" recalls Lyell's work there, as at the Temple of Serapis. "Great continents" stands for his own view.

An important consequence followed from this conclusion. As described in chapter 2 (and note particularly Figure 2.4), Lyell had explained changes in climate as resulting from the changing patterns of distribution of land and sea. These changes were caused by elevation and subsidence. If the rise and fall of the land from the sea occurred as an oscillation within narrower rather than broader boundaries, continents could be seen as less stable features of the globe. On the other hand, if, as Darwin's ideas suggested, the elevations were higher, and subsidences deeper, the distribution of land and sea would appear a more enduring feature of the earth's history. Darwin's reworking of Lyell's coral reef theory thus inevitably put Darwin at odds with Lyell's theory of climate change. Darwin was aware of this consequence of his approach but did not call it to Lyell's attention. As Darwin wrote to Hooker much later, in 1859, on a point relating to the then current glacial theory,

> But perhaps he [Lyell] unconsciously hates (do not say so to him) the view, as slightly staggering him on his favourite theory of all changes of climate being due to changes in relative position of land & water.—[145]

Darwin remained very interested in the "relative position of land & water," but his expanded view of the vertical boundaries and horizontal spaces over which the forces operated separated him from Lyell. However, he counseled himself, as he counseled others: "do not say so to him."

We opened this chapter with a discussion of Darwin's ideal of a "simple" geology. To close this chapter we note the vocabulary of a charge Darwin made against Élie de Beaumont's and d'Orbigny's views: "Is not this making Geology nice & simple for beginners?"[146] Darwin's remark was pejorative, but, underneath, one suspects a recognition of a common sympathy for the ideal. That, and an acknowledgment to himself that he was no longer a beginner and must treat the ideal with care.

CHAPTER 8

SIMPLICITY CHALLENGED

For some time to come the Glacial Theory must occupy a prominent place in Geological Investigation. The Subject appears to me the most important that has been put forth since the propounding of the Huttonian Theory & the surface of the whole Globe must be examined afresh.

WILLIAM BUCKLAND, 1840

I hope your friend will enjoy . . . his [Welsh] tour, as much as I did—it was a kind of geological novel, but your friend must have patience for he will not get a good *glacial eye* for a few days. . . .

I feel *certain* about the glacier-effects in N. Wales.—

CHARLES DARWIN TO W. D. FOX, 4 SEPTEMBER 1843

How rash it is in science to argue because any case is not one thing, it must be some second thing which happens to be known to the writer.—

CHARLES DARWIN TO T. F. JAMIESON, 6 SEPTEMBER 1861

Glaciers

Glaciers were for Darwin one of nature's grand spectacles when he first viewed them in 1833 during the H.M.S. *Beagle* voyage of 1831–1836. Only later, gradually, and through a tortuous route, did glaciers become for him a geological agency of major significance. This tortuous route of refined understanding spanned several decades. In the end his views of glaciers and glaciation would alter his ideal of a "simple" geology.

The development of glacial theory by Swiss authors has been well treated by Albert V. Carozzi, and serves as a background to my narrative. Early work in Switzerland by Ignace Venetz (1788–1859) and Jean de Charpentier (1786–1855) anchored the modern scientific understanding of glacial phenomena by arguing for the former great extension of glaciers and their role in transporting erratic boulders, and in polishing and striating rocks. Once convinced of the accuracy of their work, Louis Agassiz added the notion of ice ages. As Carozzi remarked, it took the "conjugate efforts" of Venetz, Charpentier, and Agassiz over a twenty-five-year period to overcome the old belief in the transportation of great boulders by huge water and mud currents.[1]

Of course, Venetz, Charpentier, and Agassiz had forerunners and co-contributors. On the Swiss side these included mountaineers such as Jean-Pierre Perraudin (1767–1858), as well as the botanist Karl Schimper (1803–1867), who coined the term "*Eiszeit*" ("Ice Age"). Outside Switzerland, early authorities on glaciers included Playfair, whose book describing glacial action accompanied Darwin on the *Beagle* voyage.[2] However, Darwin was ignorant of the Swiss work until after the voyage, and did not fully engage Playfair's views on glaciers during the voyage. Herries Davies, citing the memories of James David Forbes (1809–1868), suggested that by 1827 Jameson at Edinburgh University discussed the possible former existence of glaciers in Scotland.[3] Yet Darwin did not record hearing such a lecture, and Jameson did not discuss the subject in print. Where Jameson did figure largely in later debates was as the editor of the *Edinburgh New Philosophical Journal,* which carried important articles on glaciers in the 1830s and 1840s.

As a result of his experience aboard H.M.S. *Beagle,* Darwin first viewed glaciers in the setting not of the Alps, but of Tierra del Fuego. Its climate was so wretched—windy, wet, and cold even in summer—that for all aboard the *Beagle* the months spent surveying the intricate waterways were difficult ones. Even the astonishing presence of the native Fuegians posed danger. The glaciers stood out amidst the gloom. On 29 January 1833 Darwin first saw them as one of the small surveying boats entered the north arm of Beagle Channel, named during the first (1826–1830) expedition of H.M.S. *Beagle* (Figure 8.1). Darwin wrote the following in his small pocket field notebook at first sighting: "many glaciers beryl blue *most beautiful* contrasted with snow."[4] The name for the color, "beryl blue," was drawn from *Werner's Nomenclature of Colours* by Patrick Syme (1774–1845), a book Darwin carried with him aboard the ship.[5] Syme's chart represented a traditional comprehensive view of natural history: the categories of animal, vegetable, and mineral are

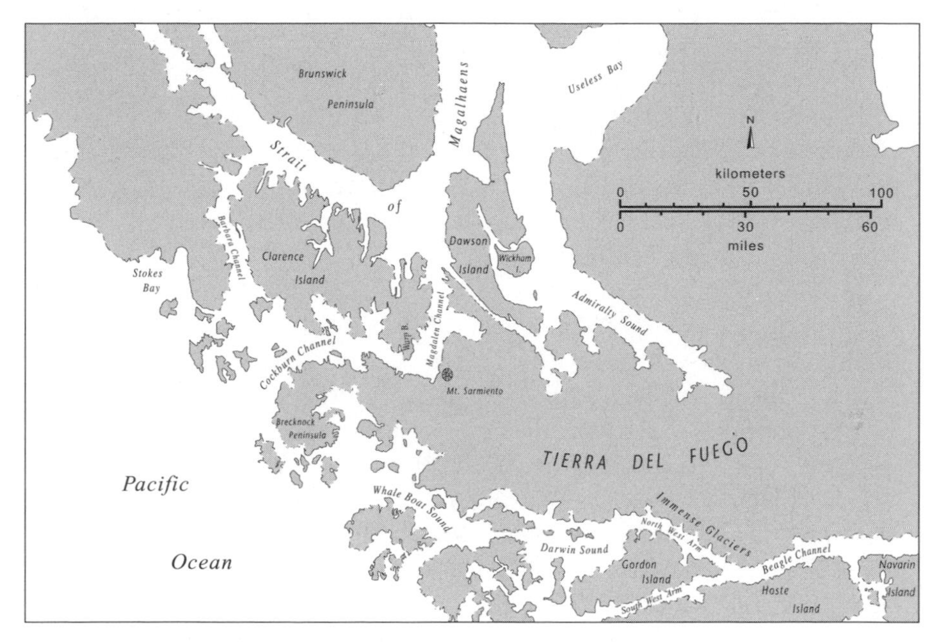

Figure 8.1. Map of Beagle Channel. Simplified version of the official chart. UMBC Cartographic Services.

linked by the possession of common colors. Darwin chose the name "beryl"—derived from beryl, the mineral—from Syme's chart (Plate 9). Presumably Darwin had left Syme's book on the *Beagle* during the expedition in the smaller boats that were used for exploration, for in the expression "beryl blue" Darwin borrowed from Syme without repeating him exactly.

Conrad Martens painted several stunning watercolors of the glaciers. He chose perspectives that included Mt. Sarmiento, which lies to the northwest of Beagle Channel. One painting was done from a sketch made on 9 June 1834, on the ship's second swing through Tierra del Fuego, from a location called Warp Bay in the Magdalen Channel facing Mt. Sarmiento (Plate 10). Darwin said of the glaciers in his notebook that their color was "blue by transmitted and reflected light," qualities that are captured in Martens's paintings.[6]

On 29 January 1833 while Darwin viewed what were labeled on the chart reproduced in Figure 8.1 as the "Immense Glaciers" of the northern arm of the Beagle Channel, he experienced danger from a glacier. In FitzRoy's words,

> We stopped to cook and eat our hasty meal upon a low point of land, immediately in front of a noble precipice of solid ice; the cliffy face of a

> huge glacier, which seemed to cover the side of a mountain, and completely filled a valley several leagues in extent.

Then a "thundering crash" shook us as the "whole front of the ice cliff" came crashing down.[7] The ice caused a wave that threatened to carry away the surveying boats. Darwin aided in their rescue. Disaster was averted. Glaciers had become more than spectacle; they had become experience.[8] Darwin was witness to the creation of icebergs.

Darwin's recorded observations on the glaciers of the northern arm of the Beagle Channel were brief but careful. With them he was continuing the tradition from the first surveying expedition, including noting where rock-bearing icebergs had been seen. (The second expedition had on board the manuscript of the records from the first expedition.) Darwin did not actually climb on any of the glaciers. In manuscript notes he made reference to the "glacier, which I was able to approach nearest to," and, later, in published work wrote, "I had no opportunity of landing on any glacier, but we passed in the Beagle and Magdalen channels within two miles of several." In his notes he remarked on "the great difference of structure in the tongue of land formed by mountain torrents & by glaciers.—" The torrents produced a "bank of pebbles; but with the glacier it is a pile of enormous boulders." He noted one boulder averaged "6 feet above the sand . . . & 30 yards in circumference." Such a large amount of material was brought down by torrents and by glaciers that he wondered why the Beagle Channel had not filled up.

In his notes, like FitzRoy, Darwin compared masses of ice falling from glaciers into the channel to "a miniature of the icebergs of the Antarctic seas." He repeated this analogy in his Diary from the voyage and added a note comparing the distribution of glaciers in the Northern and Southern Hemispheres:

> The occurrence of glaciers reaching to water's edge & in summer, in Lat: 56° is a most curious phenomenon: the same thing does not occur in Norway under Lat. 70°.—

Darwin gleaned his information on Norway's snow line, and hence glaciers, from his reading of Leopold von Buch's *Travels through Norway and Lapland*, a work that accompanied him on the voyage.[9]

In Darwin's longer set of geological notes on Tierra del Fuego there is a rich interplay of observation and hypothesis, which he later drew from in his published work. For example, he remarked,

> I cannot more accurately describe the appearance of the cliffs around Navarin Island [in Tierra del Fuego], than by the remark, which, at the

> time, I entered in my note-book, "that a vast debacle appeared to have been suddenly arrested in its course."[10]

For background, Darwin began with a good short treatment of Tierra del Fuego by Phillip Parker King (1793–1856), commander of the *Adventure* and *Beagle* on the first surveying expedition of 1826–1830. King's 1831 article identified what he called a transverse section of the continent in the sharply differing rock types in Tierra del Fuego: granite and greenstone on the west, clay-slate in the center, and recent formations on the eastern portion.[11] Similarly, while in the northern arm of the Beagle Channel, Darwin—building on King's work—reserved his closest attention for rock types, and particularly for the junction between the micaceous rocks and those of clay-slate. In Darwin's geological specimen notebooks one can trace these interests, such as where he noted for specimen 952, "Mica Slate. with Garnets: grand chain in North arm of Beagle Ch[a]." At this time Darwin did not collect boulders simply for their shape, or for their similarity to, or distinctness from, the surrounding rock. Later, in 1834 at the Santa Cruz River, he did, as indicated by his description in his specimen notebook of specimens 1994 and 1995 as "*Immense* blocks."

As we focus on Darwin's interest in glaciers while he was at Tierra del Fuego, we must remind ourselves that at that time his most keenly felt theoretical interest lay with working out the relation of cleavage and stratification there. Toward the end of the voyage he summarized his views on the relation of the laminated structure of the slates to the form of the land, views he later developed in print.[12] Considerably further down the line in his hierarchy of interests were his observations on glaciers and on what he termed "alluvium."

Darwin grouped all the superficial loose deposits, including boulders, under the heading "alluvium." There are several surprising aspects to his extensive contemporary notes on alluvium in Tierra del Fuego. First, his later-used mechanism of icebergs transporting rock appears only briefly as a possibility. It is not the dominant idea of the notes. Indeed, there is little association of ice of any kind, whether glacial or in icebergs, with his accounts of the alluvium.[13] Second, Darwin was not a rote Lyellian. For example, as mentioned earlier, in notes he used the term "debacle" to describe forces sufficient to move great bodies of rock. Third, Darwin expressed frustration with his subject of alluvium in his notes, especially during his second visit to Tierra del Fuego in 1834, after he had seen the distribution of large lava blocks in the area of the Santa Cruz River. For example, in 1834, after considering two alternative hypotheses on how material could be transported from the west to the east in Tierra del Fuego, whether by the action of ordinary tides and currents, or by sudden rushes of water, he remarked in his notes, "It is clear my data are insufficient to come to any satisfactory conclusions.—"[14] On the same page of notes he added that the ocean was very deep at some points off the east coast of Tierra del Fuego: 660 feet about a mile off Pt. S. Anna, and no sounding at 1536 feet east of Cape Froward in midchannel. This depth made water transport for rocks difficult to imagine. Overall, Darwin was aware that the subject of the transportation of rocks was a matter of interest among geologists. In his notes he mentioned FitzRoy having received inquiries on the subject from Lyell.[15] At some point—possibly as a later addition—he also referred in his voyage notes to Buckland's work on diluvial valleys, suggesting a relevance for the interpretation of the valley of the river Santa Cruz.[16] Beyond that, Darwin's notes on alluvium in Tierra del Fuego are difficult to summarize. At one point in 1834 he ventured the following claim: "Generally however the Boulders may be attributed to the alluvium.—"[17] But this was a thin assertion. From his *Beagle* period notes, it is clear that Darwin believed these were not settled questions.

Darwin's 1837 Manuscript

In 1837, three years after Darwin left Tierra del Fuego for the last time, he spent seven months transforming his private journal from the voyage into a book. As usual, careful of his expenditure of time, he noted later, "From March 13th to end of September entirely employed in my journal."[18] Anticipating speedy publication, which did not in fact occur, Darwin sent his text to the publisher and began receiving proofs back for correction in late August 1837.[19] The three years from 1834 to 1837 had been full of developments for Darwin and for the field of geology as it related to glaciers. With regard to superficial deposits, following his last visit to Tierra del Fuego he had read volume 3 of Lyell's *Principles of Geology,* which used the imported term "erratic blocks."[20] That phrase became part of Darwin's everyday vocabulary, which it had not been earlier even though the it appeared elsewhere in his library aboard ship, as in De la Beche's *Geological Manual.* After the voyage, as he was writing up his results, Darwin also largely stopped using Buckland's term "diluvium" for descriptive purposes.

Darwin in 1837, the prospective author, was a different man from Darwin in 1834, the explorer and field naturalist. As we have seen, his ambition as a geologist had solidified around the topic of crustal motion: elevation and subsidence. In the spring of 1837 he had also privately adopted an evolutionary hypothesis. Thus, as he was writing up his journal for publication, even seemingly unrelated issues often had ties, sometimes hidden, to his larger views.

An interesting feature of Darwin's work is its timing in relation to the new hypothesis on glaciers coming from Switzerland. The theory of Venetz and

Charpentier probably reached Darwin first through Lyell's 1836 presidential address to the Geological Society of London, where Lyell noted that

> M. Charpentier has lately proposed another theory which he informs us is merely a development of one first advanced by M. Venetz. The alpine blocs . . . were not carried by water, for had that been the case the largest would be either in the Alpine valleys or near the base of the great chain, and we should find their size and number diminish as we recede from their original point of departure. But the fact is otherwise, many of the blocks on the Jura . . . being of the largest dimensions. . . . According to this hypothesis . . . the erratic blocks are monuments of the greater magnitude and extent of the ancient glaciers under a different configuration of the surface.[21]

However, even though Darwin did know the Venetz-Charpentier version of the glacial hypothesis before writing up his voyage account, he, like Lyell, did not adopt its conclusions. Lyell himself used his next presidential address to the Geological Society (1837) to insist that formations "usually called diluvial" (no mention of erratic boulders) were produced "by a prolonged submersion of the land, the level of which has been greatly altered at periods very modern in our geological chronology." He went on to note in regard to the Swiss situation,

> I now believe that by far the great part of the dispersion of transported matter has been due to the ordinary moving power of water, often assisted by ice, and cooperating with the alternate upheaval and depression of land.[22]

Darwin's presentation of material related to the glaciers of Tierra del Fuego took two forms in the main body of the text for his *Journal of Researches.* In chapter 11, he recounted much of the material drawn from his Diary. (The color of the glaciers was now described as "beryl-like blue," rather than "beryl blue"—a minor change.) He compared the Beagle Channel to the valley of Loch Ness in Scotland. In so doing, he was drawing an analogy between the landscape of Tierra del Fuego and that of other locales around the world. Tierra del Fuego had become his model for understanding a certain set of climatic, geographical, and geological circumstances.

In chapter 13 of the book, Darwin presented an analytical view of what he had seen in Tierra del Fuego. His analysis had two parts. First, he now adopted the notion that icebergs had deposited erratic blocks, particularly the angular ones, seen across the eastern region of Tierra del Fuego. He wrote,

> When the land was elevated, the fragments of rock would be found deposited on the eastern side of the continent, in bands representing the ancient channels. Whether or not the hypothesis of their transport be true, such is the position of the erratic blocks in Tierra del Fuego.[23]

In his discussion Darwin also differentiated sharply between semirounded boulders, as he had found near Port Famine, for which he was not willing to offer an explanation, and angular boulders. When, toward the end of the voyage, in his Red Notebook, he did turn to De la Beche's account of erratic blocks, he criticized him for not drawing "sufficient distinction" between angular and rounded boulders.[24] In any case, for Darwin the presence of angular erratic blocks in eastern Tierra del Fuego now had an explanation.

The second portion of Darwin's analysis pertained to climate. His approach can be traced to the work of Humboldt, as interpreted by Lyell. Humboldt's article entitled "On Isothermal Lines, and the Distribution of Heat over the Globe," published in its English version in 1820–1821, was Lyell's chief theoretical source in writing the section on climate in the first volume of the *Principles of Geology*.[25] Lyell began his treatment of climate by summarizing Humboldt's work. The key ideas were that "zones of equal warmth . . . are neither parallel to the equator nor to each other"; that "the same mean annual temperature may exist in two places which enjoy very different climates, for the seasons may be nearly equalized ['insular' climates] or violently contrasted ['excessive' climates]"; and that determining these differences were "a multitude of circumstances, among the principal of which are the position, direction, and elevation of the continents and islands, the position and depth of the sea, and the direction of currents and of winds."[26] Following Humboldt, Lyell also noted the moderating effect of the ocean on climate, as in the Southern Hemisphere with its greater proportion of ocean to land.

Darwin adapted this Humboldtean/Lyellian view of climate to serve his account of the distribution of erratic boulders worldwide. In his *Journal of Researches*, he asked rhetorically,

> [W]hat are the circumstances in the southern hemisphere that produced such results [that is, of glaciers descending to the sea]? Must we not attribute them to the large proportional area of water; and do not plain geological inferences compel us to allow, that during the epoch anterior to the present, the northern hemisphere more closely approached to that condition, than it does now?

Darwin had only to turn to the second volume of Lyell's *Principles of Geology* to see a fold-out colored "Map shewing the extent of surface in EUROPE which has been covered by Water since the commencement of the deposition of the older TERTIARY strata." A pictorial depiction of a partly submerged Europe was readily at hand (see Figure 2.4). Darwin invited his readers to join him in thinking about "what is actually taking place in the southern hemispheres, only transporting in imagination each part to a corresponding latitude in the north." Darwin then brought his analogy to bear on the most famous site for erratic boulders in Europe, the Alps, and the Jura. He posited that icebergs charged with granite "would be floated from the flanks of Mont Blanc, and then stranded on the outlying islands of the Jura."[27]

While the hypothetical image of "islands of the Jura" would have startled Venetz and Charpentier, Darwin did not refer directly to their work in the main text of his *Journal of Researches.* However, in that work Darwin did juxtapose the situation of glaciers in Switzerland to that in Tierra del Fuego. He wrote,

> Every one has heard of the mass of rubbish propelled by the glaciers of Switzerland, as they slowly creep onwards. In the same manner in Tierra del Fuego, on a still night the cracking and groaning of the great moving mass may be distinctly heard. The same force, which is known to uproot whole forests of lofty trees, must, when grating over the surface, tear from the flanks of the mountain many huge fragments of rock.[28]

In his argument he also made one point that they could all share: that it was philosophically acceptable to posit the existence of glaciers where they do not exist at the present time. But their understanding of glaciers remained distinct and incompatible.

I suggest that, on Darwin's side, something of that incompatibility owed itself to his initial approach to glaciers primarily as spectacle. To him, glaciers were to be seen. They reflected light; they transmitted light. Yet, in emphasizing the visual, Darwin failed to recognize the bulk and weight of glaciers. In his summary remarks to his *Journal of Researches,* note how he separated glaciers from those spectacles that have geological power. He listed six "remarkable spectacles": the stars of the Southern Hemisphere, the waterspout, the "glacier leading its blue stream of ice in a bold precipice overhanging the sea," a lagoon island formed by coral, an active volcano, and an earthquake. Of the six spectacles on the list, he emphasized the coral island, the volcano, and the earthquake, because of their "intimate connexion with the geological structure of the world."[29] The land-bound glacier retained its interest primarily as spectacle. The glacier became effective as a geological agent only as a parent to icebergs. On its own it was not possessed of geological agency. To transfer glaciers from the category of spectacle to the category of agency would take Darwin several years.

A few factors affected Darwin's experience of glaciers. Certainly one must consider his disadvantage as a traveler who could not return to the same site year after year. Also, his firsthand experience with glaciers was primarily a visual one. As noted earlier, he did not climb onto any glacier; he observed them from a distance, albeit a short one. In this Darwin differed from the early Swiss glaciologists, and from most later British ones as well.[30] Another possible explanation for Darwin's limited sense of the power of glaciers was that within the division of labor aboard ship, he operated in the context of the natural historical tradition. Kinds of things mattered: animal, vegetable, and mineral. This was a world still consistent with Syme's color chart. Darwin's specimen notebooks reflect his collecting practice in honoring this traditional division. His work was systematic, but it was not particularly mathematical. He was not the keeper of extended series of numerical measurements aboard ship, though he did use these measurements. And he did not take the measurements himself from which weight might have been inferred. In contrast, it was King on the first surveying mission (1826–1830) who provided an estimate of 3500 feet for the line of perpetual snow in the Strait of Magellan.[31] One can at least speculate that had Darwin been more intimately involved in measuring the bulk of glaciers or, in this case, of the related matter of snow lines, he might have moved more quickly and easily to consider the potential force of glaciers as geological agents.

Darwin's 1838 Addendum

Agassiz and Darwin were contemporaries. They devoted themselves to the very largest questions in geology and natural history. However, their personal experience with existing glaciers did not overlap at the obvious point. Unlike many of his British contemporaries, Darwin never visited Switzerland. (Agassiz did see the glaciers at the Strait of Magellan in 1872, a year before his death, but he never wrote about the experience.)[32] It was Agassiz who first entered Darwin's world—in print. As already mentioned, in April 1838 the *Edinburgh New Philosophical Journal* carried a translation of Agassiz's famous Neuchâtel address of July 1837. It is not known exactly when or how Darwin first read the address, but by 27 October 1838 he had written a passionate attack on Agassiz's work, and that of Venetz and Charpentier. From the publishing point of view, this attack appeared under odd circumstances. Darwin's volume in the *Narrative* from the *Beagle* voyage had been set in print but held until FitzRoy was finished with his own work. As time went on, Darwin found material he wished to add to his text, and in the fall of 1838 he wrote a series of extended notes, or addenda, for his volume. I mention the circumstance because it may account for the invective tone of Darwin's remarks, which were likely published without the benefit of any solicitous editorial eye.

In an addendum to his *Journal of Researches* Darwin declared that the problem of the placement of erratic boulders had been solved by the theory of iceberg transport. In this context, he attacked the new glacial theory with passion. Fueling his passion, but hidden from view, was undoubtedly his unannounced belief in transmutation, and his gradualist view of species extinction, both of which differed sharply from the views Agassiz presented in his Neuchâtel address. In his attack in the *Journal of Researches,* Darwin accepted as factual the description of moraines and of the "polished and scratched surface of the rocks" provided by Venetz, Charpentier, and Agassiz. But he sought to accommodate these facts by his notion of iceberg transport. He claimed (p. 619) to make only two assumptions in doing so: (1) that an arm of the sea extended between the Alps and the Jura during the period when the area had a more tropical character, and (2) that the elevation of Switzerland, whenever it took place, was slow and gradual. In retrospect, it is clear that numerous assumptions were embedded within these assumptions, leaving Darwin (and Lyell's) position on climate and erratic placement exposed. Thus, for example, Lyell assumed that the "proportion of dry land to sea continues always the same." Of his own theory, derived from Lyell's, Darwin credited his use of strong analogies, and his theory of a climate change "shown by reasoning, independent of the existence of erratic blocks, to be probable in a high degree." His statement on glacial theory in the addendum to his *Journal of Researches* concluded with a condescending understatement: "whether this is the case with the theory of M. Agassiz, I leave the reader to decide."[33]

In challenging Agassiz's views Darwin was, probably without his knowing it, taking on a most worthy opponent. In 1838 Agassiz had already visited Britain twice, in connection with his work on fossil fish, and was well known to a number of prominently placed British geologists, among whom were Lyell, Sedgwick, and Buckland (who was in fact Agassiz's guest in Switzerland in 1838). Thus, Agassiz was ideally situated to become a spokesperson for the new glacial hypothesis.

As an indication of the general state of knowledge at the time on the question of erratic boulders, here are Sedgwick's comments to Agassiz in an important and richly detailed letter of 5 March 1838. Agassiz had written to him requesting an opinion of his 1837 Neuchâtel address. The English translation of the address had in fact not yet appeared, but Sedgwick offered his own opinion on the problem of erratic boulders:

> On the subject of the erratic blocks of Switzerland it strikes me that no one can possibly account for them without the aid of the carrying power of ice. Without knowing what it is, I am, therefore, favourably disposed

> towards at least a part of your hypothesis. A great deal of evidence, both positive and negative, has been advanced in favour of the iceberg theory. For example, Mr Darwin has shown that throughout South America erratic blocks are found within the limits of latitude where glaciers are, or may have been, down to the level of the sea; and that they are wanting in the tropical latitudes, where ice could never have existed near the sea level. In England . . . , we have a most interesting series of erratic blocks. I don't think the iceberg theory can be applied to them, because they go in almost all directions, and not towards any prevailing point of the compass, and because they follow the exact line of water-worn *detritus* and comminuted gravel. Such blocks I attribute to currents produced during the periods of elevation and unusual violence. There are many instances of rocks grooved deeply, and partially rubbed down, by the currents of what we formerly called *diluvium,* a word which is passing in some measure out of use in consequence of the hypothetical abuse of the term by one school of geologists. There are very fine examples of this kind near Edinburgh. Stones transported in this way are always rounded by attrition, and in every question about the origin of erratic blocs we ought to regard their condition (viz. whether rounded or not), as well as their geographical relation to the parent rock.[34]

In this passage Sedgwick took note of the strengths and weaknesses of the iceberg explanation for the distribution of boulders, citing Darwin along the way. (Knowledge of Darwin's work presumably came to him through Henslow.) He noted that erratic boulders "go in almost all directions" in England and, in his opinion, were best explained by currents produced during periods of unusual violence. Sedgwick also remarked on the decline of currency in the term "diluvium," though he obviously thought that the term had value if not abused.

As indicated by the comment in Sedgwick's letter, Darwin was also a well-regarded spokesman for the iceberg hypothesis. He had witnessed the descent of glaciers to the sea, and the presence of rock-bearing icebergs in the Beagle Channel. By extensive reading, he had investigated the distribution of erratic boulders worldwide. He was well equipped to serve as the explicator of Lyell's views on the subject of boulders. Like Agassiz in 1838, he was also at a high point in self-confidence and ready to defend his views.

Yet, in all remarks in the *Journal*'s addendum, glaciers themselves are rather offstage: climate, erratic boulders, icebergs—these are Darwin's chief concerns in 1838. But, within the year, Darwin did begin to acknowledge a view of glaciers distinct from his own. In 1838 he did field research at Glen Roy in Scotland, where erratic boulders also came into play. In his long article stemming from that research, Darwin touched only briefly on the subject of glaciers. But when he did, he included his own characterization of them—"glaciers, the parents of icebergs"—and the second view, "glaciers of the Alps" that he can now imagine, not descending to some former sea, but as "appendages on a greater mass of snow accumulated on far loftier chains."[35] The beryl-blue glaciers of Tierra del Fuego were no longer the only models for glaciation. By 1839 Darwin had allowed some geological role for glaciers that did not reach down to the sea. He had begun to acknowledge the glaciers of Switzerland.

Darwin's 1837–1842 Research in Shropshire, Scotland, and Wales

Darwin's geological field researches abruptly diminished in quantity after his return to England from the *Beagle* voyage. There was the occasional pursuit of an interesting topic, as with his research on the role of earthworms in making topsoil, a subject that his cousin Elizabeth Wedgwood characterized as the "Maer Hypothesis" from its Wedgwood origins.[36] But never again did Darwin seek to survey a whole new area geologically, as he did routinely during the voyage. Instead, his major efforts were directed toward processing and publishing work already completed in the field.

His limited fieldwork in geology resulted from deteriorating health, though his marriage and new commitment to work on the species question were also factors. In mid-1838 he reminded himself that he could travel more easily if he did not marry:

> If *not* marry | Travel. Europe, yes? | America????
> If I travel it must be exclusively geological United States, Mexico Depend on health & vigour & how far I become Zoological
> If I dont travel.—Work at transmission of Species—[37]

Darwin did in fact marry the next year, 1839. And his reference to "health & vigour" indicates that he was already questioning his own physical prowess, though he was still a young man. As it turned out, after the voyage he never again traveled outside the British Isles, so that his plans for overseas research trips failed. With that in mind, what geological fieldwork Darwin did manage to do after the voyage stands out all the more.

As documented by his extant field notes, Darwin did some geological fieldwork while visiting Shropshire in the summer of 1837.[38] He was at Shrewsbury for about a week spanning the end of June and the beginning of July.[39]

He had a longer visit to Maer Hall in Staffordshire and to The Mount, his father's home in Shrewsbury (Shropshire), from 25 September to 21 October, when he may have continued this work, though I am aware of no dated extant notes to that effect. He also undertook an extended and carefully planned geological field trip the next year. He had intended the trip for July or August but moved it up for reasons of health. He wrote to his cousin Fox on 15 June,

> I have not been very well of late, which has suddenly determined me to leave London earlier than I had anticipated. I go by the steam-packet to Edinburgh.—take a solitary walk on Salisbury crags & call up old thoughts of former times then go on to Glasgow & the great valley of Inverness,—near to which I intend stopping a week to geologise the parallel roads of Glen Roy,—thence to Shrewsbury, Maer for one day, & London for smoke, ill health & hard work.—[40]

In all, Darwin's tour lasted from 23 June through 1 August and included all the stops he had described to Fox. Manuscript material from this trip includes notes from Darwin's one day at Salisbury Crags, and a field notebook from his "eight good days" at Glen Roy and a folder of notes entitled "Alluvium Shropshire 1838 July."[41] Darwin's next, and last, major geological field trip was four years later. He noted in his Journal in June 1842 that he "Went to Maer, June 15th to Shrewsbury; & on 18th to Capel-Curig, Bangor, Carnarvon to Capel-Curig; altogether ten days, examining glacier action."[42]

Before discussing the two publications that emerged from these field trips, we might ask whether there were common themes connecting them. First, it is instructive to consider what these trips were not about. Darwin did not join his colleagues in the Geological Society in exploring the older strata. The one group of notes he did write regarding British strata is devoted largely to showing how their origin might be explained by movements of elevation and subsidence.[43] Second, his ventures were largely into familiar territory: Wales, which he knew from boyhood and from his explorations with Sedgwick, and the environs of Edinburgh, with which he was familiar as a student of Jameson. Glen Roy was new, though still in Scotland. Third, the themes that connect these trips are a concern with superficial deposits—once called "diluvium"—and with superficial shaping of the land, as with the parallel roads of Glen Roy. These were themes from the voyage. Before elaborating on them, acknowledgment should be made of Darwin's one day at Salisbury Crags.

Darwin intended to "take a solitary walk on Salisbury crags & call up old thoughts of former times." Was he at the age of twenty-nine already simply nostalgic for his student days at Edinburgh? Probably not. Rather, he was revisiting a site that had once been incorrectly interpreted to him in a field lecture by Jameson who had presented a Wernerian interpretation of the site. Darwin would have returned to view the site through Huttonian eyes.

Salisbury Crags is presently regarded as an igneous sill. It is situated on the north side of Arthur's Seat, an extinct volcano. Darwin did not use the word "sill" in his notes from the day, though he described the horizontality of the igneous bed in such a way as was consistent with the term "sill." Arthur's Seat he described as an "old submarine volcano" and compared some of its rocks to those he had seen in the Andes. Darwin's notes also referred to a description of the site published in 1812.[44] An additional feature of interest in the notes is Darwin's statement that in judging the origin of the Salisbury Crags, "There can be only 4 theories proposed." While the theories are not of sufficient interest to recapitulate them here, Darwin's listing of mutually exclusive alternatives is similar in form to the one he used in reasoning about the origin of the parallel roads of Glen Roy.

In order to evaluate these individual researches on superficial deposits, it is necessary to place them in two contexts: one provided by the tradition represented by Cuvier, Buckland, and Agassiz; the other represented by Lyell.

Darwin's interest in superficial deposits was common to geologists in the 1830s and 1840s, but there was an additional factor underlying his interest that not all shared, at least not to the same degree. These deposits contained remains of numerous extinct animals, as Buckland's research had emphasized. All geologists were necessarily involved in the interpretation of these remains, but when Darwin became a transmutationist in the spring of 1837, the question acquired greater urgency. Were the extinct creatures replaced by a "new creation"? Or were the extinct creatures replaced by the species that were their transmuted descendants? Those working in the tradition of Cuvier claimed the former. Agassiz was one of them. In describing the effect of the great refrigeration that had come to the earth signified by glaciers, Agassiz drew an unforgettable image of the effect of the great cold: "Death enveloped nature in its winding-sheet." Later in his address he suggested that

> the diminution of the temperature of the globe may be expressed by the following line:—
>
> 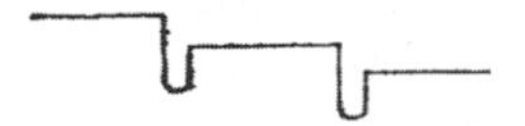
>
> Thus, the epoch of extreme cold which preceded the present creation, has only been a passing oscillation of the temperature of the globe. . . . It was attended by the disappearance of animals of the diluvian epoch of

> geologists, as the mammoths of Siberia still attest. . . . There was thus a complete separation between the existing creation and those which have preceded it; and, if the living species sometimes resemble in our apprehension those which are hid in the bowels of the earth, it nevertheless cannot be affirmed that they are regularly descended from them.[45]

For a transmutationist, present species are by definition descended from their temporal predecessors. This was the very point that Agassiz could not affirm. By their differing views of extinction and its aftermath with respect to life, these two young men, Agassiz and Darwin, were set apart from each other.

If the most profound resonance attending the question of superficial deposits was (in Agassiz's terminology) "physiological," the most significant, purely geological aspect pertained to the understanding of forces shaping the surface of the land. Herries Davies has suggested that Hutton's understanding of the wearing away of continents, and Playfair's deep appreciation of the erosional powers of rivers were insufficiently taken up by their immediate geological successors. He also has suggested how, instead, Lyell's views of the primacy of marine erosion held the day.[46]

This point is particularly apropos of Darwin, who followed Lyell in emphasizing the power of the sea to shape the land. After considering and rejecting Playfair's views emphasizing the power of rivers, Darwin wrote while in Chile,

> The ultimate conclusion which I <draw> have come to—Is that primarily the lines of elevation determine the figure of a Continent; secondarily that a gradually *retreating* ocean models the elevated points; smooths with so called Diluvium some of its asperities, determines the directions of the great slopes—&<sometimes> <not infrequently> even sometimes excavates the < . . . > valleys.—Thirdly we have rivers with aid of floods & subterranean movements modifying. altering & generally deepening those lines & depressions left by the preexisting ocean.—[47]

Obviously, as shown by his several marks of deletion, Darwin was struggling with these ideas, particularly on the question of the relation of rivers to their valleys. (In contrast, Playfair had written that "rivers have, in general, hollowed out their valleys."[48]) In the end for Darwin, elevation and the work of the "retreating ocean" were the primary agents for the surface of the land. He was thereby prone to favor an agency for explaining the distribution of diluvium/alluvium that relied on the action of the sea. Meteorological phenomena—rain, wind—were not candidates as primary geological agencies, nor were rivers, nor, initially, were glaciers.

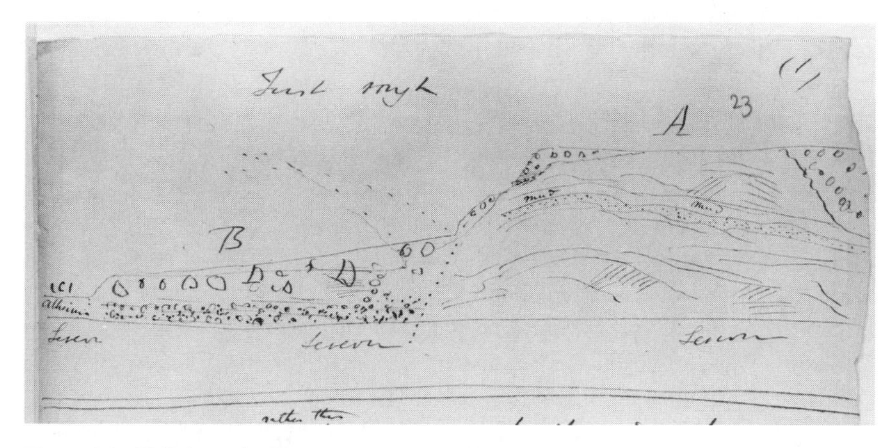

Figure 8.2. Cliff above the Severn River. From DAR 5:B23–B29. By permission of the Syndics of Cambridge University Library.

There were thus two preferences that Darwin brought to treatment of superficial deposits: for a noncatastrophic explanation of species extinction, and for a ranking of the sea as the most important factor in shaping the surface of the land. These preferences had developed during the course of the voyage, and remained strong in Darwin's mind after it.

Let us begin with the two sets of notes entitled "Alluvium," one set made in July 1837, the other in July 1838. The first set is longer and begins with a drawing of a cliff, on his estimation 100 feet high or more, above the Severn River. Heavy vegetation in the area hinders the search for precise locations. Beds of "alluvium" are clearly labeled on Darwin's sketch (Figure 8.2). In his notes Darwin used the terms "alluvium," "gravel," "scattered irregular fragments of rock," "blocks more angular than in gravel beds and often of larger size," and "shingle." He did not use the phrase "erratic boulders" or the term "diluvium." In studying the beds Darwin carefully looked for shells and recorded a number from one of the beds, but not the other. In interpreting the beds Darwin also used the vocabulary commonplace to British stratigraphic geology ("conformable" and "unconformable") and of its classifications ("New Red Sandstone"). In his 1837 notes he observed that in explaining the origin of a particular bed of pebbles, he had rejected a "deluge, which I first felt inclined to invoke" in favor of an explanation that depended on the "currents of the sea." His did so, he said, because a deluge "would have cut the red sand unconformably."

Darwin's notes on Shropshire alluvium dated 1838 are briefer, but useful in showing Darwin's development since the previous year. He remained interested in whether beds possessed marine shells and in the composition, size, and shape of superficial rocks, but now a new agency was introduced: transportation of superficial matter by glaciers that had once bordered the sea. After comparing a formation resembling a "red earthy mass" to a similar one in the Pampas, he suggested that "I do not doubt in all these cases it will turn out to be estuary mud, with glaciers acting.—"[49]

The issue of alluvium also figured prominently in notes from Glen Roy, which he explored immediately before revisiting Shrewsbury in July 1838. For a map of the region of Scotland that includes Glen Roy, see Figure 8.3. Glen Roy drew Darwin's attention because of its "parallel roads." The origin of its "alluvium" was important but secondary. In Glen Roy Darwin saw an opportunity to reanalyze a landscape described by two preceding observers. Both of these men had given such complete accounts of the phenomenon of the "parallel roads" at Glen Roy that in his own eight days at the site, Darwin could rely on their work as his guide. However, he did add information, as, for example, when he remeasured elevations at various points using his mountain barometer.

John MacCulloch (1773–1835) and Thomas Dick Lauder, the authors of the two previous geological accounts of the parallel roads, differed sharply from Darwin in their initial approach to the problem. Working independently of each other at about the same time, each man was struck by the exceptional nature of the roads. MacCulloch described the roads as a "solitary phenomena" "of which no other similar example has yet been discovered."[50] Lauder expressed his interest more romantically: on an "accidental ramble through that valley, in the course of a pedestrian tour in the West Highlands" his curiosity was "much excited" by the roads. He later returned to the area "with the purpose of endeavouring to put myself in possession of all the facts I could possibly collect, regarding these curious appearances."[51] In contrast, Darwin approached the topography as similar, if more defined, to what he had seen at Coquimbo in Chile.

MacCulloch's and Lauder's articles still read extremely well. Not only did they correctly identify the roads as former lake beaches, but also their individual techniques of investigation and argument seem exemplary on many points. Their work was easily assimilated into later glacial interpretations.

For MacCulloch and Lauder the experience of the parallel roads was both romantic and scientific. MacCulloch, a friend of Walter Scott (1771–1832), noted the traditional Highlanders' interpretation of the "roads" (possibly useful for the purposes of hunting) as the work of "Fingal and the heroes of his age" and acknowledged that in the roads a spectator could well possess the "feeling that he is contemplating a work of art." Engravings of his own draw-

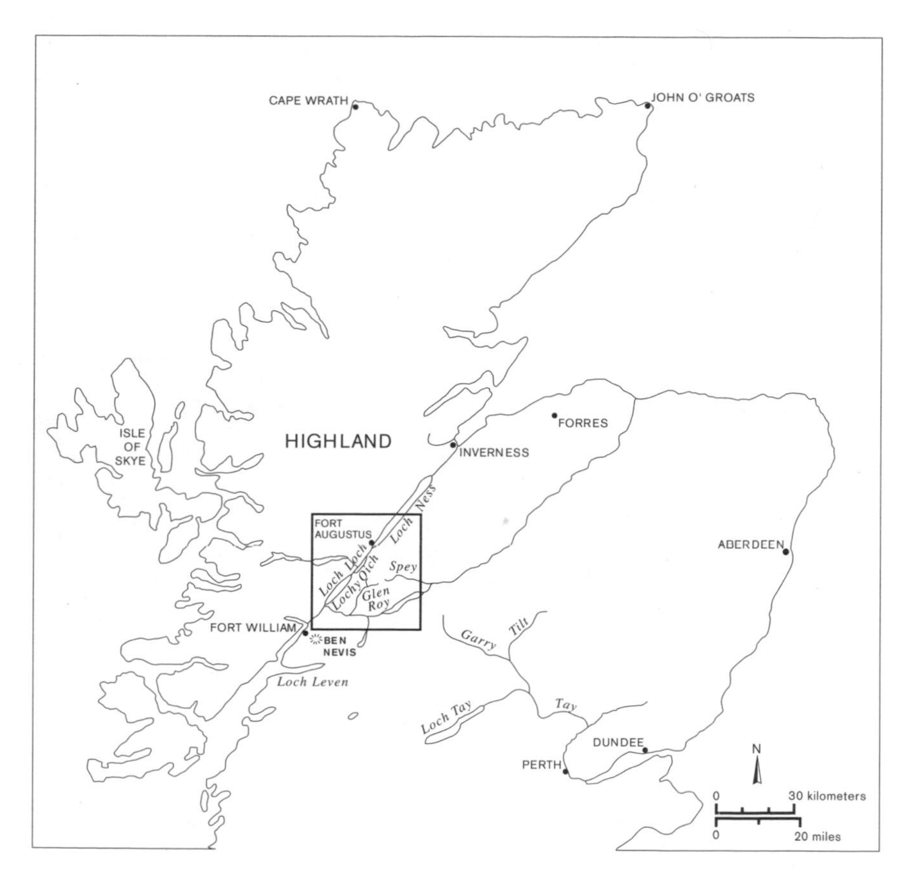

Figure 8.3. Scotland. UMBC Cartographic Services.

ings accompanied his article, and he regarded them as providing intelligibility to his paper "which words alone cannot describe."[52] Darwin modified one of MacCulloch's profile sections of the Glen Roy roads for use in his own article.

The "wild and sublime in Scottish scenery" held equal appeal for Lauder, whose sensitivity to the beauty of the landscape was shaped by the artist George Fennel Robson (1788–1833), author of *Scenery of the Grampian Mountains.* Speaking as a practical man, Robson attributed the growing interest of the "opulent classes" in Highland scenery to their "exclusion" from the Continent during the Napoleonic wars.[53] Whatever its origins, aesthetic as well as geological interest in Highland scenery continued into the next decades. A gifted artist himself, Lauder's sketches illustrate his article. One of them is reproduced as Figure 8.4. This drawing shows one aspect of the

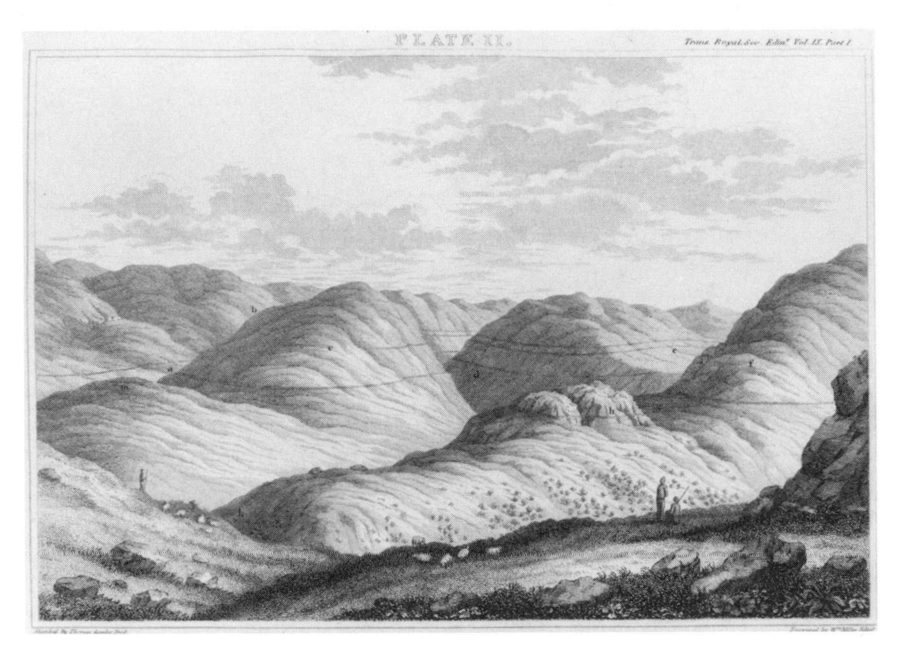

Figure 8.4. Lauder's view of the mouths of Glen Roy and Glen Spean. By permission of the Syndics of Cambridge University Library.

"roads" Lauder particularly emphasized, that is, that they encircle isolated prominences, which he took to argue for their origin as lakeshores. Lauder also compared the Lochaber region, with its steep-sided valleys, to the Alps in Switzerland, which he had visited. Lake Geneva was one of the sites he compared to those in Scotland.

On the scientific side, Lauder was not the equal of MacCulloch in describing rock types (for example, he referred one rock simply to the "primitive series"), but he was superior to MacCulloch in technical virtuosity for having brought with him a "Mr. Maclean," a civil engineer whose work was instrumental in translating the topography of Lochaber into a problem of understanding the flow of water in the region. Lauder, accompanied by Maclean, attended to both the natural beauty and the patterns of drainage of Lochaber. Both endeavors formed part of the understanding of the area's "topography," a term he used repeatedly.[54]

Darwin continued in the tradition of supplying a graphic rendering of the scenery at Glen Roy, though he was not the artist. For his article he procured a sketch drawn by Albert Way, a friend from Cambridge days. Way described it as done "with a good deal of detail" but without "Effect—that is no heavy showers passing over the middle plane of the landscape" and "no powerful results of light & shadow."[55] (This sketch is Figure 7.1.)

Let us now turn to the identity of the roads. MacCulloch keenly felt the lack of a geographical survey for the country. He did, however, measure the altitude of the roads. He focused primarily on the roads within Glen Roy itself. He found a 12-foot difference between the highest road at Glen Roy and one above it at Glen Gluoy, though it was Lauder who pressed the importance of that higher road. MacCulloch's figures that Darwin used for the three roads within Glen Roy were 972, 1184, and 1266 feet above sea level.[56]

In interpreting the roads, MacCulloch was impressive in argument and honest in admitting difficulties. He considered that the roads might have been man-made, though he thought it unlikely. He preferred the view that the horizontality of the roads suggested the action of water. He then set out three possible causes: (1) a succession of deluges caused by local elevations of the land, (2) water terraces formed as ancient riverbanks, and (3) a former lake. Almost as an afterthought—perhaps one suggested by members of his audience at the Geological Society of London—MacCulloch suggested a fourth alternative, that the lines or roads represented former sea beaches. He was clearly not impressed by the likelihood of this explanation. Rather he favored the interpretation that the roads of Glen Roy represented the beaches of a former lake that had subsided at three different intervals. He acknowledged the difficulty in explaining the creation and removal of barriers damming the lake, but suggested that the action of rivers might have played a role.[57]

In his treatment of the problem of the roads, Lauder cut to the chase. He put forward the lake theory as the best available explanation for their origin. His explanation was an advance over MacCulloch's in that, with the assistance of an engineer, he had worked out a plausible history of the drainage pattern for the area. "For the area" is a significant phrase since Lauder, more than MacCulloch, went beyond Glen Roy proper in his study. Lauder summarized his conclusions regarding the drainage of the lakes, whose beaches formed the presently visible "roads," in a series of diagrams, as well as in words. The diagrams are reproduced as Figure 8.5. Lauder thus outlined a temporal pattern for the successive production of the several "roads" or shorelines, and used visual illustrations to make his point. (In its graphic form his work brings to mind Darwin's later verbal and visual treatment of the temporal sequence of development of coral atolls.) Like MacCulloch he faced the problem of explaining the removal of the barriers damming up the lake or lakes. His explanation was a convulsive movement of the land, though he (like MacCulloch) was suitably aware that "it is much easier to suppose the existence of former barriers, than to discover the means which operated in their removal."[58]

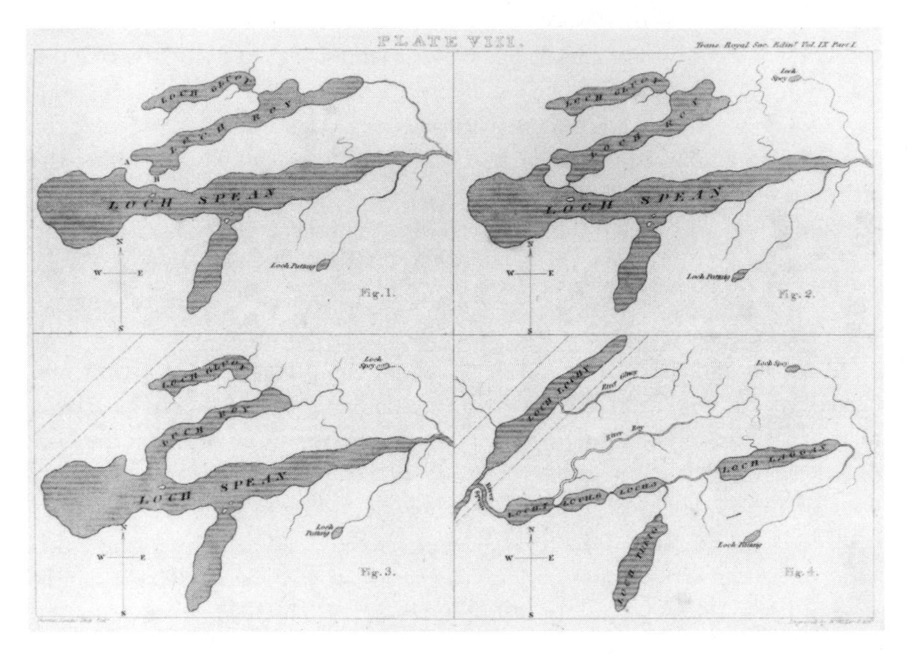

Figure 8.5. Lauder's theory of the formation of the roads of Glen Roy: 1. primary state of all the lakes, 2. state of Loch Roy after its first subsidence, 3. state of matters after the second subsidence of Loch Roy, 4. change produced by the opening of the Great Glen of Scotland. By permission of the Syndics of Cambridge University Library.

When Darwin entered Glen Roy in 1838, he had MacCulloch's and Lauder's articles firmly in mind and probably in hand. His field notebook on Glen Roy refers routinely to their observations and arguments. But he also brought to Scotland the memory of his South American experience, and that led him to favor the action of the sea in forming the roads. His predilection for a marine interpretation is apparent from the notes from Thursday the 28th of June, his first day in the field. He began with discussion of the "buttresses of alluvium" at the upper end of Loch Dochart. Of them he concluded, "Rivers could not have deposited it. Barrier of lake very lofty, & no trace of it; to the Sea more probable." Similarly, later in the week, he noted, "In all cases <I urge> deposition marine—because if not chain of lake[s] & if so there would be barrier[.]—[59]

In the roughly twenty-year interval separating Darwin's field trip to Glen Roy from those of MacCulloch and Lauder, superficial deposits had received critical scrutiny. Empirically, MacCulloch's and Lauder's descriptions of what they both termed "alluvium" were superb. They described the shape of superficial boulders and their placement. Lauder, for example, noted that above the highest shelf was "bare moorish soil" like that "covering any other mountain," while below that line were "large depositions of alluvial clay, sand, rounded pebbles, and gravel." He also noted the lack of "marine exuviæ" in the valleys.[60]

Darwin published his Glen Roy work within months after his fieldwork. He argued for a marine interpretation of the "roads" of the Glen Roy region. His chief weapon against MacCulloch's and Lauder's lacustrine interpretations was their lack of an adequate explanation for the barriers that must have dammed the lakes. Darwin wrote, "It is a startling assumption to close up the mouth of even one valley by an enormous imaginary barrier; to do this with all would be monstrous."[61] Once he had dismissed "these views which cannot be admitted" (an exclusionary principle he was later to regret), he discussed his own interpretation, which drew heavily on his knowledge of elevatory movements elsewhere in the world. To account for alluvial matter at Glen Roy, he called on observations of similar sand and shingle deposits in South America. (Like MacCulloch and Lauder before him, he could not find marine remains, though he did report discovery of shells elsewhere in Scotland, as well as in Shropshire and Staffordshire.)[62]

He then addressed the subject of the "erratic boulders of Lochaber." He argued that the granite boulders he saw scattered at various sites must have come from at least five or six miles away. He next discussed the only two "worthy" theories of transportation of erratic boulders: "great debacles" and "floating ice." His limitation of the choices is interesting because at the time of his writing, Darwin was already aware of Agassiz's glacial theory. He referred to material presented in his *Journal of Researches* for support of his own views.[63]

Let us pause now for a moment to assess how matters stood with regard to Darwin's interpretation of erratic blocks as of 1839. He had a mechanism of transportal (icebergs), a theory of climate change provided by Lyell, and several postvoyage field experiences. His understanding of elevation and subsidence served as a foundation for these views. Moreover, Darwin constructed an interpretation that was in consonance with the work of other researchers, including not only Lyell but also Murchison. In 1839 Murchison published *Silurian System*, whose importance in the history of geology is universally acknowledged as establishing the identity of a major division of the geological record. What is less well known about Murchison's book is that it represented a new chapter in the treatment of superficial deposits—the question of the "diluvium." Murchison took the issue head on, assigning a new term—"drift"—to cover "coarse and sometimes far transported fragments, to which some geologists apply the word 'diluvium'." A "second class of alluvia" included "all the deposits formed in lakes and river courses," deposits from calcareous springs, and subaerial deposits.[64] Thus, Murchison in effect preserved Buckland's distinction between diluvium and alluvium, but with a new term for the first. (In his private notes, Darwin sometimes continued to use the term "diluvium" into the 1840s.[65] The newer vocabulary, however, inserted itself fairly quickly into his writing.)

In *Silurian System* Murchison devoted five chapters to the subject of superficial deposits. The Silurian region proper was covered with "local detritus only" but "passing its northern or eastern limits, we enter districts where a large portion of the accumulations are associated with others which have been transported from Cumberland, and probably even from Scotland." To these transported materials Murchison applied the term "Northern Drift." He believed these had been delivered to their present position by icebergs. The materials were thus submarine deposits. It was also in these chapters that he quoted Darwin extensively. Darwin had given him a copy of the printed but unpublished *Journal of Researches*, and Murchison made full use of it. For example, he wrote,

> But Mr. Darwin is not satisfied with showing, that the coast of former European islands were in all probability the seats of great icebergs; he pursues his argument . . . and . . . points out the *absence* of erratic blocks in the intertropical regions. . . . Thus, the conditions of the difficult problem which we have to solve are now much more fully brought before us than in any former discussion of the subject.

Murchison also incorporated material from Darwin's research trips of 1837 and 1838 into the Shropshire/Shrewsbury area. Of particular use was the marine shell Darwin had observed in the "deep gravel pits south of Shrewsbury."

There seems to have been some reciprocal influence in play on this point, for in his Glen Roy article Darwin had suggested that Murchison's discovery of seashell fragments in Staffordshire and Shropshire led him "to examine many gravel pits there . . . and that when found, the fragments [of shells] are generally exceedingly few in number and partially decayed." Darwin concluded from this that the preservation of shells "May be considered as a remarkable and not as an ordinary circumstance." (The absence or paucity of marine shells in "drift" was a circumstance that ran counter to the iceberg hypothesis, so that proponents of the hypothesis seized on any evidence, however meager, of a formerly marine environment.) In a general way, Murchison also supported the emphasis Darwin placed on the elevation and subsidence of the land.[66] Thus, as of 1839, Lyell, Murchison, and Darwin all subscribed to a similar interpretation of superficial deposits. It was a position with its own theory (iceberg transport of rock) as well as a developing vocabulary ("drift").

With regard to the introduction of new vocabulary, Lyell's article on the eastern Norfolk drift was also significant. In this article Lyell introduced the

Lowland Scots farmers' name of "till" for unstratified drift. He also compared the appearance of terminal moraines of glaciers to the accumulations "where drift ice, charged with mud, sand and blocks, melts, and the earthy materials are allowed to fall tranquilly to the bottom" of the sea.[67]

With Lyell's article we have arrived at the year 1840, a momentous one in the British Isles for the interpretation of superficial deposits. It was during this year that Agassiz did two things that changed the course of the discussion. First, he published the book *Études sur les glaciers*.[68] Second, he toured parts of Scotland, England, and Ireland in the company of several leading geologists, reinterpreting the landscape from the point of view of glacial theory.

In considering Agassiz's role in establishing glacial theory, it is useful to distinguish between the quality of a scientist's work and its impact. Venetz and Charpentier had done original and brilliant work creating glacial theory, but it was Agassiz who ultimately had the greater impact. He was the stronger presence, or, in other words, the stronger author.

Agassiz dated the preface to his book August 20, 1840. He came to Britain that fall, after he finished his book. What united the two enterprises, his book and his tour, was his command of the facts and his ability to communicate with an audience. In both his book and his field excursions Agassiz excelled in directing the attention of others to glacial features. Indeed, in his article on "Geology" for the Admiralty handbook of 1849, Darwin listed Agassiz's *Études sur les glaciers* as a book that should accompany "the voyager in the Temperate and Polar regions."[69]

It was both the content and the manner of presentation of the book that made it so valuable to the novice. While Agassiz drew most of his examples from the Swiss Alps, he generalized his conclusions. Thus, there are separate chapters on the structure, external aspects, crevasses, formation, movement, temperature, and color of glaciers. (Like Darwin, Agassiz also used the term "beryl" in his chapter on color: riverlets on the great Zermatt glacier "flow over a bed seemingly cut inside massive beryl.")[70] There are also separate chapters on moraines and ice pinnacles. On interpretive matters there are chapters on the former extent of glaciers in the Alps, and on oscillations of glaciers during historical times. There is also a chapter arguing for the existence of large ice sheets outside the realm of the Alps.

Adding to the work's value as a handbook on the study of present and past glacial activity are its thirty-two plates. While all the studies we have cited in this chapter relied on drawings and diagrams of high quality, Agassiz's are unique in that they supplied explanatory overlays. For example, one of Agassiz's plates shows a view of the Viesch glacier. The separate overlay draws attention to specific features of the landscape. Figure 8.6 shows this overlay. Figure 8.7 shows the scene to which it refers. Agassiz also supplied notes to

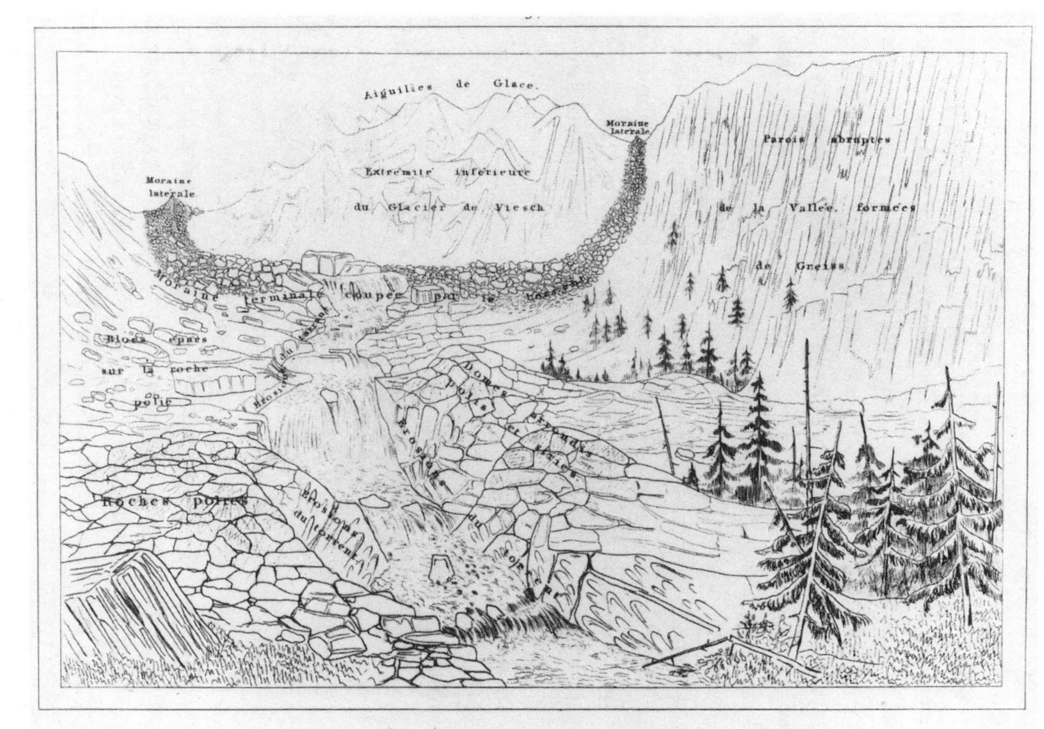

Figure 8.6. Glacier de Viesch (overlay). From *Études sur les glaciers* by Louis Agassiz.

Figure 8.7. Glacier de Viesch. From *Études sur les glaciers* by Louis Agassiz.

each plate. For Plate 9 he drew attention to such features as the "rounded domes which were polished and striated by the movement of the ice" and the manner in which the "end moraine grades without interruption into the moraine." Such written description, together with the overlays and engravings, enhanced the utility of his book. Darwin owned a copy of it and valued it. The book taught him how to recognize signs of former glaciation in local landscapes.

Not content to reach an audience beyond his native Switzerland through the medium of the written word alone, Agassiz actively proselytized in person for the glacial interpretation. He invited foreign geologists to visit him in Neuchâtel, which Buckland did in 1838. He also advanced the glacial theory abroad. A factual notice entitled "On the Polished and Striated Surfaces of the Rocks Which Form the Beds of Glaciers in the Alps" was read on his behalf before the Geological Society on 10 June 1840. Later, in the fall of that year, in a truly remarkable penetration of the British geological establishment, he presented work in person both at the Glasgow meeting of the British Association for the Advancement of Science (in September) and at the regular meeting of the Geological Society of London (in November).

Agassiz's paper to the British Association pertained to the Swiss situation, while his paper of 4 November to the Geological Society reflected his just-completed fieldwork in Britain. In between the dates of those two meetings he toured areas in Scotland and England in the company of Buckland and Lyell, assisting them in reassessing over one hundred familiar British sites. The overall theme of the 1840 fieldwork inspired by Agassiz can be taken from Buckland's exclamation that appears at the head of this chapter: the idea that, because of the new glacial theory, "the whole Globe must be examined afresh." They reported on this fieldwork at the November and early December meetings of the Geological Society.[71] Agassiz's tour of Britain in 1840 has been well described by a number of scholars, among them Boylan, Lurie, and Herries Davies in works already cited. While space does not permit reiteration of the many interesting aspects of Agassiz's tour, those pertaining to Darwin need to be addressed.

During the second half of 1840 Darwin was in poor health. He did not attend the Glasgow meeting of the British Association, nor did he accompany Agassiz on any fieldwork. In 1840 he was unable to attend many meetings of the council nor, presumably, the ordinary meetings of the Geological Society. (Attendance at ordinary meetings was not taken during this period.) Indeed, Darwin resigned from the council on 25 March 1840; at its request his name remained on the list until the next anniversary.[72] Yet it is important to recall that 1839 had brought Darwin to the forefront of geological research by virtue of his publications on South America, including the *Journal of Re-*

searches, and on Glen Roy. Now, one year later, the subject matter of these publications was of paramount interest among British geologists. Thus, although he seems not to have been present at the debates over Agassiz's glacial hypothesis, Darwin could not have helped being involved with the subject. (In later life Darwin noted that he had once met Agassiz; he mentioned having been "charmed with him." This meeting may have taken place in 1846 when the British Association convened in Southampton.[73])

The first immediate effect on Darwin of the debate came in a report in the newspaper the *Scotsman* dated 9 October 1840. The report carried Agassiz's Glasgow address, as well as early news of his fieldwork in Scotland. It also included an extract of a letter by Agassiz to Robert Jameson, who had forwarded it to the *Scotsman* since the letter had arrived just past the deadline of the current issue of the *Edinburgh New Philosophical Journal,* which he edited. In his letter to Jameson, Agassiz remarked that, as he had expected and had indeed announced at Glasgow, he found "remote traces of the action of glaciers," and the nearer he approached to the peaks of the mountains, including Ben Nevis, the more pronounced such indications became. He further claimed,

> The parallel roads of Glen Roy are intimately connected with this former occurrence of glaciers, and have been caused by a glacier from Ben Nevis. The phenomenon must have been precisely analogous to the glacier-lakes of the Tyrol.[74]

Jameson forwarded Agassiz's letter to the *Scotsman* with the suggestion that it "gives what may be considered the true explanation of the parallel roads of Glen Roy."

Agassiz expanded this point in his paper of 4 November to the Geological Society:

> Another class of phaenomena with glaciers, is the forming of lakes by the extension of glaciers from lateral valleys into a main valley; and M. Agassiz is of opinion, that the parallel roads of Glen Roy were formed by a lake which was produced in consequence of a lateral glacier projecting across the glen near Bridge Roy, and another across the valley of Glen Speane. Lakes thus formed naturally give rise to stratified deposits and parallel roads, or beds of detritus at different levels.[75]

In his Geological Society paper Buckland supported Agassiz's interpretation, while forbearing "to dwell on the phænomena of parallel terraces, though . . . convinced that they are the effects of lakes produced by glaciers."[76] It is fair to say that Darwin must have experienced that sense of having received a direct hit from Agassiz's, Jameson's, and Buckland's remarks.

Yet the news out of Scotland was not all bad for Darwin. In the article devoted to Agassiz from the *Scotsman,* Darwin was praised for his researches in the Southern Hemisphere. The author of the newspaper report believed that Darwin's researches supported Agassiz's claim that the "climate of this country at a former, but, geologically speaking, a recent epoch, was much colder than it now is." The author continued,

> [A]nd that such a state of things is consistent with the course of nature, is shown by the fact that Mr Darwin found glaciers reaching down to the level of the sea on the west coast of Chili, in latitude 46 degrees, that is, 11 degrees nearer the equator than Ben Nevis.

Glaciers that came down to the sea were very much still part of the picture. Agassiz himself credited them in his address to the Geological Society. He pointed out that

> the study of the phænomena of glaciers in different latitudes, as well as at different altitudes, together with the examination of their different effects where in contact with the sea, will introduce many modifications in the consideration of analogous phænomena in countries where glaciers have disappeared.[77]

Agassiz then went on to credit floating icebergs for the transport of blocks from Sweden to the coast of England. He did not, however, credit any of the supporters of the iceberg thesis by name, though the association of Lyell, Murchison, and Darwin with the hypothesis would have been understood by all present.

During the famous hours-long debate that concluded the Geological Society's meeting of 18 November, Whewell used Darwin's observations on the Southern Hemisphere to challenge rather than affirm Agassiz's ideas. Whewell also brought up the philosophical point that geologists were being presented with a series of alternative choices: "If we do not allow the action of glaciers, how shall we account for these appearances?" Whewell countered that it was not "within our reach at present to refer each set of phenomena in geology to its adequate cause, but that is no reason why we should receive any theory that is offered to account for it."[78] Being attuned to the physical sciences, Whewell was particularly concerned that Agassiz's claims regarding the supposed diminution of the temperature of the earth be addressed. Whewell also wondered "where we should obtain mountains as *fulcra* for glaciers." Agassiz in his speech to the Geological Society had adduced sheets of ice resembling those presently existing in Greenland to explain deposits of unstratified gravel, but Whewell was obviously not ready to employ an analogy drawn from so distant and unfamiliar a place. For Whewell, as of November 1840, the existence of mountains was a necessary condition for glacial activity.

By the end of Agassiz's glacial season in Britain, Darwin's career in geology had been effectively recharted. Thereafter, he was on the defensive regarding his explanation of the parallel roads of Glen Roy. He was also charged, as were his peers, with examining landscapes afresh in the search for signs of glacial action. Increasingly, this charge came to be one of discriminating among various possible causes—icebergs, running water, rivers, glaciers, subaerial effects—in interpreting the origin of superficial deposits. Rather lightheartedly Murchison had declared after the Glasgow meeting of the British Association that "Agassiz gave us a great field-day on Glaciers, and I think we shall end in having a compromise between himself and us of the floating icebergs!"[79] In practice, the compromise proved difficult to manage. As Boylan has shown, the council of the Geological Society actively suppressed the more wholehearted expressions of support for Agassiz's all-out glacialism. Murchison maintained his resistance to glacial interpretations, and Lyell retreated from his initial strong support. Buckland remained Agassiz's supporter (with qualifications), but when Buckland became dean of Westminster in 1845 he devoted the bulk of his time to church administration rather than to geology.

During the spring of 1841, a few months after Agassiz's British tour, Darwin sent him an admiring and conciliatory letter, noting that "I have lately enjoyed the pleasure of reading your work on Glaciers, which has filled me with admiration."[80] Surprisingly Darwin also sent along a copy of his own *Journal of Researches*, while apologizing for the manner in which he referred to Agassiz's work on glaciers in the book. Agassiz seems not to have responded to Darwin's overture, and the two men never established an ongoing correspondence. In any case, Agassiz's views were now firmly planted in Darwin's mind. Further, Agassiz's highly visual treatment of the subject was especially valuable to Darwin, who, for reasons of health, was unable to travel far. Initially Darwin used Lyell's copy of Agassiz's book, but he seems to have purchased his own, since there is a copy of it in Darwin's library.

Also in March 1841 Darwin corresponded extensively with Lyell on the subject of Glen Roy and on glacial topics generally. Darwin wrote, "I think I have thought over [the] whole case without prejudice & remain firmly convinced they are marine beaches.—"[81] Darwin had his detailed work on the roads to support his argument. Agassiz's fieldwork was sketchy by comparison. At the same time, Darwin continued to challenge Agassiz on his interpretation of Swiss phenomena. Citing Agassiz against himself, Darwin stated that ordinary glaciers could not have transported the large angular erratic blocks to the Jura (smaller, rounded boulders should be expected), a premise that led Darwin to conclude, in arguing the case to Lyell, that

> if an hypothesis is to be introduced, the sea is much simpler.—floating ice seems to me to account for every thing as well as, and some[times] better, than the solid glaciers.—The hollows, however, formed by the ice-cascades, appear to me strongest hostile fact,—though certainly, as you said, one sees hollow round cavities, on present rock-beaches.—[82]

This passage is indicative of the tenor of many of Darwin's remarks during the period: a holding to the "simpler" ideas of his own geology, while assimilating new vocabulary and facts associated with glacial theory.

Very much along these lines is his paper submitted to the Geological Society on 14 April 1841 and read on 5 May entitled "On the Distribution of the Erratic Boulders and on the Contemporary Unstratified Deposits of South America."[83] He remained true to his original judgment that icebergs had deposited the erratics. He also continued to press the analogy of his South American cases to those in Switzerland, arguing, for example,

> Had the space between Chiloe and the Cordillera been converted into land, the boulders, in their position with respect to their probable parent rocks, in their size and angular shape, would have resembled those on the Jura; the blocks of granite now lying between the islets, being the representatives of those which, M. Agassiz has lately shown, occur in the interior valleys of that range.[84]

Yet in this article Darwin adopted the current terminology of "erratic boulders," "till," and "moraine." He also allowed for the possibility that Agassiz's glacial hypothesis might be used to explain the case, and he was willing to argue difficulties on both sides. Further, he distinguished situations that favored one explanation over the other. Where boulders were found on land that he believed from other signs had been recently submerged—as at the head of the Santa Cruz River, where the form of the land suggested its having been molded by the sea—he argued for erratic boulders having been deposited by icebergs.

Darwin also emphasized in his article that he believed angularity was connected with transportal by icebergs; sheet ice, he thought, would cause more attrition to the rocks. Where, as in Tierra del Fuego, sheet ice was more easily imaginable, boulders presently on shore displayed less angularity, as if, he said, they had been worn down while resting on a sea beach. As an aside, it should be noted that in 1841 Darwin described the existence of furrows on a sandstone formation at Pernambuco, which he interpreted as an indication that not all furrows on rocks were glacially caused.[85]

On 20 July 1841 Charles and Mary Lyell departed from Liverpool for a year's trip to America. The Lowell Institute of Boston had invited him to give a series of geological lectures.[86] Lyell's departure left Darwin without his regular geological correspondent. Thus, for a year, we have little in the way of a detailed or intimate record of his reflections on the glacial issue, or geological topics generally. However, we do know that an important publication took place on 15 December 1841 that aroused Darwin's interest. Buckland reported on his field research in Snowdonia, the area that included Snowdon, at 3560 feet the highest mountain in Wales (Figure 8.8).

This territory was well known to Darwin from childhood, and he could picture the sites under discussion. Buckland reported finding evidence of glacial activity in the area. He cited the work of previous researchers who played important though not major roles in the glacial controversy. One was Joshua Trimmer (1795–1857), who was most famous for having identified marine remains at Moel Tryfan, a mountain "1000 ft above the sea" in North Wales. In a letter to Buckland, published by the Geological Society in 1831, Trimmer interpreted the marine shells as partial indications of the remains of "extensive diluvial action." These marine remains are presently interpreted as the result, not of diluvial action (as Trimmer suggested) or of marine submergence (as it appeared to Darwin), but of the motion of "ice-dredged marine sand" over the land.[87] As Herries Davies has emphasized, these marine remains at Moel Tryfan (whose height he gives as 1392 feet) formed a considerable obstacle to the early acceptance of a glacial hypothesis for the landscape of North Wales.[88]

Another person who figured in the interpretation of North Wales was John Eddowes Bowman (1785–1841). Having firsthand acquaintance with the appearance of glacial remains in Switzerland, Bowman set about to assess North Wales with his memory of Swiss features in mind. His open-minded search for glacial features in Wales is indicated by the title of his article, published in the last year of his life: "On the Question, Whether There Are Any Evidences of the Former Existence of Glaciers in North Wales?"[89] In his search Bowman believed that he could find no unequivocal signs of the work of former glaciers. From the point of view of the effect of Bowman's work on the subsequent history of glacial exploration in North Wales, Bowman's overall conclusion was soon overridden, but he did draw attention to the subject. Especially interesting from the standpoint of Darwin's work is that Bowman visited the area that Darwin would make his own. Bowman described it as follows:

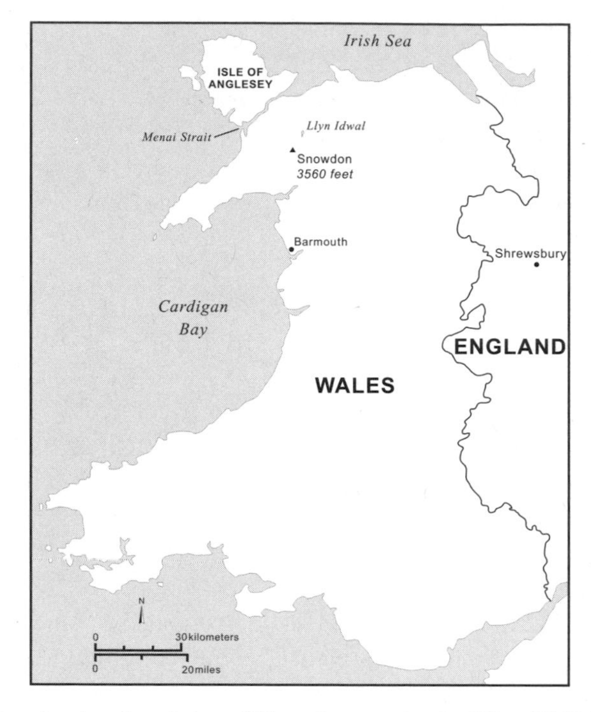

Figure 8.8. Map showing the relation of Shrewsbury to sites in Wales. UMBC Cartographic Services.

> The spot is well known to tourists as one of the wildest and most alpine in North Wales, where the torrents from Llyn Ogwen and Llyn Idwal, on issuing from their respective lakes are precipitated over the rocky barrier that forms the head of Nant Francon.

Bowman did find rounded bosses of rock at the site but interpreted them as having been formed by stone and ice carried by the torrents flowing from the lakes. Darwin would visit the site the next year and assign a glacial interpretation to the same phenomena.

Given Bowman's well-posed questions regarding glaciation, and Trimmer's evidence of marine remains at high elevations, a person of exceptional competence was needed to enter North Wales to assess the situation. Buckland was this person. Accompanied by the mining engineer Thomas Sopwith (1803–1879), Buckland investigated Snowdon and its environs in October 1841, not long after Bowman's visit. (Buckland did not visit the area around Llyn Ogwen and Llyn Idwal cited by Bowman, though he did note Bowman's work in other areas, and agree with portions of his interpretation.) Buckland concluded that the seven principal valleys of Snowdonia displayed those phenomena that Agassiz had identified as indicative of glacial action: "obtusely rounded, dome-shaped rocks—furrowed, fluted, striated and polished surfaces—and the occasional accumulation of mounds of detritus with boulders resting on their surface."[90]

Presuming that others would follow in his footsteps, Buckland described the valleys in the order in which they "may be most conveniently visited by persons approaching Snowdonia by the Holyhead road." In the valley of Conwy he noted striations in the rock, similar to those near Edinburgh. In the valley of Llugwy he noted dome-shaped rocks, over which he believed glaciers had passed. He described the valley of the Ogwyn as one of the several sources of glaciers leading to the valley of Nant Francon. The valley of the Gwyrfais he associated with a "field of ice." For the valley of the Nantel or Lyfai he noted polished surfaces with striations in the direction of the valley, a usual sign of glacial motion. In interpreting the region, Buckland relied on three agencies: (1) a version of traditional diluvialism (an "overwhelming stream of water" or a "rush of water" or "violent inundations"), (2) glaciers, and (3) icebergs, as described by Lyell in his Norfolk paper of 1840. In closing, Buckland observed that he would not comment "on the general subject of Diluvium and Drift"—the first his own term, the second Murchison's—but was ready to consider both the "action of glaciers" and "bodies of drifting water"—a rather hybrid expression seeming to include either floods or marine submergence. Whatever his blurring of categories, Buckland's main point was to assign phenomena to their proper causes, without presuming the exclusive action of any of them.

Buckland's treatment of the subject of the "Glacia-Diluvial Phænomena" around Snowdon was bound to be of great interest to Darwin. A visit was called for. Despite his ill health, Darwin journeyed to North Wales in June 1842, six months after Buckland's presentation at the Geological Society. In his Diary he noted, "May 18th. Went to Maer, June 15th to Shrewsbury; & on 18th to Capel-Curig, Bangor. Carnarvon to Capel-Curig; altogether ten days, examining glacier action."[91] In his article stemming from his ten days of field research, Darwin praised Buckland. In the first sentence he declared that he had been "guided and taught" by Buckland's work and, having "familiarized" himself with it, was led to make "a few additional observations."[92] He also titled his article in a manner that signaled its connection to Buckland. Where Buckland had written of "Glacia-Diluvial Phænomena," Darwin wrote of the "Ancient Glaciers of Caernarvonshire" and of the "Boulders Transported by Floating Ice." He conjoined the two interpretations with a conjunction rather than a hyphen, but the intent was the same. However, Buckland's diluvial alternative—"inundation" Darwin called it—was a matter to be brushed off by referring to Murchison's chapters on drift from *Silurian System,* which established that seashell-bearing deposits (similar, Darwin believed, to those in Wales) were of submarine origin.[93]

Where, then, did Darwin go in Wales and what did he see? His route overlapped that of Buckland, but he did not climb Snowdon or make his way around its surrounding valleys. Darwin could not walk and climb as formerly. Yet he hoped to see clear evidence of glacial action. This he did most dramatically at Cwm Idwal and its environs (Plate 11). He had visited the site in 1831 and was thus returning to a familiar place. Presumably he was also aware of Bowman's article and of his assessment of the site. "Cwm" is the Welsh word for a small rounded valley. As interpreted by present-day glacial theory it is a "cirque"—that is, "a deep recess in a mountainous upland formed by the erosive process of a small, individual glacier." This definition is drawn from Kenneth Addison, a current authority on the region, who has also noted that Cwm Idwal is "the most accessible site to offer a microcosm of montane glacial landforms."[94]

From both Darwin's field notes of his trip and from his published article, we know that his observations at Cwm Idwal and its environs were decisive in placing him firmly within the glacial camp. In his field notes, for example, he remarked that the moraines at Cwm Idwal were "unequivocal" and "far more perfect" than the "great ones" mentioned by Buckland, that therefore it was with "extreme pleasure to those who have read Agassiz etc = thanks to Buckland" to recognize in these lines of moraines the "retreat of glacier, in its last stage."[95]

Particularly striking to Darwin was a large boulder, broken into four pieces, that he observed on the shore of Lake Idwell. ("Lake Idwell" is Darwin's spelling; the lake is presently known by its Welsh name, "Llyn Idwal.") He described it as lying on "a low mound of detritus, probably once a terminal moraine."[96] He interpreted the scene as "an example of a boulder broken, as described by Charpentier and Agassiz, into pieces, from falling through a crevice in the ice." He noted,

> The boulder now consists of four great tabular masses, two of which rest on their edges, and two have partly fallen over against a neighbouring boulder. From the distance, though small in itself, at which the four pieces are separated from each other, they must have been pitched into their present position with great force; and as the two upright thin tabular pieces are placed transversely to the gentle slope on which they stand, it is scarcely possible to conceive that they could have rolled down from the mountain behind them; one is led, therefore, to conclude that they were dropped nearly vertically from a height into their present place.

Darwin did not note which texts of Charpentier and Agassiz he was drawing on, but the following quotation from Charpentier makes a clear point:

> The internal mass of a glacier consists of ice or rather of frozen snow in a pure state, without any mixture of earth or stones. When blocks fall through a fissure to the bottom of the glacier, they are rolled or pushed forwards. . . . When . . . a block falls quite near the lower end of a glacier through a fissure to the bottom, and at a time when the glacier is retiring, it remains nearly at the same point and in the same position which it occupied when it fell . . . the blocks mentioned above . . . are split up in their whole length.[97]

This boulder, as with the moraines at the south end of Lake Idwell, complementing those noted by Buckland east of Lake Ogwyn, confirmed Darwin as a glacialist. An example of the sorts of moraines that Darwin was viewing is shown in Plate 12 and Figure 8.9.

Another demarcating feature of glaciers that was especially important to Darwin were boss- or dome-formed rocks. In his article Darwin wrote of the phenomenon:

> The rocky and steep barrier over which the ice from the amphitheatre of Lake Idwell flowed into the valley of Nant Francon, presents from its summit to its very foot (between 400 and 500 feet), the most striking examples of boss or dome formed rocks; so much so, that they might have served models for some of the plates in Agassiz' work on Glaciers.[98]

In his field notes for Wales Darwin had remarked that "this structure required practice [to see]." Once Darwin had adjusted his vision to recognize boss- or dome-formed rocks, it became perhaps his most valued device for identifying the former presence of glaciers.

Like Agassiz and Buckland, Darwin did not draw attention to certain other features of the landscape that would announce the former presence of glaciers to a present-day observer at the site—namely, the cirque form of Cwm Idwal and the classic U-shaped valley of Nant Francon. At an important point in his essay, Agassiz did not mention U-shaped valleys and cirques as proofs of glaciers.[99] (The term "amphitheatre" was the usual term of choice for both Buckland and Darwin for what would today be referred to as a "cirque." Neither man used the phrase "U-shaped valley.") Thus, the shapes of rocks, rather than the larger profiles of landforms, were the most useful indicators of glacial passage.

Figure 8.9. View toward Llyn Idwal. Photograph by the author.

Boss- or dome-shaped rocks became key as Darwin attempted to distinguish surfaces affected by the passage of glaciers from those not so affected. Before visiting North Wales in June 1842, Darwin already possessed a view of the history of the area that posited a period of marine submergence during which the shells reported by Trimmer at Moel Tryfan, and observed by himself and others in the environs of Shrewsbury, had been deposited. In his 1842 trip he visited Moel Tryfan. While there, he did not observe any features, as bossed rocks, that signified the passage of glaciers. He did note the features of a mountain south of the upper lake of Llanberis, where the lower portions consisted of convex domes or bosses of naked rock and the upper portions were less naked and where the jagged end of the slaty rocks projected through the turf in irregular hummocks with "no smooth bosses, no scored surfaces, no boulders."[100] He believed that the mountains at this period "must have formed islands, separated from each other by rivers of ice, and surrounded by the sea." As glaciers met the sea, icebergs would also be created, which in turn would deposit their cargo of rocks as they melted. The present appearance of the North Wales landscape thus seemed now to Darwin to require for its explanation the action of former glaciers (as at Cwm Idwal), of floating icebergs (at Moel Faban), and, at periods, marine submergence (as at Moel Tryfan).

Adopting a glacial hypothesis, even in part, required Darwin to change his former views, to see things differently. He observed himself making that change. In his field notes he commented, when at Cwm Idwal looking at the moraines, "Eleven years ago, spent day here saw nothing. nor sh^{d} have if not pointed out.—therefore this notice."[101] He was referring to his earlier visit to Cwm Idwal in 1831, where in his notes he wrote mainly on lithology. He expressed the same idea but in a more expansive fashion in writing to Fitton (a substitute for the absent Lyell): "Eleven years ago, I spent a whole day in the valley, where yesterday every thing but the Ice of the Glacier was palpably clear to me, and I then saw nothing but plain water, and bare Rock."[102] He conveyed a sense of the scene still more dramatically in his *Autobiography*, where he remarked that "a house burnt down by fire did not tell its story more plainly than did this valley. If it had still been filled by a glacier, the phenomena would have been less distinct than they now are."[103]

These more dramatic retellings, to Fitton and in his *Autobiography*, omit from his notes the phrase "if not pointed out": it is important to remember Darwin's new eye had been trained by his reading of Agassiz and Buckland. Moreover, even in North Wales in 1842, he noted later to his cousin Fox that a traveler "will not get a good *glacial eye* for a few days." He said this in regard to Murchison, who "*rushed* through North Wales the same autumn & could see nothing except the effects of rain trickling over the rocks!"[104] So those two elements—having the glacial indicators pointed out, and patience in developing one's powers of discrimination—were key. Once that was done, the mind could fill in the rest.

Darwin was quite satisfied now to be counted among the glacial proponents. "Have you seen last New Eding. Phil.," he wrote to Lyell, "it is ice & glaciers almost from beginning to end."[105] Darwin was among the "ice & glaciers" contributors to the issue. In December 1842 he wrote to his cousin Fox that he was more inclined toward Switzerland than toward any other destination abroad. Nine months later he again wrote to Fox, this time more realistically: "Whenever I give myself a trip, it shall be, I think, to Scotland to hunt for more parallel roads."[106] His different choices for travel, neither accomplished for reasons of health, also signal his dual interests—in continuing to explore glacial sites, and in furthering his initial interpretation of the parallel roads of Glen Roy. One of the charms of his 1842 trip to North Wales was that it convinced him that he could accommodate both glaciers and marine submergence. As he wrote to Fitton while still in North Wales,

> These glaciers have been grand agencies; I am the more pleased with what I have seen in N. Wales, as it convinces me that my views, of the distribution of the boulders on the S. American *plains* having been affected by floating Ice, are correct. I am also more convinced that the

> valleys of Glen Roy & the neighbouring parts of Scotland have been occupied by arms of the Sea, & very likely, (for on that point I cannot *of course* doubt Agassiz & Buckland) by glaciers also.[107]

All identifiable agencies had their application.

After 1842 Darwin took no extensive geological field trips. However, several later publications indicate his continuing interest in sorting through and assigning the proper explanation to superficial deposits and issues surrounding glaciers and icebergs. Titles are suggestive: "Extracts from Letters to the General Secretary, on the Analogy of the Structure of Some Volcanic Rocks with That of Glaciers" (1845), "On the Transportal of Erratic Boulders from a Lower to a Higher Level" (1848), and "On the Power of Icebergs to Make Rectilinear, Uniformly-Directed Grooves Across a Submarine Undulatory Surface" (1855). (Full citations are given in the bibliography.) The first publication was composed of extracts of letters Darwin had written to James David Forbes, an authority on glaciers.

In closing this section, mention must also be made of Darwin's short span of fieldwork done on 20–22 August 1874 in Southampton. There he studied superficial deposits, focusing particularly on fragments of gravel that presently stand upright. These upright fragments reminded him of a similar phenomenon he had seen at home in Down where flints "as long and thin as my arm" stand upright in the "stiff red clay."[108] He associated both the upright fragments of gravel at Southampton and the upright flints at Down with differential movements of the underlying beds. In the case of the Southampton gravel Darwin believed that as lower beds of frozen snow melted—the snow being associated with interstratified drift—the "elongated pebbles would have arranged themselves more or less vertically." Darwin's views were published by James Geikie (1839–1915), to whom Darwin had written on the subject after reading what appears to have been an early release of the second edition of Geikie's *The Great Ice Age, and Its Relation to the Antiquity of Man* (1877). Geikie published Darwin's comments in *Prehistoric Europe.*[109] Clearly, the subject of superficial deposits, including their shape and position, remained of interest to Darwin well after the 1840s.

The Effect of the Glacial Theory on Darwin's Ideal of a "Simple" Geology

Glacial theory affected Darwin on a psychological and methodological level, causing him to check his enthusiasm for pure theory, even in areas removed from the subject of glaciers. In 1840 he noted that he now "set less value on theoretical reasoning in geology"; later he also referred to the "sin of speculation" and of having "burned my fingers pretty sharply in that way."[110] His newly found wisdom did not prevent Darwin from continuing to theorize: he composed two drafts of what would become the *Origin of Species* in the early 1840s. What was new was a diffidence about the certainty that he might attach to his own conclusions. Thus, for example, as the *Origin* moved toward publication, he was cautious in estimating its lasting value, fearing that he might have been a monomaniac pursuing a foolish doctrine. To Huxley he spoke of his "poor rag of a hypothesis"; to Lyell he wrote that "thinking of the many cases of men pursuing an illusion for years, often & often a cold shudder has run through me & I have asked myself whether I may not have devoted my life to a phantasy."[111]

His diffidence seems to have begun with his experience of the unhappy fate of his Glen Roy paper. On a number of occasions Darwin severely criticized his work at Glen Roy, most notably in a passage from his *Autobiography:*

> This paper was a great failure, and I am ashamed of it. Having been deeply impressed with what I had seen of the elevation of the land in S. America, I attributed the parallel lines to the action of the sea; but I had to give up this view when Agassiz propounded his glacier-lake theory. Because no other explanation was possible under our then state of knowledge, I argued in favour of sea-action; and my error has been a *good lesson to me never to trust in science to the principle of exclusion.*[112]

I have drawn attention to the word "trust" to suggest that the passage, as well as Darwin's over-twenty-year involvement with the issue of Glen Roy's parallel roads, is best understood if one focuses on that aspect of thought where confidence and belief reside. The phrase "principle of exclusion" is Darwin's own wording for the method of reasoning that was later called by philosophers of science "eliminative induction." This method requires a listing of all plausible alternative explanations, and the successive elimination of all but one explanation, which is then accepted as the true explanation.[113] In his original work on Glen Roy, Darwin believed he had identified all plausible explanations; he did not consider the glacial hypothesis.

Darwin never abandoned the use of his "principle of exclusion" method. His either/or approach to species mutability was, after all, an argument of similar logical form. Moreover, he was quick to strike out explanations that seemed to him unacceptable: diluvialism, for example, did not long remain on his list of possible explanations for geological phenomena including Glen Roy. Yet Glen Roy alerted him to the limits of the method. In a letter that anticipates the paragraph in his *Autobiography*, Darwin wrote in 1861 to the Scottish agriculturist and geologist Thomas Francis Jamieson (1829–1913):

> What a wonderful record of the old icy lakes do these shores [at Glen Roy] present! . . . I have been for years anxious to know what was the truth, & now I shall rest contented, though ashamed of myself.—*How rash it is in science to argue because any case is not one thing, it must be some second thing which happens to be known to the writer.*—[114]

Sorting out the best explanation for the origin of the so-called parallel roads of Glen Roy was a protracted process for Darwin and other midcentury geologists. Darwin's vacillating response to the glacial interpretation of Glen Roy is indicated in a letter to Buckland:

> I should much like to hear yours and Agassiz's opinion on the parallel roads, though I believe I know its outline—I cannot give up the sea, after thinking over many points of minor detail in the country, though, I am very sure, if your theory had occurred to me during the first two days of my examination I should have given up their marine and *ordinary* lacustrine origin at once.[115]

Darwin's rather surprising confession ("if your theory had occurred to me") underscored his opinion, expressed variously over the years, that observing and theory-testing went hand in hand. As he wrote in 1860, "I have an old belief that a good observer really means a good theorist."[116]

Darwin's commitment to the Glen Roy problem exceeded that of other writers on the subject because it arose in the context of his pursuit of a simple geology. In their proximity to the sea and in their horizontality, the roads provided a measurable indicator of crustal elevation analogous to what he had observed in South America. He would have preferred that the roads were of greater extent, and in 1843, confident that he could maintain a glacial theory for North Wales and a marine one for Glen Roy, he observed to his cousin Fox, "Whenever I give myself a trip, it shall be . . . to Scotland to hunt for more parallel roads."[117] As the prospect of his own future work in the area dimmed, he looked for expert assistance. There were two instances where his efforts at collaboration are instructive, the first in the 1840s and the second in the 1860s.

In 1846 Leonard Horner, at that time president of the Geological Section of the British Association for the Advancement of Science, received from Darwin a proposal for an official survey of the area around Glen Roy. Darwin wrote,

> The Parallel Roads of Glen Roy in Scotland have been the object of repeated examination, but they have never hitherto been levelled with

> sufficient accuracy. Sir T. Lauder Dick procured the assistance of an engineer for this purpose, but . . . it was impossible to ascertain their exact curvature, which . . . appeared equal to that of the surface of the sea. Considering how very rarely the sea has left narrow & well-defined marks of its action at any considerable height on the land . . . it appears highly desirable, that the roads of Glen Roy, should be examined with the utmost care during the execution of the Ordnance Survey of Scotland.

Darwin concluded his proposal with the promise that the work of the Ordnance Survey at Glen Roy would elucidate "one of the most important problems in geology,—namely the exact manner in which the crust of the earth rises in mass."[118] The British Association did recommend that such a survey of the roads be made, but (as Darwin rather anticipated) the Survey was "swamped" with work, and the job was not executed until much later. Nonetheless, Darwin's instructions reveal the reasons for his focus on Glen Roy.

Between Darwin's first and second interventions there was an interlude when others took the lead. The first was David Milne (1805–1890), who went to Glen Roy in support of Darwin's marine interpretation and came away opposing it. Milne revitalized the lake interpretation of the roads by discovering a hitherto unnoticed feature of the landscape. This was a watershed at the head of Glen Glaster, which branches off Glen Roy near its bottom. As Milne wrote in the version of his paper prepared for formal publication, the existence of this watershed "established . . . that the whole of the 4 shelves of Lochaber are coincident with water-sheds."[119] Thus, the question of the pattern drainage of the lakes was laid to rest. It could be shown at what point each lake would have drained, leaving the pattern of shelves or roads visible as they are today. However, Milne did not adopt the notion that ice had blocked the outlets to the lakes; he believed the barriers had been composed of such local materials as gravel, clay, or other detrital matter. Milne did not have a glacial eye.

The effect on Darwin was immediate. Milne had attacked what he described as the interpretation favored by Darwin, Lyell, and Horner. Darwin responded by drafting a rejoinder to Milne's points, initially intended for the *Scotsman* but later withdrawn, in which he accepted Milne's discovery but argued that it largely favored the Agassiz-Buckland interpretation.[120] The question of glacial evidence was still outstanding, however, as were a number of smaller points, most of which Darwin believed favored the marine interpretation. Darwin's rejoinder to Milne ended with a list of unanswered questions, in effect a list of directives for a future researcher to pursue.

Another researcher, who in fact accompanied Milne to Glen Roy, was

Robert Chambers (1802–1871), author of the anonymously published *Vestiges of the Natural History of Creation* (1844). (Chambers's authorship of *Vestiges* was suspected within Darwin's circle.) From Darwin's point of view, Chambers was a somewhat annoying shadow figure or *doppelgänger* as regards to Glen Roy, as he was with regard to evolution. Darwin, however, did correspond with Chambers on the subject of Glen Roy. In his book *Ancient Sea-Margins* (1848), Chambers rejected the interpretation of Milne, his fellow traveler, and sided with Darwin.[121] Indeed, Darwin's interpretation of Glen Roy, and his interest in sea-formed terraces, was the intellectual centerpiece of Chambers's book. A fold-out colored map of Glen Roy accompanied the book. What probably struck Darwin in reading it was how well Chambers understood the thematics of a simple geology as applied to Glen Roy: "Mr. Darwin came fresh from the mobile continent of South America, and pronounced these vales to have once been filled with arms of the sea." However, Chambers reworked Darwin's understanding of oceanic subsidence to argue for a worldwide drop in sea level rather than a rise in land.

A contemporary diluvialist interpretation of Glen Roy by George Mackenzie (1780–1848) also should be noted. Mackenzie was associated with Milne and Chambers, though the three men came to differing conclusions about Glen Roy. Darwin never mentioned Mackenzie's work, perhaps because he spoke disparagingly of Darwin's interpretation of Glen Roy, and the "English geologists" who favored it, without adding any useful new facts.[122] More surprisingly, Darwin did not discuss the potentially useful ideas of James Thomson (1822–1892) regarding the "viscidity of glaciers" as arguing in favor of the Agassiz-Buckland interpretation of the roads.[123] Thomson had taken his interest in glacial structure and movement from James David Forbes, who had also figured in Darwin's 1845 publication "On the Analogy of the Structure of Some Volcanic Rocks with That of Glaciers."

In the early 1860s Darwin again sought, this time successfully, to direct research at Glen Roy that he was unable to do himself. While Darwin had long hoped that Lyell would tackle the Glen Roy problem, it devolved on Jamieson, who became the Fordyce lecturer on agricultural research at the University of Aberdeen in 1862.

The context for studying Glen Roy in the 1860s was far different from what it had been twenty years earlier. Geologists were more accepting of glacial theory, as, for example, Andrew Crombie Ramsay of the Geological Survey (promoted to senior director of the Survey for England and Wales in 1862), whom Darwin attempted unsuccessfully to recruit to study Glen Roy in 1859.[124] It may seem odd that Darwin would address someone of Ramsay's stature to accept a fieldwork assignment. However, by 1859, even before publication of the *Origin* that November, Darwin was a revered figured within

British geology. In February 1859 Darwin received the Wollaston Medal, which was the highest honor bestowed by the Geological Society of London. Darwin reacted to notification of the award with an enthusiastic run of triple exclamation points.[125]

In addition to strong personal ties among the core group of high-level geologists, which included Darwin, the institutional resources for the study of geology were greater in the 1860s. The existence of the Geological Survey put many more geologists into the field. The work of the Ordnance Survey was also useful. While the full Ordnance Survey of the Glen Roy area was not published until 1874, Jamieson drew on the work of its surveyors in providing reliable figures for elevations above sea level when he was evaluating previous measurements taken at Glen Roy. Thus, for example, in 1862 Jamieson wrote to Lyell that the Ordnance surveyors had measured Glen Roy's Makoul pass at 850 feet above mean sea level. That number was consistent with an earlier measurement, made by the engineer employed by Chambers, of 847 feet above the sea for the lowest Glen Roy road. The nearly identical level of the pass was "a coincidence . . . too close to be accidental" and suggested to Jamieson that the lowest Glen Roy road "coincides with this outlet & has been determined by it."[126]

Finally, the temporal ordering of the geological past was better understood in the 1860s than it had been in the 1840s. From Jamieson's comments on Glen Roy alone, it is clear that the accent was on interpolating periods of land submergence and land glaciation, rather than on simply studying the structure of the roads themselves. As Jamieson wrote to Lyell in 1861 in the context of discussing Glen Roy, "The Pleistocene, or as you abjure the word, the Newer Pliocene, period was evidently of vast duration & probably we shall require yet to interpolate here & there a new chapter in its history."[127] That the "here & there a new chapter" would include treatment of the first records of human presence in Scotland only added to the interest. Lyell had indeed sought to work out the Glen Roy problem as part of his efforts at the time to correlate the glacial period or periods in Scotland with human artifacts. Of course, the argument of Darwin's *Origin* loomed over all such discussions.[128]

In August 1861 Jamieson ventured into Glen Roy, fortified by Lyell's and Darwin's encouragement and queries. Before his departure he was already in doubt of the marine explanation of the roads, though he took that to be the dominant view. On his return from his trip he stated his full adoption of the glacial theory of the roads to Darwin:

> I returned a few days ago from a trip to Lochaber . . . and now hasten to present you with some of the results of my visit . . . finding the marks

of ice action so plain over the whole district I cannot help thinking that Agassiz hit upon the true solution of the problem when he pronounced these marks to be the effect of glacier-lakes.[129]

Jamieson made much of the exact horizontality of the roads. Had they been marine in origin, he believed they would not have been so level, owing to the fluctuations of the tides. He was also ready to adduce a "great Greenlandic covering of ice" in describing the pattern of at least one glaciation in the area. Jamieson's provisional narrative for the Lochaber area was "this great Greenland-state of things" followed by a marine submergence of unknown amount and then "a set of glaciers after this submergence sufficient for our Glen Roy lakes." Jamieson thus set aside Agassiz's Ben Nevis glacier in favor of praising Agassiz's notion of glacial dams.[130]

Both Darwin and Lyell challenged Jamieson on a number of points regarding the glacial theory, and Jamieson, raising one of his own (the difficulty of imagining how an ice dam might hold back a lake), was respectful. Darwin also sent Jamieson his papers dealing with Glen Roy. Eventually, after a round of questioning from Lyell, Darwin outlined to him what he now took to be the sequence of events at Glen Roy:

> I suppose whole valley of Glen Spean filled with ice; then water would escape from outlet at Loch Spey & the highest shelf would be first formed. Secondly ice began to retreat, & water would flow for short time over its surface; but as soon as it retreated from behind hill marked Craig Dhu, where the outlet on level of 2^{d} shelf was discovered by Milne, the water would flow from it, and the *second* shelf would be formed. This supposes that a vast barrier of ice still remains, under Ben Nevis, along all the lower part of the Spean. Lastly I suppose the ice disappeared, everywhere along L. Laggan, L Treig & Glen Spean, except close under Ben Nevis, where it still formed a Barrier, the water flowing out at level of *lowest* Shelf, by the pass of Muckul at head of L. Laggan.—This seems to me to account for everything. It presupposes that the shelves were formed towards close of Glacial period.—[131]

After this point Darwin acceded to the glacial view of the Glen Roy roads, never to waver.

Jamieson returned to Glen Roy the next summer, July 1862, for another round of fieldwork. He was quite aware that "the case has been so much contested that the geological world will not readily adopt any explanation," and he rather hoped that someone with, say, Ramsay's stature would enter the field.[132] Thus, he took great pains with his work, publishing a portion of it in

1862, and the full report in 1863. In the full report Jamieson reviewed with respect all previous work, including the diluvial hypothesis, before making the case for the glacial explanation. He checked previous observations, crediting some, questioning others. He identified more signs of glaciation in the landscape than had any previous observer. He also postulated more precisely than anyone had before the likely positions of former glaciers—or "ice-streams" as he preferred to describe them—with regard to the roads: "grant then these two ice-streams, one in the Great Caledonian Valley and the other at Glen Treig, and the problem of the parallel Roads can be solved." He also postulated, in accordance with ideas that Ramsay was then developing, that Lake Trieg, a "profound pool," was of glacial origin.

Jamieson emphasized the role played by the weight of the ice in shaping the surface of the land. It was an idea he would continue to develop, as in an 1865 article where he argued for the depression of land in Scotland having been caused by the weight of glacial ice, with a rebounding of land following the last retreat of the glaciers. This sense of the great weight of glaciers or ice streams stemmed from what Jamieson referred to as the "enormous thickness" of the ice he believed to have covered the region.[133]

Jamieson's map of the Glen Roy region, taken from his 1863 article, captured the essence of his interpretation (Figure 8.10). Red lines signified the roads. Red arrows marked out outlets of the old lakes. Red shading marked old deltas of the lakes. Gold shading indicated areas of glacial moraines. (Darwin wrote to Lyell in 1862, "The moraines opposite L. Treig are obviously very important."[134]) Blue markings indicated the direction of glacial striae. Jamieson's solution ended the life of the marine theory of the origin of the roads. It was also a vindication of the original lake theory, though water was now believed to drain toward the east rather than toward the west. Jamieson's solution was also, of course, a vindication of Agassiz's and Buckland's initial judgment on Glen Roy.

Jamieson's views were promulgated to a wide audience by Lyell in his *Geological Evidences of the Antiquity of Man* (1863).[135] As the *Saturday Review* described it, this work was "a trilogy, the constituent elements of which should be headed respectively, Prehistoric Man, Ice, and Darwin."[136] This review suggests the conjunction of issues in Lyell's mind, and for much of the public, as of 1863. As is well known, Lyell's book disappointed Darwin for not coming far enough along on transmutation, but in writing to Lyell about the book Darwin also needled him for not writing as enthusiastically as he might of the glacial explanation of Glen Roy: "I wonder you did not add a vivid sentence . . . at the end of Glen Roy, on our almost still seeing the glacier lakes."[137]

With the glacial reinterpretation of Glen Roy secure, Darwin lost a key ex-

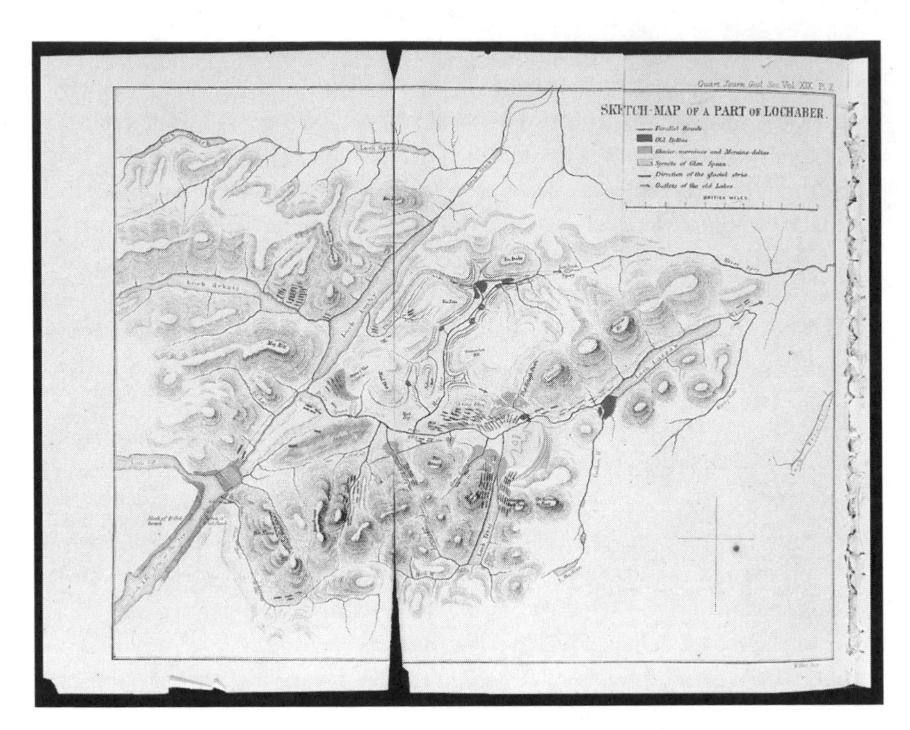

Figure 8.10. Jamieson's map of Lochaber. Courtesy of the Library of Congress.

emplar for his original simple geology. The shape of the roads could not be used as an indicator of the pattern of the rise of the crust of the earth. The glacial solution for Glen Roy also added to the case for jettisoning Lyell's explanation of the origin of climate. Already in the late 1830s and 1840s Darwin had departed from Lyell in imagining large-scale movements in the rise and fall of land. This led him to presume that continents were relatively enduring. By the 1840s he opposed those who posited land bridges to accommodate problems in explaining the geographical distribution of species. Once he had accepted the likelihood of a colder period in Europe prior to the present day—initially associated with the presence of icebergs as well as of glaciers—he also integrated it quickly into his emerging species theory. In his 1844 and 1856–1858 drafts of his species theory, as well as in the *Origin*, he used this change in climate to explain the presence of Arctic vegetation on mountaintops in Europe.[138] He believed that as the glacial cold receded, Arctic plants remained only in the coldest, highest locations, lower elevations being tenanted by plants newly arriving from temperate areas.

Darwin's departure from Lyell on the issue of climate is apparent from his correspondence. To Hooker he wrote, after sending him the chapter on the geographical distribution of species from the manuscript for the *Origin*, "I grieve to say that I feel compelled to disbelieve in the Lyellian view that [the] Glacial Epoch is connected with position of continents."[139] On receiving Hooker's reply, which is not extant, he answered more fully:

> I demur to what you say, that we change climate of the world to account for "migration of bugs flies &c", we do nothing of sort; for we rest on scored rocks, old moraines, arctic shells & mammifers.—I have no theory whatever about cause of cold, no more than I have for cause of elevation & subsidence; & I can see no reason why I sh^{d} not use cold, or elevation or subsidence to explain any other phenomena, such as of distribution.—I think if I had space & time, I could make pretty good case against any great continental changes since Glacial epoch, & this has mainly led me to give up the Lyellian doctrine as insufficient to explain all mutations in climate.—

A few months later, as the *Origin* was on the eve of publication, Darwin wrote again to Hooker:

> With respect to migration during the Glacial period, I think Lyell quite comprehends, for he has given me a supporting fact. But perhaps he unconsciously hates (do not say so to him) the view, as slightly staggering him on his favourite theory of all changes of climate being due to changes in relative position of land & water.—

In the 1840s and 1850s Darwin had to choose between explaining all the major geological phenomena in one comprehensive theory and accepting the arguments for the existence of a worldwide glacial period. Clearly by 1859 he had chosen the latter. As he wrote to Ramsay in 1862, "The shelves are a magnificent record of the Glacial period.—"[140]

In the 1860s a new possible explanation for glacial periods emerged. James Croll (1821–1890) posited that changes in the earth's orbit had been the cause of glacial periods in the earth's history. Croll's views were summarized in *Climate and Time in Their Geological Relations* (1875).[141] The image of the earth as a self-sustaining enclosed unit was thereby erased. While Darwin had always left the door open for extraterrestrial influences on the earth, as with his early interest in meteorites, his simple geology had focused on events and forces within and on the surface of the earth. By the 1860s glacial theory required a step away from that focus. In thirty years, from the 1830s to the 1860s, glacial theory had transformed geological science, and with it Darwin's initial idea of a simple geology. Lyell's theory of climate was, for Darwin, dead. Glaciers as well as the sea might shape the land. Jamieson's notion of land rebounding following the melting of glaciers suggested a new mechanism for elevation. Nothing in geology was left untouched by glacial theory, least of all Darwin's understanding of the earth's crust.

CHAPTER 9

GEOLOGY AND SPECIES

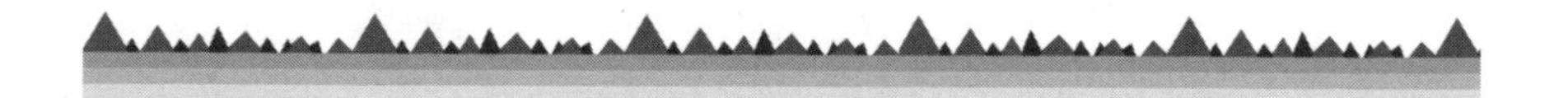

The Legacy of a Book

The *Origin of Species* cast a long shadow. It ushered in the present era in paleontology when an evolutionary perspective is routinely brought to bear on the interpretation of fossils. Yet, paleontology, and geology in general, played a subordinate role in the book itself. The chapters devoted to geology come late in the argument, well after the case has been made for natural selection. In the first four chapters Darwin developed the main points to his argument, well demarcated by his concise chapter titles: "Variation under Domestication," "Variation under Nature," "Struggle for Existence," and—the seminal chapter—"Natural Selection." Chapters 5–8 were various in subject: "Laws of Variation," "Difficulties on Theory," "Instinct," and "Hybridism." Following these, come two geological chapters—"On the Imperfection of the Geological Record," and "On the Geological Succession of Organic Beings"—and an additional two chapters that bear on geology, both entitled "Geographical Distribution." The final chapters wrap up the argument with a discussion of a variety of topics, including embryology, and a recapitulation and conclusion.

The message of Darwin's chapter on the imperfection of the geological record was a negative one, that geologists should not expect every stage in the development of a species to have been preserved in the fossil record. Chapter 10, on geological succession, was also cautionary as to what it claimed re-

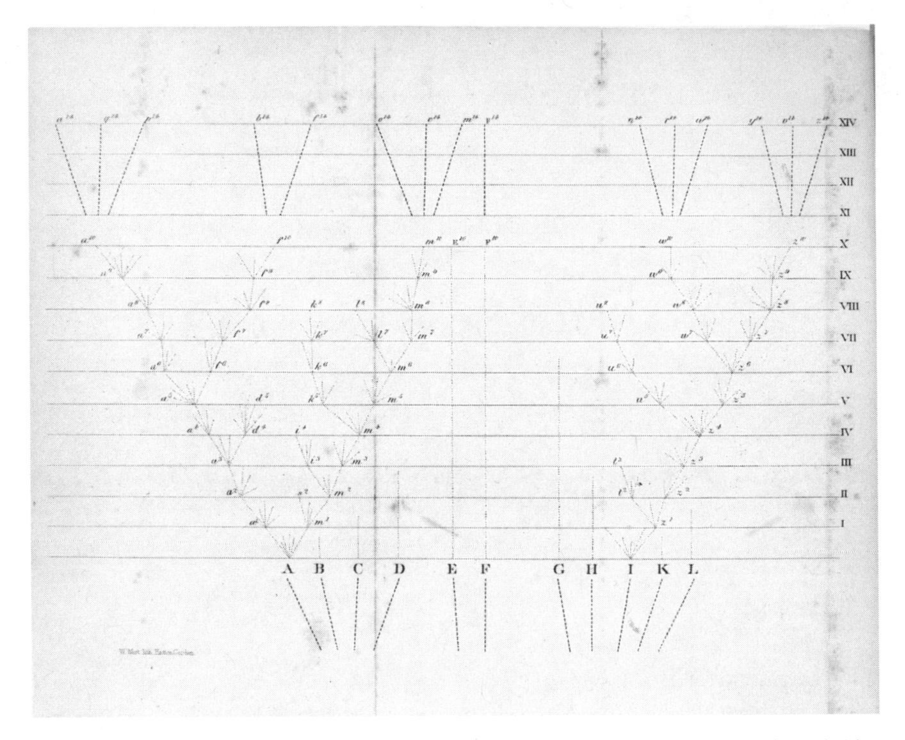

Figure 9.1. Diagram from *On the Origin of Species*. By permission of the Syndics of Cambridge University Library.

garding the possibility of tracing the evolutionary descent of species through the fossil record. The single diagram, which accompanied the chapter on natural selection represented an idealized model for the formation and descent of species without providing any labels, either of species or of geological strata (Figure 9.1). As he sent the diagram in to his publisher, Darwin remarked, "It is an odd looking affair, but is *indispensable* to show the nature of the very complex affinities of past & present animals.—"[1]

Yet paleontology, and geology generally, played a larger role in the development of Darwin's species theory than would appear from the *Origin*. Without attempting to be comprehensive—for the literature on Darwin is vast—my goal is to document that claim. This chapter will focus primarily on Darwin's view of geology and species during the *Beagle* voyage. Chapter 10 will carry the story from his adoption of a transmutationist hypothesis in 1837 to his publication of the first edition of the *Origin* in 1859.

Darwin's Formulation of the Species Question during the *Beagle* Voyage

Charles Darwin's grandfather, Erasmus Darwin, described his goal as to enlist "imagination under the banner of science."[2] In keeping notes during the *Beagle* voyage, Charles Darwin had a quite different charge in view: to collect and observe—in other words, to develop the factual basis of natural history. There is no record from the voyage, so far as I am aware, of Charles Darwin's reflections on his grandfather's imaginative speculations on transmutationist ideas, or, for that matter, on those entertained by the French naturalist Jean Baptiste de Lamarck (1744–1829) or by the zoologist Robert Edmond Grant (1793–1874), whom Darwin knew at Edinburgh. The views of all three men were known to Darwin before the voyage, though he did not have aboard ship any of his grandfather's writings or any of Lamarck's transmutationist works, nor did he correspond with Grant. In the narrow confines of the *Beagle* library, reference books won out over philosophy. Transmutationist views, however, were sufficiently common as to be expressed, often in rebuttal, in various texts that were part of the ship's library, volume 2 of Lyell's *Principles of Geology* being a case in point.[3]

Yet an unrecorded presence may still be a presence. In regard to Darwin's parallel series of geological and zoological notes from the voyage, it is noteworthy that they were not entirely private. To some degree all the collections aboard ship, and notes about them, fell under the domain of the captain. He, and therefore others of his choosing, were potential readers. This circumstance may help to explain the fact that in his *Beagle*-period notes, Darwin often posed provocative opinions as questions. (Nonetheless one should not make FitzRoy of the *Beagle* days as the biblical literalist he later became.) When Darwin did adopt a transmutationist position in 1837, he was in the gratifying position of vindicating, by virtue of a greater command of factual material, a position taken by his grandfather. He could be a radical, but a conserving radical, preserving and extending an intellectual tradition already established within the family. When in later life Charles Darwin reviewed his grandfather's work, he said he was "much disappointed, the proportion of speculation being so large to the facts given."[4] By comparison, his own work from the *Beagle* onward resulted in an enormous quantity of fact, one measure of which is the thousands of specimens collected and commented on.

Understanding the nature and extent of Darwin's zoological work during the voyage has recently become considerably easier by virtue of the recent publication of his zoology notes and specimen lists. In addition to providing transcriptions of Darwin's texts, Richard Keynes has also provided current identifications for a number of the specimens.[5] His work on Darwin's zoological collection joins surveys by Porter of Darwin's plant collections and by Smith of Darwin's insects.[6] By helpfully publishing the specimen lists—which are literally lists—together with more narrative notes, Keynes also underlined the centrality of specimens to Darwin's work in zoology during the voyage. Whereas in geology Darwin turned to more essay-like discussions of theoretical issues as the voyage proceeded, in zoology he remained focused on individual specimens. His brief theoretical remarks—or, more often, questions—appear within entries devoted to description of an animal and its habits.

Before turning to Darwin's notes from the voyage, I should like to place on record, for future reference, Darwin's familiar statement from his *Autobiography* concerning the progress of his belief on transmutation.

> During the voyage of the *Beagle* I had been deeply impressed by discovering in the Pampean formation great fossil animals covered with armour like that on the existing armadillos; secondly, by the manner in which closely allied animals replace one another in proceeding southwards over the Continent; and thirdly, by the South American character of most of the productions of the Galapagos archipelago, and more especially by the manner in which they differ slightly on each island of the group; none of these islands appearing to be very ancient in a geological sense.[7]

This statement was written over forty years after his experiences on the voyage; there is nothing so compact and oriented toward transmutation in his contemporary notes. However, one can reconstruct Darwin's frame of mind by examining the categories he used in his notes that would pertain logically to transmutation. I will consider the zoological notes first and then the geological.

The word "transmutation" does not appear in the zoological or geological notes. The word "evolution" occurs once in the zoological notes: to describe the spiraling flight of the frigate birds at the Galápagos Islands, "showing their perfect skill in evolutions when many are darting at the same floating morsel.—"[8] Overall in the zoological notes, Darwin's most useful categories of analysis were taxonomic. On the taxonomic scale, from kingdom at the top, to species and varieties at the bottom, with the important exception of a number of invertebrates, including corals, Darwin's interest lay toward the bottom of the scale. Usually, as would make sense for a field naturalist, Darwin worked on the species level. Defending his field experience, and that of his local informants, he wrote in February 1834, "Whatever Naturalists may say, I shall be convinced from such testimony as Indians & Gauchos that there are two species of Rhea in S. America."[9] Toward the end of the voyage in 1835 at the Galápagos he refined his vocabulary to include the phrase "slight variations" as in speaking about the tortoises: "It is said that slight variations in the form of the shell are constant according to the Island which they inhabit.—"[10] In sum, the species/variety distinction was largely where Darwin focused his attention during the voyage, and the eventual yield to him was great, namely, the circumstance of geographical isolation as a means for creating new varieties, or, even, species.

In one category of organisms—the invertebrates—Darwin routinely raised questions relating to higher taxonomic groupings than species and variety. This was often simply a matter of trying to answer the question regarding a specimen: what is it? Even in his first scientific paper, delivered while a student at Edinburgh, Darwin reported that the so-called ova of the bryozoan *Flustra foliacea* were in actuality larvae of the leech *Pontobdella*.[11] Questions of identity also occur with some frequency in Darwin's zoology notes. For example, he puzzled over zoea, which is a distinctive larval stage of large crustaceans such as crabs and shrimps. He also struggled over the classification of several "Corallina" specimens, wondering whether "I must rank these beings as belonging to the Vegetable rather than animal world.—" These specimens were eventually identified as coralline algae, a classification that justified Darwin's initial hesitation.[12] Coralline algae also were discussed in Darwin's *Structure and Distribution of Coral Reefs,* where he described them as belonging "to one of the lowest classes of the vegetable kingdom."[13] Thus, in the instance of some specimens associated with coral reefs, there was a change in designation from the animal to the vegetable kingdom. Another instance where a major change in classification figured among *Beagle* species were the barnacles, a number of which Darwin studied in detail during the voyage, and which were being transferred, during that time, by several investigators (though not Richard Owen) from the Mollusca to the Crustacea. After the voyage Darwin contributed to this transfer, using, as did others, developmental stages as criteria in classification.[14]

Overall Darwin's work with the invertebrates led in turn to continued interest in generation and development. While Darwin was too much of an imperfectionist with regard to the fossil record to attempt to establish parallels between states of development and the stratigraphical record, as did Agassiz, on occasion he did draw on developmental analogies in an attempt to understand the advent of new species. Thus, in February 1835, in a passage to be quoted later in this chapter, he toyed with the idea that species might be propagated in a fashion analogous to the budding of plants. M. J. S. Hodge has aptly referred to Darwin as a "life long generation theorist"; similarly Phillip R. Sloan, working in a related vein, has investigated what he terms the young

Darwin's "invertebrate research program."[15] Finally, we should make the rather obvious point that the study of generation and development was inherently easier to do for plants or invertebrates than with vertebrates because specimens of the former were readily available and easier to handle. This meant that large biological questions could be addressed more directly.

Darwin also used such metaphors as "chain of Nature" and "scale of Nature" (as in "What is this animal? Where does it come in the scale of Nature?"). Geographical categories appear in the notes with frequency: "centre of creation," "aboriginal," and, not to overlook the obvious, the phrase "geographical distribution" itself. The phrase and category "the economy of nature" appears along with subsidiary concepts: "stations," "habitations," and "fecundity."[16]

For Darwin, "adaptation" was a term of useful, indeed often startling employment, as in his remark in 1834 at the island of Chiloé, when he was considering the possible specificity of lice to different peoples: "Man springing from one stock according [to] his *varieties* having different parasites.—[It leads one to many reflections]."[17] Finally, a word should be said about Darwin's frequent employment in his notes of the terms "Nature" and "Creator" or "created." Darwin's usage was in accord with that of his peers as, for example, in the English version of Cuvier's *Animal Kingdom*. The editor of the work, Edward Griffith (1790–1858), wrote that the word "nature" encompassed "the laws regulating those beings which collectively compose the universe." He suggested that in this sense "we are accustomed to personify nature, and to use its name for that of its Creator."[18] In his notes, Darwin, like Griffith, used the terms "Nature" and "Creator" interchangeably. At this time Darwin did not use the term "creationist"; the word seems to have entered his vocabulary with his "Sketch of 1842."[19]

We turn now to Darwin's geological notes. In these notes the most frequently used category is indicated by the phrase "organic remains." In importance it was the counterpart of the "species/variety" category of the zoological notes. The phrase "organic remains" was a commonplace one.[20] As Darwin employed it, the phrase carried the methodological charge to search for and record the presence of fossils in every formation. This practice was in accord with standards then being established in British geology. So well known had the standards become that in February 1832 at Bahia in Brazil, Darwin would criticize Humboldt for failing to note organic remains in a formation in Columbia:

> These <<Tertiary>> formations are much the same as those described by Humboldt in Columbia: he also mentions a fresh-calcareous tufa, but . . . does not specify the organic remains by which it may be recognized.[21]

Identifying the fossils in strata was becoming imperative for good geological research.

From the first year of the voyage onward, Darwin had in mind larger issues surrounding the general notion of "organic remains." In September and October of 1832 after working through the beds at Bahía Blanca (in present-day Argentina), he commented,

> I have been this particular in describing these beds, in which the organic remains occurred.—for the comparison of formations in different parts of the world which contain animals of equal grade in the chain of nature seems at present to be much wanted in Geology.—[22]

Significantly, Darwin used the metaphorical term "chain of nature" here as he had done in his zoological notes. The "chain of nature" phrase suggests a rank order connecting higher and lower. The theme of comparing fossils in different parts of the world is also expressed in his notes from the Falkland Islands. There, to his surprise, Darwin found organic remains in sandstone interlaced with highly laminated clay-slate. He thought this one of the "oldest (or most inferior) formations ever to be found fossiliferous." He remarked that

> it will therefore be preeminently interesting to compare these fossils with those of a similar epoch in Europe: to compare how far the actual species agree & the comparative number of individuals of each sort.—[23]

In effect, Darwin was proposing a census of species and of individuals within species, an interest that fostered his later direct concern with populations. To finish the story on the Falkland fossils, it should be mentioned that on Darwin's return home, Murchison placed them in the lower part of his newly established Silurian System.[24]

Associated with the category of organic remains were the subjects of geographical distribution and of extinction. These topics were important to Darwin in their own right, but also as elements from which he assembled hypothetical but concrete explanatory narratives for the regions he visited. I emphasize the concreteness of these stories because Darwin can seem abstract in his work on species, as in the genealogical diagram from the *Origin* (see Figure 9.1). In constructing these narratives, Darwin noted that he was positing sequences of events rather than dating them. Thus, in writing about the formation of beds at Bahía Blanca, which he visited the first year of the voyage, he commented,

> Subsequent to this elevation, the present order of things commenced, the sea has continued wearing away some points & in others heaping up banks, thus <wearing> creating a line of coast. best adapted for its own motions.—It is certain that this period is a long one, from the breadth & number of the parallel dunes.—Having no date to calculate the time <in> which one is formed it is impossible even to conjecture the absolute time[.]—[25]

Darwin would employ the phrase "the present order of things commenced" again, and throughout his career he would remain drawn to the subject of "absolute time."

As a geologist Darwin was interested in imaging a sequence of land movements that might have created the present distribution of land animals. For example, of Navarin Island, which forms the southern bank of a portion of the Beagle Channel in Tierra del Fuego, Darwin remarked in 1833 that

> we find Guanaco, Foxes & Mice; it is highly probable that these animals passed over from the main land, before the Beagle channel had broken through the bed of Alluvium.—[26]

Darwin was also interested in the distribution of major groupings of species, as, for example, in the fact that South America was home to the guinea pig.

Although Darwin brought a geologist's appreciation of the significance of land movements to the subject of geographical distribution, the underpinning to his approach to the subject of distribution went back, through Lyell, to the idea of "centres of creation" developed by Carl Linnaeus (1707–1778). Lyell employed this idea to explain, among other things, the appearance of the distribution of vegetation on a group of islands, where the "original isle was the primitive focus, or centre, of a certain type of vegetation, where, in the surrounding isles, there would be a smaller number of species, yet all belonging to the same group." This would have the effect of giving "an appearance of *centres* or *foci* of creation . . . as if there were favourite points where the creative energy has been in greater action than in others."[27]

Regarding species extinction, Darwin's attention during the voyage was focused on the *processes* that might produce extinction more than on the identity of the extinct animals. Proper identification of the bones found would come in London. The provisional nature of his field identifications can be noted in his opening comments on specimens collected at Bahía Blanca in September and October 1832:

> [T]he number of *fragments* of bones of quadrupeds is exceedingly great:—I think I could clearly trace 5 or 6 sorts.—the head of one very

> large animal (with singular anterior cavity) has 4 large square hollow molar teeth; perhaps it may be the Megalonyx; the lower jaw & one molar tooth of some smaller animal., I conjecture one of Edentata & perhaps allied to the Armadilloes: the molar teeth of some large animal. (Rodentia?) 741 - - - - 744: bones of some smaller quadruped. like deer[.][28]

The run of specimen notes listing bones in numerical order has similar comments. Again, in the specimen notebook the accent was on noting the type of bed in which the bone, often fragmentary, was found. There is some anatomical description (as for specimen 730, "Head of the Femur"), but identification of parts was clearly not the primary task at hand. The one diagram in the notes on the 1832 collection was not of a bone, but rather of a hypothetical reconstruction of the process of formation of the plain wherein bones were found.[29] Darwin seems not to have had with him aboard ship a copy of Cuvier's classic text on fossil mammals, *Recherches sur les ossemens fossiles de quadrupèdes,* and had the identification of bones been his goal during the voyage, this work would have been the necessary atlas. Substantial portions of Cuvier's summary views were available in Griffith's English text. On shipboard Darwin also had access to encyclopedia entries on the various extinct animals, including those drawn from the *Dictionnaire classsique d'histoire naturelle* edited by Bory de Saint-Vincent.[30]

Darwin's collections of South American vertebrate fossils formed the basis for the part of his *Zoology of the Voyage of H.M.S. Beagle* entitled *Fossil Mammalia,* with analysis and description of specimens by Owen and a separate "Geological Introduction" by Darwin describing the sites where the specimens were found and, if possible, the conditions and relative date of their deposition. As a side point, we should mention that in *Fossil Mammalia* Darwin did not retain his specimen numbers. That, combined with the fragmentary condition of many of the organic remains, makes it difficult to associate each numbered specimen as listed in his geological notebooks with a published description. Further complicating matters, a number of specimens were destroyed during World War II when the Royal College of Surgeons was bombed. The surviving specimens are in the possession of the Natural History Museum in London, formerly known as the British Museum (Natural History).

There is a curious and untold story pertinent to the voyage years that concerns the animal emblematic of the Cuvierian tradition: the *Megatherium* ("huge beast"). Cuvier had assigned this large fossil animal to the same family as the present-day sloths, which are relatively small animals.[31] By the 1830s the fossil vertebrates were as well known to the public as they were to specialists. As the 1830s began, what were lacking in Britain were actual specimens of the *Megatherium*. It was Darwin's good fortune that the *Beagle*'s route took him to the site of the original finds of the *Megatherium,* where he was successful in collecting bones. These preceded him home and were exhibited by Buckland and Clift, curator of the Hunterian Museum at the Royal College of Surgeons, at the meeting of the British Association for the Advancement of Science held at Cambridge in June 1833. Henslow wrote to Darwin with enthusiasm about the bones.[32] However—and this was known to Darwin apparently as early as 1832—another prominent British collector, the diplomat Woodbine Parish, had already taken the "greater part" of a skeleton back to England. Parish presented an account of this and two other specimens of *Megatherium* at the 13 June 1832 meeting of the Geological Society; the following week, Buckland discussed the most complete of the specimens in detail at the British Association. In his notes from 1832 Darwin quoted from a newspaper report on Parish's presentation to the Geological Society.[33]

The curious aspect to the *Megatherium* story relates what would eventually be shown to be a false interpretation of its external covering. In the 1830s it was believed that "its skin, thick and . . . ossified, was divided into a number of polygonous scales."[34] Or, in Buckland's words in his 1832 article:

> A further peculiarity [in the *Megatherium*] consists in the fact of its sides and back having been armed with a coat of mail like the armadillos, which also obtain their food by the act of continual digging in the ground; this coat of mail exceeds an inch in thickness. The Professor suggested his opinion, that one use of the bony armour is to prevent the annoyance which this class of animals would feel, without some such protection, from the constant presence of sand and dirt with which the act of digging and scratching for their daily food would otherwise fill their skins.[35]

Buckland's remarks reflect his keen interest in relating structure to the manner of life of an animal—in short, a well-developed paleontological perspective. These remarks were known to be in Darwin's hands by March 1834.

Earlier, in 1832, Darwin had collected a number of osseous plates similar to those described both in Griffith's English edition of Cuvier's book and in Buckland's article. His contemporaneous notes read,

> At Punta Alta the only organic remain I found in the Tosca (excepting mere particles of shells) was a most singular one: it consisted in an extent of about 3 feet by <4> 2 covered with thick osseous polygonal plates; forming together a tesselated work: it resembles the case of Armadillo on a grand scale: these plates were double. or an interval of few inches between them.—With it was only a fragment [of?] joint of extremity.—[36]

In his specimen list Darwin noted for number 735 "Pentagonal osseous plates" and for specimens 736, 737, and 738, "Fragments of the latter: Is it a sort of hide?" He also noted for similar plates numbered 807 and 808, "These are said to belong to the Megatherium: Can it belong to the animal of which (822) is the lower jaw[?]"[37]

In 1832 Darwin hesitated regarding the association of the plates with *Megatherium,* but the next year he referred to the osseous plates simply as "Megatherium hide." He also noted that the cases of present-day armadillos are often found separate from their bodies, which might account for plates not being found together with bones. He recorded that the eighteenth-century traveler Thomas Falkner (1707–1784) had associated the polygonal plates with armadillos:

> I myself found the shell of an animal, composed of little hexagonal bones, each bone an inch in diameter at least; and the shell was near three yards over. It seemed in all respects, except it's [*sic*] size, to be the upper part of the shell of the armadillo; which, in these times, is not above a span in breadth.[38]

While Darwin was writing up his narrative of the *Beagle* voyage in 1837, all of his fossil bones were with Owen at the Royal College of Surgeons, and he omitted the question of the interpretation of the osseous plates. However, in notebook entries of 1837 Darwin continued to refer to the *Megatherium* as possessing a hide.[39]

In 1839 Owen at the Royal College of Surgeons reassigned the large osseous plates to an extinct animal he named the *Glyptodon,* which represented a new genus in the order Edentata. Central to Owen's work was the nearly entire specimen of the animal located by Parish, who had sent a drawing of the specimen, together with one of its teeth, to the Royal College of Surgeons. On the basis of the drawing and the tooth, Owen named the animal "glyptodon" or "sculpted tooth." Parish had announced the find in his book *Buenos Ayres and the Provinces of the Rio de la Plata* published in 1838.[40] Then in March 1839, in a masterpiece of synthesis and argument, Owen showed not only that the "portions of tesselated armour transmitted to Europe . . . testify to their having formed part of the structure of an edentate animal, widely different from, and much smaller than, the *Megatherium,*" but also that "the opinions . . . in favour of the *Megatherium* . . . rest on no better ground than the mere fact of bony armour of some gigantic quadruped and the skeleton of the *Megatherium* having been discovered in the same continent."[41]

In effect, Owen had stripped the *Megatherium* of its osseous plates. Darwin's specimens of *Megatherium* bones were listed on Owen's chart of the known remains of the animal. Owen used the list to argue the point that no osseous plates had been found at the same location as *Megatherium* bones.

Eventually the Royal College of Surgeons received the entire *Glyptodon* specimen recovered under the direction of Parish. Portions of it are listed in its museum catalogue of 1845.[42] Illustrated there was *Glyptodon clavipes*. The plate also bears the telling interpretive heading "Gigantic Extinct Armadillo" (Figure 9.2). Owen's illustration of *G. clavipes* recalls Darwin's initial assessment of his specimens of large fossil osseous plates: "it resembles the case of Armadillo on a grand scale." The *Glyptodon*, as Owen treated it, fitted easily into a belief in descent. Quite probably Darwin did intend to invoke that view when he wrote in his *Autobiography* of "great fossil animals covered with armour like that on the existing armadillos."

When in April 1840 Owen published on the osseous plates that Darwin brought home, he assigned some to a related genus *Hoplophorus* rather than to *Glyptodon* on the basis of the pattern displayed by the plates.[43] Since Owen could not assign all of Darwin's specimens of plates and associated bones to one genus, the section of *Fossil Mammalia* describing the specimens was burdened with the general title "Description of Fragments of Bones, and of Osseous Tesselated Dermal Covering a Large Edentata."[44] More conjecturally, it is possible that Darwin had reasons of his own for not making more of the tight fit Owen had asserted between the *Glyptodon* and the modern armadillos. In his autobiographical statement, for example, Darwin did not mention the word "*Glyptodon*." Nor did he do so in the second edition of the *Journal of Researches*, published in 1845.[45] This may have been intentional.

In the interval separating his return to England from Owen's publication on *Glyptodon*, Darwin had adopted a transmutationist hypothesis. In considering descent in the notebooks, he did not insist that a direct ancestor be found to connect extinct fossil species to their living descendants. Rather, he suggested that one should expect to find extended family relationships. The *Glyptodon*-armadillo relationship, when it was suggested, may have seemed to set too high a standard for a fit, or—in the vocabulary of Owen's 1839 article—an "affinity."

Two quotations from the notebooks suggest the direction of Darwin's views in the 1837–1839 period:

> We may look at Megatheria, armadillos & sloths as all offsprings of some still older type. . . .
>
> Now according to my view. in S. America parent of all armadilloes might be brother to Megatherium.—uncle now dead.[46]

Darwin was cautioning himself against tracing a direct line of descent between extinct and living species. While at the time he wrote these entries he believed

Pl I
GIGANTIC EXTINCT ARMADILLO (Glyptodon clavipes Ow)
Length of the Carapace 5ft 7in

Figure 9.2. Gigantic extinct armadillo. Courtesy of the Library of Congress.

that the *Megatherium* possessed dermal armor, his hesitation seems to have continued even after the *Glyptodon*-armadillo relation was known. Thus, in his 1844 "Essay" on the species question, he wrote of present and former animals in South America as "intimately related" and remarked on "the numerous enormous animals of the megatheroid family, some of which were protected by an osseous armour like that, but on a gigantic scale, of the recent armadillo," but he did not invoke the tight *Glyptodon*-armadillo example.[47] It remained in the background, a surety of descent for those inclined to the transmutationist hypothesis.

Questions arose regarding the *Megatherium*'s hide. In his Bridgewater Treatise of 1836, Buckland, following Cuvier, showed the *Megatherium* with a hide formed by osseous plates.[48] Owen challenged Buckland, arguing that the *Megatherium*'s "toes were developed to sustain and wield long and compressed claws, such as form the compensating weapons of defence of the hair-clad Sloths and Anteaters."[49] By the close of the decade the *Megatherium* hide had been assigned to a new animal that showed striking "affinity" (Owen's term) to the present-day armadillo. Darwin figured in the public debate because his specimens of *Megatherium* bones were studied and prized, and because he had collected osseous plates believed to be part of the *Megatherium* hide. Owen, who was then emerging as the successor to Clift and Buckland, resolved the question of the *Megatherium* hide by his identification of the *Glyptodon*. Inadvertently he also added to Darwin's then-forming case for transmutation. In Darwin's file on paleontology, there is an undated slip on which he reminded himself that Owen's paper on the *Glyptodon* "must be studied."[50] The question of the *Megatherium* hide proved a high-stakes endeavor for Owen and for Darwin, though each was to pursue the subject of affinity in a different direction. A trace of that endeavor appears in Darwin's autobiographical remarks on the pathway he followed to transmutation.

Let us now return to the chronology of the voyage. Darwin worked with the great fossils on the eastern side of South America in 1832, the first year of the voyage. As time went on, he sought to integrate his understanding of the geological history of the area with the distribution of fossils, and with the distribution of present-day species. In doing so, he often thought in terms of particular animals. In 1834, sometime before mid-April, he reviewed his geological notes taken to that point in the voyage and wrote a ten-folio essay, "Reflection on reading my Geological notes."[51] In it his schematic presentation of the history of life on the eastern side of South America is not transmutationist, but it is highly sequential. The key passage reads,

> May we conjecture that these [elevations] . . . began with greater[(x)] strides, that rocks from seas too deep for life . . . were rapidly elevated & that immediately when within a proper depth. life commenced. . . . The elevations **rapidly** continued; land was produced. on which great quadrupeds lived: the former inhabitants of the sea perished . . . the present ones appeared.—
>
> [(x)]Perhaps the first opening of the N & S. crack in the crust of the globe. forming the Cordilleras

The tantalizingly enigmatic phrase "life commenced" would seem, in this context, to refer to the belief, common before the dredging of ocean floors was possible, that no life could exist at great depths. Equally striking in Darwin's remarks is the sequence: elevation of land from the sea (as Darwin's footnote suggested, perhaps caused by the opening in the crust of the globe that formed the Andes) and the production of the land where the "great quadrupeds lived."[52]

Darwin then speculated on the origin of species in the Pampas and Patagonia in the period since the deposition of superficial layer of reddish argillaceous earth called "Tosca." He wrote,

> As Patagonia has risen from the waters in so late a period, it may be interesting to consider whence came its organized being[s].—I have conjectured the absence of trees in the fertile Pampas & rich valleys of B[anda] Oriental. to be owing to no Creation having taken place subsequently to the formation of the superior Tosca: with respect to the Plants. I know nothing.—The . . . larger animals being of easier knowledge. best deserve observation.—We will consider only those animals South of the Plata.—

Using the reference works he had with him aboard ship, as well as his own field knowledge, Darwin then listed the indigenous animals and their ranges: the skunk, the otter, the common fox, the jaguar, the puma, and so forth. He noted that all but two (*Cavia Patagonica,* the Patagonian cavy, and *Dasypus pichiy,* an armadillo) have sufficiently wide geographical ranges that their presence south of the Rio de la Plata could be explained without invoking a new "Creation."[53]

Some time after the initial draft of his "Reflections," Darwin joined FitzRoy in traveling up the Santa Cruz River in a trip that lasted from 18 April to 8 May 1834. After the trip he read volume 3 of Lyell's *Principles of Geology,* which led to further thought and more notes, written largely on the versos of the original text of his "Reflections." Fundamentally Darwin was already a Lyellian in his approach to the question of species extinction. Like Lyell, he believed that species died out singly rather than in batches. This belief attuned him to questions of population, for as the numbers of individuals in a species dwindled, the possibility of the extinction of a whole species increased. However, in volume 3 of the *Principles* Lyell presented a principle of interpretation for the fossil record that Darwin challenged or, at least, queried. Lyell suggested dating recent strata by comparing the proportion of extant to extinct species of shells. As an example he suggested that

> if strata should be discovered in India or South America, containing the same small proportion of recent shells as are found in the Paris basin, *they* also might be termed Eocene. . . .
>
> There might be no species common to the two groups; yet we might infer their synchronous origin from the common relation which they bear to the existing state of the animate creation.[54]

In his copy of Lyell's book, Darwin added the following trenchantly: "if the rate of change is everywhere the same." In a similar comment in his annotations to his "Reflections" essay, he wrote that establishing the age of Patagonian beds with respect to Europe could only be done by studying the relative proportion of recent shells, but he continued, "This rests on the supposition that species become extinct in [the] same ratios over the whole world."[55] Here we see Darwin pulling away from Lyell.

In February 1835 Darwin returned to the subject of species extinction in an essay on the subject.[56] He was puzzled by the extinction of a large fossil he had found at the port of San Julián in what is presently Argentina. The fossil, clearly of an extinct vertebrate—Mastodon (?)," he queried—seemed to have been deposited quite recently in geological terms, that is, after the "*re*-modelling into steps of what at first must appear the grand (so called) diluvial covering of Patagonia." Again he puzzled over what might have caused the extinction of the animal, for he believed that nothing in the geological landscape suggested that the vegetation could have changed drastically in the period leading up to its demise. In his essay Darwin also commented on an 1831 publication by Buckland concerning fossil elephants of an extinct species found in cliffs of frozen mud on Eschscholtz Bay in what is today Alaska. He took issue with Buckland, who was also struggling with the "perplexing question" of climate in relation to the extinct elephant of the Arctic.[57] Buckland had concluded that the bones of modern animals found with those of the extinct elephant had fallen on the beach, and were not coeval. He believed the climate had changed in the interval between the life of the elephant and that of modern animals in the area.

In his notes Darwin argued against Buckland's interpretation, wanting instead to see the superficial gravel of Eschscholtz Bay as similar to what he now interpreted as marine depositions in South America. To explain the position of Buckland's elephant, Darwin adduced what he termed his own "not very satisfactory explanations," including "the chance that the animals did not require quite the temperature of the torrid zone." In short, Darwin, following his own Lyellian interpretation of South America, resisted Buckland's conclusions regarding the elephant bones of Alaska. His arguments in the February 1835 essay up to this point were not distinctly new for him. However, they do show him engaged with what was then current literature, including what he referred to as Buckland's "beautiful paper."

There was, however, something quite new in Darwin's February 1835 essay. In it he posed a hypothesis: perhaps species are born and die, as do individuals. The Italian geologist Giovanni Battista Brocchi (1772–1826) suggested this idea, only to be rejected by Lyell in the *Principles of Geology.* In February 1835 Darwin revived the idea as a point for speculation, having recently been drawn to compare the duration of the life of a species to that of an individual by observing the grafting of apple trees in Chiloé.[58] ("Posit the opposite" is one strategy for theory change Lindley Darden has suggested; Darwin's move here is a good example of that strategy in action. Darwin rejected Lyell's rejection.)[59] In the same essay Darwin then went on to suggest that births and deaths must be balanced, for otherwise "the number of inhabitants has varied exceedingly at different periods," which would be "in contradiction to the fitness which the Author of Nature has now established." In suggesting a constancy in number of inhabitants for the world, Darwin was appealing to a natural theological sense of order, but deploying it in a new direction: to require that the birth of species be accounted for.

The February 1835 essay is interesting on a number of grounds. It shows Darwin as a discoverer in the empirical sense: he landed at the site in January 1834 and found what he referred to in his contemporary notes as "large bones," which would become "Mastodon (?)" in the essay. He also imagined that the "fossil animal, whilst in a like manner [to that of the present-day Guanaco] crossing the ancient bay, fell into one of the muddy creeks, and was there buried"—an interesting observation in the light of recently found footprints of the animal.[60] (Still later Owen would identify the bones as a new species, *Macrauchenia patichonica,* but that development came after the voyage.) The document also shows Darwin's maturity as a participant in what were then ongoing discussions in British geology.

It is now 1835 in our account. In September and October of that year Darwin explored the Galápagos archipelago. He was on or nearby the following four islands on the following dates: Chatham (now San Cristóbal), 17–22 September; Charles (now Floreana or Santa María), 23–27 September; Albemarle (now Isabela), 29 September–3 October; and James (now Santiago or San Salvador), 8–17 October. A rich literature has developed around Darwin's trip to these islands, most recently Edward Larson's broad study of the history of their exploration and the precise tracing of Darwin's route by Greg Estes, K. Thalia Grant, and Peter R. Grant.[61] Important as those weeks on the islands were to Darwin, we should not pluck them out of the context of the voyage in its totality. If, for example, Darwin had visited the islands at the beginning of the voyage, rather than near its end, his thought on species creation and extinction, and on the desirability of relating both to geological processes, would not have been as developed. By September 1835, he had become attuned to the subject of the durability of species.

Darwin was also looking forward to the islands as a prime site for new collections. Thus, he wrote to his family,

> I am very anxious for the Galapagos Islands,—I think both the Geology & Zoology cannot fail to be very interesting.—With respect to Otaheite [Tahiti], that *fallen* paradise, I do not believe there will be much to see.—[62]

"Fallen"—the double meaning aside—would have meant to him "used up" as regard to the collector seeking novelty.

On the Galápagos Islands Darwin expected to find novel species, "novel" in the sense that the islands had never been subject to systematic collection by a trained naturalist and could be expected to yield a rich harvest. There was also an open question as to whether they would be similar or identical to species on the mainland. In his contemporary notes made at Charles Island Darwin phrased the inquiry as follows:

> 26th & 27th [September] I industriously collected all the animals, plants, insects & reptiles from this Island.—It will be very interesting to find from future comparison to what district or "centre of creation" the organized beings of this archipelago must be attached.—[63]

The Galápagos Islands did prove to provide a range of possible answers to that question, both at the time of Darwin's visit and later.

Among the interesting animals were the tortoises, which gave their name to the islands. The giant tortoises were not found on the South American continent, though the relevant edition of Cuvier's *Animal Kingdom* connected them, "perhaps" by "some analogy," with the Indian tortoise, "*Testudo Indica*," whose carapace measured three feet long on average.[64] In his contemporary notes Darwin did not use a scientific name for the tortoise, or speculate on its possible connections with other tortoises in the world. Nor did he suggest anywhere in his contemporary notes that he believed the Galápagos tortoise to be unique to the islands. Later, in his book on the voyage, with the backing of his authority on reptiles, Thomas Bell (1792–1880), Darwin did suggest the Galápagos Islands as the probable point of origin for all the world's giant tortoises.[65]

Overall, Darwin associated Galápagos species with nearby land, namely, South America. As he said in his zoology notes, and later emphasized more generally in his published account, "The Ornithology is manifestly South American.—"[66] But in what is one of the best-documented cases of empirical discovery in science, Darwin also recorded patterns of variation among species on the islands that surprised him and that, ultimately, pushed him across the line to a transmutationist position. I use the word "discovery" intentionally, for Darwin clearly learned something that he had not known before from on-site inspection of the Galápagos fauna and flora. I am making this point since the pendulum has possibly swung too far in the opposite direction: at first, Darwin's experience in the Galápagos Islands was overemphasized as the turning point in his arrival at a transmutationist position. Now that most readers have learned that Darwin did not become an evolutionist until 1837, the Galápagos experience is possibly credited with too little.

The mockingbirds stand out in importance among the animals and plants he observed and collected on the islands. On the evidence of his contemporary notes, one piece of new information stands out as essential. It was given to him by Nicholas O. Lawson (dates unknown), a British subject who was at that time interim governor of the penal colony that had been established on Charles Island in 1832 by the Republic of Ecuador. There was an element of serendipity in the timing of Darwin's contact with Lawson, for they met on Charles Island, the second of Darwin's four stops, which allowed him time to factor Lawson's remarks into his expectations for subsequent stops.

The contemporary evidence for Darwin's view of the mockingbirds begins with an entry in his field notebook, written on Chatham Island, probably on one of the last two days he was there, that is, either 21 or 22 September, when he recorded in his Diary, "We slept on the sand-beach, & in the morning after having collected many new plants, birds, shells & insects, we returned in the evening on board.—" In his field notebook he commented, "The Thenca [mockingbird] very tame & curious on these Islands. I certainly recognise S. America in Ornithology:- would a botanist. 3/4 of plants in flower."[67] In context Darwin's phrase "in these Islands" suggests that Darwin was presuming a single species on all the islands; at this point in his travels he had explored only Chatham.

Darwin's next stop was on Charles Island. Here he met Lawson, who, as an Englishman in South America in the 1830s, was a familiar sort of figure, though not so prominent as to be mentioned in Mulhall's *English in South America*. In 1830 Lawson, who had already visited the islands to assess their fertility, imported domestic animals to Charles Island. In 1838 he was still on the island, continuing to give scientific and navigational information to visitors, as he had done for FitzRoy and Darwin.[68] An Ecuadorian encyclopedia article estimates that the population of Charles Island fell from 300 in 1835 to 12 in 1851, owing to poor government, which it attributes to persons other than Lawson.[69] Whatever the difficulties of supervising a penal colony, Lawson behaved flawlessly toward his unexpected guests. FitzRoy wrote of Lawson, "At his table we found the welcome of a countryman, and a variety of food quite unexpected in the Galápagos Islands." For Darwin the yield was even greater.

Lawson was a keen observer of nature. He commented on the decline in the number of tortoises, due to exploitation of them for food by residents of Charles Island and crews of passing whaling ships. More importantly, he brought home to Darwin the difference in shapes of tortoise shells according to the island of origin of the tortoise. On this point Darwin's earliest-known statement, made while on the Galápagos or shortly afterward, is as follows:

> It is said that slight variations in the form of the shell are constant according to the Island which they inhabit—also the average largest size

> appears equally to vary [pencil insert] according to the locality.—Mr Lawson states he can on seeing a Tortoise pronounce with certainty [pencil insert] from which island it has been brought.—[70]

Presumably Lawson elaborated on this opinion in conversation with Darwin, giving details of his method of identification, but either Darwin did not record Lawson's remarks or his notes are missing. What Lawson would have pointed out to Darwin was the distinction among a number of groups, most strikingly the difference between those with dome-shaped and those with saddle-shaped carapaces. On Darwin's first stop, at Chatham Island, he would have seen the first. On his second stop, at Charles, he would have seen the second, either as living animals or as remains, for Lawson used the "large terrapin shells . . . to cover young plants, instead of flower pots."[71]

The tortoises of Charles Island are believed to have become extinct sometime after the *Beagle*'s visit; records of successful "turpining" expeditions to the island by American whaling vessels in 1831–1868 extend through 1837 but not afterward.[72] The existence of the distinctive saddle-back tortoises at Charles Island had been pointed out by the American naval officer David Porter (1780–1843) as early as 1815. Noting that it differed from the species of tortoise on another island, Porter wrote accurately of the Charles Island tortoise: "The form of the shell is elongated, turning up forward in the manner of a Spanish saddle."[73] Unfortunately, Porter's book was not part of the *Beagle*'s library; had it been, Darwin would have been cued earlier to study the tortoise variants. In any case, Darwin later remarked that the tortoises brought home aboard the *Beagle* were too small "to institute any certain comparison."[74] The tortoises thus do not appear in the published *Zoology* from the voyage.

How then are we to interpret the initial effect of Lawson's comment on Darwin? My reading of Darwin's text, together with his subsequent action while on the Galápagos, is that Lawson's comment was *a* trigger, and possibly *the* trigger, that allowed Darwin to capture and note the situation among the Galápagos mockingbirds. Here he was greatly aided by the circumstance that on the very island—Charles—where he met Lawson, he also encountered mockingbirds that differed from those on Chatham Island, which he had just visited.

The Charles Island mockingbird differs noticeably from the Chatham mockingbird by the possession of what one recent observer described as its "large breast spots, like a bikini."[75] The Charles mockingbird is now regrettably extinct on the main island but survives on Champion and Gardner islets off its shore. In collecting the Charles Island mockingbird, Darwin did record its island of origin in his contemporary notes. Similarly when he collected the mockingbirds on the other two islands, he explored and noted their island of origin. In his published *Zoology,* he wrote, "I fortunately happened to observe, that the specimens which I collected in the two first islands we visited, differed from each other, and this made me pay particular attention to their collection."[76] In his contemporary notes, he commented,

> This birds [*sic*] which is so closely allied to the Thenca of Chili (Callandra of B. Ayres) is singular from existing as varieties or distinct species in the different Is^{ds}.—I have four specimens from as many Is^{ds}.—These will be found to be 2 or 3 varieties.—Each variety is constant in its own Island.—[Note (a) the Thenca of <Chatham Is^{d}> Albemarle Is^{d} is the same as that of Chatham Is^{d}.—] This is a parallel fact to the one mentioned about Tortoises.[77]

This statement has key ingredients for establishing the contemporaneous nature of Darwin's discovery: the citation of it as a fact, like that regarding the tortoises, and their mention of there being different varieties or species assigned according to island. Darwin's phrase regarding each variety as constant according to its own island does not, in actuality, apply to the tortoises, though from the context it is clear that he was repeating what he had heard from Lawson, rather than what he had observed personally.

To underline the enduring nature of Darwin's discovery, let us compare his initial conclusions to current knowledge with respect to the Galápagos mockingbirds. Figure 9.3 indicates Peter R. Grant's current opinion regarding the birds. He recognizes four separate species, one on Hood Island (Española), which Darwin did not visit. The Charles mockingbird is still recognized as distinct from the Chatham mockingbird. This key empirical aspect of Darwin's observation has remained constant from 1835 to the present.

The Charles Island mockingbird is the most distinct. Beyond that point, current judgments do not correspond precisely with those Darwin made in 1835, or those made by the ornithologist John Gould (1804–1881) in 1837. In 1835 Darwin judged the Albemarle and Chatham forms to be the same. Gould regarded them as distinct, and they continue to be regarded as distinct today. Gould, however, grouped the mockingbirds of James Island with those of Albemarle, which, as Grant's diagram suggests, would not be done today.[78] Thus, there was partial but not complete correspondence between Darwin and Gould's judgments and those of the present day. This may be read as error or, perhaps better, as a level of messiness inherent to the process of discovery. It is also the case that discussions are ongoing as to the appropriate taxonomy for the birds.[79]

The Galápagos species continued to occupy Darwin's mind during the voy-

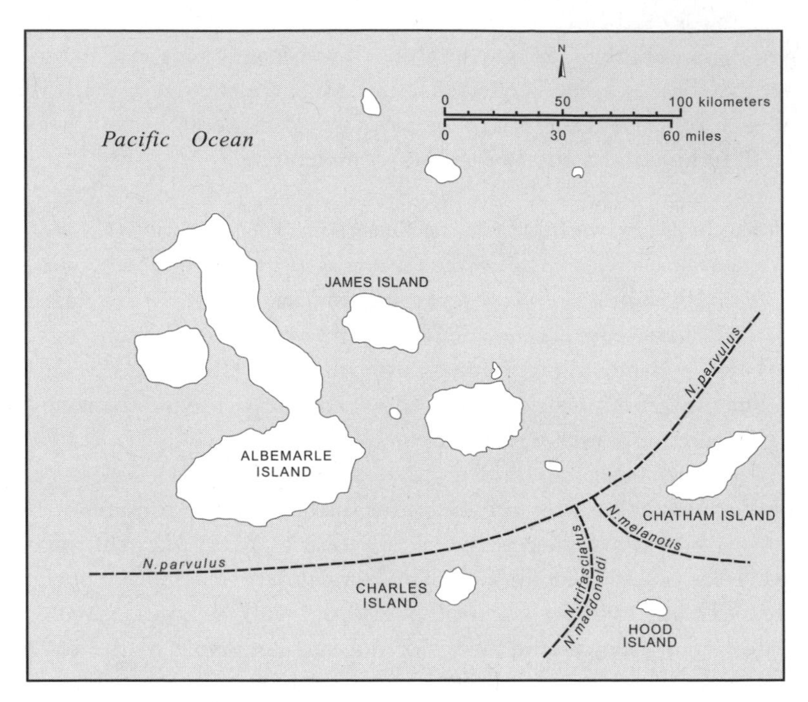

Figure 9.3. Distributional map for species of Galápagos mockingbirds. UMBC Cartographic Services, redrawn courtesy of Peter R. Grant and Princeton University Press from *Ecology and Evolution of Darwin's Finches.*

age. To his mentor Henslow he wrote of his Galápagos collections, "I shall be very curious to know whether the Flora belongs to America, or is peculiar. I paid also much attention to the Birds, which I suspect are very curious.—"[80] In July 1836 the *Beagle* called at the island of St. Helena in the Atlantic Ocean; he noted it as "possessing an unique Flora, this little world, within itself"—a metaphor he was later to apply to the Galápagos archipelago when he referred to it as "a little world within itself."[81] The Galápagos species as a whole—that "little world"—remained of special interest, as a model for island fauna and flora. He also continued to develop his thoughts on the mockingbirds. As he prepared his collections for examination by the experts at home, he went through his Zoology Notes and his specimen catalogues, assembling separate lists of specimens. In part this was a clerical task, with his servant and copyist Syms Covington assisting him. But in the course of constructing these lists, Darwin added new material and paused to reflect on the significance of some specimens. Of particular importance was his reflection on the Galápagos mockingbirds, for he added a passage that is tantalizing and yet so brief that it has permitted varying interpretations. The passage reads,

> When I see these Islands in sight of each other, & <but> possessed of but a scanty stock of animals, tenanted by these birds, but slightly differing in structure & filling the same place in Nature, I must suspect they are only varieties. . . . If there is the slightest foundation for these remarks the zoology of Archipelagoes will be well worth examining; for such facts <would> undermine the stability of Species.[82]

In my reading of the passage I fasten on the phrase "undermine the stability of Species," as Darwin for the first time raised the question explicitly of loss of fixity in species. This represented a step beyond the "birth and death" language he employed in mid-1835. Further, the sorting out of mockingbirds, island by island, carried imbedded within it a possible mechanism for how the stability of species might be undermined. I read the phrase "I must suspect they are only varieties" as Darwin debating with himself questions that could be resolved only on his return to England.

The impact of other Galápagos birds was less dramatic than that of the mockingbirds. The finches were important as providing examples of new species, but they were not the critical example that led to Darwin's adoption of a transmutationist hypothesis. One wishes that the phrase "Darwin's mockingbirds" rather than the phrase "Darwin's finches" had become current; it would aid producers of various television documentaries in telling the story with the right emphasis.[83]

In the last year of the voyage many of Darwin's activities were directed toward reentering England. England met him halfway. In June 1836 he encountered a representative of British science in the person of John Herschel at the Cape of Good Hope in Africa. Reading Herschel had inspired Darwin in 1831. Now, five years later, they would possibly have touched on the subject that Herschel had addressed in a letter of 20 February 1836 written to Lyell. Herschel opened the letter to Lyell thanking him for the gift of a recent edition of the *Principles* and then immediately went to the most sensitive subject:

> Of course I allude to that mystery of mysteries the replacement of extinct species by others. . . . I cannot but think it an inadequate conception of the Creator, to assume it as granted that his combinations are exhausted upon any one of the theatres of their former exercise—though . . . we are led by all analogy to suppose that he operates through

> a series of intermediate causes & that in consequence, the origination of fresh species, could it ever come under our cognizance would be found to be a natural in contradistinction to a miraculous process.

Herschel then went on to comment that no such process of producing species was known, and, indeed, it would be difficult to distinguish whether a species was nascent or just dying out, though it was possible that both processes were going on at once—some groups "spreading" from their centers while others were "retreating to their last strong holds."[84] All that separated Herschel's and Darwin's views of 1836 was that—possibly—Darwin had intimations of how a process of the "origination of fresh species" might work from his experiences with the Galápagos mockingbirds.

Herschel's terminology, his phrase "origination of fresh species," certainly calls to mind Darwin's eventual title: *On the Origin of Species*. For the historian it also solves the nagging problem of choosing historically apt vocabulary since Darwin did not use the then current term "transmutation" or the modern term "evolution" in its current sense during the voyage. The phrase "the species question" is a historian's catchall phrase rather than one in use in Darwin's circle during the 1830s. Perhaps Herschel's letter provides the best description in contemporary language of the directions of Darwin's interest: reflections on "the replacement of extinct species by others" and on "the origination of fresh species." One final point can be made here: although Lyell argued against transmutation in the *Principles*, Herschel clearly interpreted him as having reopened the question. As Herschel wrote to Lyell, "You have succeeded . . . in adding dignity to a subject already grand by exposing to view the immense extent & complication of the problems it offers for solution."[85] The subject of species had become a set of problems awaiting solution.

To what extent Herschel and Darwin discussed the subject of species in June 1836 is unknown. Certainly Herschel's precise statements were new to him in December 1838, when he added a heartfelt "Hurrah" to his note that "Herschel calls the appearance of new species. the mystery of mysteries."[86] But the two men, with their shared regard for Lyell, were on a common track throughout the 1830s, which would certainly have been reflected in their conversations in June 1836.

CHAPTER 10

GEOLOGY AND THE *ORIGIN OF SPECIES*

I have just finished my sketch of my species theory. If, as I believe that
my theory is true & if it be accepted even by one competent judge, it will be
a considerable step in science. . . . I therefore write this, in case of my sudden death,
as my most solemn & last request . . . that you will devote 400£ to its publication. . . .
With respect to Editors.—M$^{r.}$ Lyell would be the best if he would undertake it:
I believe he w^{d} learn some facts new to him. As the Editor must be a geologist, as well
as Naturalist. The next best Editor would be Professor Forbes of London. The next best
(& quite best in many respects) would be Professor *Henslow*??. D$^{r.}$ Hooker would
perhaps correct the Botanical Part probably=he would do as Editor= Dr Hooker
would be *very* good The next Mr Strickland.—If no<ne> of these would
undertake it, I would request you to consult with Mr Lyell, or some
other capable man, for some Editor, a geologist & naturalist.

CHARLES DARWIN TO EMMA DARWIN, 5 JULY 1844

From being accustomed to look at the world under a geological point of view
I view . . . the *whole* existing Fauna & Flora as a *mere fragment;* . . .

CHARLES DARWIN TO HEWETT COTTRELL WATSON, 26 AUGUST 1855

I think geologists are more converted than simple naturalists
because more accustomed to reasoning.

CHARLES DARWIN TO ALFRED RUSSEL WALLACE, 18 MAY 1860

"In former case position, in latter time"

Within six months of his arrival back in England Darwin had become a transmutationist. In his own date book he identified March 1837 as the critical period for his adoption of a transmutationist hypothesis. His entries in his Red Notebook, where he first used such phrases as "if one species altered" or "if one species does change into another," provide the documentary evidence that corroborates that date.[1] Further, it is now generally accepted by scholars that it was his interaction with professional taxonomic experts in London that permitted him to posit a transmutationist hypothesis, for the concept of "species" was defined by such men as John Gould and Richard Owen as they went about their daily work. It was their office, not Darwin's, to name his specimens.[2] This naming was of two kinds: (1) identification of specimens of unknown character and (2) the marking off and naming of good species. For the large fossils from South America the first task was paramount; for the more familiar living animals, as the birds, the second task dominated. Both sorts of identifications figured in his adoption of transmutationism.

As of March 1837 Darwin's understanding of transmutationism had both a geographical and a temporal dimension. On the geographical side stood the forming of species (what would today be called "speciation"); on the temporal side stood descent. The two processes were tied together by the notion of succession: as one species succeeded another across space, so one species succeeded another in time. When Darwin discussed this idea in the Red Notebook, he had particular species in mind: on the geographical side most especially, the two South American rheas; on the temporal side, the fossil bones that he believed—following Owen's advice—to be an extinct guanaco. Comparing the two he said,

> The same kind of relation that common ostrich bears to (Petisse. & diff kinds of Fourmiller): extinct Guanaco to recent: in former case position, in latter time. (or changes consequent on lapse) being the relation.—[3]

The examples he chose have their own complexities, but the key point to emphasize is the comparison he drew between change over time and change over space.

The case of the ostriches was straightforward. The common ostrich was *Rhea americana*. The "Petisse" was the lesser rhea, *Pterocnemia pennata*. Darwin customarily referred to the two rheas in his field notes as "Avestruz" and "Avestruz petise" from the Spanish "*avestruz*" (ostrich) and "*avestruz petiso*" (small ostrich). The common rhea is presently found from northeastern Brazil to the Río Negro in Central Argentina, and the lesser rhea in the

Patagonian lowlands, where Darwin collected portions of a specimen, and in the high Andes of Peru, Bolivia, northern Chile, and northwestern Argentina. The lesser rhea became known as Darwin's rhea following its identification by Gould at a meeting of the Zoological Society of London on 14 March 1837. Gould was then unaware that the species had already been described in 1834 by d'Orbigny.[4] Darwin drew on the explanatory possibilities of the case of the two rheas throughout his transmutation notebooks.[5] The next term "Fourmillier" (Darwin's spelling) is more problematic since it was drawn from "Fourmilier" (antbird), though Darwin did not use the term elsewhere in his notes. More likely, he was referring to birds that he described elsewhere as *Myothera* and that are now assigned to the South American family Rhinocryptidae, a small family related to the Formicariidae or antbirds. He collected a number of specimens of these birds, which displayed patterns of geographical distribution that he associated with the rheas, that is, a pattern that would suggest a process of speciation.[6]

The phrase "extinct Guanaco to recent" has a beautifully clear meaning, at least on the surface. Darwin intended by this phrase to refer to the bones he had collected at the port of San Julián, having "no idea at the time, to what kind of animal these remains belonged."[7] After examining them, Owen had written to Lyell in a letter dated 23 January 1837 describing them as follows:

> RUMINANTIA
> Fam: Camelidae
> 2 cervical vertebrae, portions of femur, & fragments of a Gigantic Llama! as large as a Camel, but an *Auchenia* (from the plains of Patagonia)[8]

The animal Owen had described as a "Gigantic Llama" was the one, following Owen, that Darwin referred to in his Red Notebook passage as the "extinct Guanaco." His interpretation of what he took to be two llamas or guanacos was that they bore the same kind of relation to each other as did the two rheas. There was a notion of succession that, to Darwin, suggested a genetic relationship between the two species.

In a presidential address to the Geological Society of London on 17 February 1837, Lyell developed the notion of succession in regard to South American vertebrate species collected by Darwin and described by Owen. He reviewed Darwin's collection according to Owen's initial interpretations and concluded,

> These fossils . . . establish the fact that the peculiar type of organization which is now characteristic of the South American mammalia has been

> developed on that continent for a long period, sufficient at least to allow of the extinction of many large species of quadrupeds. The family of the armadillos is now exclusively confined to South America and here we have from the same country the Megatherium, and two other gigantic representatives of the same family. So in the Camelidæ, South America is the sole province where the genus Auchenia or Llama occurs in a living state, and now a much larger extinct species of Llama is discovered. . . . These facts elucidate a general law previously deduced from the relations ascertained to exist between the recent and extinct quadrupeds of Australia; for you are aware that to the westward of Sydney on the Macquarie River, the bones of a large fossil kangaroo and other lost marsupial species have been met with.[9]

Thus Darwin was able to draw not only from Owen's description of the South American bones but also from Lyell's interpretation. Further Darwin could move easily from the notion of succession to the notion of descent, though the word does not appear in the Red Notebook.

There are interesting complexities in the story of Darwin's use of the information available to him from contemporary paleontology. We have cited the case of the *Megatherium* hide. Darwin also used information that was later altered in his citation of the "extinct Guanaco."[10] Initially, in January 1837, Owen associated the fossil he named *Macrauchenia* with the order Ruminantia; before writing up the *Zoology* he placed it with Pachydermata. Owen's association of the fossil with camels was based on the vertebrarterial canal in the neck being like that in camels while his later assignment of the fossil to Pachydermata depended in large part on the structure of the fossil's astragalus or ankle bone (Figure 10.1). Owen's research notebooks on the specimen have survived and support this interpretation.

Owen's reassessment of the fossil seems to have come in the autumn of 1837, sometime before he gave the fossil its scientific name in December of that year.[11] As to present-day views, Owen's *Glyptodon* is still regarded as phylogenetically closely related to the armadillos. However, *Macrauchenia* is placed in Litopterna (an order of extinct South American ungulates), its camel-like features being only superficial. South America's llamas and guanacos are connected phylogenetically with Old World camels.[12] Thus, there is an irony that two examples of close relation that Darwin presumed to exist during the time he was keeping the transmutation notebooks are now discarded. An instance in the notebooks where Darwin cited both examples in the same sentence is from Notebook A, page 9, written in mid-1837: "Will it be supposed that the armadillos have eaten out the Megatherium [then presumed to have scutes]. The Guanaco the Camel [*Macrauche-*

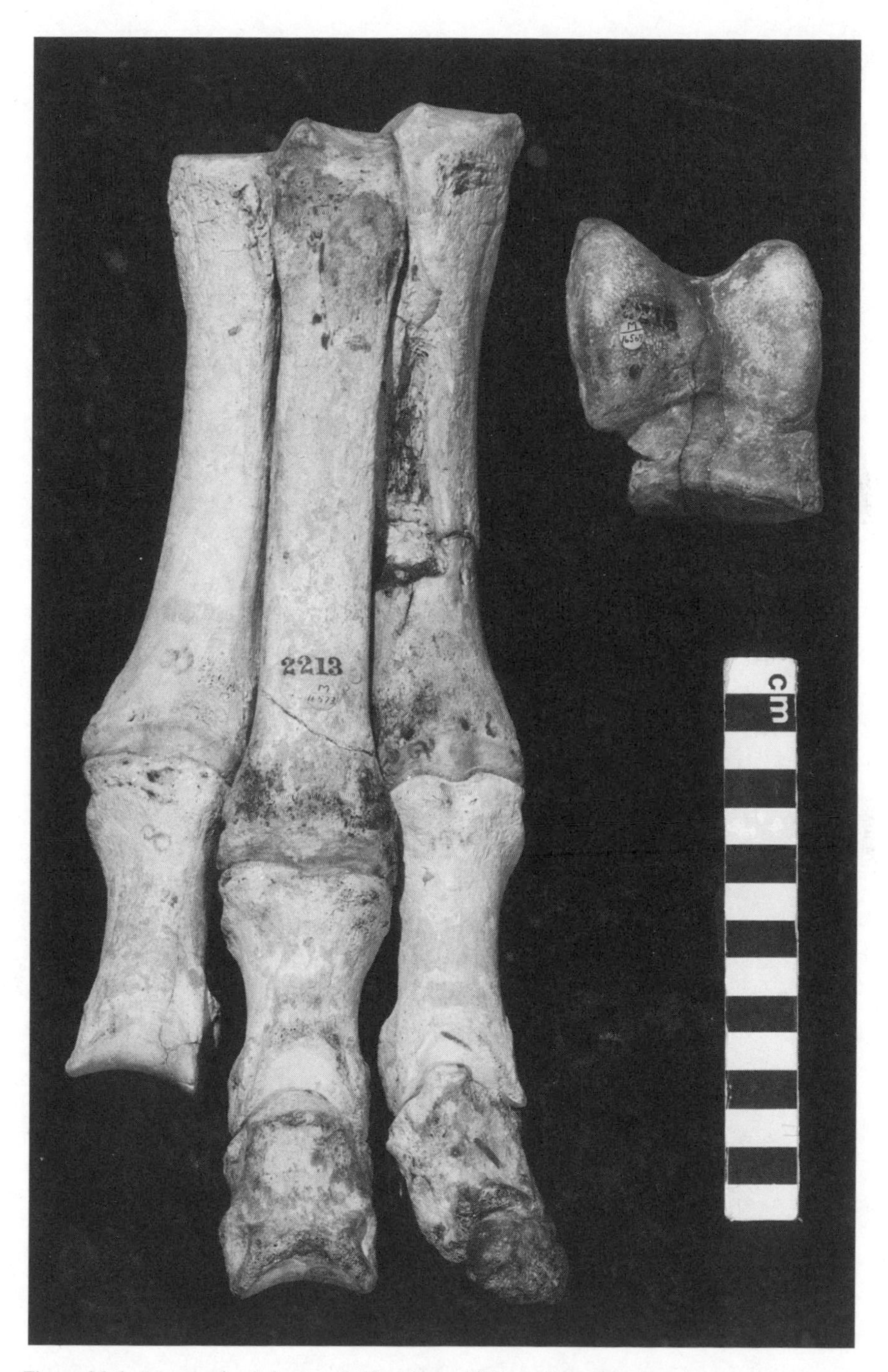

Figure 10.1. *Macrauchenia* bones. © The Natural History Museum, London.

nia]."[13] Darwin was here trying to explain why one group replaced the other.

What can be made of these ironies? First, Darwin was clearly relying on the information he could gain from specialists working on his collections. This meant Owen in the case of vertebrate fossils. Second, on the evidence of his notebook entries, Darwin did not believe it likely that he would be able to identify the fossil progenitor of any given species, so that shifts in classification would not have been fatal to his theory. Thus, in his autobiographical passage quoted earlier in this chapter, he avoided the problem of naming by referring merely to "great fossil animals covered with armour like that on the existing armadillos." The connection between the armadillos and the glyptodonts fit this description at the time he was writing his autobiography, and, indeed, it still does today. In considering lines of descent, Darwin was always cautious about identifying the "parent" of any species, as in this quotation from Notebook B:

> Cuvier objects to <tran>propagation of species, by saying, why not have some intermediate forms been discovered. between palaeotherium, megalonyx, mastodon, & the species now living.—Now according to my view, in S. America parent of all armadilloes might be brother to Megatherium.—uncle now dead.[14]

Thus, Darwin argued that one ought not expect to connect living species with their precise ancestors.

We must now make due acknowledgment of the strength of the relationship between Owen and Darwin before they broke publicly over transmutation. Not only did Owen supply Darwin with descriptions of the vertebrate fossils, he was also engaged in anatomical studies that complemented Darwin's geological interests. "Owen has been doing some grand work in morphology of the vertebrata," Darwin enthused to Hooker in 1846.[15] Later, in 1854, when he pictured an "archetype shell" of a sessile cirripede, he was using the vocabulary of archetypes that calls to mind Owen's work.[16]

Like Darwin, Owen had large views on biological subjects. This is demonstrated by his lectures of 1837 to the *Royal Society of Surgeons,* which Phillip Sloan has annotated and published. While Owen the anatomist focused on interior aspects of growth, Darwin the geologist focused on the exterior conditions of life. For example, in the spring of 1837 as Owen was writing his Hunterian Lectures and Darwin his book on the voyage, both men considered the question of the causes for the extinction of species. Darwin held ideas congruent to Owen's opinion that "species could perish by the wearing out of the contained vital energy in the germ." Thus, in Notebook B, for example, Darwin recorded that "Mr Owen suggested to me" that the production of monsters might "present an analogy to production of species."[17]

On occasion Darwin used Owen as a foil. For example, as a transmutationist Darwin had to explain the reason for the existence of the sexes. He faced this problem as he read Owen's edition of a work by John Hunter (1728–1793) in which Owen used the term "unnatural hermaphrodites." For Owen's "unnatural" Darwin substituted the term "abortive." To Darwin "each Man or mammalia being abortive hermaphrodite simplifys [*sic*] case [of the origin of the sexes] much."[18] Here Darwin took Owen's "unnatural" qualifier for "hermaphrodite" and reinterpreted it in transmutation fashion as "abortive." Nature, viewed through a transmutationist lens, could accommodate hermaphrodites.

In sum, Owen and Darwin were intellectually close in the 1830s and 1840s. They were also socially linked, as is suggested by Owen inviting the Darwins to a musical evening in 1841 (the date is uncertain) and by Owen's visit to Down House in 1848.[19]

Darwin's Pathway, 1837–1859

From 1837 onward Darwin developed his ideas on transmutation: first in a series of notebooks, then in three drafts of his theory—the "Sketch" (1842), an "Essay" (1844), and "Natural Selection" (1856–1858)—and, finally, in the *Origin* (1859) itself.[20] Between 1844 and 1859 lay a fifteen-year hiatus. Darwin was waiting for the moment when, as he wrote in the testamentary letter that prefaces this chapter, he might find "one competent judge" to accept his views. Concern with audience was paramount. He would not publish until he was sure of a positive reception among those he respected most. His own grandfather had published transmutationist views in poetic form; his contemporary Robert Chambers published transmutationist views anonymously. Neither man limited his audience; neither man turned the scientific tide.

Darwin's own approach would be to convince judges one by one—Hooker and Lyell were among his candidates. In the meantime, partly on the advice of Hooker, he sought projects that would cement his reputation among career naturalists, accommodate his by-now severe physical limitations, and result in publications that would complement his theoretical commitments. Once he had completed his work on the *Beagle*'s geology, he undertook a systematic treatment of barnacles, both living and fossil. This was followed by study of pigeon breeding.[21] Eventually, in 1858, he found his primary judge in the shape of a co-discoverer, Wallace, whose work prompted Darwin to change course, abandoning a manuscript then in progress in favor of publishing a shorter treatise, the *Origin.*

The primary texts in which Darwin developed his transmutationist ideas differ in genre: notebook entries are speculative and broad-ranging, later texts expository. While most of the subjects Darwin considered in his long career were first raised in his notebooks, it is only in the "Sketch" and "Essay" that his basic argument for evolution through natural selection is fully articulated. Still, even after Darwin articulated his views, his theory remained elastic, largely because little was known about variation and heredity and various topics associated with geology. Scholars have studied the pathway Darwin followed as he assembled his overall argument, from his adoption in March 1837 of a transmutation viewpoint, to his reading of Malthus on population in September 1838 and his subsequent construction of a theory of natural selection. Rather than review the full course of Darwin's intellectual development, we will examine several geological aspects to his species work as viewed from the period of his theoretical notebooks (1836–1839) through the first edition of the *Origin.*[22]

The Geological Record

The earth's geological record was of central interest to anyone positing transmutation. Yet what sort of evidence would it provide? On this point Darwin's position was consistent with his own earlier position, which he owed to Lyell. The metaphor of the earth's strata as comparable to a book with many of its pages torn out governed Darwin's treatment of the subject. This book metaphor was not, in fact, unique to Lyell, though Darwin consistently credited it to him. Sedgwick used the metaphor in his 1830 presidential address to the Geological Society, as when he noted that at points in the sequence of earth's formations "a leaf seems to be torn out from the volume of her history."[23] Herschel favored linguistic metaphors: "Words are to the Anthropologist what rolled pebbles are to the Geologist—Battered relics of past ages."[24] But Darwin looked to Lyell in making the point:

> Lyell's excellent view of geology, of each formation being merely a page torn out of a history, & the geologist being obliged to fill up the gaps.—[25]

This view recurred in the *Origin,* with Herschel's metaphor of language attached:

> For my part, following out Lyell's metaphor, I look at the natural geological record, as a history of the world imperfectly kept, and written in

> a changing dialect; of this history we possess the last volume alone, relating only to two or three countries. Of this volume, only here and there a short chapter has been preserved; and of each page, only here and there a few lines. Each word of the slowly-changing language, in which the history is supposed to be written being more or less different in the interrupted succession of chapters, may represent the apparently abruptly changed forms of life, entombed in our consecutive, but widely separated, formations.[26]

The continuity of the book analogy between Darwin's entries in Notebook D and the *Origin*, together with the elaboration of the metaphor in the *Origin*, suggests how useful it was to him.

Still, in some torn books a whole chapter may be present here and there. In Notebook E he faced the difficulty—or opportunity—squarely:

> My very theory required each form to have lasted for its time: but we ought in the same bed if very thick to find some change in upper & lower layers.[27]

He finished off by asking, "[D]oes not Lonsdale [curator and librarian at the Geological Society] know some case of change in vertical series[?]" He was also pleased when current research in geology seemed to offer instances supporting his position. For example, in Notebook E he reviewed an 1839 work by Sedgwick and Murchison entitled "Classification of the older stratified rocks of Devonshire and Cornwall" that he believed described a "beautiful case, showing the gradation from one grand system to another." He then indicated what use he would make of the material:

> The argument must be thus put, shall we give up whole system, of transmut., or believe that time has been much greater, & that systems, are only leaves out of whole *volumes.*—[28]

He scored passages in Sedgwick and Murchison's article that emphasized their point that "the zoological groups of Devonian rocks are all of characters intermediate between those which mark the Carboniferous and Silurian epochs."

The broadest geological question from Hutton onward was the eternality of the earth: Did the earth have a beginning in time? Or, did it show no vestiges of a beginning? Throughout the notebooks Darwin clearly presumed that the earth did have a beginning in time, for he speculated as to its age. However, there are remnants of Huttonianism even here, for he consistently argued for a very old earth. His transmutationist views led him "to believe the world older than *geologists* think."[29] In a later notebook he railed against his fellow geologists:

> I utterly deny the right to argue against my theory [of transmutation], because it makes the world far *older* than what Geologists, think: it would be doing, what others but fifty years since [did] to geologists. . . . Being myself a geologist, I have argued to myself, til I can honestly reject such false reasoning[.][30]

In drawing out the imagined age of the earth, Darwin retained aspects of the uniformitarian assumption of a virtually eternal earth. He was also loyal to the anti-progressionism of Lyell in that he did not invoke the image of a cooling earth to account for changes in species. Thus, with Huttonian skepticism as to the existence of evidence from the distant geological past, he argued, "The bottom of the tree of life is utterly rotten & obliterated in the course of ages.—"[31] In thinking that, Darwin was aware that some of his geologist colleagues would disagree.

In addition to all these features of Darwin's residual Lyellian anti-progressionism, another aspect, usually overlooked in treatments of Darwin's views on species, runs through the transmutation notebooks. Darwin valued the idea that there was some stability or constancy in life on earth. Sometimes he spoke as if the number of species on earth remained constant, as in the following statement: "Twelve of the contemporaries must have left no offspring at all, so as to keep number of species constant.—"[32] Sometimes it was the sheer quantity of life that was stable:

> The *quantity of life* on planet at different periods, depends,—on relations of desert, open ocean, &c this probably on long average, equal quantity, 2^{d} on relations of heat & cold. therefore probably fewer now than formerly.—*The number of forms depends* on the external relations (a fixed quantity) & one subdivision of stations & diversity—.—this perhaps on long average equal.—[33]

This thread of speculation derived from a philosophical orientation toward uniformity.

Descent

We come now to the subject of descent. The word came somewhat late to Darwin's vocabulary. It first appeared in the transmutation notebooks toward the end of Notebook B, where Darwin remarked of two "inhabitants of the Tropics" that they had a "real relationship"—"viz descent."[34] In the course of his notebook speculations on descent, Darwin developed several new ideas. Lyell's schema that species were real but impermanent in nature was the foundation on which he built—even in reaction—as suggested by his commentaries in his own copies of the fifth and subsequent editions of the *Principles of Geology*.[35] Extending beyond Lyell's purview, however, were his own experiences in the field, as modulated by the judgments of the professional naturalists at home regarding his collections. While both Lyell and Darwin read widely and deeply in the literature of natural history and geology, Darwin's firsthand zoological field experience gave him the greater authority in considering the species question. It was not until Lyell began his own species notebooks in the 1850s that he approached Darwin's grasp of the fundamental issues.[36]

Darwin, like Lyell and Linnaeus before him, continued to focus on species as the primary unit of analysis. Indeed, he reflected on his own method in that regard:

> If species made by isolation; then their distribution (after physical changes) would be in rays—from certain spots.—**Agrees with old Linnæan doctrine & Lyells. to certain extent**[37]

While, as Ernst Mayr and others have pointed out, Darwin wrestled with the concept of species his entire life, he always believed that transmutation took place at the species level.[38] A sign of that wrestling occurs in Notebook C. In attempting to set up a general principle whereby species could be distinguished from varieties, he drew on his own experience with South American ornithology. Because the South American rheas had contiguous ranges ("interlock" was his word), they should be considered good species; for some of the continental mockingbirds he would "till neutral ground [be] ascertained, call them varieties." This led him to the general conclusion that a "species is only a fixed thing with reference to other living being.—"[39]

Related to Darwin's focus on species was his emphasis on extinction as the force that had shaped life. As he wrote in Notebook C, "The more I think, the more convinced I am, that *extinction* plays greater part then [*sic*] *transmutation*.—"[40] Here was the geologist speaking, tutored, as was all of Darwin's generation, in the tradition of Cuvier. As a corollary to the emphasis on extinction, Lamarck's approach to transmutation was ruled out. Lamarck had discounted the possibility of extinction and had posed the alternative explanation that species had transmuted.[41] In contrast, Darwin's vision placed extinction at the center. Even before he came to Malthus, and to the (yet-unnamed) notion of natural selection, death was recognized as the great pruner and shaper.

What, then, was Darwin's understanding—one almost automatically says "picture"—of the history of life? The metaphor he used was the traditional one of the "tree of life." But he altered it to accommodate his geological views even at first use. As he wrote in Notebook B, "The tree of life should perhaps be called the coral of life, base of branches dead; so that passages cannot be seen.—"[42] His freehand drawing of this idea is shown in Figure 10.2. To this

Figure 10.2. First sketch from Notebook B. By permission of the Syndics of Cambridge University Library; produced by UMBC Cartographic Services.

Darwin added another thought: "Is it thus fish can be traced right down to simple organization.—birds—not." He offered no further comment in words on his choice of examples. He then followed this thought with a variation of his previous drawing. In his accompanying drawing (Figure 10.3) the dotted

Figure 10.3. Second sketch from Notebook B. By permission of the Syndics of Cambridge University Library; produced by UMBC Cartographic Services.

line suggests the unknowable genealogy of birds, the solid line the (hypothetically) more knowable genealogy of fish. This diagram is *a* or, perhaps, *the* progenitor to the diagram that eventually appeared in the *Origin*.

The drawing is highly abstract. With it Darwin could imagine how the descent of species could have occurred without having to commit himself to any particular pathway. Moreover, as a design it could be transferred from one taxonomic level to another. In his initial discussion of the diagram in the *Origin*, he let letters A through L stand for species while in a later discussion he let the same letters stand for genera. As he stated in that later discussion, he was most interested in illustrating the question of "genealogical *arrangement*" rather than the degree of difference among descendants.[43]

In contrast to Darwin's view, consider the diagrams produced by his contemporaries. In his *Philosophie zoologique*, Lamarck pictured a movement from one grouping toward another in a clearly identified pathway for the transformation of species (Figure 10.4).[44] On the English side, Phillip Sloan has identified in Owen's notes of 1827 a hand-drawn diagram by Joseph Henry Green (1791–1863) of the "ascending scale" of animal life that "illustrates a genetic and branching arrangement similar to those found in Lamarck's work."[45] These schemes were quite different from Darwin's, labeled rather than unlabeled as to identity of organism, focused on the higher levels of taxonomy rather than species or genera, and trained on the ascent from simplicity to complexity rather than on purely genealogical descent.

With his transmutationist views and abstract notion of descent Darwin also had a different agenda from most of those working in paleontology and stratigraphy. He had received concrete help in coming to his earliest understanding of transmutation from Owen's identifications of *Beagle* specimens. But, once he had become a transmutationist, Darwin did not look first to paleontology and stratigraphy for his primary material in understanding transmutation. Rather, he looked to plant and animal breeding. In a textual way, this is signified by his starting a notebook he called "Questions and Experiments." In this notebook, begun in mid-1839, Darwin sought to record the opinions of breeders and agriculturists regarding variation and inheritance.[46]

While Darwin was pursuing his research on variation and inheritance, Sedgwick and Murchison were working out the stratigraphy of the lower part of the fossiliferous geological record, and in 1841 Phillips divided the history of life into three eras—Palaeozoic, Mesozoic, and Cainozoic, names that still stand.[47] They did not share Darwin's belief that the bottom of the tree of life had been obliterated in the course of ages. There may also have been some resentment that Darwin had not done the hard slogging in working out the earliest stages of the fossil record. In any case, these men would be among the least sympathetic early readers of the *Origin*. It would be the next generation of stratigraphers and paleontologists, the redoubtable Thomas Henry Huxley (1825–1895) among them, who would embrace evolutionary conclusions.[48]

Yet Darwin did not ignore the paleontological approach. In his barnacle studies there is systematic discussion of the geological history of fossil forms

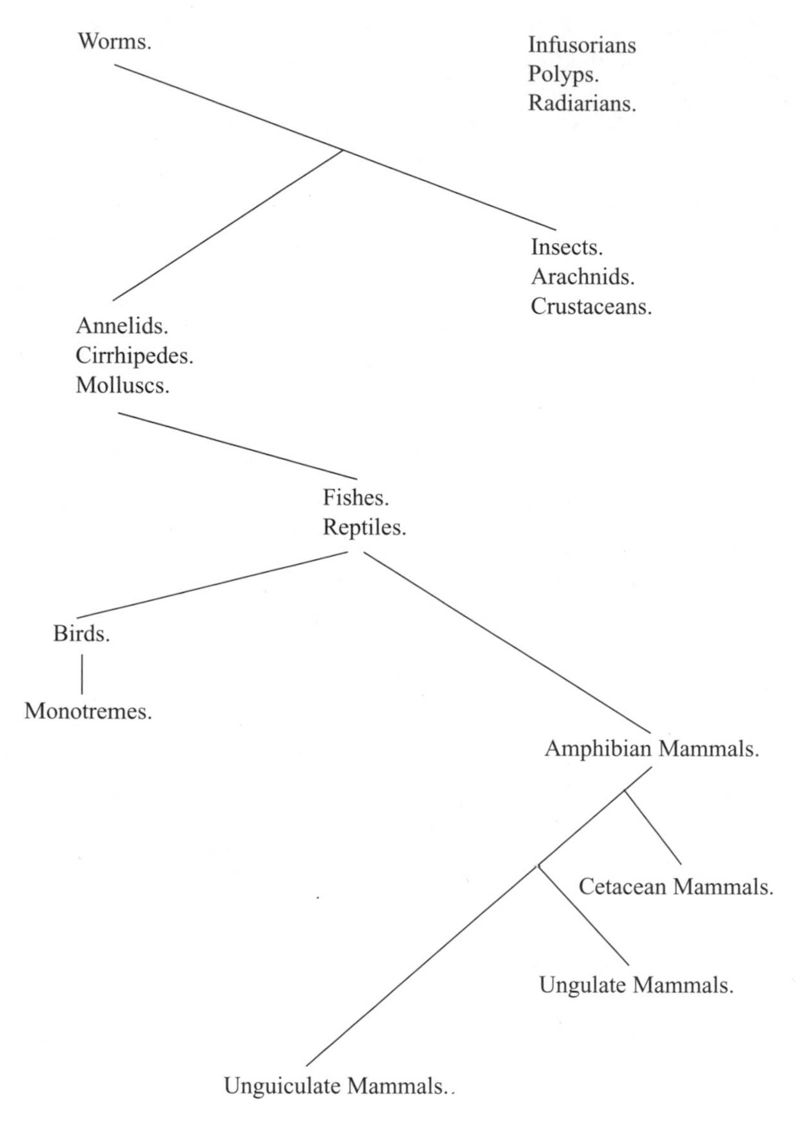

Figure 10.4. Lamarck's diagram. UMBC Cartographic Services.

found in Great Britain. While Darwin made no attempt to draw a phylogenetic tree for the evolution of the barnacles, there are remarks in the text that carry a transmutationist implication, as, for example, in his reference to "Pollicipes,—the oldest known genus, from which, in one sense, all ordinary Cirripedes, both sessile and pedunculated, seem to radiate."[49] Also, as Stott has shown in detail, Darwin's study of barnacles was particularly cued to the study of evolutionary gradation as regard to gender: from normal barnacles, which were hermaphroditic, through separate sex barnacles.[50] In his studies of pigeons, Darwin was similarly cued to the issue of descent. It was the summing up of slight variations by breeders that had, over time, produced the present array of domesticated pigeons, which "have descended from the rock-pigeon (*Columba livia*)."[51] On the pigeons he noted that he was following the common opinion of naturalists.

In addition to his own work, Darwin often noted whether paleontological facts were "for" or "against" his theory. Thus, for example, he rejoiced in *Archaeopteryx*. He wrote to Dana in 1863: "The fossil Bird with the long tail & fingers to its wings . . . is by far the greatest prodigy of recent time. It is a grand case for me; as no group was so isolated as Birds."[52] Darwin kept score: *Archaeopteryx* was one for his side. He was also alert to instances where a branching scheme seemed applicable to the fossil record. Thus, for example, he was pleased when the paleontologist John William Salter (1820–1869), who prepared specimens for exhibition in the Museum of Practical Geology,

> showed me . . . the Spirifers [a large group of articulate brachiopods] of Devonian, Lower & Upper Carboniferous formations arranged . . . after my diagram in the "Origin", & it astonished me what a beautiful branching gradation he made by intercalating the varieties & species according to geological age.—[53]

Having failed to persuade Salter to translate his visual display of specimens glued on a board into a publication, Darwin approached another expert on fossil brachiopods, Thomas William St Clair Davidson (1817–1885), for assistance. As Darwin wrote to Chambers, who had encouraged him in this direction, Davidson had "seen Salters [*sic*] table of species grouped like a tree."[54] Darwin hoped that Davidson would supply such a tree, though, in the end, he did not. In any event, Darwin desired to see "descent with modification" "exemplified and worked out in some detail, and with some single group of beings."[55]

On occasion Darwin was also interested in speculating concerning actual lines of descent. In 1860, following publication of the *Origin,* he and Lyell exchanged a series of letters on the subject. Lyell clearly still thought of Darwin's views as implying a notion of progressive development in the sense that he believed that a major group, say, mammals, could arise independently at more than one place and time. Darwin disagreed with that view, and to suggest patterns of development he thought might have occurred, he wrote to Lyell sketching out two alternatives (Figure 10.5):

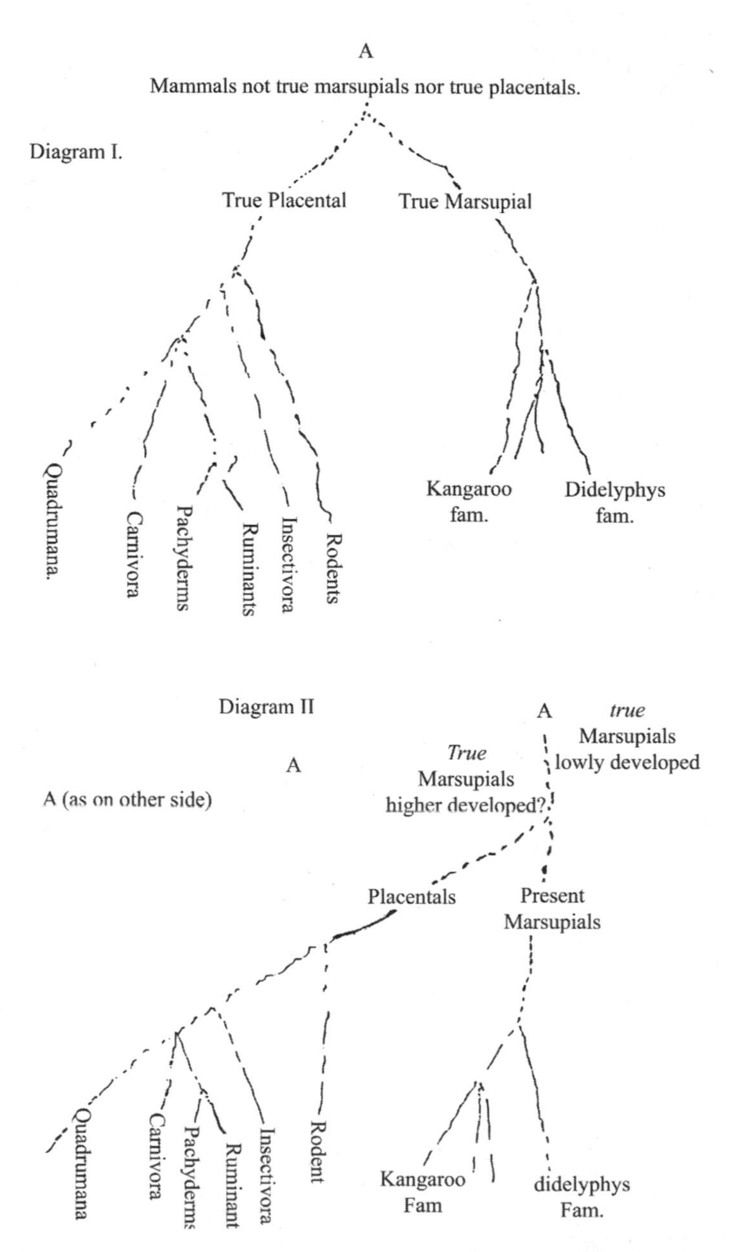

Figure 10.5. Darwin's conjectural diagrams on animal development. By permission of the Syndics of Cambridge University Library and Cambridge University Press; produced by UMBC Cartographic Services.

> I enclose 2 diaggrams [*sic*] showing the sort of manner I *conjecture* mammals have been developed: . . . I have not knowledge enough to choose between these two diagrams; if the Brain of Marsupials in embryo closely resembles that of placentals, I sh$^{d.}$ strongly prefer no$^{r.}$ 2. . . . As a general rule I sh^{d} prefer no^{r} I. diagram.[56]

What these two diagrams show is that as of the 1860s, Darwin was deeply interested in genealogy in the broadest terms and prepared to speculate, at least in private, on the subject.

Classification and Divergence

During the period of the theoretical notebooks (1837–1839), with his notion of descent in hand, and with his useful—and geologically inspired—image of the coral of life, Darwin faced two related problems. One was understanding the present and historic work of classification as a human enterprise. On this he made considerable progress. The other was accounting for the number and kinds of organisms in the world, both presently and in the past. On this issue he made some progress during the notebooks period, but continued to struggle with the issue of "divergence" after this period. On classification Darwin firmly came down on the side of descent as the basis of a natural arrangement: "We now know what is the natural arrangement, it is the classification of . . . relationship; latter word meaning descent."[57] In the course of keeping Notebook C, he developed this point, which allowed him to separate genealogical relationship—for which he reserved the term "affinity"—from physical resemblance that did not have a genealogical basis.

Once that distinction was established in his mind, Darwin turned to examine the Quinarianism of William Sharp MacLeay (1792–1865). In this system of classification the animal kingdom was divided into five "circles of affinity" connected by smaller "osculant" groups (Figure 10.6).[58] While never an advocate of the system, Darwin on occasion used its vocabulary of "inosculation" and "osculant groups" and was aware of its influence among a number of London systematists. In Notebook C Darwin came to view the MacLeay circles as an artifact stemming from the difficulty of representing a classification system in a two-dimensional manner. He noted the difficulty "in arranging animals" on paper as one would a "drying plant, all brought in one plane."[59] What Darwin had in mind here is that if one could view MacLeay's "circles of affinity" from above, they would appear as the tops of cones. The actual genetic parent to those species represented on the circle would thus be

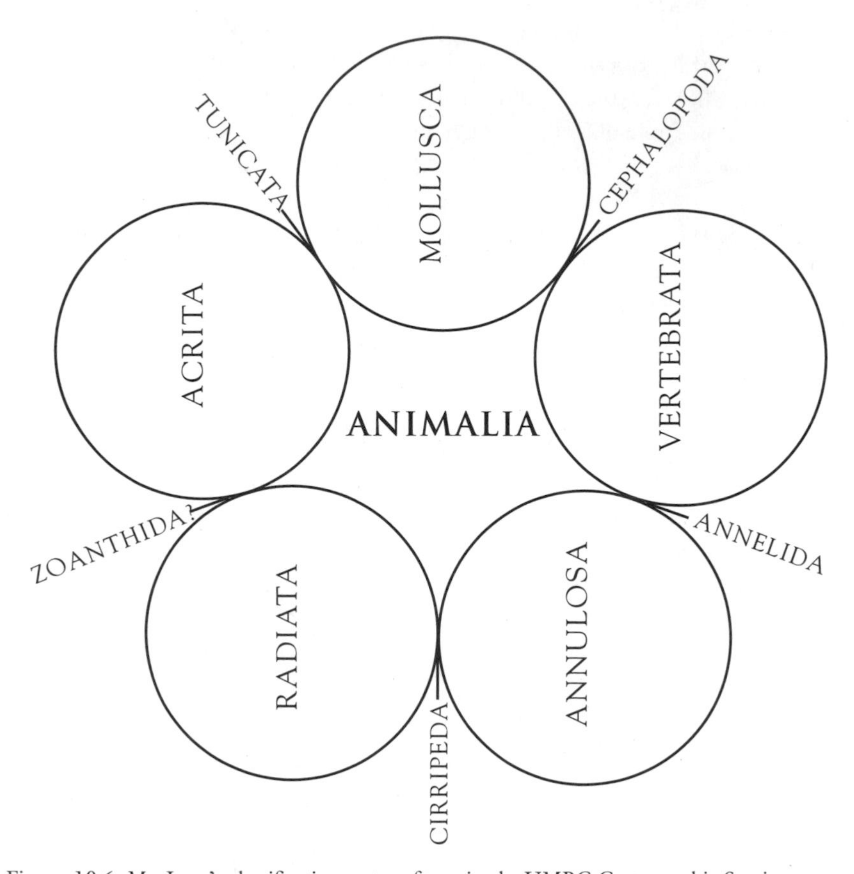

Figure 10.6. MacLeay's classification system for animals. UMBC Cartographic Services.

below the surface at the point of the cone. As Darwin wrote, "[T]he central twigs dying, affinities would be . . . in broken circles."[60]

Quinarianism aside, Darwin did seek accommodation for his new understanding of species with his fellow naturalists. One person he pursued was George Waterhouse (1810–1888), who had described mammals and insects from Darwin's *Beagle* collections. Darwin courted Waterhouse with a gift of a volume by Lyell ("I am so determined to make you a geologist") and was explicit in 1843 in describing his own goals for a natural system of classification:

> According to my opinion (which I give every one leave to hoot at, like I should have, six years since, hooted at them, for holding like views) classification consists in grouping beings according to their actual *rela-*

> *tionship,* ie their consanguinity, or descent from common stocks. . . . To me . . . the difficulty of ascertaining true relationship ie a natural classification remains just the same, though I know what I am looking for.—[61]

Waterhouse did not take Darwin's bait. He did not assert a genealogical foundation to a natural system of classification. Rather, he responded with his own definition:

> Well, I *will* explain what I am aiming at when I attempt to arrange a group in a natural manner—or, in other words, when I attempt a Natural Classification—I think a number of animals may be so arranged as to display (symbolically by their relative positions) a great many known facts relating to their structure, habits, and geographical distribution; . . . but as the term "natural" certainly has a very vague meaning when thus used, I have no objection to apply the word "*useful*" instead[.]—[62]

Waterhouse's emphasis on the operational aspect of classifying what is "useful" proved the way out of the dilemma.

Darwin himself could not outline the descent of species. Indeed, in his exchange with Waterhouse he emphasized how few forms had been preserved geologically:

> [Y]ou will admit fossil & recent beings all come into one system.— . . . It w[d] take a Chapter to argue, how probable it is that Geology has never revealed & never will reveal, more than one out of million forms, which have existed.

Darwin went on in his next letter to argue, provocatively, that

> if every organism, which had ever lived, or does live were collected together . . . a perfect series would be presented, linking all, say the Mammals, into one great, quite indivisible group[.]—[63]

In his reply Waterhouse did not address Darwin's geological argument. As his "Observations on the Classification of the Mammalia" suggested, he remained close to a schematizing notion that used circles in representing nature.[64] And as indicated by his exchanges with Darwin in correspondence, Waterhouse was searching for rules that would guide him in classification, though he was not a literal Quinarian.

Darwin and Waterhouse were separated on principle. Yet, in their view of what constituted good practice in natural history, they were much closer. Waterhouse was eager, as he said, to see classifications unite "a great many known facts" with regard to organisms. Darwin was no different, and in 1843 when Waterhouse was seeking employment at the British Museum, he praised him for his "skill & accuracy in discriminating species" and as superior to himself in the "Zoological requirements of a naturalist."[65] Darwin was no doubt here recalling Waterhouse's settling of some of his own very pointed questions as he reviewed his "Animals" list from the voyage. "Are the various specimens of mice, which I have collected varieties or species?" Darwin had asked. In February 1837 Waterhouse answered him: there were nineteen new species of mice among the specimens Darwin had given him.[66]

The Darwin-Waterhouse connection provides a good illustration of the pragmatic nontheoretical manner in which the issue of species could be handled in Britain by the 1840s. A number of historians have converged on the point that institutional as much as intellectual factors shaped the issue. The point of dispute was the naming of species. By what rules were species to be named and who was to name them? Darwin's own practice was clear after his return from the voyage: he approached naturalists associated with established institutions and deferred to their judgment. Doing so meant subordinating defining what species are—their meaning—to the practice of assigning them scientific names. Calling and defining were kept distinct operations. John Beatty has argued this point convincingly.[67] Further, by the mid-1840s the traditional practice of assigning Latin names using the Linnaean system of binomial nomenclature had triumphed over challenges from reformers and radicals of various persuasions, among them Quinarians and Lamarckians.

The victory of the Linnaean system of naming was instantiated in a document published as a British Association "Report of a Committee appointed 'to consider of the rules by which the Nomenclature of Zoology may be established on a uniform and permanent basis.'"[68] The committee was chaired by Hugh E. Strickland and included Darwin among its members, and, seemingly at Darwin's suggestion, Waterhouse. Gordon McOuat has provided a full analysis of the committee's report.[69] Darwin questioned Strickland's insistence on priority as the standard for assigning names, but this was a relatively minor matter. Overall the two men were in agreement. McOuat has also shown that, in a related move, the zoologist John Edward Gray (1800–1874) effectively made the British Museum the governing institution in the naming of species because of its practice of providing duplicates of specimens to provincial museums.[70] London thereby won over the provinces and achieved at least parity with centers on the continent.

These developments represented a triumph of British pragmatism. For example, as McOuat recounted, Gray at the British Museum set up a system whereby the scientific name of each species was listed on a separate sheet of paper, rather than being recorded in a bound catalogue, "so that at any future time the leaves may be separated and bound in any other form." The key idea was, "Systems change. Species endure as the ground of discourse."[71] McOuat also argued that species, rather than specimens, as at the Royal College of Surgeons, were key to the British Museum's arrangement of its collections. Similarly, the retention of Latin binomial nomenclature furthered the goal of international cooperation. As Strickland observed in 1843, "Let us hope that these efforts to produce uniformity in the scientific language of zoology will tend to facilitate intercourse between the naturalists of all countries."[72]

As they impinged on Darwin, who was a key participant on the Strickland committee and who remained a trusted figure within museum circles, these developments directed him toward a more utilitarian, or what McOuat has described as a "conventionalist"—or even "cynical"—attitude toward the species concept in the *Origin*. Mayr had also noted such a movement.[73] Yet, though Darwin was more openly speculative in his transmutation notebooks than elsewhere concerning the possible causes behind the formation of species, throughout his career he focused on the species/variety end of the taxonomic scale. As we have seen, this was true in his zoological notes from the voyage.

After the voyage, Darwin also depended on and deferred to the authority of professional zoologists in the naming of his specimens. The treatment of the mockingbirds by Gould is a case in point. There is a thread connecting his practice from his earlier to his later works. The very nature of the transmutation notebooks did not require that Darwin pursue one course at the expense of others. In his dealings with the specialists working on his collections he could be a conventionalist; in his more private speculations he could emphasize the search for the causes of similarity of organisms—that is, he could think more biologically.

One constant in Darwin's thinking was avoiding Lamarck's explanation of transmutation. While Darwin occasionally entertained the idea that there was some progressive tendency driving transmutation, he never embraced it. There was no inherent directionality to change in species. He tended to attribute the diversity of life by adducing changes in climate:

> Speculate on multiplication of species by travelling of climates & the backward & forward introduction of species.—[74]

But if climate was the driver, then its effects were presumably reversible. As he put it in Notebook E,

> I doubt not [that] if the simplest animals could be destroyed, the more highly organized ones. would soon be disorganized to fill their places.—The Geologico-geographico changes must tend sometimes to augment and sometimes to simplify structures[.][75]

This led to the question as to whether the exact same species might be formed twice.

Lyell's cyclicism in the early 1830s suggested that the same species could arise twice, but by 1838 in his publication of *Elements of Geology,* he explicitly denied the possibility. Darwin noted, "Lyell has remarked species never reappear when once extinct.—"[76] Lyell's opinion aside, whether the same species, or forms, could come into existence more than once was a question that continued to be debated through the 1850s. Darwin thought not, but transmutationism was sometimes associated with such a belief.[77]

Again distinguishing himself from Lamarck, Darwin asserted, "My theory leaves quite untouched the question of spontaneous generation.—"[78] He debated this point with his confidant Hooker in the 1850s, arguing that

> there was nearly as much difference between trying to find out whether species of a genus have had a common ancestor & concerning oneself with the first origin of life, as between making out the laws of chemical attraction & the first origin of matter.[79]

While bracketing the subject of spontaneous generation, Darwin did speculate on the early history of life. In Notebook E he suggested,

> In early stages of transmutation, the relations of animals & plants to each other would rapidly increase, & hence number of forms. once formed would remain stationary, hence all present types are ancient.[80]

This line of thinking enabled Darwin to imagine the building of interrelationships among different forms of life—"animals and plants to each other"—and thus to account for both the multiplicity of species and the presence of many still-extant groups in early portions of the fossil record. It was thereby an explanation for what in the notebooks he called "diversity" and what he would later call the "principle of divergence."[81] It took considerable time, until perhaps as late as the second half of 1856, before Darwin fully formulated the principle.[82]

The question of explaining the multiplicity of species on the earth was integral to Darwin's theory, particularly after his adoption of selection in 1838. As numerous commentators have suggested, after he adopted selection as the core of his theory, Darwin deemphasized geographical isolation as an explanation for the creation of new species. One reason for this shift in his thinking that has not been suggested thus far, that I am aware of, pertains to his geology. After Darwin had completed his book on the origin and distribution of coral reefs, he leaned more to a belief in the long-term stability of continents than he had formerly. In 1856 he noted his belief "that continents as continents, and oceans as oceans, are of immense antiquity."[83] His attention was thereby directed toward consideration of what might happen to species on continents and away from consideration of what might happen to species on islands.

The Primacy of Geology

In the mid-1840s and the 1850s Darwin engaged in disputes regarding geology and species with Edward Forbes and Hooker. Darwin regarded both men as close colleagues, as indicated by his naming them as possible editors of his 1844 "Essay." These were "in-house" disputes. Hooker, a botanist, had become a transmutationist by 1859. Although Edward Forbes was never a transmutationist (he died in 1854 at the age of thirty-nine), his younger brother David Forbes (1828–1876), also a geologist, was an early supporter of the *Origin*.[84] In Darwin's disputes with both Forbes and Hooker the question arose of which branch of natural history should be preeminent in evaluating claims regarding the past distribution of species. In each case, Darwin argued in favor of geology, according to his own best understanding of the subject at that time.

Darwin was a Lyellian, as was Edward Forbes. Yet, as we have suggested already, Darwin's interpretation of the origin of coral islands had led him to regard them as products of long-term subsidence and, thus, of great antiquity. Correspondingly, continents were also seen as governed by long-term motion. As compared to Lyell, Darwin in effect had become a stabilist. Thus, when in 1846 Forbes wrote to Darwin regarding a possible land bridge—which he referred to as "Atlantis"—connecting the Azores with the Iberian peninsula and North Africa, his message fell on deaf ears (Figure 10.7).[85] Darwin deemed the idea too hypothetical and speculative. Forbes's hypothesis that the exposed Miocene formations were the remains of a former landmass did not persuade him.

Darwin was appalled at the ease with which Forbes's claims were accepted by a number of their colleagues, who, he believed, were ready to move too quickly to posit land bridges to explain current patterns of distribution among plants and animals. As Darwin wrote to Hooker in 1856, "Indeed even one

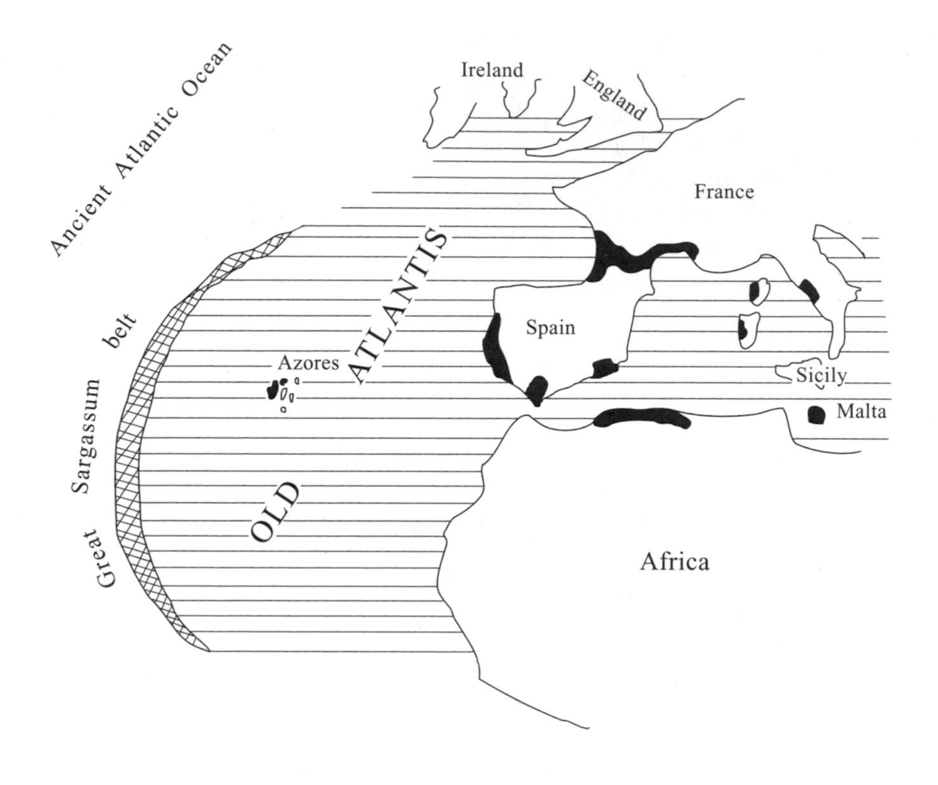

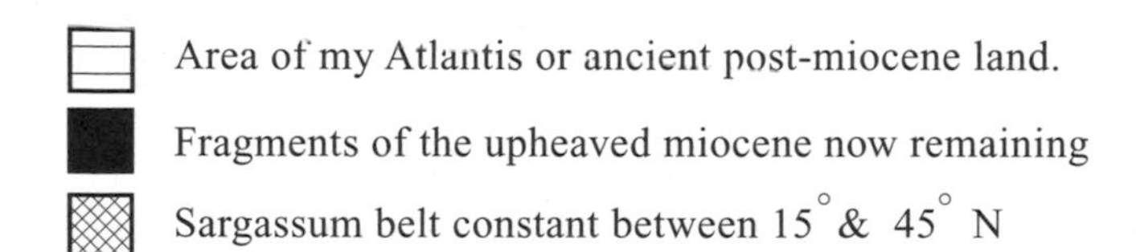

Figure 10.7. Forbes's Atlantis. By permission of the Syndics of Cambridge University Library and Cambridge University Press; simplified and reproduced by UMBC Cartographic Services.

continental extension is an awful gulp to me. I never made a continent for my Coral Reefs.—"[86] In favor of his own predilection against land bridges, he also argued that "not a fragment of secondary or palæozoic rock has been found on any isl[d]. above 500 or 600 miles from a mainland.—"[87]

Rather than posit land bridges Darwin preferred to consider alternative explanations for distribution. Chief among these was a possibility that Forbes had minimized: "natural transport."[88] Indeed, Darwin came to fault Forbes on this point: "I cannot avoid thinking that Forbe's [*sic*] Atlantis was an ill-service to Science, as checking a close study of means of dissemination."[89] Collecting information on species dispersal did indeed become one of Darwin's favored occupations as time went on. His study took place against a background of suspicion of continental extensions. His geology did not favor them.

Darwin's commitment to the primacy of geology among the natural historical sciences was also evident in his exchanges with Hooker over the origin of coal. Their discussion began in the mid-1840s and continued intermittently for many years. In the mid-1840s Hooker began to supplant Lyell as Darwin's adviser and sounding board. "You are my public," Darwin wrote to Hooker in 1860, suggesting how deeply he valued their correspondence.[90]

In the Red Notebook, in an entry written during the *Beagle* voyage, Darwin had speculated that coal might be formed on the open coast. He imagined that "much vegetable matter from thickly wooded mountains, probably chiefly leaves" might collect in basins, as those between the island of Mocha and the mainland.[91] The vegetable matter could eventually become coal. In this scenario the environment for the formation of coal would be salt or brackish water, rather than freshwater. Darwin found a disciple for his views on coal in Edward William Binney (1812–1881), a Manchester solicitor and paleobotanist.

Building on Darwin's speculations regarding subsidence in the *Journal of Researches,* as well as on his own researches in the coalfields in the region around Manchester, Binney argued for the "probable" marine origin of coal. He also tied this argument to Darwin's work in Patagonia:

> It was owing to the observations of Mr. Charles Darwin, on the coast of Patagonia, that geologists were first presented with a series of phenomena of the gradual rising of land, it then being in a state of repose, for a considerable period, and again rising. . . . Upon first reading his work, I immediately saw a series of phenomena, the reverse of which, I have long been convinced had taken place during the formation of our beds of Coal, and that, in all probability, they were the opposite of what was taking place on other parts of the earth's crust at that time.[92]

Coalfields were formed during periods of repose, which allowed for the growth of trees; these beds were later submerged in some sort of marine setting, a shallow ocean being Binney's preferred solution. For some time Darwin had also believed that *Sigillaria,* a characteristic plant of coal formations, had been aquatic, and he was pleased to know that someone else had come to a similar conclusion.[93] He joked to Hooker the next year: "be a good boy & make Sigillaria a submarine seaweed."[94]

Hooker sent a negative reply within the week. Darwin recoiled from the "savage" tone of Hooker's letter. In his letter to Darwin, now missing, Hooker seems to have linked *Sigillaria* with tree ferns, which he and Darwin both agreed were terrestrial plants. In attempted rebuttal Darwin pointed out that Hooker himself had suggested that the affinities of *Sigillaria* were obscure.[95] Binney's paper and the exchange with Darwin presaged a major paper by Hooker, who was then serving as botanist to the Geological Survey, on the general subject of the vegetable productions of the Carboniferous period. Hooker thoroughly reviewed the complex evidence regarding *Sigillaria,* visiting Binney's collection of specimens at the Manchester Museum in the process. In his article Hooker compared the growth of ferns to the possible growth of the extinct *Sigillaria,* but noted that this "does not necessarily indicate a close botanical affinity between them."[96] The argument between Darwin and Hooker over the origin of coal continued for over twenty years, with the marine hypothesis becoming less persuasive as time went on.[97]

For our purposes it is Darwin's approach to the problem of the origin of coal that is of paramount importance. His comments to Hooker in the letter of 6 May 1847 are to the point:

> Could a Botanist tell from structure alone that the mangrove family, almost or quite alone in dicots:, could live in the sea— . . . ? Is it a safe argument, that because algæ are almost the only, or the only, submerged sea-plants, that formerly other groups had not members with such habits; . . . but I am forgetting myself, I want only to some degree to defend myself, & not burn my fingers by attacking you.—*The foundation of my letter, . . . though I daresay you will think it absurd, is that I would rather trust, cæteris paribus, pure geological evidence than either Zoolog. or Botan. evidence:* I do not say that I w^{d} sooner trust *poor* geolog. evidence than *good* organic: I think the bases of pure geological reasoning is simpler, (consisting chiefly of the action of water on the crust of the earth, & its up & down movements) than bases drawn from the difficult subject of affinities & of structure in relation to habits.[98]

For Hooker, the botanist, the key question was the affinities of the plants found as fossils in the coal. But for Darwin the subject of the origin of coal began with what he already knew about the crust of the earth. For him the dispute with Hooker was not so much about the coal, or even *Sigillaria,* as it was about elevation and subsidence. Indeed—though this point is implicit rather than explicit in Darwin's comments—if coal were formed in a shallow marine setting, the presence of coal could be a marker for the repose of the crust or for subsidence. Geology mattered most.

The Origin of Variability

From his notebooks through the *Origin* Darwin was aware of his ignorance of the causes of variation. In Notebook B he wrote of "Unknown causes of change . . . Volcanic isl^{d}.—Electricity"; in the *Origin* he wrote, "Our ignorance of the laws of variation is profound."[99] (A similar situation obtained for inheritance.) Yet in a long interval between the *notebooks* and the *Origin* there was an interesting shift in the weight he placed on possible causes for variability. When he first adopted transmutation, he had in mind physical circumstances as inducing variation, hence, his remark in Notebook B: "animals, on separate islands, ought to become different if kept long enough . . . apart, with slightly differen[t] circumstances . . . Galapagos Tortoises, Mocking birds."[100] "Slightly different circumstances" played the causative role. In the course of arriving at the theory of natural selection, and becoming increasingly interested in domestic productions, he came to believe he needed to interpret the notion of slightly differing conditions more broadly. The unsettling effects of domestication on the reproductive system of animals seemed an additional likely cause of increased variability.

In Darwin's "Sketch" of 1842 and his "Essay" of 1844, this approximately equal balancing of causes for variability is prominent.[101] In Darwin's *Origin* it is not. Darwin's former interest in positing a close correlation between variability in organisms and geological change has receded. Why this was so can be explained by changes in his geological opinion and by the nature of his own ongoing empirical researches. As we have seen in the previous section, Darwin regarded geology as primary. He was reluctant to embrace the continental extensions, as suggested by Forbes, that were not founded on independent geological evidence. His own research on coral reefs had also made him in the 1840s more of a continental stabilist than he had been in the 1830s. By the late 1850s he was less inclined than formerly to adduce the direct effect of geological change to account for variability in organisms, though the possibility was retained, as indicated in the draft and published versions of the *Origin*.[102]

From 1846 to 1854, as Darwin classified barnacles, he became increasingly secure in his belief in the immense variability among organisms in nature. As he wrote under the subject of "Variation" in his volume on the living species of sessile barnacles, "Not only does every external character vary greatly in most of the species, but the internal parts very often vary to a surprising degree."[103]

In 1855, once he was "quite done" with the barnacles, and as he was preparing to write his book "*for & versus* the immutability of species," he turned to breeding pigeons, since "I want to get the young of our domestic breeds to see how young, & to what degree, the differences appear."[104] As is well known, his work on pigeons played a prominent role in the *Origin*. Following his work breeding pigeons he noted in the *Origin* that "those characters which are mainly distinctive of each breed, for instance the wattle and length of beak of the carrier, the shortness of that of the tumbler, and the number of tail-feathers in the fantail, are in each breed eminently variable."

The effect of Forbes's unacceptable continental extensions conjoined with demonstrable variability within species, as established by the test cases of barnacles and pigeons, encouraged Darwin to reduce the role of geology in accounting for variability in organisms. In the *Origin*, geological change played a lesser role as an agent for promoting variation than it had in the early formulations of his theory.

It should also be noted that in his rebalancing efforts in accounting for variability, Darwin did not wish to promote "mere chance" as the active agent.[105] To have done so would have seemed an ad hoc argument to many of his contemporaries. Moreover, it would have added to his difficulties in explaining why some species have changed little over time.

The Rate of Change for Species

From his earliest to his latest comments on time as related to species, Darwin deferred to geology. Yet, there were so many bracketed issues in geology—the question of the heat of the earth for one, the question of extraterrestrial influences on the earth for another—that geology did not provide ready answers to the question of the rate of species change. Moreover, Darwin was not consistently Lyellian in his approach. Thus, as noted in chapter 5, during the voyage Darwin correlated the movement of the native peoples in South America to the rise of the Andes, which, for a gradualist, suggested a rather compressed understanding of the time involved in the rise of mountains. Early on Darwin also voiced skepticism of Lyell's method of dating formations using the relative proportion of recent shells, commenting, "This rests on the supposition that species become extinct in same ratios over the whole world."[106] Another example of Darwin's thinking unconventionally for a gradualist came just as he became a transmutationist in March 1837. In his Red Notebook he suggested that "if one species does change into another it must be per saltum—or species may perish."[107]

Zoology and botany were alternative candidates to geology as the most fundamental of the natural historical sciences. A life force could be seen as driving the transmutation process. As Darwin sought to persuade Hooker and Lyell of the plausibility of his theory of natural selection, it is clear from their correspondence how much they both associated the course and rate of transmutation with properties interior to organisms themselves, and how Darwin, in correspondence, sought to convince them of the merits of his own alternative view. Hooker believed that a transmutationist position entailed change by species over time, with species having a sort of "expansive power" to adjust to altered conditions.[108] Darwin argued that, instead, time was "important only [in] so far as giving scope to selection."[109]

A key exchange between Darwin and Lyell over the rate of change in species came in 1860, while Lyell was reading the *Origin* with care. The starting point for their discussion was such statements as Darwin's: "Species of different genera and classes have not changed at the same rate, or in the same degree."[110] Lyell pressed Darwin on such claims: if such islands as the Canaries and Madeiras have existed since the Miocene, would not natural selection have had time to modify one of the vertebrate animals into a primitive mammal (where none had existed previously)? "Or if not, does it not give us a sort of scale for the vast geological time required for such a transmutation. No Galapagos saurian having risen to a mammal."[111] Darwin countered that he could not "conceive [of] any existing reptile being converted into a Mammal," though he admitted that Lyell's line of attack was a good one—why natural selection had not done more.[112]

Still troubled, Lyell responded that nine-tenths of the world's species are immutable, and that when change comes, they must die.[113] Darwin agreed with the unchanged nature of most species in principle, though he approached the question differently.[114] On one point they found easy agreement: time for diversification. Lyell wrote that "the time required for change is longer than was supposed."[115] Darwin agreed, emphasizing that "*a long lapse of time*" was necessary for natural selection to do its work, and adding for good measure, "But oh what work there is before we shall understand the genealogy of organic beings.—"[116]

Darwin's reference to "a long lapse of time" is out of step with some current work in evolutionary theory. The evolutionary biologist Peter R. Grant has measured evolutionary change among Galápagos finches in as short a time span as decades. He has also observed that Darwin had the opportunity to witness a similar change at his home in the village of Downe. In the winter of 1854–1855, Darwin observed freezing weather so severe that it killed four-fifths of the birds on his grounds.[117] Had Darwin studied which birds died and which survived, he might have been able to see natural selection in action, much as Peter Grant and his wife Rosemary Grant have done. Evolution occurring in nature on so short a time scale as to be susceptible to human measurement lay outside even Darwin's imagination.

The theory of punctuated equilibria, elaborated by Niles Eldredge and Stephen Jay Gould, is a second example of a current interest within evolutionary biology that centers on the subject of rates of change in species.[118] As Frank H. T. Rhodes showed, Darwin did on occasion entertain views compatible with the theory of punctuated equilibria, which emphasizes "rapid and episodic events of speciations" set against more fundamental continuity ("homeostatic equilibria").[119] As Rhodes suggested, relying in part on Darwin's correspondence with the paleontologist Hugh Falconer (1808–1865), Darwin was aware of the longevity of most species and could on occasion entertain a view that resonates with the current theory of punctuated equilibria.

In finding within Darwin's body of work some similarity to current theory, one must bear in mind several points. First, as Hooker said, Darwin's views tended to be "elastic."[120] The immediate context of Hooker's remark related to progression, but Hooker's point was applicable to any number of subjects. Darwin responded to Hooker with good-humored male banter: "But what a villain you are to heap gratuitous insults on my *elastic* theory; you might as well call the virtue of a lady *elastic,* as the virtue of a theory accommodating in its favours."[121]

On the question of rate of change, the "elasticity" of Darwin's theory could allow consideration of relatively quick change combined with stasis. Equally, it could accommodate a more gradual image of change. Second, the vocabulary of evolutionary biology was only just beginning to form. For example, Darwin did not have at hand the term "speciation" to describe the process that formed species, let alone the current distinction between "allopatric speciation" (occupying separate or nonoverlapping geographic areas) and "sympatric speciation" (occupying the same or overlapping geographical areas). Third, as numerous commentators have suggested, Darwin did move away from his earliest mechanism for species change, which was premised on geographical isolation, toward using natural selection to explain not only adaptation but also the origin of species. As a pithy reminder of that fact, one can cite Darwin's reply in 1858 to his eldest son William Erasmus Darwin (1839–1914), who had just read his father's *Journal of Researches* and had enquired about the Galápagos Islands. Darwin replied,

> Your feat of so soon finishing my Journal is very grand. . . . With respect to Galapagos, I suppose *all* the productions came, many of them very long ago, from America; & hence their general American character; but that they have since been modified by my principle of Natural Selection.—[122]

Lost is any mention of island-by-island assortment of mockingbirds or tortoises; explanation of change in species was subsumed under the general heading of "Natural Selection."

Why Darwin was so insistent on natural selection to explain the origin of species as well as the origin of adaptation has many possible explanations. I have already suggested the role played by Darwin's increasingly stabilist view of continents, which made such situations as at the Galápagos Islands seem exceptional rather than usual as sites for the proliferation of species. But there are other possible explanations. One not yet offered, that I am aware of, has to do with the desirability of constructing a theory that would resemble Newton's theory of gravitation in drawing together disparate phenomena under a single set of laws. After the *Origin* was published, Darwin remarked on some parallels between his work and Newton's, and his comrade-in-theory Alfred Russel Wallace drew an even sharper comparison:

> [The *Origin*] will live as long as the 'Principia' of Newton. It shows that nature is . . . a study that yields to none in grandeur and immensity. *The cycles of astronomy or even the periods of geology will alone enable us to appreciate the vast depths of time* we have to contemplate in the endeavour to understand the slow growth of life upon the earth.

In another letter Wallace wrote:

> Never have such vast masses of widely scattered and hitherto quite unconnected facts been combined into a system and brought to bear upon the establishment of such a grand and new and simple philosophy.[123]

Whatever the Newtonian connection, the dignity of Darwin's long argument did direct attention away from settings, such as emerging islands, where change had already been observed in its beginning stages, and thus might be relatively sudden in geological terms. Wallace's "vast depths of time" were to become key to understanding the "slow growth of life upon the earth."

"On the Lapse of Time"

In the *Origin* Darwin headed one section "On the Lapse of Time."[124] In it he estimated how much time had passed since the latter part of the Secondary period. His estimate was a "far longer period than 300 million years."[125] This estimate aside, it is striking what Darwin was *not* doing in the *Origin*. He was not providing an estimate for the age of the earth. Nor was he providing an estimate on the age of the entire stratigraphical record. He wanted only to argue against the view that "time will not have sufficed for so great an amount of organic change" as was suggested by his theory.[126] In short, his was a defensive maneuver.

To whom did he turn for support in this project? The nature of his inquiry allowed him to sidestep the question of the age of the earth. Thus, he did not address the questions of whether the earth was cooling and, if so, at what rate. This would have been to revive the ideas of Buffon or De la Beche.[127] Nor was Darwin compelled to estimate the age of the stratigraphic column. Otherwise he might have opened a correspondence with Sedgwick, or with Phillips, who was already interested in the subject.[128] In "On the Lapse of Time" Darwin mentioned three authors aside from Lyell: Hugh Miller; James Smith, known as "Smith of Jordanhill" (1782–1867); and Andrew Crombie Ramsay, whom we have already discussed in regard to glaciers. Miller and Smith were cited as observers of the "slowness with which the rocky coasts are worn away."[129] Miller's book *Testimony of the Rocks* (1857) had recently captured popular attention.[130] Smith of Jordanhill, a yachtsman, was known as an authority on the change of levels in land and sea on the Scottish coast. Lyell and Ramsay were Darwin's prime guides to the treatment of time in the *Origin*.

While Lyell did not claim time to be infinite, he did insist that it was, in Darwin's words, "incomprehensibly vast."[131] Lyell also supplied the description of the key geological site that Darwin analyzed in arriving at a quantitative expression for the passage of time. This was the region of the Weald in southeastern England. It is an oval-shaped area bound by a chalk escarpment and lies between the North and South Downs, an area now agricultural but whose forests once supported a local iron industry.[132] Lyell included a map of the area in the *Principles of Geology* and spent a chapter describing it. He identified five formations he thought had originated as horizontal strata formed at the bottom of the sea. He believed that these strata had been forced upward into the form of a dome, whose crown was cut off afterward. The presently visible beds were the result of the denudation of the strata that formed the crown. As Lyell put it,

> [The] quantity of denudation or removal by water of vast masses which are assumed to have once reached continuously from the North to the South Downs is so enormous, that the reader may at first be startled by the boldness of the hypothesis.[133]

With customary grand style Lyell then went on to assure his reader that once "sufficient time" was allowed for the gradual rise of strata, the waves and currents of the ocean might indeed accomplish the denudation, with no assistance from a diluvial wave. However, Lyell gave no quantitative expression for the time required for the denudation of the Weald.

In the *Origin* Darwin did offer a quantitative estimate: "306,662,400 years; or say three hundred million years." He arrived at his figure by a simple comparison based on a hypothetical rate for the process of denudation: if the sea can degrade a 500-foot-high cliff one inch a century, then how much time will it take for the sea to erode formations 1100 feet thick extending over twenty-two miles? Darwin received his 1100-foot estimate of thickness of the original Weald formations from Ramsay.[134]

Darwin first learned of Ramsay through his 1846 article entitled "On the Denudation of South Wales and the Adjacent Counties of England."[135] This classic work went beyond Lyell's treatment of denudation to emphasize the process of marine planation.[136] Darwin recognized in Ramsay's treatment of Wales an approach allied to the one that he had adopted for South America, particularly in regard to an appreciation for the work of crustal movements in explaining landforms. He wrote Ramsay a warm letter with a copy of his own just-published book.[137] With this exchange of letters Darwin and Ramsay began an intellectual exchange that would benefit, and challenge, each of them. In 1846 Darwin was a well-published author. Ramsay envied his station in life and his "command of his own time."[138] He was the slightly younger man but already of senior rank at the Geological Survey, where he had the available facts of British geology at his fingertips. In 1848 Ramsay was invited to Down House to talk science, and asked to bring a map of South America with him. Following the visit, Darwin drew on Ramsay's knowledge repeatedly. In the course of writing the *Origin,* Darwin requested information from Ramsay on the depths of deposits in Britain from the Cambrian upward. Ramsay replied the next day, "[F]or I know you never act without good reason, & make good use of any information that can be given."[139] In the *Origin* Darwin cited the information Ramsay gave him on the maximum thickness of Palæozoic, Secondary, and Tertiary strata in Britain, as well as his data on the Weald.[140]

Darwin's reason for requesting this information was to convert figures for maximum thickness of strata to figures for the passage of time. Such conversions posed implicit questions for Ramsay as well, though not ones he believed to be answerable, at least for the moment. Ramsay had ended his 1846 article with a disclaimer: "As we estimate time, it is vain to attempt to measure the duration of even small portions of geological epochs."[141] But the questions could not be brushed aside indefinitely, and in the *Origin* Darwin converted Ramsay's number for the average thickness of Weald deposits into an estimate of temporal duration. Before leaving Ramsay, we should mention that in the 1860s he supported Darwin on evolution and reinterpreted Wealden morphology in fluvialist terms.[142]

The *Origin* was published in November 1859. By the end of the year, a reviewer had seized on Darwin's 300-million-year estimate for the denudation of the Weald as improbable. As Burchfield summarized the reviewer's objections,

> [Darwin] had uncritically assumed that a cliff 1,000 feet high would be eroded only half as fast as one 500 feet high; he had neglected undercutting, differing hardness of strata, and the fact that erosion would occur on both sides of the valley at once; and he had begun with a gross assumption of the rate of erosion in the first place.[143]

Darwin responded swiftly to his critic. Within weeks, for the *Origin*'s second edition, he suggested that the rate of denudation be estimated at two or three inches per century, which would reduce the period to 100 to 150 million years. He dropped the estimate entirely from the third and subsequent editions.[144] Privately, he took satisfaction that some geologists, such as John Jukes (1811–1869), director for Ireland of the Geological Survey, thought his first estimate reasonable.[145] But Darwin also warned Lyell, who was preparing to assign dates to the passage of time during the glacial ages: "Having burnt my own fingers so consumedly with the Wealden, I am fearful for you; . . . take care of your fingers; to burn them severely, as I have done, is very unpleasant."[146] To Asa Gray, he wrote simply, "In fact geologists have no means of gauging the infinitude of past time."[147]

Encased as it was in the book of the moment, Darwin's estimate for the denudation of the Weald served to focus greater attention on the question of the duration of past time. As a subject, time transgressed disciplinary boundaries. Geology, physics, astronomy, chemistry, all played a role in its solution. The question also pressed scientists toward quantification. In the first forty years after the *Origin* was first published, the quest for accurate estimates of the extent of past time was located with the stratigraphers, notably John Phillips, who held the chair in geology at the University of Oxford, and with the physical scientists, most prominently William Thomson (1824–1907), Lord Kelvin as of 1892. Phillips and Thomson had long been suspicious of the scientific merits of uniformitarian geology in regard to the treatment of time. From the mid-1860s through the 1890s their more modest non-Lyellian views of the duration of past time prevailed.

With the discovery of radioactivity at the turn of the century the quest took a new turn, for the phenomenon suggested both a method of measuring the age of rocks and a new source of heat for the sun, and hence, an earth with a longer possible history of habitation. By the middle of the twentieth century the age of stratigraphic record and the age of the earth had been established, and the question of the extent of past time broadened to include the age of the universe. The literature on this subject is rich.[148] Here we will carry the story only through the years immediately following publication of the first edition of the *Origin*.

Among geologists Darwin's chief protagonist on the subject of time was

Phillips, who in 1859 was president of the Geological Society of London. Their differences were several. Darwin was a transmutationist; Phillips was not. Darwin was Lyellian in his approach to geology; Phillips was the literal heir of the tradition of stratigraphy pioneered by his uncle, William Smith. Darwin's approach to paleontological questions was highly abstract; Phillips tallied the presence of named groups. On the subject of time Darwin drew on a deistic interpretation of the world; Phillips favored a Christian interpretation. Darwin was (comparatively) a gradualist in his view of geological processes; Phillips believed that geological processes could work rapidly.[149]

Phillips reviewed the *Origin* in his 17 February 1860 presidential address to the Geological Society. He accused Darwin of "abuse of arithmetic" in his calculation for the denudation of the Weald and suggested alternative computations. His estimates ranged from a low value of 12,000 years to a high value of 1,332,000 years, depending on the assumptions involved. Both numbers fell short of Darwin's "inconceivable number of 306,662,400 years."[150] Phillips had already wrestled with the question of varying assumptions affecting estimates for lapses of time in his 1855 *Manual of Geology*. In that work he had also questioned Lyell's interpretation of the Weald, so in 1860 he was already familiar with the subject that Darwin had raised.[151] Phillips next confronted the issues raised by Darwin's *Origin* when he gave the Rede Lecture at the University of Cambridge in May 1860. In the lecture he put forward his own quantitative estimates for the duration of time represented by the stratigraphic record, the longest of which was 95,904,000 years.[152] This was the first time that Phillips had put forward a figure for the entire stratigraphic column. The question of duration was now put permanently in play within the discipline.[153]

At this point William Thomson entered the lists. From the 1840s onward he had believed that the Huttonian/Lyellian approach to geological time was inconsistent with a proper physical understanding of the dissipation of heat within the solar system.[154] His work in thermodynamics had led him to view time as irreversible and limited. Darwin's *Origin* provoked him to examine the issue of time publicly. He began with a paper on the sun's heat prepared for the September 1861 meeting of the British Association and reworked for fuller publication in the March 1862 issue of *Macmillan's Magazine*. Thomson ridiculed Darwin's 300,000,000-year estimate for the denudation of the Weald and suggested that it seemed "most probable that the sun has not illuminated the earth for 100,000,000 years, and almost certain that he [the sun] has not done so for 500,000,000 years."[155]

In June 1862 Thomson began consulting with Phillips on the subject of geological time. Phillips's 96-million-year estimate for the formation of the earth's strata was in the same range as Thomson's 1863 estimate of 98 million years for the consolidation of the earth's crust.[156] As scholars have described in detail, estimates for the age of the crust of the earth varied considerably in the last forty years of the nineteenth century, but the figure of 100 million was frequently cited.

The point for evolutionists was sufficiency rather than any absolute figure for the age of the earth or of its strata. In 1878, four years before his father's death, George Darwin wrote to Thomson:

> I think what my father said about time is quite justifiable from a biological point of view; 100 or 200 million years is "incomprehensibly vast"—even a million is not conceivable in the way a hundred is. I have no doubt however that if my father had had to write down the period he assigned at that time, he w[oul]d have written a 1 at the beginning of the line & filled the rest up with 0's. Now I believe that he cannot quite bring himself down to the period assigned by you, but does not pretend to say how long may be required. I fail to see the justice of your remark that a few hundred million years would be insufficient to allow of transmutation of species by nat[ural] selection. What possible datum can one have for the rate at which it has or can work?[157]

George Darwin's point remained pertinent. Among the various figures then under discussion, no one value for the age of the crust was conclusive one way or the other for the theory of evolution through natural selection.[158] In broadest terms, the significance of Darwin's treatment of time in the *Origin* was that it brought urgency to the questions of the age of the stratigraphical column and of the earth itself. The *Origin of Species* cast a long shadow.

CONCLUSION

Charles Darwin's grand adventure as a geologist took place during the circumnavigation of the earth by H.M.S. *Beagle* (1831–1836). By the 1830s it was understood that geologists were required to travel beyond Europe to secure the foundations of their science. Political circumstances at the time encouraged such voyages to be undertaken and, concomitantly, large questions in geology to be asked, questions that touched on a number of aspects of the general culture, including religion and art. At home, geologists, including Darwin's former teachers, provided him with a receptive audience for his findings; the British government sponsored publication of his specialized researches conducted during the voyage; and the general public welcomed his overall account of the voyage, including its geology.

Because of ill health Darwin's years as a geological traveler ended much too soon, but in a poetically satisfying rounding out of his experience, he ended his travels where he had begun them—in Wales. His first major effort in the field was with Sedgwick in North Wales in 1831. Eleven years later, when he was thirty-three, his fieldwork in the same region in search of evidence of former glaciation demonstrated his ability to see a familiar landscape with fresh eyes. This was his last major geological work in the field.

In his middle years Darwin contributed toward the subdiscipline of paleontology with his work on fossil barnacles. Toward the end of his life he resumed study on the action of earthworms in forming soil. He had discussed the subject in 1837 before the Geological Society of London, and in 1881 it was the subject of his last book.[1] Throughout his life, whether he was climbing in the Andes, trekking through Wales, working at his desk on fossil specimens, or examining the soil in his garden, Darwin's "noble science of Geology" retained its allure.[2]

But if geology was a continuing strand of Darwin's work and, consequently, of his identity, what was his impact on the science? Was it major or minor? A part of his general impact on science, or distinct? Summaries of his numerous discrete contributions to the field have been made in this book, but judgment on the largest questions needs to be offered as a concluding statement.

Before the *Origin,* Darwin was known to the world of science as a naturalist and geologist. His closest professional association had been with the Geological Society of London, and the majority of his publications to 1859 were geological in whole or in part. The primary reason why he is not well known as a geologist today is that the *Origin of Species,* read as a biological text, has outshone all else. However, there are secondary reasons for the oversight. Darwin did not do his major work within geology's core paradigm, which was then stratigraphy. Rather he worked off geology's "list," which was composed of a number of subjects other than stratigraphy. His range was great; some thirty topics from the structure of the earth to the formation of rocks are covered in one notebook alone.[3] Indeed Darwin's very range can allow him to seem a contemporary to present-day geologists who take the time to become familiar with his work. There is very little in present-day geology, from understanding cleavage and foliation to speculation concerning the interior of the earth, that Darwin did not address, either in print or in his notes.

In addition to great range as a geologist, Darwin was also ambitious. He believed that it would be possible to create a "simple" geology based on an understanding of the vertical motions of the earth's crust, elevation and subsidence. While full achievement of that goal eluded him, its scale is worthy of note. Only recently, with plate tectonics, has a similarly large-scale theory enjoyed success. Part of Darwin's scheme for a simple geology has survived close to the form in which he imagined it: his theory of the structure and distribution of coral reefs, which emphasized the role of subsidence in the creation of coral islands, most strikingly atolls. *Coral Reefs* held a special place in Darwin's estimation, and, alone of the three parts of his geological trilogy from the voyage, he rewrote it substantially when it appeared as a second edition in 1874.

The coral reef theory received respect upon publication, and quick converts in Lyell and Dana. Yet geological opinion in the decades immediately following Darwin's death was divided on the merits of the theory. In his Rede Lecture of 1909 Archibald Geikie concluded with the regret that "the overwhelming evidence [is] in favour of elevation rather than depression [subsidence] among many oceanic islands."[4] However, John Wesley Judd reminded his readers of Darwin's wish in 1881 that "some doubly rich millionaire would take it into his head to have borings made in some of the Pacific and Indian atolls, and bring home cores for slicing from a depth of 500 or 600 feet." Judd also saw support for Darwin's theory in the Funafuti expeditions, which accomplished borings of nearly twice that (1114 feet).[5] Eventually Darwin's theory was fully confirmed empirically. In 1952 the American geologist Harry S. Ladd (1899–1982) struck basalt beneath a layer of about 1300 meters of carbonates on Enewetak Atoll. Beside the bore hole Ladd erected a sign: "Darwin was right!"[6]

A similar delay occurred in corroborating Darwin's petrological insight regarding the origin of diversity in igneous rocks. His insight awaited integration into a tradition of research that could yield empirical results, though this time in a laboratory rather than a field setting. Igneous petrology has a place in its history for Darwin, not as proffering a universal solution to the problem of the diversity of igneous rocks, but as suggesting "possible explanations for some igneous phenomena."[7]

But in one area—evolution—Darwin did offer a universal solution that has held to this day. As Hewett Cotrell Watson wrote in 1859, in the flush of his first reading of the *Origin,* "You are the greatest Revolutionist in natural history of this century, if not of all centuries."[8] Allowing for rhetorical hyperbole on Watson's part, his overall point must be allowed. Geology both contributed to the revolution establishing evolutionary theory and was affected by it.

Of course, the *Origin* provoked as many questions as it answered. Immediately on its publication Darwin's provisional dating of the fossil record was contested. Further, his strong claim regarding the imperfection of the fossil record has continued to be controversial. Since 1859 there have also been inconsistencies in several of his opinions. John Wesley Judd, for example, was troubled by the incongruence between Darwin's ideas of elevation and subsidence and his reliance on the permanence of continents and ocean basins. Judd expressed his surprise to Darwin who, "looking at me with a whimsical smile," said,

> I have seen many of my old friends make fools of themselves, by putting forward new theoretical views or revising old ones, after they were sixty years of age; so, long ago, I determined that on reaching that age I would write nothing more of a speculative character.[9]

As Darwin's reticence suggested, there would be much working out of theory for geologists to do. In sum, as a book the *Origin* continued to stand on its own for later generations, including our own, though, like many classic works, in ways that invited continued inspection and debate.

To close we should return to where we began, with Geikie's and Judd's assessments. Following Whewell, they divided the realm of geology into uniformitarian and catastrophist sects. They counted themselves fellow members with Lyell and Darwin among the uniformitarians, whose long time scale for geology and (eventual) evolutionary presumption seemed well established by 1909. Yet, from the vantage point of the present day, the nineteenth-century catastrophists also had their victories. Glacial theory, a catastrophist claim, stood, with transmutationism as monumental accomplishments. On occasion, Hugh Falconer and Joseph Hooker paired aspects of the two rhetorically as being of similar importance, and nineteenth-century geology indeed appears unthinkable in retrospect without the theories of evolution and of ice ages.[10] Further, with Darwin's manuscripts as well as his published work in hand, we now know that it was glacial theory that derailed Darwin's development of his simple geology in the late 1830s and early 1840s. Thus, the careers of Agassiz and Darwin, of catastrophism and uniformitarianism, were intertwined. Both made major contributions to the science of geology.

As for Darwin, in an early notebook he recorded his ambitions for a simple geology and for understanding the transmutation of species. Comparatively, he was more successful in establishing one than the other, though his tectonic ideas survive in part. Yet his commitment to the discipline that embraced both indicates why as a young man he so confidently asserted of himself, "I a geologist."

ABBREVIATIONS

A	Charles Darwin, Notebook A
Autobiography	Nora Barlow, ed., *The Autobiography of Charles Darwin* (London: Collins, 1958)
B	Charles Darwin, Notebook B
C	Charles Darwin, Notebook C
CD	Charles Darwin
Correspondence	Frederick Burkhardt et al., eds., *The Correspondence of Charles Darwin,* 13 vols. (Cambridge: Cambridge University Press, 1985–2002)
CP	Paul H. Barrett, *The Collected Papers of Charles Darwin* (Chicago: University of Chicago Press, 1977)
CR	Charles Darwin, *The Structure and Distribution of Coral Reefs. Being the First Part of the Geology of the Voyage of the Beagle* (London: Smith, Elder and Co., 1842; 2d ed. 1874)
D	Charles Darwin, Notebook D
DAR	Darwin Archive, Manuscripts Department, Cambridge University Library
Diary	Richard Darwin Keynes, ed., *Charles Darwin's "Beagle" Diary* (Cambridge: Cambridge University Press, 1988)
E	Charles Darwin, Notebook E

GSA	Charles Darwin, *Geological Observations on South America. Being the Third Part of the Geology of the Voyage of the Beagle* (London: Smith, Elder and Co., 1846)
JR	Robert FitzRoy, ed., *Narrative of the Surveying Voyages of His Majesty's Ships Adventure and Beagle,* vol. 3: *Journal and Remarks. 1832–1836,* by Charles Darwin (London: Henry Colburn, 1839). Also issued separately as *Journal of Researches into the Geology and Natural History of the Various Countries Visited by H.M.S. 'Beagle' . . . 1832–1836* (London: Henry Colburn, 1839)
JR (1845)	Charles Darwin, *Journal of Researches into the Natural History and Geology of the Countries Visited during the Voyage of H.M.S. Beagle round the World, under the Command of Capt. Fitz Roy, R.N.*, 2d ed. (London: John Murray, 1845)
M	Charles Darwin, Notebook M
Marginalia	Mario A. Di Gregorio and N. W. Gill, eds., *Charles Darwin's Marginalia* (New York: Garland, 1990)
Notebooks	Paul H. Barrett, Peter J. Gautrey, Sandra Herbert, David Kohn, and Sydney Smith, eds., *Charles Darwin's Notebooks, 1836–1844: Geology, Transmutation of Species, Metaphysical Enquiries* (London and Ithaca: British Museum [Natural History] and Cornell University Press, 1987)
Origin	Charles Darwin, *On the Origin of Species by Means of Natural Selection, or the Preservation of Favoured Races in the Struggle for Life* (London: John Murray, 1859)
RN	Charles Darwin, Red Notebook
VI	Charles Darwin, *Geological Observations on the Volcanic Islands Visited During the Voyage of H.M.S. Beagle, together with Some Brief Notices of the Geology of Australia and the Cape of Good Hope. Being the Second Part of the Geology of the Voyage of the Beagle* (London: Smith, Elder and Co., 1844)
Zoology, Fossil Mammalia	Charles Darwin, ed., *The Zoology of the Voyage of H.M.S. Beagle, under the Command of Captain Fitzroy, during the Years 1832 to 1836,* part I: *Fossil Mammalia,* by Richard Owen (London, 1838)

Symbols Used in the Transcriptions of Darwin's Notebooks

< >	Darwin's deletion
<< >>	Darwin's insertion
bold type	Darwin's later annotation
[ ]^{CD}	Darwin's brackets
[ ]	Editors' brackets

NOTES

Preface

1. *Quarterly Journal of the Geological Society of London* 15 (1859):xxiii.

2. John Wesley Judd, "Darwin and Geology," in Albert Charles Seward, ed., *Darwin and Modern Science* (Cambridge: Cambridge University Press, 1909), 337–384; Archibald Geikie, *Charles Darwin as Geologist* (Cambridge: Cambridge University Press, 1909); Sandra Herbert, "Remembering Charles Darwin as a Geologist," in Roger G. Chapman and Cleveland T. Duval, eds., *Charles Darwin, 1809–1882: A Centennial Commemorative* (Wellington, New Zealand: Nova Pacifica, 1982), 231–237.

3. Charles Lyell, *Principles of Geology, Being an Attempt to Explain the Former Changes of the Earth's Surface, by Reference to Causes Now in Operation,* 3 vols. (London: John Murray, 1830–1833).

Chapter 1. "I a geologist"

1. Charles Darwin, "*Life*. Written August—1838," in *Correspondence,* 2:438. Throughout this work Darwin's orthographic choices (as here in "illdefined") are quoted as they appear in the manuscript or as they are transcribed in the published sources unless otherwise noted.

2. Ibid., 2:439, 440.

3. *Notebooks,* 528–529 (M:39–40).

4. Darwin, "*Life,*" in *Correspondence,* 2:439–440, for all quotations in this paragraph. For an interpretation that traces Darwin's self-critical and deferential manner to his unresolved bereavement following his mother's death, see John Bowlby, *Charles Darwin: A New Life* (New York: W. W. Norton, 1990), 58–79; for a challenge to that view, see Janet Browne, *Charles Darwin: Voyaging* (New York: Alfred A. Knopf, 1995), 18–22. On Darwin's 1819 holiday in Wales, see Peter Lucas, "'Three Weeks Which Now Appears Like Three Months': Charles Darwin at Plas Edwards, 1819," *National Library of Wales Journal* 32 (2001):133–146.

5. Darwin, "*Life,*" in *Correspondence,* 2:439.

6. *Autobiography,* 44.

7. Anonymous, *The Wonders of the World* (Dublin: B. Smith, 1825).
8. Ibid., 162.
9. The opportunity for schooling varied greatly. After his matriculation at the University of Cambridge, Darwin became a member of a minuscule minority. The number of students admitted annually to Cambridge colleges averaged 440 from 1820 through 1829. Even by 1914, only 1 percent of the relevant age group was entering any university. Primary-level education became compulsory, and hence universal, in the 1870s and free after 1891. Still, some opportunities for schooling existed earlier. By 1840, about half of English brides and two-thirds of English grooms were sufficiently literate to sign the marriage register. Yet an 1844 naval handbook "made no mention of reading and writing requirements for enlisted personnel except for gunners." See Gillian Sutherland, "Education," in Francis Michael Longstreth Thompson, ed., *The Cambridge Social History of Britain 1750–1950,* 3 vols. (Cambridge: Cambridge University Press, 1990), 3:138, 157, 122; and David F. Mitch, *The Rise of Popular Literacy in Victorian England* (Philadelphia: University of Pennsylvania Press, 1992), 152, 176, and, for the quoted material, 39 (n. 75).
10. Robert E. Schofield, *The Lunar Society of Birmingham: A Social History of Provincial Science and Industry in Eighteenth-Century England* (Oxford: Clarendon Press, 1963), 35–59, 193–218. For a vibrant new account of the group, see Jenny Uglow, *The Lunar Men: Five Friends Whose Curiosity Changed the World* (New York: Farrar, Straus and Giroux, 2002).
11. Browne, *Charles Darwin: Voyaging,* 34.
12. *Autobiography,* 45–46; Browne, *Charles Darwin: Voyaging,* 28–32.
13. Emma Darwin to CD, February 1839, in *Correspondence,* 2:172.
14. *Autobiography,* 46.
15. L. S. Dawson, *Memoirs of Hydrography,* 2 vols. (Eastbourne: Henry W. Keay, 1885), part 1, 47, 48; part 2, 8.
16. George Stephen Ritchie, *The Admiralty Chart: British Naval Hydrography in the Nineteenth Century* (London: Hollis and Carter, 1967), 3, 195.
17. John Lynch, *The Spanish American Revolutions, 1808–1826,* 2d ed. (New York: W. W. Norton, 1986), 1.
18. Richard Graham, *Independence in Latin America: A Comparative Approach,* 2d ed. (New York: McGraw-Hill, 1994), 138.
19. Archibald Day, *The Admiralty Hydrographic Service, 1795–1919* (London: Her Majesty's Stationery Office, 1967), see map following p. 48 of "Principal British Hydrographic Surveys to 1829."
20. CD to Catherine Darwin, 14 February 1836, in *Correspondence* 1:490.
21. Recollection of Philip Gidley King. See CD to Caroline Darwin, 2–6 April 1832, in *Correspondence,* 1:221 n. 2.
22. Darwin to Frederick Watkins, 18 August 1892, in *Correspondence,* 1:261.
23. *Diary,* 90.
24. John A. Phillips and Charles Wetherell, "The Great Reform Act of 1832 and the Political Modernization of England," *American Historical Review* 100 (April 1995):414; quotation from Robert K. Webb, *Modern England from the Eighteenth Century to the Present,* 2d ed. (New York: Harper and Row, 1980), 197.
25. Erasmus Alvey Darwin to CD, 18 August 1832; W. D. Fox to CD, 30 June 1832; CD to J. S. Henslow, 18 May–16 June 1832, in *Correspondence,* 1:259, 245, 238.
26. For a biography emphasizing Darwin's political situation, see Adrian Desmond and James Moore, *Darwin: The Life of A Tormented Evolutionist* (New York: Warner Books, 1991); also, regarding Darwin's aloofness from the immediate crush of current political events, see Sandra Herbert, "Essay Review," *Isis* 84 (1993):116.
27. *Correspondence,* 1:128–136, 2:37–39.
28. *Autobiography,* 95: "As for myself I believe that I have acted rightly in steadily following and devoting my life to science. I feel no remorse from having committed any great sin, but have often and often regretted that I have not done more direct good to my fellow crea-

tures. . . . I can imagine with high satisfaction giving up my whole time to philanthropy, but not a portion of it; though this would have been a far better line of conduct."

29. William H. Goetzmann, *Exploration and Empire: The Explorer and the Scientist in the Winning of the American West* (New York: W. W. Norton, 1966), xi.

30. Bruno Latour, *Science in Action: How to Follow Scientists and Engineers through Society* (Cambridge: Harvard University Press, 1987), 219, 232.

31. *Notebooks,* 449 (E:173): "Mr Greenough on his Map of the World has written Mastodon found at Timor.—thinks he has seen specimen at Paris Museum.—"Also see *Correspondence,* 7:324 n. 5.

32. Joseph Banks to Jacques Julien Houton de Labillardière, 6 June 1796, quoted in Gavin de Beer, *The Sciences Were Never at War* (London: Thomas Nelson, 1960), 55. On Banks's genuine internationalism, within limits, see John Gascoigne, *Science in the Service of Empire: Joseph Banks, the British State and the Uses of Science in the Age of Revolution* (Cambridge: Cambridge University Press, 1998), 157–165.

33. Jacques Julien Houton de Labillardière, *Relation du voyage à la recherche de 'La Pérouse' . . . pendant les années 1791, 1792, et pendant la 1ère et la 2ème année de la République françoise.* 2 vols. + atlas (Paris: H. J. Jansen, 1800), 1:v–xvi.

34. For a classic treatment emphasizing the unity of Enlightenment thought with regard to both science and society, see Ernst Cassirer, *The Philosophy of the Enlightenment,* trans. Fritz C. A. Koelln and James P. Pettegrove (Princeton: Princeton University Press, 1951).

35. Dawson, *Memoirs of Hydrography,* 2:14. Italics added.

36. Caroline Darwin to Sarah Elizabeth Wedgwood, 5 October 1835, in *Correspondence,* 1:505.

37. Harry Woolf, *The Transits of Venus: A Study of Eighteenth-Century Science* (Princeton: Princeton University Press, 1959), 182–197.

38. John C. Beaglehole, "James Cook," *Dictionary of Scientific Biography,* 3:396.

39. Glyndwr Williams, "New Holland to New South Wales: The English Approaches," in Glyndwr Williams and Alan Frost, eds., *Terra Australis to Australia* (Melbourne: Oxford University Press Australia, 1988), 117–159.

40. Woolf, *Transits of Venus,* 167.

41. Charles to Caroline Darwin, 30 March–12 April 1833, in *Correspondence,* 1:302.

42. Beaglehole, "James Cook," 3:397.

43. John C. Beaglehole, *The Exploration of the Pacific,* 3d ed. (Stanford: Stanford University Press, 1966), 9.

44. Quoted in Glyndwr Williams and Alan Frost, "*Terra Australis:* Theory and Speculation," in Williams and Frost, *Terra Australis to Australia,* 28.

45. CD to Caroline Darwin, 12 November 1831, in *Correspondence,* 1:179; Louise Crossley, *Explore Antarctica* (Cambridge: Cambridge University Press, 1995), 29.

46. Bernard Smith, "Cook's Posthumous Reputation," in *Imagining the Pacific in the Wake of the Cook Voyages* (New Haven: Yale University Press, 1992), 240.

47. Gananath Obeyesekere, *The Apotheosis of Captain Cook: European Mythmaking in the Pacific* (Princeton: Princeton University Press, 1992); and, in response, Marshall Sahlins, *How 'Natives' Think, about Captain Cook, for Example* (Chicago: University of Chicago Press, 1995).

48. Gillian Beer, "Travelling the Other Way," in Nicholas Jardine, James Secord, and Emma Spary, eds., *Cultures of Natural History* (Cambridge: Cambridge University Press, 1996), 322–337. Sandra Herbert, "The Place of Man in the Development of Darwin's Theory of Transmutation: Part I," *Journal of the History of Biology* 7 (1974):223–232; Desmond and Moore, *Darwin,* 132–135; Browne, *Charles Darwin: Voyaging,* 234–253.

49. "Preparatory Map for the Lapérouse Expedition," in Marie-Hélène Tesnière and Prosser Gifford, eds., *Creating French Culture: Treasures from the Bibliothèque nationale de France* (New Haven: Yale University Press, 1995), 355.

50. Ultimately some material from the voyage appeared in Alessandro Malaspina, *Viaje*

Pólitico-Científico Alrededor del undo por las Corbetas Descubierta y Atrevida . . . desde 1789–1794 (Madrid: Impr. de la viuda e hijos de Abienzo, 1885). Also see Sandra Herbert, ed., *The Red Notebook of Charles Darwin* (London and Ithaca: British Museum [Natural History] and Cornell University Press, 1980), 77, 137.

51. Alexander von Humboldt, *Personal Narrative of Travels to the Equinoctial Regions of the New Continent, during the Years 1799–1804,* trans. Helen Maria Williams, 7 vols. in 9. (London: Longman, 1814–29).

52. Kurt-R. Biermann, "Alexander von Humboldt," *Dictionary of Scientific Biography,* 6:549–555; also Michael E. Hoare, "The Forsters and Cook's Second Voyage 1772–1775," in Walter Veit, ed., *Captain James Cook: Image and Impact* (Melbourne: Hawthorn Press, 1972), 109.

53. Humboldt, *Personal Narrative,* 1:xv.

54. Browne, *Charles Darwin: Voyaging,* 42. O. A. Bushnell, "Aftermath: Britons' Response to News of the Death of Captain James Cook," *Hawaiian Journal of History* 25 (1991):7.

55. Helmut de Terra, *Humboldt: The Life and Times of Alexander von Humboldt, 1769–1859* (New York: Alfred A. Knopf, 1955), 76–85; "Studies of the Documentation of Alexander von Humboldt," *Proceedings of the American Philosophical Society* 102, no. 2 (December 1958):566; Arthur P. Whitaker, "Alexander von Humboldt and Spanish America," *Proceedings of the American Philosophical Society* 104, no. 3 (June 1960):317–322. On the career of Humboldt's traveling companion in South America, see Eduardo G. Ottone, "The French Botanist Aimé Bonpland and Paleontology at Cuenca del Plata," *Earth Sciences History* 21 (2002):150–165.

56. Nicholaas Rupke, "Humboldt's Fame," presented at the History of Science Society meeting, 29 October 1995, Minneapolis.

57. For Williams's role in spreading a feminized version of the French Revolution, see Gary Kelly, *Women, Writing, and Revolution, 1790–1827* (Oxford: Clarendon Press, 1993), 30–79.

58. *Correspondence,* 1:52 n. 3, 167, 157, 120 n. 2.

59. Ibid., 1:163, 204.

60. *Autobiography,* 68.

61. *Correspondence,* 1:120.

62. W. H. Brock, "Humboldt and the British: A Note on the Character of British Science," *Annals of Science* 50 (July 1993), 365.

63. Humboldt, *Personal Narrative,* 1:3.

64. Ibid., 1:43.

65. Ibid., 2:19–20.

66. Ibid., 1:153–154.

67. Ibid., 1:197–198.

68. Ibid., 1:198.

69. Ibid., 1:iii.

70. Ibid., 1:230.

71. Susan Faye Cannon, *Science in Culture: The Early Victorian Period* (New York: Dawson and Science History Publications, 1978), 104. Cannon's was the first to capture a large group of activities in nineteenth-century science under the rubric of "Humboldtian" science.

72. For a perceptive recent reading of Humboldt's intent, see Michael Dettelbach, "Global Physics and Aesthetic Empire: Humboldt's Physical Portrait of the Tropics," in David Philip Miller and Peter Hanns Reill, eds., *Visions of Empire: Voyages, Botany, and Representations of Nature* (Cambridge: Cambridge University Press, 1996), 258–292. Dettelbach suggests that Humboldt was the first to use extensively and systematically the techniques of isoline cartography (p. 261). Thrower defines isotherms as "lines of equal average temperature" in discussing Humboldt, but also credits the English astronomer Edmond Halley (1656–1742) with establishing thematic quantitative cartography. See Norman J. W. Thrower, *Maps and Civilization: Cartography in Culture and Society* (Chicago: University of Chicago Press, 1996), 129, 95–100. See also Thrower's short list of isograms, pp. 247–248.

73. Alexander von Humboldt to CD, 18 September 1839, in *Correspondence,* 2:425–426. Translation by the editors.

74. *Correspondence,* 2:426.

75. Ibid., 2:230 n. 4.

76. CD to Alexander von Humboldt, 1 November 1839, in *Correspondence,* 2:239–240.

77. On Darwin's unsuccessful prodding, see *Correspondence,* 2:237, 348, 408.

78. Francis Beaufort, "Memorandum," 11 November 1831, in Robert FitzRoy, ed., *Narrative of the Surveying Voyages of His Majesty's Ships Adventure and Beagle,* 3 vols. + appendix to vol. 2. Vol. 2: *Proceedings of the Second Expedition, 1831–1836 under the Command of Captain Robert Fitz-Roy* + appendix, 2:40.

79. *Correspondence,* 1:162–163.

80. FitzRoy, *Narrative,* 2:24, 26, 30, 32.

81. Ibid., 2:32, 33.

82. Richard I. Ruggles, "Geographical Exploration by the British," in Herman R. Friis, ed., *The Pacific Basin: A History of Its Geographical Exploration* (New York: American Geographical Society, 1967), 253–255.

83. In the original of the chart the landmasses are edged in color. This copy is one held by the Library of Congress, which also holds a full-sized version of the 1826 edition. At the moment I am working on the presumption that FitzRoy would have had the large version in his chart room, though I have yet to find a written reference to it. I wish to acknowledge the assistance of Gary Fitzpatrick of the Geography and Map Division at the Library of Congress in calling my attention to this chart.

84. For material from Beaufort's "Memorandum," see FitzRoy, *Narrative,* 2:23, 39, 28, 38; for Lyell's explanation of the origin of coral reefs, see Lyell, *Principles of Geology,* 2:283–301. Volume 2 of the *Principles* was not published until January 1832. Beaufort's remarks in his "Memorandum" of the previous November suggests that he was already familiar with Lyell's position.

85. FitzRoy, *Narrative,* 2:36. Also see Sandra Herbert, "Charles Darwin as a Prospective Geological Author," *British Journal for the History of Science* 24 (1991):168 n. 26.

86. Day, *Hydrographic Service,* 44–67.

87. "Report of the Committee on Which the Foregoing Voyage was Ordered," in William Webster, *Narrative of a Voyage to the Southern Atlantic Ocean, in the Years 1828, 29, 30, Performed in H.M. Sloop Chanticleer, under the Command of the Late Captain Henry Foster,* 2 vols. (London: R. Bentley, 1834), 2:370.

88. Webster, *Narrative of a Voyage to the Southern Atlantic Ocean,* 2:215. My colleague Stephen G. Brush has provided a description of Baily's method and aims in what follows.

The "period" (the time to go back and forth) of a pendulum (small massive object, "bob," suspended by a massless string of length L) is

$$T = \pi\sqrt{(L/g)}$$

where g = acceleration of gravity. The "seconds pendulum" used in geodetic measurements was designed so that T = 1 second at a particular place. The same pendulum was then taken to another place and the period was measured (that is, how many vibrations in a mean solar day, compared to about 60 × 60 × 24 [86,400 times per day at the first place]). The ratio of periods would give the ratio of values of the square root of g at those two places.

If the two places have different latitudes, then one must correct for the difference in centrifugal force (which reduces g by about 2 percent at the equator where the rotation speed is greatest). For an idealized homogeneous earth the surface would be an oblate spheroid (flattened at the poles, bulging at the equator because of rotation). For this surface, one can calculate g as a function of latitude taking account of (1) the variation in distance from the center of the earth to its surface [the R in Newton's law of gravity $1/R^2$] and (2) the variation in centrifugal force. Deviations from homogeneity, such as local density variations, would produce local variations from the calculated g.

The work of the *Chanticleer* on worldwide variation in gravity suggests that its tasks were located more in the realm of pure science compared to the more applied hydrographical work done aboard the *Beagle*. The astronomer Baily, in assessing the expedition's findings, drew what were intriguing but, in the long run, inadequate conclusions from the data. He suggested (p. 220) that "the force of gravity seems to be greater in islands situate at a distance from the main land than it is on continents." Baily puzzled over this, noting that islands, being for the most part volcanic, were formed of dense material, yet they were surrounded by water, which is of lower density. Possibly, as Brush has suggested to me, Baily was looking too much for causes that would produce density variations only at the surface of the earth. In any case, the question of gravity variation at the earth's surface was resolved with the modern theory of isostasy. For the debate as it stood at midcentury, see Stephen G. Brush, *Nebulous Earth: The Origin of the Solar System and the Core of the Earth from Laplace to Jeffreys* (Cambridge: Cambridge University Press, 1996), 158–162. Also see Victor F. Lenzen and Robert P. Multhauf, "Development of Gravity Pendulums in the Nineteenth Century," *Smithsonian Institution National Museum Bulletin* 240 (1966):301–347.

89. Webster, *Narrative of a Voyage to the Southern Atlantic Ocean,* 2:370. A committee of the Royal Society of London drew up the instructions for the voyage. Davis Gilbert, the president of the society, served as the committee's chair. William Fitton, president of the Geological Society of London, and William Herschel, president of the Royal Astronomical Society, were also members of the committee. For its report, see pp. 369–382. Bailey was also a member of the committee.

90. Browne, *Charles Darwin: Voyaging,* 179. Also, FitzRoy, *Narrative,* 2:17, and, for the discussion of FitzRoy's procedures and results on the 1831–1836 voyage, see the Appendix, 325–352.

91. FitzRoy, *Narrative,* Appendix, 345.

92. Alfred Friendly, *Beaufort of the Admiralty: The Life of Sir Francis Beaufort, 1774–1847* (London: Hutchinson, 1977), 256.

93. FitzRoy, *Narrative,* 2:20.

94. Browne, *Charles Darwin: Voyaging,* 227.

95. Ibid., 229. According to the "How Much Is That" Web site, £1800 in 1831 would have the purchase power of £87,072 in 2001.

96. Martin J. S. Rudwick, *The Great Devonian Controversy: The Shaping of Scientific Knowledge among Gentlemanly Specialists* (Chicago: University of Chicago Press, 1985), 461.

97. The scholarship is put to full use in Browne, *Charles Darwin: Voyaging,* 202–210.

98. *Correspondence,* 1:162 n. 2.

99. Ibid., 1:312, 315 n. 1.

100. *JR,* 608.

101. For a heated exchange of views on the subject, see Paolo Palladino and Michael Worboys, "Science and Imperialism," *Isis* 84 (1993):91–102, and Lewis Pyenson, "Cultural Imperialism and Exact Sciences Revisited," *Isis* 84 (1993):103–108. To some extent the difference in views corresponds to a difference in focus. Palladino and Worboys are concerned with "the problem of imperialism, of why and how one people dominates another" (p. 91); Pyenson is concerned primarily with narrower questions concerning the practice of science in imperial settings. For an important treatment of the "problem" of imperialism, see Norman Etherington, "Reconsidering Theories of Imperialism," *History and Theory* 21, no. 1 (1982):1–36. Etherington separates nineteen-century overseas colonial expansion from the imperialism of the 1895–1914 period.

102. Richard Koebner and Helmut Dan Schmidt, *Imperialism: The Story and Significance of a Political Word, 1840–1960* (Cambridge: Cambridge University Press, 1964). For a survey of principal authors writing on imperialism, see Roger Owen and Bob Sutcliffe, *Studies in the Theory of Imperialism* (London: Longman, 1972), 331–376.

103. In January 1830 FitzRoy remarked, "There may be metal in many of the Fuegian mountains, and I much regret that no person in the vessel was skilled in mineralogy, or at all

acquainted with geology. . . . [I inwardly resolved] that if ever I left England again on a similar expedition, I would endeavour to carry out a person qualified to examine the land; while the officers, and myself, would attend to hydrography." FitzRoy, *Narrative,* 1:385.

104. For a review of the literature regarding imperialism that includes discussion of its relation to the growth of geographical knowledge, see Patrick Wolfe, "Imperialism and History: A Century of Theory, from Marx to Postcolonialism," *American Historical Review* 102 (April 1997):378–420, especially pp. 409–410. Also see the important collection of articles in Roy MacLeod, ed., *Nature and Empire: Science and the Colonial Enterprise. Osiris* 15 (2000), of which the following is particularly pertinent in its general approach: Sverker Sörlin, "Ordering the World for Europe: Science as Intelligence and Information as Seen from the Northern Periphery," 51–69.

105. A definition proposed by David Fieldhouse and quoted in Robert A. Stafford, *Scientist of Empire: Sir Roderick Murchison, Scientific Exploration and Victorian Imperialism* (Cambridge: Cambridge University Press, 1989), 1.

106. *Correspondence,* 1:306 n. 6. For the history of the dispute, which included an American role that set the Rosas government in Buenos Aires at odds with the United States, see H. S. Ferns, *Argentina* (London: Ernest Benn, 1969), 87–88, 253–260. Ferns's assessment accurately predicted the sort of war that did in fact occur in 1982 over the islands.

107. Beaufort, "Memorandum," in FitzRoy, *Narrative,* 2:30.

108. The analysis holds that there was a continuity between informal empire, based on trade, that developed in the mid-nineteenth century and the formal British Empire of the late-nineteenth century. See John Gallagher and Ronald Robinson, "The Imperialism of Free Trade," *Economic History Review,* 2d ser., 1 (1953):1–15. The key passage in the article (p. 13) reads, "British policy followed the principle of extending control informally if possible and formally if necessary. . . . The usual summing up of the policy of the free trade empire as 'trade not rule' should read 'trade with informal control if possible; trade with rule when necessary'." For a favorable view of the applicability of the Robinson-Gallagher thesis in regard to Latin America, see Richard Graham, "Robinson and Gallagher in Latin America: The Meaning of Informal Imperialism," in William Roger Louis, ed., *Imperialism: The Robinson and Gallagher Controversy* (New York: Franklin Watts, 1976), 217–221. For the alternate view that "British enterprise in Latin America before 1914 operated independently of diplomatic promotion or assistance," see D. C. M. Platt, "Economic Imperialism and the Businessman: Britain and Latin America before 1914," in Roger Owen and Bob Sutcliffe, eds., *Studies in the Theory of Imperialism* (London: Longman, 1972), 296. For an early statement of the "informal empire" thesis as related to Argentina, see H. S. Ferns, "Britain's Informal Empire in Argentina, 1806–1914," *Past and Present* 4 (November 1953):60–75. For a recent account basically supporting that thesis in regard to Chile and Brazil, as well as Argentina, see P. J. Cain and A. G. Hopkins, *British Imperialism: Innovation and Expansion, 1688–1914* (London: Longman, 1993), 276–315. For a survey of key actors, see Michael G. Mulhall, *The English in South America* (Buenos Ayres: Standard Office, 1878).

109. Robert A. Stafford, "Geological Surveys, Mineral Discoveries, and British Expansion, 1835–1871," *Journal of Imperial and Commonwealth History* 12 (1984), 5–32; Stafford, *Scientist of Empire;* James A. Secord, "King of Siluria: Roderick Murchison and the Imperial Theme in Nineteenth-Century British Geology," *Victorian Studies* 25 (1982):413–442. More generally, see Stafford, "Scientific Exploration and Empire," in William Roger Louis, ed., *The Oxford History of the British Empire,* vol. 3: *The Nineteenth Century,* ed. Andrew Porter (Oxford: Oxford University Press, 1999), 294–319. See also Patrick Wyse Jackson, ed., *Geological Travellers* (New York: Pober Publications, in press).

110. Stafford, "Geological Surveys," 7 and Appendix, "Overseas Geological Surveys Undertaken or Proposed, 1835–1871," 23–24.

111. Syms Covington recorded a naming in his journal entry for 5 March 1834: "Wullia, or Buttons land; the place where the Indians were put on shore, & native place of Jemmy Button; here the Indians are very numerous. Near here is Mount 'Darwin' the highest mountain

in Tierra del Fuego. here the ship was surrounded with canoes the <whole> short time we were here (bartering)[.]" Syms Covington Journal, 1831–1836, Mitchell Library, Sydney. Given that the Fuegians would likely have had a name for the mountain, the designation "Mount Darwin" formed part of an overlay to a previously established pattern of naming (as "Buttons land" for "Wullia"), as Covington's quotation marks ('Darwin') perhaps indicated.

112. Adam Sedgwick, "Presidential Address [18 February 1831]," *Proceedings of the Geological Society of London* 1 (November 1826–June 1833):298. In a volume of great value for the theme of expansion in the service of science, see in particular Suzanne Zeller, "The Colonial World as a Geological Metaphor: Strata(gems) of Empire in Victorian Canada," *Osiris* 15 (2000):85–107.

113. William Daniel Conybeare, "Report on the Progress, Actual State, and Ulterior Prospects of Geological Science," *Report of the British Association for the Advancement of Science* (1833):413.

114. Owen to Murchison, 12 March 1849, quoted in Stafford, "Geological Surveys," 14.

115. George Gaylord Simpson, *Discoverers of the Lost World: An Account of Some of Those Who Brought Back to Life South American Mammals Long Buried in the Abyss of Time* (New Haven: Yale University Press, 1984), 205–208. See also Maria Margaret Lopes and Irina Podgorny, "The Shaping of Latin American Museums of Natural History, 1850–1990," *Osiris* 15 (2000):108–118.

116. J. C. Beaglehole, *The Life of Captain James Cook* (London: Adam and Charles Black, 1974), 236–247; Dawson, *Memoirs of Hydrography,* part 1, 121. Further wreckage from Lapérouse's expedition was found in 1962. See Norman Douglas and Ngaire Douglas, eds., *Pacific Islands Yearbook,* 16th ed. (North Ryde, Australia: Angus and Robertson, 1989), 517. On general background, see also Keith R. Benson and Philip F. Rehbock, eds., *Oceanographic History: The Pacific and Beyond* (Seattle: University of Washington Press, 2002).

117. *CR,* Appendix, *"A Detailed Description of the Reefs and Islands in the Coloured Map, Plate III,"* 151–205.

118. Ibid., 191.

119. DAR 32.1:72 [volume no.: folio].

120. Ibid., 37.1:385. Darwin worked the information into *GSA,* 71.

121. Stafford, *Scientist of Empire,* 144.

122. David Oldroyd, *Thinking about the Earth: A History of Ideas in Geology* (Cambridge: Harvard University Press, 1996), 120. Oldroyd points out that had the geology of the globe first been worked out in New Zealand, no geologist would have placed a boundary between the Cretaceous and the Tertiary, for sedimentation continued in that part of the world whereas there was a major break in Europe. Also see pp. 120–130 for discussion of the bureaucratic structures for the geological mapping of the world, much of it dominated by English-speaking geologists.

123. Charles Darwin, "Geology," in John F. W. Herschel, ed., *A Manual of Scientific Enquiry; Prepared for the Use of Her Majesty's Navy: and Adapted for Travellers in General* (London: John Murray, 1849), 156–195; also in *CP,* 1:227–250.

124. CD to Caroline Darwin, 28 April 1831, in *Correspondence,* 1:122. See Stuart Max Walters and E. Anne Stow, *Darwin's Mentor: John Stevens Henslow, 1796–1861* (Cambridge: Cambridge University Press, 2001), 78–107.

125. CD to Caroline Darwin, 29 April 1836, in *Correspondence,* 1:496.

126. Erasmus Alvey Darwin to CD, 5 March 1823, in ibid., 1:7.

127. Browne, *Charles Darwin: Voyaging,* 97, 123–124. *Correspondence,* 1:76, 89, 102, 110.

128. J. S. Henslow to CD, 20 November 1831, in *Correspondence,* 1:183.

129. CD to J. S. Henslow, 11 July 1831, in ibid., 1:126.

130. *Origin.* J. D. Hooker to CD, 2 July 1860, in *Correspondence,* 8:270 and pp. 590–597, Appendix VI, "Report of the British Association meeting in Oxford, 26 June—3 July 1860."

131. *Correspondence,* 1:539.

132. Information courtesy of E. S. Leedham-Green, Assistant Keeper, University Archives, University of Cambridge.

133. John F. W. Herschel, *A Preliminary Discourse on the Study of Natural Philosophy* (London: Longman, 1831). On Darwin's reading of Herschel, see *Correspondence,* 1:118 n. 2; Walter F. [Susan Faye] Cannon, "John Herschel and the Idea of Science," *Journal of the History of Ideas* 22 (April–June 1961):215–239; and Michael Ruse, "Darwin's Debt to Philosophy: An Examination of the Influence of the Philosophical Ideas of John F. W. Herschel and William Whewell on the Development of Charles Darwin's Theory of Evolution," *Studies in History and Philosophy of Science* 6 (June 1975):159–181.

134. William Whewell, *History of the Inductive Sciences from the Earliest to the Present Times,* 3 vols. (London: J. W. Parker, 1837).

135. Cambridge University Calendar for 1831 as cited in *Correspondence,* 1:112.

136. Herschel, *Preliminary Discourse,* 131.

137. Ibid., 108–110, 211–17.

138. Ibid., 38.

139. Ibid., 160.

140. Ibid., 216–217.

141. John F. W. Herschel, *Results of Astronomical Observations Made during the Years 1834, 5, 6, 7, 8, at the Cape of Good Hope; Being the Completion of a Telescopic Survey of the Whole Surface of the Visible Heavens, Commenced in 1825* (London: Smith, Elder and Co., 1847).

142. Herschel, *Preliminary Discourse,* 100.

143. *Correspondence,* 1:112 n.3. Darwin placed tenth on the list of 178 of those who took the pass degree. On the examination and degree system at the University during these years, see D. A. Winstanley, *Early Victorian Cambridge* (Cambridge: Cambridge University Press, 1940), 65–71, and Sheldon Rothblatt, *The Revolution of the Dons: Cambridge and Society in Victorian England* (London: Faber and Faber, 1968), 181–190.

144. CD to W. D. Fox, 15 February 1831, in *Correspondence,* 1:118.

145. *Autobiography,* 68.

146. Herschel, *Preliminary Discourse,* 262.

147. Browne, *Charles Darwin: Voyaging,* 118–132.

148. *Autobiography,* 45; Erasmus Darwin to CD (then nicknamed "Bobby"), 14 November 1822, in *Correspondence,* 1:3–5.

149. *Autobiography,* 52.

150. Ibid., 53. According to Joan Eyles, Jameson did move toward the igneous interpretation of trap rock, though there is a discrepancy between Darwin's memory of Jameson's views in 1826 and the evidence Eyles cites. See Joan Eyles, "Robert Jameson," *Dictionary of Scientific Biography,* 7:69–71.

151. W. D. Conybeare and William Phillips, *Outlines of the Geology of England and Wales, with an Introductory Compendium of the General Principles of that Science and Comparative Views of the Structure of Foreign Countries, Part I* (London: William Phillips, 1822), 443.

152. Charles Daubeny, *A Description of Active and Extinct Volcanos; with Remarks on Their Origin, Their Chemical Phænomena, and the Character of Their Products, As Determined by the Condition of the Earth during the Period of Their Formation* (London: William Phillips, 1826), figure on p. 452. On p. 451 Daubeny stated that his table was to represent gradation of opinion, and that it did not express exactly present views of individuals named.

153. James A. Secord, "The Discovery of a Vocation: Darwin's Early Geology," *British Journal for the History of Science* 24 (1991):139.

154. Secord, "The Discovery of a Vocation: Darwin's Early Geology," 134–138. Paul Barrett also briefly made this point. See Paul H. Barrett, "The Sedgwick-Darwin Geologic Tour of North Wales," *Proceedings of the American Philosophical Society* 118, no. 2 (1974):146.

155. Robert Jameson, *Manual of Mineralogy: Containing an Account of Simple Minerals, and Also a Description and Arrangement of Mountain Rocks* (Edinburgh: Archibald Constable, 1821); *Marginalia,* 432–440.

156. Annotation to Jameson, *Manual of Mineralogy,* 127, listed in *Marginalia,* 434.

157. The scale of relative hardness has come to be known as the Mohs scale. John G. Burke, "Friedrich Mohs," *Dictionary of Scientific Biography,* 9:447–449, outlines Jameson's and Mohs's close association as fellow students of Werner but suggests that Mohs's overall system of classification, based primarily on crystal form, hardness, and specific gravity, was not well received by most mineralogists.

158. Jameson, *Manual of Mineralogy,* 335, 345. "Mountain," as in the phrase "mountain rocks," derives from the German term "*Gebirge*"—literally meaning "mountain"—which Werner adapted to his own use. Ospovat translated the term as "rock mass" or "rock formation" when used in a geological sense. He also noted that in 1786 the English equivalent of *Gebirge* was ground, as meaning "the rock in which a mineral vein or bed is found." See Abraham Gottlob Werner, *Short Classification and Description of the Various Rocks,* ed. and trans. Alexander M. Ospovat (New York: Hafner, 1971 [original text 1786]), 98.

159. Jameson, *Manual of Mineralogy,* 341.

160. Ibid., 342. Secord, "Discovery of a Vocation," 137, reproduces an example of Darwin's marginalia in Jameson's book.

161. Jameson, *Manual of Mineralogy,* 444–466.

162. *Autobiography,* 57–58.

163. Phillip Reid Sloan, "Darwin's Invertebrate Program, 1826–1836: Preconditions for Transformism," in David Kohn, ed., *The Darwinian Heritage* (Princeton: Princeton University Press, 1985), 71–120. Rebecca Stott, *Darwin and the Barnacle* (New York: W. W. Norton, 2003), 1–41.

164. John Stevens Henslow, "Geological Description of Anglesea," *Transactions of the Cambridge Philosophical Society* 1 (1822):359–452.

165. University of Cambridge Archive, vol. 39.17.1, Woodwardian Professor Guardbooks.

166. *CD,* "Journal," in *Correspondence,* 1:539. Italics added.

167. Browne, *Charles Darwin: Voyaging,* 138. I am also indebted to Nancy Mautner for discussion on these points.

168. *Autobiography,* 60. Secord, "Discovery of a Vocation," 142–144.

169. Adam Sedgwick, *A Syllabus of a Course of Lectures on Geology* (Cambridge: J. Hodson, 1821), 2d ed., 1832; 3d ed. (at the University Printers), 1837. In the course of three editions Sedgwick's outlines went from 45 to 57 to 66 pages. In the 1821 version Sedgwick covered a discussion of basic geological terms (stratification, formations, etc.) and a detailed listing of Primitive, Transitional, Secondary, and Tertiary rocks. He clearly stated his opposition to sacred theories of the earth, while employing the category of diluvial detritus. In the 1832 version of the lectures, the influence of Charles Lyell is discernible. In addition to topics discussed in the first edition, Sedgwick also discussed a number of topics that would be of interest to Charles Darwin, including the parallel roads of Glen Roy (p. 6), and the "remarkable effects of earthquakes—elevation and depression of sea coasts, &c" (p. 7).

170. David R. Oldroyd, "Adam Sedgwick and Lakeland Geology (1822–24)," *Proceedings Thirtieth International Geological Congress* 26 (1997):197–204.

171. Nicolaas A. Rupke, *The Great Chain of History: William Buckland and the English School of Geology, 1814–1849* (Oxford: Clarendon Press, 1983), 51–63.

172. Brian P. Dolan, "Governing Matters: The Values of English Education in the Earth Sciences, 1790–1830" (Ph.D. dissertation, University of Cambridge, 1995), 45–69. Since nicknames are a sign of acceptance, it is interesting that at Cambridge, Clarke was known as "Stone" Clarke, "to distinguish him from 'Bone' Clark the anatomist and 'Tone' Clarke the Professor of Music." Peter Searby, *A History of the University of Cambridge,* vol. III: *1750–1870* (Cambridge: Cambridge University Press, 1997), 20 n. 26.

173. Quoted in John Willis Clark and Thomas McKenny Hughes, eds., *Life and Letters of the Reverend Adam Sedgwick,* 2 vols. (Cambridge: Cambridge University Press, 1890), 1:380.

174. James A. Secord, *Controversy in Victorian Geology: The Cambrian-Silurian Dispute* (Princeton: Princeton University Press, 1986), 63.

175. "Wedgwood's sketch of strata uncovered in the digging of the Trent and Mersey Canal, from a letter to Thomas Bentley, 1767" reproduced in Schofield, *Lunar Society,* facing p. 96.

176. Browne, *Charles Darwin: Voyaging,* 29.

177. CD to J. S. Henslow, 11 July 1831, in *Correspondence,* 1:125. The Severn River runs through Shrewsbury.

178. Ibid.

179. CD to C. T. Whitley, 12 July 1831, in *Correspondence,* 7:466.

180. CD to C. T. Whitley, 10 August 1828, in ibid., 7:465. Whitley was Reader in natural philosophy and mathematics at Durham University, 1833–1855.

181. CD to J. S. Henslow, 11 July 1831, in *Correspondence,* 1:125. On the design of the instrument, see Darwin, "Geology," 160: "One of the simplest clinometers is that constructed by the Rev. Prof. Henslow: it consists of a compass and spirit-level, fitted in a small square box; in the lid there is a brass plate, graduated in a quadrant of 90 degrees, with a little plumb-line to be suspended from a milled head at the apex of the quadrant. The line of intersection of the edge of the clinometer, when held horizontally, with the plane of the stratum, gives its strike, range, or direction; and its dip or inclination, taken at right angles to the strike, can be measured by the plumb-line. In an uneven country, it is not easy without the clinometer to judge which is the line of greatest inclination of a stratum; . . . A flat piece of rock representing the general slope can usually be found, and by placing a note-book on it, the measurement can be made very accurately." In the photograph the spirit-level lies to the right of the compass face. The plumb, which Darwin referred to in his letter to Henslow as a "heavy ball," was stored in a small recess just to the left of the sighting wire on the compass. See also Solene Morris and Louise Wilson, *Down House: The Home of Charles Darwin* (London: English Heritage, 1998), 20, and for photographs of other instruments aboard the *Beagle.*

182. Michael B. Roberts, "Darwin at Llanymynech: The Evolution of a Geologist," *British Journal for the History of Science* 29 (1996):469–478.

183. D. R. Oldroyd, personal communication, mid-1990s.

184. CD to W. D. Fox, 9 July 1831, in *Correspondence,* 1:124. These maps are stored in DAR 265. On them, see Michael R. Roberts, "I Coloured a Map: Darwin's Attempts at Geological Mapping in 1831," *Archives of Natural History* 27 (2000):69–79. For the accompanying notes to one of the maps, see Sandra Herbert and Michael B. Roberts, "Charles Darwin's Notes on His 1831 Geological Map of Shrewsbury," *Archives of Natural History* 29 (2002):27–29.

185. *Autobiography,* 69–70.

186. Peter Lucas, "'A Most Glorious Country': Charles Darwin and North Wales, Especially His 1831 Geological Tour," *Archives of Natural History* 29 (2002):26, and pp. 6 and 11 for reproductions of pencil sketches of Welsh scenery by Darwin's cousin Fanny Wedgwood (1806–1832). On several of those earlier visits, see Peter Lucas, "'Three Weeks Which Now Appears Like Three Months': Charles Darwin at Plas Edwards, 1819," *National Library of Wales Journal* 32 (2001):133–146; and "Jigsaw with Pieces Missing: Charles Darwin with John Price at Bodnant, the Walking Tour of 1826 and the Expeditions of 1827," *Archives of Natural History* 29 (2002):359–370. On the Welsh landscape, see John Davies, *The Making of Wales* (Cardiff: Cadw, 1996).

187. Barrett, "Sedgwick-Darwin Geologic Tour," 163 (quoting Sedgwick on Darwin's "man"), and p. 149 quoting the statesman Robert Lowe (1811–1892), on the 14-pound hammer possessed by Darwin. While a mallet-like 14-pound hammer was a possibility, it would not have been standard choice. By way of comparison, the heaviest of the thirty-three historically significant geological hammers in the inventory of the Sedgwick Museum in Cambridge is 3 pounds, 13 ounces, as measured by the curator of the Museum Mr. Rod Long. According to Dr. Graham Chinner, on purely geological grounds a mallet-like hammer would have been counterproductive, smashing to useless bits the highly cleaved slates and grits of North Wales, as Sedgwick likely would have advised Darwin. There are a number of additional reasons to call

it into question. Initially the event may have been misobserved. Lowe, while already a fine classical scholar, was not a geologist and would not likely have been familiar with the tools of the trade. Further, Lowe, an albino, had sight in only one eye, and with that eye his vision was limited. Then too, his reminiscence was written in 1876, forty-five years after the event, during which interval details may well have lost their clarity. Another possibility is that the numeral may have been a typographical error. Lowe typed the reminiscences himself. He had taken up typewriting to overcome his poor penmanship, caused by his partial blindness, but he found the "art of striking the right letter quite as difficult as writing with a pen." See A. Patchett Martin, ed., *Life and Letters of the Right Honourable Robert Lowe, Viscount Sherbrooke,* 2 vols. (London: Longman, 1893), 2:397; 1:19–20 for the quotation from the typescript; 1:201–207 for Lowe's (and his wife's) high regard for Darwin; and 2:397 on Lowe's typewriting. Finally, Martin may have mistranscribed Lowe's remark.

188. Adam Sedgwick to Roderick Murchison, 13 September 1831, in Clark and Hughes, eds., *Life and Letters of the Reverend Adam Sedgwick,* 1:378. The editor notes that the term "secondaries" included the carboniferous rocks, as was usual at that time.

189. The map appears as the frontispiece to Conybeare and Phillips, *Outlines of the Geology of England and Wales.* For a schematic reproduction of the Welsh portion of the Greenough map, see Secord, *Controversy in Victorian Geology,* 52.

190. Conybeare to Sedgwick, 24 April 1828, in Clark and Hughes, *Life and Letters of Reverend Adam Sedgwick,* 1:324.

191. Ibid., 325.

192. CD to Thomas McKenny Hughes, 24 May 1875, in Clark and Hughes, *Life and Letters of Reverend Adam Sedgwick,* 1:380. For the complete text of Darwin's field notes, see Paul H. Barrett, "The Sedgwick-Darwin Geological Tour of North Wales." Darwin changed from "we" to "you" to "I" in varying portions of his field notes from the trip, indicating when he was working with Sedgwick and when separately.

193. Michael B. Roberts, "Darwin's Dog-leg: The Last Stage of Darwin's Welsh Field Trip of 1831," *Archives of Natural History* 25 (1998):66–67.

194. Barrett, "The Sedgwick-Darwin Geologic Tour of North Wales," 155. Secord, *Controversy in Victorian Geology,* 58.

195. Adam Sedgwick, "On the Origin of Alluvial and Diluvial Formations," *Annals of Philosophy,* n.s. 9 (1825):243: "To such materials as these the term diluvial (indicating their formation by some great irregular inundation) is now applied by almost all the English school of geologists."

196. Barrett, "The Sedgwick-Darwin Geological Tour of North Wales," 160. Oldroyd, "Adam Sedgwick and Lakeland Geology (1822–1824)," 200, credits the local Lakeland guide and geologist Jonathan Otley (1766–1856) with distinguishing bedding, jointing, and cleavage-features in the Lakes and with conveying his knowledge to Sedgwick, who acknowledged the debt.

197. Secord, "Discovery of a Vocation: Darwin's Early Geology," 146–148. Secord has provided a map (p. 146) tracing Darwin's route against the background of a version of the Greenough map.

198. Sedgwick to Murchison, 13 September 1831, in Clark and Hughes, *Life and Letters of Reverend Adam Sedgwick,* 1:378.

199. Secord, "Discovery of a Vocation: Darwin's Early Geology," 146.

200. Darwin to Hughes, 24 May 1875, in Clark and Hughes, *Life and Letters of the Reverend Adam Sedgwick,* 1:380–381.

201. Barrett, "Sedgwick-Darwin Geologic Tour," 162.

202. Lucas, "Charles Darwin and North Wales." Roberts, "Darwin's Dog-leg," and, with revised chronology and the suggestion that Darwin may also have gone on to Dublin with Sedgwick, "Just before the Beagle: Charles Darwin's Geological Fieldwork in Wales, Summer 1831," *Endeavour* 25 (2001):33–37. With their deep knowledge of North Wales—Lucas stressing literary and topographical evidence, Roberts the geological—the authors have re-

opened the question of the 1831 field trip to great effect, deftly employing new information to show how familiar Darwin already was with North Wales, which made Sedgwick's instruction all the more readily useful. The date of Darwin's arrival in Barmouth is now established as 23 August on the evidence of the journal of the Lowe brothers (Robert and Henry Porter [1810–1887]), then being tutored in mathematics by Whitley. Lucas, "Charles Darwin and North Wales," p. 3.

203. Emma Darwin to W. D. Fox, 8 December 1863, in *Correspondence,* 11:690.

204. Lucas, "Charles Darwin and North Wales," 7.

205. Adam Sedgwick to CD, 18 September 1831, in *Correspondence,* 1:157. Of the books Sedgwick cited, those not appearing in references to this chapter include Jean François d'Aubuisson de Voisins, *Traité de géognosie,* 2 vols. (Strasbourg: F. G. Levrault, 1819), and Robert Bakewell, *An Introduction to Geology: Comprising the Elements of the Science in Its Present Advanced State, and All the Recent Discoveries; with an Outline of the Geology of England and Wales,* 3d ed. (London: Longman, 1828). William Lonsdale (1794–1871) was the curator and librarian of the Geological Society of London from 1829 to 1841.

206. CD to J. S. Henslow, 9 September 1831, in *Correspondence,* 1:149.

Chapter 2. Geology

1. The magnitude of its achievements has been recognized by historians of geology and discussed in detail. Building on an older and still valuable scholarship, several recent studies have assessed the field in ways that support each other on fundamental points. Among book-length treatments: Gabriel Gohau, *A History of Geology,* rev. and trans. Albert V. Carozzi and Marguerite Carozzi (New Brunswick: Rutgers University Press, 1990); Rachel Laudan, *From Mineralogy to Geology: The Foundations of a Science, 1650–1830* (Chicago: University of Chicago Press, 1987); Roy Porter, *The Making of Geology: Earth Science in Britain, 1660–1815* (Cambridge: Cambridge University Press, 1977); Mott T. Greene, *Geology in the Nineteenth Century: Changing Views of a Changing World* (Ithaca: Cornell University Press, 1982); and Oldroyd, *Thinking about the Earth.*

2. Thomas S. Kuhn, "The History of Science," *International Encyclopedia of the Social Sciences* 14 (1968):77. Turning around the issue of the generality of Darwin's presentation, one can emphasize the logical coherence of the *Origin.* On this score, see Ernst Mayr, *One Long Argument: Charles Darwin and the Genesis of Modern Evolutionary Thought* (Cambridge: Harvard University Press, 1991). In interpreting the *Origin,* one can also emphasize Darwin's role as a theorist who operated in natural historical disciplines that traditionally have not been regarded as rich in theory, because they were largely nonmathematical in content. Here see Sandra Herbert, "The Place of Man in the Development of Darwin's Theory of Transmutation: Part II," *Journal of the History of Biology* 10 (1977):157–178. On the coherence of the entire corpus of Darwin's work, see Michael T. Ghiselin, *The Triumph of the Darwinian Method* (Chicago: University of Chicago Press, 1969).

3. *CR, VI, GSA.*

4. Listed in *CP.* Papers are cited in their original form in the bibliography.

5. Porter, *Making of Geology,* 202.

6. Martin J. S. Rudwick, *The Meaning of Fossils: Episodes in the History of Palaeontology* (London: Macdonald, 1972), 213; Sandra Herbert, "Darwin the Young Geologist," and Martin J. S. Rudwick, "Darwin and the World of Geology (Commentary)," in David Kohn, ed., *The Darwinian Heritage* (Princeton: Princeton University Press, 1985), 503, 511–512.

7. *Autobiography,* 77.

8. I borrow the phrase from George Malcolm Young, *Portrait of an Age: Victorian England,* ed. G. K. Clark (London: Oxford University Press, 1977), 18.

9. *Correspondence,* "The books on board the *Beagle,*" 1:553–557.

10. CD to Susan Darwin, 3 December 1833, in ibid., 1:359.

11. CD to Catherine Darwin, 22 May–14 July 1833, in ibid., 1:314.

12. Henslow to CD, in ibid., 1:327.

13. In the bibliography, an asterisk indicates the books on geology that are known to have formed part of the Beagle's library.

14. CD to Henslow, 24 July–7 November 1834, in *Correspondence,* 1:399.

15. Porter, *Making of Geology,* 203.

16. William Phillips, *An Elementary Introduction to the Knowledge of Mineralogy: Comprising Some Account of the Characters and Elements of Minerals; Explanations of Terms in Common Use; Descriptions of Minerals, with Accounts of the Places and Circumstances in Which They Are Found; and Especially the Localities of British Minerals,* 3d ed. (London: William Phillips, 1823), sect. a, p. i.

17. Ibid.

18. *Encyclopædia Britannica,* 6th ed., 20 vols. (Edinburgh: Archibald Constable, 1823), 9:550.

19. Alexandre Brongniart, "On the Relative Position of the Serpentines (Ophiolites), Diallage Rocks (Euphotides), Jasper, &c., in Some Parts of the Apennines," in Henry De la Beche, ed., *A Selection of the Geological Memoirs Contained in the Annales des Mines, Together with a Synoptical Table of Equivalent Formations, and M. Brongniart's Table of the Classification of Mixed Rocks* (London: William Phillips, 1824), 161.

20. Henry Thomas De la Beche, *Sections and Views Illustrative of Geological Phænomena* (London: Treuttel and Würtz, 1830), 70–71. This book is not known to have been aboard the *Beagle.* I have used it for the expressive diagram, and as coming from a geological author with whom Darwin became acquainted after the voyage through their common participation in the Geological Society of London.

21. Erasmus Darwin, *The Botanic Garden, A Poem, in Two Parts.* Part 1: *The Economy of Vegetation.* Part 2: *The Loves of the Plants.* (Part 1, 1791 1st edition; Part 2, 1790 2d edition.) (London: J. Johnson, 1791–1790), 1:facing p. 65 of "additional notes."

22. Conybeare and Phillips, *Outlines of the Geology of England and Wales,* ii. Italics added.

23. Phillips, *Mineralogy.*

24. John Playfair, *Illustrations of the Huttonian Theory of the Earth* (Edinburgh: Cadell and Davies, 1802), 515.

25. George Greenough, *A Critical Examination of the First Principles of Geology* (London: Longman, 1819), 234–235.

26. J. M. Rodwell to Francis Darwin, 8 July 1882, cited in *Correspondence,* 1:125 n. 2.

27. Laudan, *From Mineralogy to Geology,* 66. Also see David Oldroyd, *Sciences of the Earth: Studies in the History of Mineralogy and Geology* (Aldershot, England: Variorum, 1998).

28. Laudan, *From Mineralogy to Geology,* 88. Ospovat has characterized the German miners' (and Werner's) use of the term '*Gebirgsart*" as follows: "A German miner always speaks of being in a *Gebirge* [mountain] as soon as he goes below the surface of the earth, whether the surface is mountainous or plain. . . . For example, if on his way down into the earth the miner encountered two sandstones widely separated from each other by layers of other kinds of rocks, he would give each of the sandstones a different and distinct name, as, for example, first sandstone and Rother Todter sandstone. Altogether, then, the German miner used three criteria in designating a rock as a *Gebirgsart:* method of work; extent of the deposit; and location. . . . Where the miners used work methods, Werner used structure and texture, from which, he believed he could determine the mode of formation of the rock and to a certain extent the time of its formation. . . . Werner also relied upon the relative position of rocks, considering this the most important clue to the time of the rock's formation." See Alexander M. Ospovat, "Reflections on A.G. Werner's 'Kurze Klassifikation'," in Cecil J. Schneer, ed., *Toward a History of Geology* (Cambridge: MIT Press, 1969), 250–251. Oldroyd has suggested the term "genetic" rather than "historical" to describe Werner's theory. Oldroyd, *Thinking about the Earth,* 101–102.

29. Greene, *Geology in the Nineteenth Century,* 19–68.

30. Laudan, *From Mineralogy to Geology,* 102–112.

31. Alexander von Humboldt, *A Geognostical Essay on the Superposition of Rocks, in Both Hemispheres* (London: Longman, 1823). This English-language edition is the one cited in DAR 32.1 57^{v}, even though Darwin owned the second French edition of 1826. See *Marginalia,* 415; *Correspondence,* 1:561. Darwin could read French, but if there was an English translation available, he usually found his way to it.

32. Humboldt, *Geognostical Essay,* sect. B, p. 1.

33. Ibid., 461–464; Sedgwick, *Syllabus of a Course of Lectures on Geology* (1821), 6. For Lyell's later attack on this view, see Lyell, *Principles of Geology* 3:145: "Some writers have attempted to introduce into their classification of geological periods an *alluvial epoch,* as if the transportation of loose matter from one part of the surface of the land to another had been the work of one particular period."

34. Georges Cuvier, *The Animal Kingdom Arranged in Conformity with Its Organization . . . with Additional Descriptions of All the Species Hitherto Named, and of Many Not Before Noticed,* by Edward Griffith and others, 16 vols. (London: Whittaker, 1827–1835). Supplementary volume 11 on the fossils by Edward Pidgeon has the table as a fold-out frontispiece. The completed publishing history of this work is described in Charles F. Cowan, "Notes on Griffith's *Animal Kingdom of Cuvier* (1824–1835)," *Journal of the Society for the Bibliography of Natural History* 5 (1969):137–140. Darwin had at least some volumes from the edition aboard the *Beagle.*

35. Humboldt, *Geognostical Essay,* 465–477. The system was known as pasigraphy. The British Survey used it later for keying maps. For treatment of the tradition in which Humboldt worked, see W. R. Albury and D. R. Oldroyd, "From Renaissance Mineral Studies to Historical Geology, in the Light of Michel Foucault's *The Order of Things,*" *British Journal for the History of Science* 10 (1977):187–215.

36. De la Beche, *Sections and Views,* vii.

37. Henry Thomas De la Beche, ed., *A Selection of the Geological Memoirs Contained in the Annales des Mines, Together with a Synoptical Table of Equivalent Formations, and M. Brongniart's Table of the Classification of Mixed Rocks* (London: William Phillips, 1824); De la Beche, *Geological Manual* (London: Treuttel and Würtz, 1831).

38. V. A. Eyles, "Henry De la Beche," *Dictionary of Scientific Biography* 4:9–11.

39. J. J. d'Omalius d'Halloy, "Memoir on the Geographical Extent of the Formation of the Environs of Paris," in De la Beche, *Selection of the Geological Memoirs,* 9. Italics added.

40. Alexandre Brongniart, "On the Zoological Characters of Formations, with the Application of These Characters to the Determination of Some Rocks of the Chalk Formation," in De la Beche, *Selection of the Geological Memoirs,* 236.

41. Ibid., 237.

42. De la Beche, "A Synoptical Table of Equivalent Formations," in De la Beche, *Selection of the Geological Memoirs,* fold-out table facing p. xxii. An updating, reworking, and simplification of this table appeared in De la Beche, *Geological Manual,* pp. 38–39. In the original table of 1824 De la Beche had three columns, each designating the "English," "French," and "German" equivalent formations. In the 1831 table the national titles are dropped in favor of names assigned by authors. De la Beche's four columns read "Improved Wernerian," "Conybeare," "Omalius d'Halloy, 1830," and "Brongniart, 1829." The original table is more detailed in its use of French and German, as well as English, names for formations. For a reproduction of the 1824 *Selection of the Geological Memoirs* table, see Laudan, *From Mineralogy to Geology,* 160; for reproduction of a portion of the 1831 *Geological Manual* table, see Secord, *Controversy in Victorian Geology,* 31.

43. On defining an "English" as opposed to a "British" grouping of geologists, see Rupke, *Great Chain of History,* 3–27. "English geology"—a phrase of contemporary usage—was, in Rupke's view, represented most purely by Buckland and Sedgwick with their lecture courses at Oxford and Cambridge, respectively. In my view, Rupke's characterization is useful for this period since it allows one to acknowledge the separation between geological traditions of England and Scotland, though for some figures, such as Lyell, who studied under Buckland but was

more sympathetic to Huttonianism, it is not helpful. Secord also correctly suggests recognition that the Cambridge group of geologists were more mathematically sophisticated and hence oriented toward the study of structure than were their Oxford counterparts. See his *Controversy in Victorian Geology*, 63–68.

44. Conybeare and Phillips, *Outlines of the Geology of England and Wales*, vi–viii.

45. Ibid., vii.

46. Ibid., xxviii.

47. Constant Prévost, "*Diluvion*," in J. B. Bory de Saint-Vincent, ed., *Dictionnaire classique d'histoire naturelle*, 17 vols. (Paris: Rey and Gravier, 1822–1831), 5:508–509.

48. Georges Cuvier, *Discours sur les révolutions de la surface du globe*, 3d French ed. (Paris: G. Dufour, 1825), 288–289.

49. William Phillips, "Preliminary Notice," in Conybeare and Phillips, *Outlines of the Geology of England and Wales*, n.p.

50. Greenough, *Critical Examination of the First Principles of Geology*, 1.

51. Hugh S. Torrens, "Patronage and Problems: Banks and the Earth Sciences," in R. E. R. Banks, B. Elliott, J. G. Hawkes, D. King-Hele, and G. Ll. Lucas, eds., *Sir Joseph Banks: A Global Perspective* (Kew: Royal Botanic Gardens, 1994), 49–75.

52. V. A. Eyles, "George Bellas Greenough," *Dictionary of Scientific Biography* 5:518–519; Joan M. Eyles, "William Smith," *Dictionary of Scientific Biography* 12:486–492. The title of Smith's map was *A Delineation of the Strata of England and Wales, With Part of Scotland.*

53. The minutes of the meeting of 7 January 1820 note that Greenough's map was then to cost £1720.0.0 prior to publication. The retail price of the map was to be £6.6; the trade price £4.14.6; and the price to the society £4.2.6. Geological Society of London Archives. Assessing the relationship between Smith's map and Greenough's map (or, in effect, the society's map) has been a challenge to historians. L. R. Cox argued that Greenough first intended a map on Wernerian principles but altered his approach once having visited Smith in 1808 and viewed his work. See L. R. Cox, "New Light on William Smith and His Work," *Proceedings of the Yorkshire Geological Society* 25 (1942):41–43, 78. For further support of the claims to priority of Smith and other mineral surveyors, see Torrens, "Patronage and Problems," 59–69.

54. William Buckland, *Reliquiæ Diluvianæ; or Observations on the Organic Remains Contained in Caves, Fissures, and Diluvial Gravel, and on Other Geological Phenomena, Attesting the Action of an Universal Deluge* (London: John Murray, 1823). See Darwin's instruction to himself to (fully) "Read this work—" in *Notebooks*, 413 (E:60).

55. William Buckland, "On the Occurrence of Remains of Elephants, and Other Quadrupeds, in the Cliffs of Frozen Mud, in Eschscholtz Bay, within Beering's Strait, and in Other Distant Parts of the Shores of the Arctic Seas," in F. W. Beechey, *Narrative of a Voyage to the Pacific and Beering's Strait . . . 1825, 26, 27, 28*, 2 vols. (London: Henry Colburn and R. Bentley, 1831), 2:331–356. Darwin's personal copy of Beechey's book was the 1832 American, or "Yanky Edit" as he called it. This edition did not have the Buckland appendix. *Notebooks*, 34 (RN:45).

56. Bakewell, *Introduction to Geology;* d'Aubuisson de Voisins, *Traité de géognosie*.

57. Arthur Birembaut, "Jean-François d'Aubuisson de Voisin," *Dictionary of Scientific Biography* 1:327. Also Laudan, *From Mineralogy to Geology*, 107–112.

58. David R. Stoddart, ed., "Coral Islands by Charles Darwin," *Atoll Research Bulletin* 88 (December 1962):18–19.

59. De la Beche, *Geological Manual*, 140–143; Lyell, *Principles of Geology*, 2: 283–301.

60. John Michell, "Conjectures Concerning the Cause, and Observations on the Phænomena of of Earthquakes; Particularly of That Great Earthquake of the First of November 1755, Which Proved So Fatal to the City of Lisbon, and Whose Effects Were Felt As Far As Africa, and More or Less throughout Almost All Europe," *Philosophical Transactions of the Royal Society of London* 51 (1760):566–634; Daubeny, *Description of Active and Extinct Volcanos;* George Poulett Scrope, *Considerations on Volcanos: The Probable Causes of Their Phenomena, the Laws Which Determine Their March, the Disposition of Their Products, and Their Connexion*

with the Present State and Past History of the Globe; Leading to the Establishment of A New Theory of the Earth (London: William Phillips, 1825).

61. Scrope, *Considerations on Volcanos,* 214.

62. Thrower, *Maps and Civilization,* 95, and, on the tradition, 95–105.

63. *Correspondence,* 1:239 n. 3, 308 n. 4, and 381 n. 5 for information regarding the date of Darwin's receipt and reading of the three volumes of the *Principles.*

64. Leonard Wilson, *Charles Lyell: The Years to 1841* (New Haven: Yale University Press, 1972), 314.

65. *Autobiography,* 77, 101. Italics added.

66. CD to Leonard Horner, 29 August 1844, in *Correspondence,* 3:55.

67. Lyell, *Principles of Geology,* 1:65, 140.

68. See, for example, the extended passage in ibid., 1:461–462.

69. Charles Lyell to Mary Horner, 17 February 1832, quoted in Wilson, *Charles Lyell,* 344.

70. Lyell, *Principles of Geology,* 1:1.

71. Martin J. S. Rudwick, "The Strategy of Lyell's *Principles of Geology, Isis* 61 (1970):5–33. See also his introduction to the reprint of Lyell, *Principles of Geology,* 3 vols. [1830–1833] (Chicago: University of Chicago Press, 1990–1991), 1:vii–lviii.

72. Lyell, *Principles of Geology,* 1:167.

73. For discussion of Lyell's range of sources and a reconstructed bibliography to the first edition, see Lyell, *Principles of Geology,* vol. 3, bibliography compiled by Martin J. S. Rudwick (1991 reprint of first edition of 1833), 113–160.

74. Lyell, *Principles of Geology,* 1:150.

75. Lyell, *Principles of Geology,* 1:30.

76. Lyell, *Principles of Geology,* 1:31, where Lyell cited only "Quirini." The name is obscure, but Rhoda Rappaport identified the author as Giovanni Quirini or "Querini," in *When Geologists Were Historians, 1665–1750* (Ithaca: Cornell University Press, 1997), 107. Rudwick identified the author as Giorgio Quirini in the bibliography (p. 149) to his 1991 reprint edition of Lyell, *Principles of Geology,* iii.

77. Lyell was not the first British geologist not to adopt the term "diluvium." Scrope had preceded him, and De la Beche was cautious. But Lyell was the one to mount the attack, contra Buckland, on "loose and slightly-consolidated strata of gravel and sand, and which are usually called diluvian formations. . . ." Lyell, *Principles of Geology,* 1:144.

78. Wilson, *Charles Lyell,* 278.

79. Lyell, *Principles of Geology,* 1:81.

80. Ibid., 1:83.

81. Ibid., 1:125. Also see Dov Ospovat, "Lyell's Theory of Climate," *Journal of the History of Biology,* 10 (Fall 1977):317–339.

82. Lyell, *Principles of Geology,* 1:104–124.

83. Ibid., 1:145.

84. Ibid., 2:1.

85. See Oldroyd, *Thinking about the Earth,* xxv, on "primary" as "an obsolete term, formerly used to designate what is now called the Precambrian; then extended to include the Palaeozoic; and then used to refer to what is now called the Palaeozoic." Also see p. 82, for development of the term by Giovanni Arduino (1735–1795).

86. Lyell, *Principles of Geology,* 3:364.

87. Ibid., 3:viii.

88. Ibid., 3:395. Martin J. S. Rudwick, "Charles Lyell's Dream of a Statistical Palaeontology," *Palaeontology* 21, pt. 2 (May 1978):225–244.

89. Lyell, *Principles of Geology,* 3:chap. 8.

90. Ibid., 3:chaps. 24, 19.

91. Ibid., 3:383, quoting [George Poulett Scrope] *Quarterly Review* 86 (October 1830):464. For the identity of the reviewer as Scrope, see Wilson, *Charles Lyell,* 273–277, 303.

92. Lyell, *Principles of Geology,* 3:347–348.

93. *Correspondence,* 1:399.

94. CD to Henslow, 24 July–7 November 1834, in ibid., 1:399.

95. CD to Henslow, 12 August 1835, in ibid., 1:462.

96. William Whewell, "[Review of] *Principles of Geology,* Vol. II, by Charles Lyell," *Quarterly Review* 47 (1832):126.

97. William Buckland, "On the Fossil Remains of the Megatherium, Recently Imported into England from South America," *Report of the British Association for the Advancement of Science* (1833):104–107; Conybeare, "Report on the Progress, Actual State, and Ulterior Prospects of Geological Science."

98. CD to J. S. Henslow, March 1834, in *Correspondence,* 1:370.

99. The standard history of the society in its first century is Horace B. Woodward, *The History of the Geological Society of London* (London: Geological Society, 1907). See pp. 286–296 for a list of presidents and their addresses; the 1843–1845 addresses were not published. In recent years, research in the records of the Geological Society was facilitated by the work of its archivist. See John C. Thackray, "The Archives of the Geological Society of London," *Earth Sciences History* 3 (1984):3–8. On characterizing the field, see also Rudwick, *Great Devonian Controversy,* 17–60, and Secord, *Controversy in Victorian Geology,* 14–38. For a survey of the field of British geology, laterally, from the point of view of the year 1835, see Martin J. S. Rudwick, "A Year in the Life of Adam Sedgwick and Company," *Archives of Natural History* 15 (1988):243-268.

100. The linkage of England and Wales was usual, as is suggested by the title of Conybeare and Phillips's book, and by the title of Greenough's map. (Much of Wales remained relatively ungeologized, however.) Edinburgh remained the center of geological science in Scotland in the 1830s and 1840s. Jameson, still active at the university, was then editor of the *Edinburgh New Philosophical Journal,* which was important during the period for, among other things, publishing translations of the work of Continental geologists, including those writing on glaciers. Geological study in Ireland was well represented from the 1790s onward, and when the Geological Society of Dublin was formed in 1832, it rested on an already distinguished past. The influence of the metropolis of London grew as the Geological Survey of Ireland, born in 1845, operated under the aegis of the Geological Survey of Great Britain, directed by Henry De la Beche, a key figure in the Geological Society of London. See Gordon L. Herries Davies, *North from the Hook: 150 Years of the Geological Survey of Ireland* (Dublin: Geological Survey of Ireland, 1995), 1–48.

101. Simon Knell, *The Culture of English Geology, 1815–1851: A Science Revealed through Its Collecting* (Aldershot, England: Ashgate, 2000). David Oldroyd has also emphasized to me the activities of the Royal Geological Society of Cornwall, founded in 1814, and of the Highland and Agricultural Society of Scotland. The Highland and Agricultural Society published important geological work on the Midland Valley of Scotland in the 1840s. For a London acknowledgment of Scottish and Irish endeavors, see George Bellas Greenough, "Presidential Address [20 February 1835]," *Proceedings of the Geological Society of London* 2 (November 1833–June 1838):154–155.

102. John C. Thackray, *To See the Fellows Fight: Eye Witness Accounts of Meetings of the Geological Society of London and Its Club, 1822–1868* (Standford in the Vale, England: British Society for the History of Science, 2003), 153.

103. Martin J. S. Rudwick, "The Foundation of the Geological Society of London: Its Scheme for Co-operative Research and Its Struggle for Independence," *British Journal for the History of Science* 1 (1963):325–355.

104. Jack Morrell and Arnold Thackray, *Gentlemen of Science: Early Years of the British Association for the Advancement of Science* (Oxford: Clarendon Press, 1981), 453.

105. Ibid., 460–466.

106. Roderick Impey Murchison, *The Silurian System, Founded on Geological Researches in the Counties of Salop, Hereford, Radnor, Montgomery, Caermarthen, Brecon, Pembroke, Mon-*

mouth, Gloucester, Worcester, and Stafford; with Descriptions of the Coal-Fields and Overlying Formations (London: John Murray, 1839).

107. Rudwick, *Great Devonian Controversy,* 201.

108. Cannon, *Science in Culture,* chap. 2.

109. Isaac Todhunter, *William Whewell: An Account of His Writings with Selections from His Literary and Scientific Correspondence,* 2 vols. (London: Macmillan, 1876), 1:32. Clark and Hughes, *Life and Letters of Reverend Adam Sedgwick,* 1:264–265.

110. Hugh S. Torrens, "Geology in Peace Time: An English Visit to Study German Mineralogy and Geology (and Visit Goethe, Werner and Raumer) in 1816," in Bernhard Fritscher and Fergus Henderson, eds., *Toward A History of Mineralogy, Petrology, and Geochemistry* (Munich: Institut für Geschichte der Naturwissen-schaften, 1998), 147–175.

111. Rudwick, *Great Devonian Controversy,* 412. The five presidents were Buckland, Greenough, Lyell, Murchison, and Sedgwick. The five nonpresidents were Robert Austen (later named Godwin-Austen) (1808–1884), De la Beche, John Phillips (1800–1874), Thomas Weaver (1773–1855), and David Williams (1773–1855).

112. See Rudwick, *Great Devonian Controversy,* and Secord, *Controversy in Victorian Geology* for the month-by-month details of the relations of the Sedgwick and Murchison in the 1830s.

113. Wilson, *Charles Lyell,* 361.

114. Rudwick, *Great Devonian Controversy,* 72 (Buckland's appointment as canon of Christ Church was made in 1825); Thackray, *Gentlemen of Science,* 538. (Sedgwick was appointed to a prebendary stall at Norwich Cathedral in 1834.)

115. *Punch*'s "Geology of Society" cartoon of 1841 is reproduced in Secord, *Controversy in Victorian Geology,* 35.

116. Hugh S. Torrens, "Patronage and Problems: Banks and the Earth Sciences," 49–75, and, more generally, the collected essays in his *The Practice of British Geology, 1750–1850* (Aldershot, England: Ashgate, 2002).

117. D. T. Moore, J. C. Thackray, and D. L. Morgan, "A Short History of the Museum of the Geological Society of London, 1807–1911, with a Catalogue of the British and Irish Accessions, and Notes on Surviving Collections," *Bulletin of the British Museum (Natural History) Historical Series* 19 (1):54. Also see John Challinor, "Thomas Webster," *Dictionary of Scientific Biography* 14:210–211.

118. CD to Henslow, 9 July 1836, "I am going to ask you to do me a favor. I am very anxious to belong to the Geolog: Society. I do not know, but I suppose it is necessary to be proposed some time before being balloted for, if such is the case, would you be good enough to take the proper preparatory steps. Professor Sedgwick very kindly offered to propose me, before leaving England: if he should happen to be in London, I daresay he would yet do so.—" *Correspondence,* 2:499.

119. Woodward, *History of the Geological Society of London,* 268. The admission certificate is from the archives of the society. Henslow wrote out a nomination for Darwin by hand, rather than using the printed form. The printed form stated that at least one of the signers should be personally acquainted with the candidate.

120. List of Officers, *Proceedings of the Geological Society of London* 2 (November 1833–June 1838):356.

121. Certificate of Candidature of Charles Darwin, Royal Society of London Archive.

122. David Allen, *The Naturalist in Britain: A Social History,* 2d ed. (Princeton: Princeton University Press, 1994), 151–152.

123. Roderick Impey Murchison, "Presidential Address [15 February 1833]." *Proceedings of the Geological Society of London* 1 (November 1826–June 1833):445. Murchison was referring to *Icthyosaurus platyodon.* On Mary Anning, see Woodward, *History of the Geological Society of London,* 115. On the early participation of women in geology, also see Michele L. Aldrich, "Women in Paleontology in the United States, 1840–1960," *Earth Sciences History* 1 (1982):14–22.

124. Woodward, *History of the Geological Society of London,* 118. Woodward commented that Graham's article was the last by a woman to appear in the *Transactions* for forty years. See Maria Graham, "An Account of Some Effects of the Late Earthquakes in Chili [*sic*]. Extracted from a letter to Henry Warburton, Esq.," *Transactions of the Geological Society of London,* 2d ser., 1 (1824):413–415.

125. See George Greenough, "Presidential Address [21 February 1834]," *Proceedings of the Geological Society of London* 2 (November 1833–June 1838):56–69. Remarkably, Maria Graham defended her observations, and her credibility, in *A Letter to the President and Members of the Geological Society of London, in Answer to Certain Observations Contained in Mr. Greenough's Anniversary Address of 1834* (London: T. Brettel, 1834). On Graham, and on the significance of the dispute for Lyell and Darwin, see Sandra Herbert, "Les divergences entre Darwin et Lyell sur quelques questions géologiques," in Yvette Conry, ed., *De Darwin au darwinisme: science et idéologie* (Paris: J. Vrin, 1983), 70.

126. Martina Kölbl-Ebert, "Mary Buckland (née Morland), 1797–1857," and "Charlotte Murchison (née Hugonin), 1788–1869," *Earth Sciences History* 16 (1997):33–38, 39–43; and, in an excellent comprehensive survey, "British Geology in the Early Nineteenth Century: A Conglomerate with a Female Matrix," *Earth Sciences History* 21 (2002):3–25.

127. For Charles's visits with the Lyells in the company of the Horners, or with the Horners themselves, see *Correspondence,* 1:532; 2:12, 13, 36, 131, 132, 166. On the relationship of the Lyells and the Horners with Darwin, see Wilson, *Charles Lyell,* 433–460. Once engaged, Emma Wedgwood, wittily if a shade triumphally (though she was kindly advising Charles to refrain from the insult of renting a house opposite the Horner family), referred to the Horner daughters as the "Horneritas." Sensitive to possible slights, Emma had already reminded Charles, following their engagement, that he should call on Mrs. Horner "pretty soon I think." (*Correspondence,* 2:131, 123). I thank Leonard Wilson for information on the Horner daughters.

128. *Correspondence,* 2:12.

129. Reported in CD to Emma Wedgwood, 30 November–1 December 1838, in ibid., 2:133.

130. Woodward, *History of the Geological Society of London,* 242–244. The "Misses Horner" (unnamed), as well as Lady Lyell, were present as Horner and Lyell attempted to integrate the society. The (temporary) permission for women to attend meetings involved a move from Somerset House to Burlington House. In 1863, the society voted to return to Somerset House, a move that ended the addition of women guests to the audience.

131. Henrietta Litchfield, ed., *Emma Darwin: A Century of Family Letters,* 2 vols. (London: John Murray, 1915), 1:62.

132. I thank Michael Roberts for Fanny Owen's birth and death dates. The romance between Fanny and Charles is suggested in volume 1 of the *Correspondence.* Darwin's initial approach to marriage seems to have been utilitarian and not specific to Emma, though, by the time of their marriage, the attachment was secure. (See Darwin's remark in favor of a "nice soft wife" in "This is the Question Marry/Not Marry" of July 1838 in *Correspondence,* 2:444.)

133. Emma Wedgwood, Diary 1825, University of Keele Archive W/M 1158.

134. Emma Wedgwood to CD, 3 January 1839, in *Correspondence,* 2:157. For the paper, read to the Royal Society of London on 7 February 1839, see *CP,* 1:87–137. Charles thus apparently met Emma's deadline.

135. Emma Wedgwood to CD, 9 January 1839, in *Correspondence,* 2:163. If Emma showed no interest in attending geological meetings, in 1837 Charles's older sister Caroline (1800–1888) had regretted that her sex would prevent her from hearing Lyell's presidential address. The anniversary addresses were customarily attended by ranking members of society, who were not themselves geologists. (*Correspondence,* 2:7.)

136. Rudwick, *Great Devonian Controversy,* 25. For a violation of the norm against publication of reports of discussion, note the episode described in Paul N. Pearson and Christopher

J. Nicholas, "Defining the Base of the Cambrian: The Hicks-Geikie Confrontation of April 1883," *Earth Sciences History* 11 (1992):76–79.

137. Murchison, "Presidential Address [1833]," 447.

138. On current practice, which does not have a set assignment of meaning in relation to color, see John Barnes, *Basic Geological Mapping,* 3d ed. (Chichester: John Wiley, 1995), 97: "Choose colours with care and with due consideration for tradition. In general, use pale colours for rock units which cover wide areas and strong colours for rocks with limited outcrop, such as thin beds and narrow dykes. . . . Relate your colours to rock mineralogy."

Murchison regretted that the French and English geologists were unable to unite on a standard for coloring maps in order that "geological maps might be a sort of book written in a universal language." He noted, however, that "we must recollect that the principle of their colouring was decided and put into execution long before the publication of Mr Greenough." Roderick Impey Murchison, "Presidential Address [18 February 1842]," *Proceedings of the Geological Society of London* 3 (November 1838–June 1842):667.

139. Thomas S. Kuhn, *The Structure of Scientific Revolutions,* 2d ed. (Chicago: University of Chicago Press, 1970), 109.

140. Paul Hoyningen-Huene, *Reconstructing Scientific Revolutions: Thomas S. Kuhn's Philosophy of Science* (Chicago: University of Chicago Press, 1989), particularly "The Functions of Paradigms in the Sense of Exemplary Problem Solutions," 154–163.

141. For the classic assessment, to which we will return in chapter 9, see John C. Greene, "The Kuhnian Paradigm and the Darwinian Revolution in Natural History," in Duane H. D. Roller, ed., *Perspectives in the History of Science and Technology* (Norman: University of Oklahoma Press, 1971). On Kuhn's overall schema in relation to geology, and in particular Lyell, see Alberto Elena, "The Imaginary Lyellian Revolution," *Earth Sciences History* 7 (1988):126–133.

142. Smith was valorized by Sedgwick in his 1831 presidential address of the society for "having been the first, in this country, to discover and teach the identification of strata, and to determine their succession by means of their imbedded fossils." See Eyles, "William Smith," 12:488–489. On Smith's achievements, see Hugh S. Torrens, "Timeless Order: William Smith (1769–1839) and the Search for Raw Materials 1800–1820," in C. L. E. Lewis and S. J. Knell, eds., *The Age of the Earth: From 4004 BC to AD 2002* (London: Geological Society, 2001), 61–83. On the relative claims to originality of English and French geologists regarding the use of fossils in stratigraphy, see Martin J. S. Rudwick, "Cuvier and Brongniart, William Smith, and the Reconstruction of Geohistory," *Earth Sciences History* 15 (1996):25–36.

143. On scientific concepts in relation to scientific disciplines, see Stephen Toulmin, *Human Understanding* (Oxford: Clarendon Press, 1972), 133–260, and note his summary remark (p. 166): "*Every concept is an intellectual micro-institution.*"

144. On the guarded, though (from an historian's point of view) close relation of Kuhn's "paradigm" to his own notion of "research programme," see Imre Lakatos, *Philosophical Papers: The Methodology of Science Research Programmes,* ed. John Worrall and Gregory Currie (Cambridge: Cambridge University Press, 1978), 91.

145. Thackray in Moore, Thackray, and Morgan, "Short History of the Museum of the Geological Society of London," 53.

146. Ibid., 55.

147. Eyles, "William Smith," 487–488.

148. William Whewell, "Presidential Address [16 February 1838]," *Proceedings of the Geological Society of London,* 2 (November 1833–June 1838):633, 643; "Presidential Address [15 February 1839]," 3 (November 1838–June 1842):76, 92.

149. William Henry Fitton, "Presidential Address [20 February 1829]," *Proceedings of the Geological Society of London* 1 (November 1826–June 1833):115–116. Henry Thomas De la Beche, *A Tabular and Proportional View of the Superior, Supermedial, and Medial (Tertiary and Secondary) Rocks,* 2d ed. (London: Truettel, 1828).

150. William Whewell, "Presidential Address [1838]," 636.

151. Murchison, "Presidential Address [1842]", 649.

152. Greenough, "Presidential Address [1835]," 170.

153. John Thackray, "Charles Lyell and the Geological Society," in D. J. Blundell and A. C. Scott, eds., *Lyell: The Past Is the Key to the Present* (London: Geological Society, 1998), 18.

154. Ibid., 18.

155. William Henry Fitton, "Presidential Address [15 February 1828]," *Proceedings of the Geological Society of London* 1 (November 1826–June 1833):59. Italics added.

156. On the identification and naming of these formations, see Rudwick, *Great Devonian Controversy,* and Secord, *Controversy in Victorian Geology.* For Murchison's work at coaching Whewell for the 1838 and 1839 addresses, only partly responded to, see Rudwick, pp. 233–235, 268–275. Murchison wanted Whewell to attack De la Beche; Whewell declined to do so, but praised Murchison and Sedgwick. The lower boundary of the Silurian remained contentious until nearly the end of the nineteenth century.

157. See the classic paper Thomas Webster, "On Some Freshwater Formations on the Isle of Wight, With Some Observations on the Strata over the Chalk in the Southeast Part of England," *Transactions of the Geological Society of London* 2 (1814):161–254. The beginnings of the correlation of the geology of England and France can be traced back to the "prize-winning dissertation on the former junction of England with France (1753)." Rappaport, *When Geologists Were Historians,* 199. See also Kenneth L. Taylor, "Nicolas Desmarest," *Dictionary of Scientific Biography* 4:70–73.

158. Whewell, "Presidential Address [1838]," 633.

159. Whewell, "Presidential Address [1839]," 83–84.

160. Ibid., 80.

161. Roderick Impey Murchison, "Presidential Address [17 February 1843]," *Proceedings of the Geological Society of London* 4 (November 1843–April 1845):149.

162. Secord, *Controversy in Victorian Geology,* 311. For a treatment of the literature on the stratigraphical paradigm up to 1982, see Herbert, "Darwin the Young Geologist," 494–500.

163. On Mantell's contributions to paleontology, see Dennis R. Dean, *Gideon Mantell and the Discovery of Dinosaurs* (Cambridge: Cambridge University Press, 1999), especially 76–85 (on the *Iguanodon,* Mantell's greatest discovery), 268–279 (on Mantell's growing understanding of stratigraphy), and 243–251 (on the eventual rift between Mantell and Owen).

164. In the interest of economy of citation, only abbreviated references to the various addresses are given: *Proceedings of the Geological Society of London* 1 (1816–1833):127–131 [1829]; 197, 204–205 [1830]; 305, 271–279 [1831]; 372–373 [1832]; 440–441, 444–446 [1833]. Vol. 2: 51 [1834]; 174–175 [1835]; 369–373, 371 [1836]; 508–523 [1837]; 624–626 [1838]. Vol. 3: 85–92 [1839]; 238–244, and, relatedly, 244–247 [1840]; 498–509 [1841]; 640–652 [1842]. Vol. 4:139–150 [1843].

165. The date and source of origin for paleontology in Britain has been a matter of debate. For the growth of paleontology per se, Rudwick has also emphasized the primacy of the French tradition; see his *Meaning of Fossils,* 101–163, and, more recently, his edition of selections from Cuvier's writings, *Georges Cuvier, Fossil Bones, and Geological Catastrophes: New Translations and Interpretations of Primary Texts* (Chicago: University of Chicago Press, 1997), 264–267. On Buckland as paleontologist, see Rupke, *Great Chain of History,* 130–148; also see Dean, *Gideon Mantell.*

166. Murchison, "Presidential Address [1843]," 149–150.

167. On Murchison's siding with Sedgwick over Lyell on the interpretation of the fossil record regarding life, see Secord, *Controversy in Victorian Geology,* 292.

168. William Buckland, "Presidential Address [21 February 1840]," *Proceedings of the Geological Society of London* 3 (November 1838–June 1842):239.

169. Henry T. De la Beche, "Instructions for the Local Directors of the Geological Surveys of Great Britain & Ireland [22 May 1845]," British Geological Survey Archives, GSM1/4, fol. 57. I thank James Secord for a copy of this document.

170. Crosbie Smith, "Geologists and Mathematicians: The Rise of Physical Geology," in *Wranglers and Physicists: Studies on Cambridge Physics in the Nineteenth Century,* ed. P. M. Harman (Manchester: Manchester University Press, 1985), 73–83.

171. Whewell, "Presidential Address [1838]," 645.

172. Ibid., 644.

173. Roderick Impey Murchison, "Presidential Address [17 February 1832]," *Proceedings of the Geological Society of London* 1 (November 1826–June 1833):376.

174. Greenough, "Presidential Address [1834]," 54.

175. Whewell, "Presidential Address [1838]," 644–645.

176. William Buckland, "Presidential Address [19 February 1841]," *Proceedings of the Geological Society of London* 3 (November 1838–June 1842):509–516, subtitled "Geological Dynamics.—Glacial Theory."

177. Whewell, "Presidential Address [1838]," 645.

178. Whewell, "Presidential Address [1839]," 85.

179. James A. Secord, "The Geological Survey of Great Britain as a Research School, 1839–1855," *History of Science* 24 (1986):224.

180. Greenough to De la Beche, 11 November 1836, quoted in Rudwick, *Great Devonian Controversy,* 181. Darwin's "one volume" referred to his then-current plan to publish his geological observations from the voyage in a single volume. Darwin intended to "emigrate to London, where I can complete my geology." CD to Henslow, 30–31 October 1836, in *Correspondence,* 1:512. Also see Martin J. S. Rudwick, "Charles Darwin in London: The Integration of Public and Private Science," *Isis* 73 (1982):186–206.

181. On De la Beche's suspicion of Lyell, see Martin J. S. Rudwick, "Caricature as a Source for the History of Science: De la Beche's Anti-Lyellian Sketches of 1831," *Isis* 66 (1975):534–560.

182. Eyles, "George Bellas Greenough," 5:519.

183. Leopold von Buch to Murchison, 20 April 1846, Murchison Manuscript, Geological Society of London Archives, quoted in Secord, "King of Siluria," 414–415.

184. *Correspondence,* 1:327–328.

185. *CP,* 1:3–16. For Lyell's commentary on Darwin's work, see Charles Lyell, "Presidential Address [19 February 1836]," *Proceedings of the Geological Society of London* 2 (November 1833–June 1838):367–369, 378, 389.

186. Charles Lyell "Presidential Address [17 February 1837]," *Proceedings of the Geological Society of London* 2 (November 1833–June 1838):505–506. Darwin had presented his report to the society on 4 January 1837. *CP,* 1:41–43.

187. Whewell, "Presidential Address [1839]," 97.

188. Whewell, "Presidential Address [1838]," 625, 631, 643–645. For Darwin's three 1837 papers delivered after the previous year's anniversary address, see *CP,* 1:44–53,

189. Whewell, "Presidential Address [1839]," 92–93. Whewell had reference to Darwin's 1838 paper (*CP,* 1:53–87).

190. CD to Whewell, 10 March 1837, in *Correspondence,* 2:10.

191. CD to Henslow, 14 October 1837, in ibid., 2:50–51.

192. *The Charter of the Geological Society of London: Instituted 1807; Incorporated 1826. with the Bye-Laws Adopted at the General Meeting on the 1st of May, 1827* (London: Richard Taylor, 1836), 21–22, with abbreviation.

193. See Patrick J. Boylan, "Lyell and the Dilemma of Quaternary Glaciation," in Blundell and Scott, *Lyell,* 154. Boylan has indicated how the council, which held glacial theory in disfavor, shaped the printed record in regard to proglacial views presented by Agassiz, Buckland, and Lyell at the November 1840 general meeting of the society. While Darwin was one of the secretaries at the time of the meeting, his term ended in February 1841. Therefore, he would have been out of office later in 1841 when the council considered Agassiz's and Buckland's papers for publication in the *Transactions.* Boylan notes (p. 156), "The Council clearly was un-

willing to publish anything supporting the glacial theory, and yet was reluctant to take the final step of rejecting [the proglacial papers]." Buckland formally withdrew his papers on 28 June 1842.

194. CD to Henslow, 14 October 1837, in *Correspondence,* 2:51.

195. Thackray, "Archives of the Geological Society," 4, where the numerical spans listed are narrower. Thackray's later work in the archives turned up more reports.

196. Titles are from Referee's Reports, Geological Society Archives. Buckland's report was dated 9 March 1838; Sedgwick's first report (volcanic phenomena) was undated, his second report (on raised beaches) was dated 10 July 1837.

197. Thackray, "Archives of the Geological Society," 4. *CP,* 1:49–53.

198. Referee Report 47, 9 March 1838, Geological Society of London Archives, reproduced in *Correspondence,* 2:76.

199. Adam Sedgwick, Referee Report 48 (not dated), Geological Society of London Archives, reproduced in ibid., 2:88. The date of the report was after 15 May 1838, when CD wrote to William Lonsdale at the society regarding the paper. Darwin knew Sedgwick was to review it. (Ibid., 2:87–88.) See *CP,* 1:53–86, for the version published in the *Transactions.*

200. Boylan, "Lyell and the Dilemma of Quaternary Glaciation," 154.

201. William Fitton, Referee's Report 172, 4 March 1837; Charles Lyell, Referee's Report 154 [yes], Geological Society of London Archives.

202. See Referees Reports 1829–1842, numbers 8, 35, 38, 65, and 216. (Another report, 86, of a paper by Horner, is of uncertain provenance, and possibly by Darwin, according to the archivist's list. I shall not treat it.) Geological Society of London Archives. The reports are reproduced in *Correspondence,* 2:13, 52–53, 102–103, 251, and 175. See also p. 335 for three summary "no's" from a report of 1842.

203. Charles Darwin, Referee's Report 65, 20 October 1837, Geological Society of London Archives, reproduced in ibid., 2:52–53. The paper was Johann Georg Forchhammer, "On Some Changes of Level Which Have Taken Place during the Historical Period in Denmark," *Proceedings of the Geological Society of London* 2 (November 1833–June 1838):554–556. The paper was published, with the diagram recommended by Darwin, as "On Some Changes of Level Which Have Taken Place in Denmark During the Present Period," *Transactions of the Geological Society of London,* 2d ser., 6 (1842):157–160. Darwin had met Forchhammer at the Horners. See *Correspondence,* 2:36.

204. Charles Darwin, Referee's Report 8, 7 September 1838, Geological Society of London Archives, cited in *Correspondence,* 2:102–103. The paper was Robert Alfred Cloyne Austen, "Account of the Raised Beach Near Hope's Nose, in Devonshire, and Other Recent Disturbances in That Neighbourhood," *Proceedings of the Geological Society of London* 2 (November 1833–June 1838):102–103. In the version of Austen's paper published in the *Proceedings,* no mention is made of coral reefs. Presumably in the longer version of the paper Darwin read, some use of coral reefs was made that contradicted his own notions of coral reef formation. (Austen, incidentally, was later known as Godwin-Austen.)

Chapter 3. Specimens

1. CD to Caroline Darwin, 9–12 August 1834, in *Correspondence,* 1:404.

2. *JR,* 407.

3. DAR 33:249^{v}.

4. The geological specimen notebooks, property of the Sedgwick Museum, are presently on deposit as DAR 236 in the Darwin Archive, Cambridge University Library.

5. *JR,* 598–599.

6. *GSA,* 2.

7. William Henry Fitton, "Instructions for Collecting Geological Specimens," in Phillip Parker King, *Narrative of a Survey of the Intertropical and Western Coasts of Australia . . . 1818–1822,* 2 vols. (London: John Murray, 1827), 2:623–624. I thank Graham Chinner for this ref-

erence. These ideas are developed in Sandra Herbert, "Doing and Knowing: Charles Darwin and Other Travellers," in Patrick Wyse Jackson, ed., *Geological Travellers* (New York: Pober Publications, in press).

8. William Buckland, "Instructions for Conducting Geological Investigations, and Collecting Specimens," *American Journal of Science and Arts* 3 (1821):251.

9. Alexandre Brongniart, "Notice Concerning the Method of Collecting, Labelling, and Transmitting Specimens of Fossil Organized Bodies, and of the Accompanying Rocks," *American Journal of Science and Arts* 1 (1819):73. In 1845, writing for the Geological Survey, De la Beche recommended collecting cubes with six-and-a-half-inch sides of marble and other ornamental rock suitable for building purposes. British Geological Survey Archive GSM1/4:52.

10. John MacCulloch, "On the Forms of Mineralogical Hammers," *Quarterly Journal of Science, Literature, and the Arts* 11 (1821):6. FitzRoy, *Narrative,* 2:56, described the use of a geological hammer to hurl at the tame birds on St. Paul's Rocks, despite a concern over breaking the handle, the heavier the hammer the greater the concern.

11. Advertisement, with descriptions, as it appeared in Conybeare and Phillips, *Outlines of the Geology of England and Wales.*

12. The "platypus" hammer is treated in J. E. Taylor, ed., *Notes on Collecting and Preserving Natural-History Objects* (London: Hardwicke and Bogue, 1876), 7–8, and discussed in Anne Larsen, "Equipment for the Field," in Nicholas Jardine, James A. Secord, and Emma Spary, eds., *Cultures of Natural History* (Cambridge: Cambridge University Press, 1996), 364–366.

13. CD, "Geology," in CP, 1:229. Darwin's own geological hammer, shown in Figure 3.5, was presented to Down House by Charles Darwin's grandson, Professor Charles Galton Darwin (1887–1962) when the house opened as a Museum in 1929. It is recorded in the first catalog of the Museum as Darwin's geological hammer and is very likely a hammer that saw use during the Beagle voyage.

14. Quoted from a recollection, written after CD's death, by a school mate at Shrewsbury. *Correspondence,* 1:112 n. 5.

15. Janet Browne, *Charles Darwin: The Power of Place* (New York: Alfred A. Knopf, 2002), 469.

16. Sally Newcomb, "Contributions of British Experimentalists to the Discipline of Geology: 1780–1820," *Proceedings of the American Philosophical Society* 134 (1990):161–225. On cultural aspects, see Jan Golinski, *Science as Public Culture; Chemistry and Enlightenment in Britain, 1760–1820* (Cambridge: Cambridge University Press, 1992), 236–287. The popular, rather than strictly academic, nature of British chemistry can also be inferred from the titles of some contemporary books on the subject, as, for example, John Griffin, *Chemical Recreations: A Compendium of Experimental Chemistry,* 8th ed. (Glasgow: R. Griffin, 1838). This same John Griffin served as the English translator for the straightforwardly academic work coming out of Berlin: Henry Rose, *A Manual of Analytical Chemistry* (London: Thomas Tegg, 1831).

17. Dolan, "Governing Matters," 80–85 on Mawe and the "mineralogical tourist," and 147 and 158–190 on the differences between Davy and Clarke.

18. *Autobiography,* 45–46.

19. The dates of the relevant letters from Erasmus to his brother are 25 October 1822, 14 November 1822, 8 December 1822, 5 March 1823, 18 May 1823, 10 January 1825, 17 January 1825, and 24 January 1825. *Correspondence,* 2:1–18. Unfortunately Charles's replies to the letters have not been located, and may not be extant. There is also a year-and-a-half gap from mid-1823 to early 1825 in Erasmus's side of the correspondence. For background on Charles's chemistry, see Browne, *Charles Darwin: Voyaging,* 28–32.

20. Catherine to CD, ca. June 1823, in *Correspondence,* 1:8, "How snug the Laboratory will be in Winter!!"

21. Erasmus to CD, 14 November 1822, in ibid., 1:3 ("I will then have the remainder of ye cow sent up to me, and I will spend it on chemical instruments.—"). Also, 1:5, 8 December 1822, "If the cow is not utterly consumed the next milking . . ." In 1822 Charles was thirteen, his brother Erasmus eighteen.

22. Erasmus to CD, 25 October 1825, in ibid., 1:2.

23. Among the works are William Thomas Brande, *A Manual of Chemistry,* 2d ed., 3 vols. (London: John Murray, 1821); William Henry, *The Elements of Experimental Chemistry,* 9th ed., 2 vols. (London: Baldwin, 1823); Samuel Parkes, *The Chemical Catechism,* 10th ed. (London: Baldwin, 1822); and Andrew Ure, *A Dictionary of Chemistry,* 2d ed. (London: Thomas Tegg, 1823). Editions cited are those current in the early 1820s; Ure's work was part of Darwin's library in later life and presumably was used by him as a young man.

24. Erasmus to CD, 24 January 1825, in *Correspondence,* 1:13.

25. Ibid., 1:474, 209, 489.

26. *Notebooks,* 58 (RN:115). For another instance, see *Notebooks,* 124 (A:119).

27. Phillips, *Mineralogy,* xxxvi–xxxvii, lvi. In his specimen notes on Tasmania, Darwin noted of specimen 3474, "White. compact. uneven fracture. Aluminous stone. strong smell of do [ditto]:"

28. Henry T. M. Witham, *The Internal Structure of Fossil Vegetables Found in the Carboniferous and Oolitic Deposits of Great Britain* (Edinburgh: Adam and Charles Black, 1833), 76; Albert Johannsen, *Manual of Petrographic Methods* (New York: McGraw-Hill, 1918), 572–574. On Sorby's technique, see D. W. Humphries, *The Preparation of Thin Sections of Rocks, Minerals, and Ceramics* (Oxford: Oxford University Press, 1992), 1–5. I am indebted to Graham Chinner for these references.

29. Erasmus to CD, 14 November 1822, in *Correspondence,* 1:3.

30. Herschel, *Preliminary Discourse,* 219. On the two types of goniometer, see Ure, *Dictionary of Chemistry,* 380–381. Also see Steven C. Turner, "Goniometer," in Robert Bud and Deborah Jean Warner, eds., *Instruments of Science* (New York: Garland, 1998), 290–292. On the reflective goniometer, see D. C. Goodman, "William Hyde Wollaston," *Dictionary of Scientific Biography* 14:490–491. It was William Phillips who used the reflective goniometer systematically. See his *Mineralogy.* For instructions on the use of both types of goniometer, see Henry James Brooke, *A Familiar Introduction to Crystallography; Including an Explanation of the Principle and Use of the Goniometer* (London: W. Phillips, 1823), 25–32. These pages were also published separately as an extract, with three added paragraphs of advice on use, under the title "A Description of the Principles and Method of Using the Common and Reflective Goniometers." The extract also bore the notation "These instruments are made and sold by W. Cary, No. 182, Strand."

31. CD to Henslow, 12 November 1833, in *Correspondence,* 1:352.

32. J. S. Henslow to CD, 22 July 1834, in ibid., 1:395. "Phillips's book"—the *Mineralogy* of 1823—did in fact contain measurements of crystal planes taken with Wollaston's reflective goniometer.

33. On the difficulties of the using the reflecting goniometer, and an important survey of goniometers generally, with excellent color photographs, see Ulrich Burchard, "History of the Development of the Crystallographic Goniometer," *Mineralogical Record* 29 (1998):532.

34. First Geological Specimen Notebook, for specimen 320, observation in pencil and hence presumably later in date than the original entries. Darwin listed two angular measurements for fragments of hornblende: 52°50′ and 125°. Graham Chinner remarked of Darwin's measurement that it suggests use of the contact goniometer, since the values for hornblende using the reflective goniometer ought to be 55°30′ and 124°30′ (Phillips, *Mineralogy,* 63). (A current reference work lists an angle of 55°49′.) Darwin's measurement was sufficient to differentiate hornblende from the very similar pyroxene (87° and 93°), though that could have been done by eye. The hornblende would have been in a rock, and Darwin would have broken off cleavage fragments to measure for it.

35. Dolomite must be powdered to react. Jameson referred to it as "feebly" effervescing, as compared to the "violent" effervescence when acid was poured on granulated foliated limestone. Jameson, *Manual of Mineralogy,* 2:265.

36. John Mawe, "List of Useful Articles," in *Familiar Lessons on Mineralogy and Geology* (London: Longman, 1829), facing p. 116.

37. John Griffin, *A Practical Treatise on the Use of the Blowpipe, in Chemical and Mineral Analysis* (Glasgow: Richard Griffin, 1827), 11. On the history of the blowpipe, including numerous photographs of mineralogical kits for travelers (Berzelius's own leather rollup, ca. 1821, and a pocket mineralogical compendium made by Cary, London, ca. 1827, being pictured), see Ulrich Burchard, "The History and Apparatus of Blowpipe Analysis," *Mineralogical Record* 25, no. 4 (1994):251–277. Also see Burchard, "Blowpipe," in Bud and Warner, *Instruments of Science,* 68–69; and William B. Jensen, "The Development of Blowpipe Analysis," in John T. Stock and Mary Orna, eds., *The History and Preservation of Chemical Instrumentation* (Dordrecht: D. Reidel, 1986), 123–149.

38. Jakob J. Berzelius, *The Use of the Blowpipe in Chemical Analysis, and in the Examination of Minerals,* trans. J. G. Children (London: Baldwin, 1822), 1. This work was current at the time Darwin departed on the *Beagle* but seems not to have been part of the ship's library. The reference to Berzelius on DAR 37.2:794^{v} is written in pencil and is clearly postvoyage in date.

39. Mawe, *Familiar Lessons on Mineralogy and Geology,* 4.

40. Berzelius, *The Use of the Blowpipe,* xxvi.

41. David R. Oldroyd, "Edward Daniel Clarke, 1769–1822, and His Rôle in the History of the Blowpipe," *Annals of Science* 29 (October 1972):214–234.

42. Phillips, *Mineralogy,* the then current edition being the third of 1823; or Jameson, *Manual of Mineralogy* (1821).

43. DAR 32.1:9. For discussion on the use of borax as flux, see Berzelius, *Use of the Blowpipe* (1822), 54–56; Ure, *Dictionary of Chemistry,* 238; and Burchard, "Blowpipe Analysis," 267. For treatment of the history of mineral identification in relation to chemical analysis, see David R. Oldroyd, "A Note on the Status of A. F. Cronstedt's Simple Earths and His Analytical Methods," *Isis* 65 (1974):506–512.

44. Berzelius, *Blowpipe,* 28: "*Oxidation* ensues when we heat the subject under trial before the extreme point of the flame, where all the combustible particles are soon saturated with oxygen. . . . For *reduction,* a fine beak must be employed, and it must not be inserted too far into the flame of the lamp; by this means we obtain a more brilliant flame, the result of an imperfect combustion, whose particles, as yet unconsumed, carry off the oxygen from the subject of experiment."

45. Berzelius, *Blowpipe,* table facing p. 118, which lists a number of oxides producing yellow tints in fusions with borax under specified conditions. Graham Chinner comments on Darwin's remark that it would help to know what sort of brightness the coating of rock had: if metallic, it could have been specular hematite; if brightly colored, possibly a uranium mineral (Chinner, personal communication). For finer discriminations, see Alfred J. Moses and Charles Lathrop Parsons, *Elements of Mineralogy, Crystallography and Blowpipe Analysis* (New York: Van Nostrand, 1916), 179, and George J. Brush, *Manual of Determinative Mineralogy,* rev. Samuel L. Penfield (New York: John Wiley, 1911), 148–149.

46. For a map of the area, see *Baker's Map of the University and Town of Cambridge 1830* (Cambridge: Cambridgeshire Records Society, 1998).

47. Catherine Darwin to CD, 30 October 1835, in *Correspondence,* 1:468.

48. Ibid., 1:149, 176, 405, 512.

49. John C. Thackray, "Mineral and Fossil Collections," in Arthur MacGregor, ed., *Sir Hans Sloane: Collector, Scientist, Antiquary* (London: British Museum Press, 1994), 132.

50. University of Cambridge Archive, UP 7, fol. 184 (a). Henslow remained interested in museums, shifting his attention to local institutions after he no longer lived in Cambridge full-time. See his circular "On Typical Objects in Natural History," *Report of the British Association for the Advancement of Science* [Glasgow 1855] (1856):108–110.

51. University of Cambridge Archive, CUR 55.1. The document is inscribed "The Revd J. Croft with Prof. Whewell's Comple." The transcription is taken from Robert Willis and John Willis Clark, *The Architectural History of the University of Cambridge,* 4 vols. (Cambridge: Cambridge University Press, 1886 [vols. 1–3 reprinted 1988]), 3:100.

52. Searby, *History of the University of Cambridge,* 24–28, 206–209.

53. In a postcard to T. McKenny Hughes postmarked 22 January 1897, George Howard Darwin (1845–1912) reported sending two packing cases of specimens to the geological museum at Cambridge. A reply dated 24 January 1897 noted that when the new museum [the Sedgwick] was opened, "[Y]ou will see some good old historical collections among which that made by your father will have an honourable place[.]—" Both documents are presently kept in DAR 236.

54. Duncan M. Porter, "The *Beagle* Collector and His Collections," in David Kohn, ed., *The Darwinian Heritage* (Princeton: Princeton University Press, 1985), 973–1019. Porter described the following geological contributions on Darwin's specimens. On fossil vertebrates there was a brief notice by William Buckland in *Geology and Mineralogy, Considered with Reference to Natural Theology,* 2d ed., 2 vols. in 1 (London: Pickering, 1837), 603, as well as the primary work done by Richard Owen that culminated in the volume *Fossil Mammalia* resulting from the *Beagle* voyage. On fossil invertebrates there was George Brettingham Sowerby's work on the fossil shells from the Cape Verde Islands, St. Helena, Tasmania, and South America as it appeared in *VI,* 153–160, and *GSA,* 249–264; Edward Forbes's "Descriptions of Secondary Fossils Shells from South America," in *GSA,* 265–268; William Lonsdale on six fossil corals from Tasmania in *GSA,* 161–169; and John Morris and Daniel Sharpe, "Description of Eight Species of Brachiopodus Shells from the Palaeozoic Rocks of the Falkland Islands," *Quarterly Journal of the Geological Society of London* 2 (1846):274–278. Christian Gottfried Ehrenberg described volcanic dust and ash tuff from Darwin's collections (Porter, p. 995). The botanist Robert Brown also supplied information on Darwin's silicified wood (*GSA,* 225). Darwin himself also credited Alcide d'Orbigny with identifying numerous shells (*GSA,* iv).

55. Thackray in Moore, Thackray, and Morgan, "Short History of the Museum of the Geological Society of London," 56–58.

56. On the "passion to collect," see Ruth Lord, *Henry F. du Pont and Winterthur: A Daughter's Portrait* (New Haven: Yale University Press, 1999), 185: "Collecting implies control: it provides an island of orderliness, a bulwark of reassuring, tangible ownership and self-regulated fulfillment in the midst of uncertainty." Along parallel lines, see Hooker to CD, 27 or 28 December 1862, in *Correspondence,* 10:629: "there is a much more catholic view of specimens . . . which includes a love of all [the] artistic pleasure they afford, the pleasure of classyfying by eye, & not by knowledge, the historic interest attached to them, & the pleasure of being successful in obtaining, as well as the love of *possessing,* without which much of all the rest is nothing."

57. CD to J. S. Henslow, 6 October 1836, in ibid., 1:507.

58. *GSA,* vi.

59. The Woodwardian Museum was located in what are presently the University Central Offices, next to the Senate House. Information on the Department is from Graham Chinner.

60. Alfred Harker, "Notes on the Rocks of the 'Beagle' Collection.—I.," *Geological Magazine,* n.s. 4 (March 1907):101, which contains all the quotations from Harker used in this paragraph.

61. For example, Challinor listed a first use of "andesite" (after the Andes mountains) by von Buch in 1835, and a first use for "rhyolite" by Richthofen in 1860. John Challinor, *A Dictionary of Geology,* 5th ed. (Cardiff: University of Wales Press, 1978), 10, 260.

62. Ibid., 142.

63. Maxwell R. Banks, "A Darwin Manuscript on Hobart Town," *Papers and Proceedings of the Royal Society of Tasmania* 105 (1971):13. Darwin's Tasmanian specimens numbered 3445, 3451–3454, 3456, and 3472 that he listed as greenstones, Banks reassigned to the category dolerites. For careful discussion of Darwin's experience of Hobart and its environs, see F. W. and J. M. Nicholas, *Charles Darwin in Australia* (Cambridge: Cambridge University Press, 1989), 83–104, and Maxwell R. Banks and David Leaman, "Charles Darwin's Field Notes on the Geology of Hobart Town—A Modern Appraisal," *Papers and Proceedings of the Royal Society of Tasmania* 133, no. 1 (1999):29–50.

64. CD to Henslow, 30–31 October 1836, in *Correspondence,* 1:513.

65. Constance Richardson, "Petrology of the Galapagos Islands," *Bernice B. Bishop Museum Bulletin* 110 (1933):45–67. In 1931 Constance Richardson was president of the student Sedgwick Club, and is pictured in the 1932 photograph of the club as well. Presumably her work on Darwin's specimens was done at Cambridge in 1932. In the 1996 list for the university she is listed as Sister Mary Constance. I am indebted to Graham Chinner and to Elisabeth Leedham-Green for this information. This article forms the supplement to Lawrence John Chubb, "Geology of Galapagos, Cocos, and Easter Islands," *Bernice P. Bishop Museum Bulletin* 110 (1933):1–44.

66. Cecil Edgar Tilley, "The Dunite-Mylonites of St. Paul's Rocks (Atlantic)," *American Journal of Science* 245 (1947):483–491. Tilley referred to the specimens without mentioning their numbers.

67. As of the date of this writing, the exhibit by G. A. Chinner is in the Petrology Department at the Sedgwick Museum in Cambridge; the exhibit by David Norman, prepared for a conference held in Santiago, Chile, in 1993, is on the first floor of the museum.

68. Paul N. Pearson, "Charles Darwin on the Origin and Diversity of Igneous Rocks," *Earth Sciences History* 15 (1996):49–67. Also of importance is John W. Judd's "Critical Introduction" to the 1890 reprint of *VI.* For a new, and indispensable treatment of the history of igneous petrology, building on the work of Hatten S. Yoder Jr. and others, see Davis A. Young, *Mind over Magma: The Story of Igneous Petrology* (Princeton: Princeton University Press, 2003).

69. Jameson, *A System of Mineralogy,* 3 vols. (Edinburgh: Archibald Constable, 1820), 2:172–174. See p. 172 on diallage: "Its colours are grass-green, which sometimes inclines to emerald-green, or to mountain-green." On p. 174 Jameson explained that the name "diallage," meaning "difference," derived from this mineral's single distinct cleavage, whereas other minerals with which it was confused had two distinct cleavages. The term "serpentine," which came in gradually during the sixteenth and seventeenth centuries, described "a greenish or reddish mineral magnesium silicate with hydroxyl (OH)" whose name derived "from the markings and colour suggesting a serpent's skin." Challinor, *Dictionary of Geology,* 276.

70. *JR,* 8.

71. DAR 32.1:25.

72. *VI,* 32.

73. Ibid., 125.

74. Ibid., 126.

75. Tilley, "Dunite-Mylonites of St. Paul's Rocks (Atlantic)," 483.

76. Ibid., 483. On the expeditions, see Margaret Deacon, *Scientists and the Sea 1650–1900* (Aldershot, England: Ashgate, 1997), 333–406, and Susan Schlee, *The Edge of an Unfamiliar World: A History of Oceanography* (New York: Dutton, 1973), 107–138.

77. Tilley, "Dunite-Mylonites of St. Paul's Rocks (Atlantic)," 491. For some history of interest in St. Paul's Rocks after Darwin, as well as his own analysis, see Henry S. Washington, "The Petrology of St. Paul's Rocks (Atlantic)," in *Report on the Geological Collections Made during the Voyage of the "Quest" on the Shackleton-Rowett Expedition to the South Atlantic & Weddell Sea in 1921–1922* (London: British Museum [Natural History], 1930), 126–144. Washington (p. 134) noted that the *Challenger* visited St. Paul's Rocks in August 1873. He also noted that John Murray, then editor of the *Challenger* expedition reports, arranged for a special study of St. Paul's Rocks.

78. Witham, *Internal Structure of Fossil Vegetables.* Information on Witham's technique of drawing is from Graham Chinner.

79. Lorraine Daston and Peter Galison, "The Image of Objectivity," *Representations* 40 (1992):98. Geological thin sections are not discussed but fit the argument.

80. CD to John Lubbock, [1881], in Francis Darwin, ed., *The Life and Letters of Charles Darwin,* 3 vols. (London: John Murray, 1887), 3:250.

81. Challinor, *Dictionary of Geology,* 199; use of term dated to 1880s.

82. Sergei Ivanovich Tomkeieff, *Dictionary of Petrology* (Chichester: John Wiley, 1983), 427.

83. Pearson, "Charles Darwin on the Origin and Diversity of Igneous Rocks," 56. Humboldt had come to this conclusion since basalt was often found lying over trachyte, as at Auvergne. Also see Daubeny, *Description of Active and Extinct Volcanos,* 379–429, regarding a temporal ordering for igneous productions.

84. Scrope, *Considerations on Volcanos,* 147–148.

85. DAR 37.2:786.

86. Pearson, "Charles Darwin on the Origin and Diversity of Igneous Rocks," 56. Scrope, *Considerations on Volcanos,* 86: "I am of the opinion that it would be found useful to institute a third and intermediate class of rocks, which should include those compounds in which the proportions of felspar and the ferruginous minerals are so nearly balanced that the general aspect of the rock is neither entirely felspathic nor ferruginous. . . . These rocks might be called *greystone.*"

87. DAR 37.2:770. Original reference in Pearson, "Charles Darwin on the Origin and Diversity of Igneous Rocks," 57, without the specimen numbers being listed.

88. DAR 37.2:772. Pearson, "Charles Darwin on the Origin and Diversity of Igneous Rocks," 58, noted that Darwin did not collect at the key site. However, he did note that the key site was represented by specimens like those already collected, namely, numbers 3267 and 3268.

89. I wish to thank Graham Chinner for arranging access to the collection, Stephen Laurie for locating these specimens, and Dudley Simon for photographing them. Specimen 3268 in the upper left corner of the photograph does not bear a green label. The label numbers it does bear (6204 and 6866) are recorded for specimen 3268 in the fair copy of Darwin's catalogue held at the Sedgwick Museum.

90. Richardson, "Petrology of the Galapagos Islands," 53.

91. Ibid., 48.

92. Ibid.,

93. Graham Chinner, who has gathered together the thin sections from the Darwin collection, kindly provided them for photography.

94. Alexander R. McBirney and Howel Williams, *Geology and Petrology of the Galapagos Islands* (Boulder, CO: Geological Society of America, 1969), 54.

95. Gregory Estes, K. Thalia Grant, and Peter R. Grant, "Darwin in Galápagos: His Footsteps through the Archipelago," *Notes and Records of the Royal Society of London* 54 (2000):367 n. 181.

96. *VI,* 117.

97. For a photograph of Albemarle Island specimen 3248 that contains large feldspar phenocrysts, easily visible to the naked eye, see Pearson, "Charles Darwin on the Origin and Diversity of Igneous Rocks," 57.

98. *Notebooks,* 63; Herbert, ed., *Red Notebook,* 63 n. 159; plates in *Zoology, Part III: Birds,* 60–64. Note Plate 16 (*Mimus trifasciatus*), Plate 17 (*Mimus melanotis*), and Plate 18 (*Mimus parvulus*).

99. Scrope, *Considerations on Volcanos,* 238–239.

100. Ibid., 145.

101. Ibid., 146.

102. Ibid., 146. Quoted in Pearson, "Charles Darwin on the Origin and Diversity of Igneous Rocks," 57.

103. Scrope, *Considerations on Volcanos,* 222.

104. *VI,* 118.

105. CD to Lyell, 18 September 1849, in *Correspondence,* 4:252.

106. Young, *Mind over Magma,* 126–129.

107. For treatment of the abundance of Darwin's speculations on the origin of rocks, see *Notebooks,* 18, 83–84.

108. Leopold von Buch, *Description physique des îles Canaries, suivie d'une indication des principaux volcans du globe.* With atlas. (Paris: F. G. Levrault, 1836). Darwin's reading notes appear in RN:137, 150, and A:39–40, 42 of *Notebooks.*

109. Pearson, "Charles Darwin on the Origin and Diversity of Igneous Rocks," 58.

110. Ibid., 58.

111. Ibid., 59. The original reference is to Hugh L. Pattinson, "On a New Process for the Extraction of Silver from Lead," *Report of the British Association for the Advancement of Science* (1839):50–55. See p. 54: "It only remains to consider, how it happens that lead in the act of consolidation gives up a portion of its silver to the surrounding and still fluid lead; and the most simple view of the matter is . . . that it is an instance of true crystallization, in which the homogeneous particles of lead are drawn together by virtue of their molecular attraction, to the exclusion of the foreign body, silver."

112. *VI,* 120.

113. Ibid., 119–120. Pearson ("Charles Darwin on the Origin and Diversity of Igneous Rocks," 59) pointed out that for the James Island anomaly, Darwin suggested the feldspar crystals sank, rather than rose, by proposing that "many bubbles of gas enveloped in the lower part of that flow might have lowered the specific gravity of the melt sufficiently for the feldspar to sink."

114. Alfred Harker, *The Natural History of Igneous Rocks* (London: Methuen, 1909), 310, 321. Also see Joseph Iddings, "The Origin of Igneous Rocks," *Bulletin of the Philosophical Society of Washington* 12 (1892–1894):91–128.

115. Harker, *Natural History of Igneous Rocks,* 310. For an elegantly brief history of the ideas concerning rocks and their formation, see Oldroyd, *Thinking about the Earth,* 192–223.

116. Pearson, "Charles Darwin on the Origin and Diversity of Igneous Rocks," 64.

117. Norman L. Bowen, *The Evolution of the Igneous Rocks* (Princeton: Princeton University Press, 1928). See Davis A. Young, *N.L. Bowen and Crystallization-Differentiation: The Evolution of a Theory* (Washington, D.C.: Mineralogical Society of America, 1998), 11–12: "Both of Darwin's suggested mechanisms (crystal settling and filter pressing) . . . would play a major role in Bowen's theory. . . . It should be noted that Darwin never spoke of 'differentiation' or 'filter pressing.' He did, however, speak of 'sinking,' 'rising,' and 'separation' of crystals." Filter pressing refers to Darwin's speculation that "masses of liquid could be squeezed into fractures from a mush of already extensively crystallized liquid, thus resulting in formation of an igneous body whose composition would differ from that of the original body of liquid" (p. 11). On the slow adoption of a laboratory-centered petrology, see Carl-Henry Geschwind, "Becoming Interested in Experiments: American Igneous Petrologists and the Geophysical Laboratory, 1905–1965," *Earth Sciences History* 14 (1995):47–61.

118. In his notes Darwin cited Hall and Playfair regarding the junction (DAR 38.2:906): Basil Hall and John Playfair, "Account of the Structure of the Table Mountain, and Other Parts of the Peninsula of the Cape," *Transactions of the Royal Society of Edinburgh* 7 (1815):269–278 + plates 13–15. See p. 278 ("granite does not derive its origin from aqueous deposition") for Playfair's reading of the site as an "*instantia crucis,* with respect to the two theories [Wernerian and Huttonian] concerning the formation of rocks." Darwin was clearly a Huttonian.

119. In present-day terms Graham Chinner identified these as banded cordierite hornfels, that is, a rock backed and altered by an igneous intrusion. The two specimens were part of his 1991 exhibit of Darwin specimens.

120. DAR 236.

121. DAR 38.2:908–909.

122. *Notebooks,* 114 (A:94).

123. *VI,* 150.

124. *GSA,* 165.

125. Ibid., 165.

126. Ibid., 168. Darwin's originality on this point was strongly noted by the geologist John W. Judd (1840–1916) in his "Critical Introduction" to *GSA* in the 1890 Ward, Lock, and

Company combined edition of the three parts of the geology of the voyage of the *Beagle* (Freeman item 279). Judd's introduction has been reprinted in Paul H. Barrett and R. B. Freeman, *The Works of Charles Darwin,* vol. 9 (New York: New York University Press, 1987), 7–14.

Chapter 4. The Romantic Thread

1. Roy Porter, "Gentlemen and Geology: The Emergence of a Scientific Career, 1660–1920," *Historical Journal* 21 (1978):821.
2. Nora Barlow, ed., *Charles Darwin's Diary of the Voyage of H.M.S. "Beagle"* (Cambridge: Cambridge University Press, 1933); the current edition (1988) is cited as *Diary.*
3. CD to Caroline Darwin, 25–26 April 1832, in *Correspondence,* 1:226.
4. *JR,* 29; a reworking from *Diary,* 64. Darwin's diary, which he usually referred to as his journal, has been published under the title of *Diary.* The use of the term "diary" was to avoid confusion with the published *Journal of Researches.*
5. *Correspondence,* 1:525.
6. Ibid., 2:431.
7. J. M. Herbert to CD, 1[—4] December 1832, in ibid., 1:287.
8. CD to J. M. Herbert, 2 June 1833, in ibid., 1:320. Darwin had used the full metaphor shortly before: "You expect sadly more than I shall ever do in Nat: Hist:—I am only a sort of Jackall, a lions provider; but I wish I was sure there were lions enough.—" CD to W. D. Fox, 23 May 1833, in ibid., 1:316.
9. Ibid., 2:11, 21.
10. James Paradis, "Darwin and Landscape," in James Paradis and Thomas Postlewait, eds., *Victorian Science and Victorian Values: Literary Perspectives* (New York: New York Academy of Sciences, 1981), 85. A recent work also emphasizing Darwin's indebtedness to the Romantic movement is Robert J. Richards, *The Romantic Conception of Life: Science and Philosophy in the Age of Goethe* (Chicago: University of Chicago Press, 2002), 514–554.
11. *JR,* 569, 567.
12. *JR,* 212.
13. Catherine Darwin to CD, 27 November 1833, in *Correspondence,* 1:356.
14. Caroline Darwin to CD, 28 October 1833, in ibid., 1:345.
15. *JR,* 455, 523.
16. David Kohn, "Darwin's Ambiguity: The Secularization of Biological Meaning," *British Journal for the History of Science* 22 (1989):226.
17. CD to John Murray, 21 September 1861, in *Correspondence,* 9:273.
18. Browne, *Charles Darwin: Power of Place,* 346–350.
19. CD to C. T. Whitley, 8 May 1838, in *Correspondence,* 7:469.
20. *JR,* 604.
21. Ibid., 11.
22. Ibid., 604–605; reworked from *Diary,* 444.
23. Erasmus Darwin, *The Temple of Nature; or the Origin of Society* (London: J. Johnson, 1803).
24. Quoted in Duncan M. Porter, "Darwin's Missing Notebooks Come to Light," *Nature* 291 (7 May 1981):13.
25. *JR,* 233; reworked from *Diary,* 125–127, 135.
26. *Diary,* 317, n. 1.
27. *JR,* 406, and 406–407 for the subsequent quotation.
28. All quotations in this paragraph from Paradis, "Darwin and Landscape," 86.
29. *Autobiography,* 43–44, 85.
30. Alan Frost, "Captain James Cook and the Early Romantic Imagination," in W. Veit, ed., *Captain James Cook: Image and Impact* (Melbourne: Hawthorn, 1972), 90–106.
31. J. G. Malcolmson to CD, 31 August 1839, in *Correspondence,* 2:215.
32. Stephen Gill, *Wordsworth and the Victorians* (Oxford: Clarendon Press, 1998), 16–18.

Darwin's six-volume collection of Wordsworth's poetry (the 1840 edition) is extant but privately owned, and bears numerous marginal scorings and annotations (p. 4 n. 8).

33. William Wordsworth, *The Excursion* (Oxford: Woodstock Books, 1991), 163.

34. Quoted in Paradis, "Darwin and Landscape," 97.

35. *Notebooks,* 529 (M:40).

36. Quoted in *Notebooks,* 529 n. 40–41, quotation expanded here.

37. For reproductions of Martens's work, see Richard Darwin Keynes, ed., *The Beagle Record: Selections from the Original Pictorial Records and Written Accounts of the Voyages of H.M.S. Beagle* (Cambridge: Cambridge University Press, 1979), and Richard Keynes, *Fossils, Finches and Fuegians: Charles Darwin's Adventures and Discoveries on the Beagle, 1832–1836* (London: HarperCollins, 2002). The two Martens sketchbooks held by Cambridge University Library can also be viewed at www.lib.cam.ac.uk/digitallibrary.htm. Also see Elizabeth Ellis, *Conrad Martens: Life and Art* (Sydney: State Library of New South Wales Press, 1994). On the general subject of art and science during the period, see Smith, *Imagining the Pacific,* 1–39. On the narrower subject of geologically informed art, a fine recent book is Rebecca Bedell, *The Anatomy of Nature: Geology and American Landscape Painting, 1825–1875* (Princeton: Princeton University Press, 2001).

38. Ellis, *Conrad Martens,* 109, n. 4.

39. Ibid., 120.

40. For reproductions of the work of both Fielding and Turner, as well as essays describing the tradition of British landscape painting, see Andrew Wilton and Anne Lyles, *The Great Age of British Watercolours, 1750–1880* (Munich: Prestel, 1993).

41. Lecture passages quoted from Ellis, *Conrad Martens,* 76.

42. CD to Susan Darwin, 28 January 1836, in *Correspondence,* 1:483. The painting at Tierra del Fuego, developed from a sketch made 5 March 1834, was sold by Martens to CD for 3 guineas on 17 January 1836. The painting of the banks of the Santa Cruz River, developed from a sketch made 3 May 1834, was sold by Martens to CD for 3 guineas on 21 January 1836. Keynes, *Beagle Record,* 395, 397.

43. *JR,* 604.

44. Ibid., 606.

45. Ibid., 539–569.

46. Ibid., 356.

47. Ibid., 368–369; reworked from *Diary,* 292.

48. Ibid., 368–381; reworked from *Diary,* 292–304.

49. Keynes, *Beagle Record,* 199–213, 396–397. Some of Martens's work was reproduced in FitzRoy, *Narrative,* 2. See the engravings facing pp. 336, 348, 351, 352. Also of interest is the chart of the mouth of the river surveyed by the *Beagle* officers facing p. 339.

Chapter 5. A Prospective Author

1. For the Table of Contents, see *Autobiography,* 19. In the text itself (p. 116) Darwin used the more general phrase "My Several Publications."

2. Quoted in *Correspondence,* 3:397.

3. There were interesting breaks in this pattern. In 1844, after completing *VI* but before beginning *GSA,* Darwin allowed himself five months, from 13 February to 5 July, to sketch out in 230 pages his species theory. He also lingered between sorting his species notes in 1854 and actually beginning to write in 1856. Otherwise, the interval between the completion of one book and the beginning of another was short. Until 1861 Darwin also cleared his desk of one book before starting another. In that year he began work on his orchid book before finishing *The Variation of Animals and Plants under Domestication* (1868). (For publication data, see Richard B. Freeman, *The Works of Charles Darwin: An Annotated Bibliographical Handlist,* 2d ed. [Folkestone: Dawson, 1977].) In his later botanical works, the ending of the writing of one and the beginning of another was also not so sharply delineated as formerly, as in 1873 when

he began writing on fertilization and then turned to insectivorous plants. For Darwin's writing schedule, see Gavin de Beer, ed., "Darwin's Journal," *Bulletin of the British Museum (Natural History) Historical Series* 2, no. 1 (1959):10–11, 13–15, 19.

4. *Autobiography,* 136.

5. Notes on "Zoological walk" with William Kay to Portobello in Scotland, DAR 5:49–51.

6. Darwin, *Life and Letters of Charles Darwin,* 1:117.

7. Duncan Porter (*Nature* 291:13) has shown that the often-quoted sentence in the *Origin* (p. 489) beginning "It is interesting to contemplate an entangled bank . . ." derives from a passage in Darwin's plant notes from the voyage describing the "entangled mass" of the vegetation. "Entangled" was a common word in Darwin's vocabulary, appearing in his earliest notes from the voyage in the description of rocks (Field Notebook 1.4, Darwin Manuscripts, Down House). However, the word was especially prominent in his notes on Tierra del Fuego, being used not only to describe the intertwining of living and dead branches of the stunted beech trees, which grew along the banks of the Beagle Channel, but also the "entangled" hair of the Fuegians. See *JR,* 227, 231–33, 235. On the phrase "entangled bank" it is also possibly relevant that Darwin's childhood home in Shrewsbury sat on a bank directly above the Severn River. For further associations with the phrase going back to John Milton (1608–1674), see David Kohn, "The Aesthetic Construction of Darwin's Theory," in Alfred I. Tauber, ed., *The Elusive Synthesis: Aesthetics and Science* (Dordrecht: Kluwer, 1996), 19–33.

8. Quoted in CD, "Geology," in *CP,* 1:231.

9. *Autobiography,* 81; also see *Diary,* 437, for a recollection of the scene written in 1836, where no book is mentioned, though the sharpness of the scene is recalled.

10. Lyell, *Principles of Geology;* Leopold von Buch, *Travels through Norway and Lapland,* trans. John Black, notes Robert Jameson (London: H. Colburn, 1813); Humboldt, *Personal Narrative.*

11. Darwin recorded the debit as of February 1842 when 1337 copies had been sold. For the second edition in 1845 Darwin abandoned Henry Colburn, publisher of the first edition, from whom, as he reported bitterly, he had "never received one penny," for John Murray, who paid him £150 for the "entire Copyright of my Journal." *Correspondence,* 2:312; 3:169, 247.

12. *Correspondence,* 2:14–15.

13. While the exact amount Darwin spent on publishing his geology works is not known, there are some indications of expenditures at various points. On 18 May 1841 Charles wrote to his wife Emma, "2–300£ out of the [family] funds" might be required (*Correspondence,* 2:318). In January 1843 Darwin referred to a £60 expenditure on *CR,* without making it clear whether it was he or the publisher who had made the payment. The *Correspondence* offers no evidence on payments regarding *VI,* but see 3:307 n. 4 for information from Darwin's Account Book (Down House Manuscript Collection) recording a number of payments for illustrations for *GSA,* including those of 15 December 1845 and 5 July, 19 August, and 4 and 18 September 1846, totaling £47 13*s.* In addition, on 10 August 1847 CD paid a lump sum to Smith, Elder and Co. of £61 8s.; on 27 February 1848 he received £5 8s. 4d. from the same firm.

14. CD to W. D. Fox, 25 March 1843, in *Correspondence,* 2:352. Later in life CD did earn substantial income from his publications.

15. CD to John Maurice Herbert, 3 September 1846, in ibid., 3:338.

16. CD to Smith, Elder and Co., 6 June 1846, in ibid., 13:365.

17. Extant notebooks are listed in ibid., 1:545–546. Extracts from the notebooks are published in Nora Barlow, ed., *Charles Darwin and the Voyage of the Beagle* (London: Pilot Press, 1945).

18. Relative numbers of pages for the zoological and geological notes were first reported in Howard E. Gruber and Valmai Gruber, "The Eye of Reason: Darwin's Development during the *Beagle* Voyage," *Isis* 53 (1962):189. The main run of Darwin's geology notes is contained in DAR 32–38. A portion of these notes, as with the zoology notes in DAR 30–31, were kept on rule-lined paper bearing a C. Wilmot 1828 watermark. The paper shows signs of having been bound. In the course of the voyage Darwin's note-taking style evolved. Originally notes on the versos referred to the facing page; at DAR 32.1:61 he altered his practice to have notes on versos refer to the text on the corresponding rectos. Also, specimen numbers began by being integrated into the text; later they were moved out to the left margin.

19. All in DAR 41, the first two in the hand of Syms Covington, Darwin's servant and copyist, with Darwin's corrections added in his own hand. A holograph version of "Coral Islands," differing at points from the fair copy, is also in DAR 41 and has been published in Stoddart, "Coral Islands," 1–20.

20. The notebook labeled "Santiago Book," catalogued 1.18 at Down House, is unpublished in its entirety. It was begun about August 1834 and used as a field notebook through February 1835. Thereafter, until approximately midway in 1836, the notebook was used for more general speculations. On dating the notebook, see Frank J. Sulloway, "Further Remarks on Darwin's Spelling Habits," *Journal of the History of Biology* 16 (1983):367–376. The Red Notebook, which dates from late May 1836 to late May/mid-June 1837, has been published in *Notebooks,* as well as in a separate issue.

21. Daubeny, *Volcanos,* 264. On Humboldt and von Buch with regard to the Canary Islands, see Young, *Mind over Magma,* 51–52.

22. DAR 32.1:18. The underlining of "partial sinking" is a later pencil addition, as is marginal scoring that appears in the manuscript next to the second sentence quoted. Much later in life, in his *Autobiography,* Darwin wrote, "The geology of St. Jago is very striking yet simple: a stream of lava formerly flowed over the bed of the sea, formed of triturated recent shells and corals, which it has baked into a hard white rock. Since then the whole island has been upheaved. But the line of white rock revealed to me a new and important fact, namely that there had been afterwards subsidence round the craters, which had since been in action, and had poured forth lava." This white line is also referred to in DAR 32.15, and is visible in Martens's painting reproduced as Figure 5.1.

23. In the drawing, the alphabetical letters and the phrase "Level of Sea" are in ink; the remainder of the drawing is in pencil. The letter "K" refers to a dislocation. The rule-drawn diagonal line represents the dip. The encircled "Lava 1836" is a penciled notation Darwin made on a return visit to the island in 1836. Passages quoted in the discussion of the section drawing are from DAR 32.1:15–18 and versos with occasional reference to the first geological specimen notebook.

24. Harker, "Rocks of the 'Beagle' Collection," 102. The term "Limburgite" is defined in Tomkeieff, *Dictionary of Petrology,* 321, as a "dark-coloured rock, occurring as flows, sills or dykes, composed of phenocrysts of pyroxene and olive (hyalosiderite) in a groundmass of brown alkali-rich glass that contains a second generation of small crystals of the same minerals. Feldspar is absent."

25. *GSA,* 3.

26. Harker, "Rocks of the 'Beagle' Collection," 103.

27. *VI,* 3: "Numerous white balls appearing like pisolitic concretions, from the size of a walnut to that of an apple, are embedded in this deposit; they usually have a small pebble in their centres. Although so like concretions, a close examination convinced me that they were nulliporae, retaining their proper forms, but with their surfaces slightly abraded: these bodies (plants as they are now generally considered to be), exhibit under a microscope of ordinary power, no traces of organization in their internal structure."

28. Footnote 1 to the extended passage quoted reads, "In places. this rock has entangled & fixed portions of lower augitic rock.—"

29. *VI,* 9.

30. Harker, "Rocks of the 'Beagle' Collection," 103.

31. On Sedgwick's structural approach to formations, see Secord, " Discovery of a Vocation," 57–68, and, on the orientation of Cambridge geologists toward mathematics, see Smith, "Geologists and Mathematicians," 49–83.

32. DAR 32.2:125^{v}. The full quotation reads (fol. 125), "In the above described formations, after observing the crystalline & nearly pure siliceous nature of the one & highly laminated structure of the Slate. I had not the slightest expectation to find organic remains;—the whole appearance of the country had led me to suppose they belonged to that class which does not bear the signs of the coexistence of living beings with its formation. I was therefore the more surprised to find near the Settlement <to find> within the Slate; beds of <slaty>

sandstone. which (b) abounded with impressions & casts of shells.—" The footnote "(b)" is that quoted in the text as preceded by the statement (fol. 125^{v}), "From the above reasons I think this one of the oldest (or most inferior) formations which ever is fossiliferous.—the general character of the organic remains would also lead to this conclusion." In his journal Darwin remarked, "The whole aspect of the Falkland Islands, were however changed to my eyes from that walk; for I found a rock abounding with shells; & these of the most interesting geological æra.—" *Diary,* 146–147. In the spring of 1837 Murchison examined some of Darwin's fossils from the Falklands, declaring them to belong to lower Silurian rocks. *Notebooks,* 66.

33. Edward Sabine (1788–1883), a leader of the "magnetic crusade" in Britain, characterized work done on the 1831–1836 *Beagle* voyage as "amongst the most important contributions to magnetical science" and the "precursor of what British naval officers will accomplish for magnetism in the southern hemisphere." Edward Sabine, "Report on the Variations of the Magnetic Intensity Observed at Different Points of the Earth's Surface," *Report of the British Association for the Advancement of Science* [Liverpool 1837] (1838):32, 33. FitzRoy had provided unpublished data on magnetic intensity gathered during the voyage to Sabine, who incorporated it into his survey. Nathan Reingold, "Edward Sabine," Dictionary of Scientific Biography 12:49–53. FitzRoy published his data on magnetic variation in FitzRoy, *Narrative,* appendix, 65–88. Regarding Humboldt's role in stimulating magnetical work, see the comment, "Humboldt's conclusion, from his South American observations, that magnetic intensity varies from place to place, established geomagnetism as a subject particularly deserving the attention of the scientist." Robert P. Multhauf and Gregory Good, *A Brief History of Geomagnetism and a Catalog of the Collections of the National Museum of American History* (Washington, D.C.: Smithsonian Institution Press, 1987), 11.

34. *Diary,* 33.

35. John Cawood, "The Magnetic Crusade: Science and Politics in Early Victorian Britain," *Isis* 70 (1979):493–518, including p. 497: "The cosmical tradition favored by Humboldt, Arago, and Sabine derived from an astronomical approach to geomagnetism and regarded terrestrial magnetism as just one of a number of interconnected telluric or earth forces which were responsible for the phenomena manifest in or on the earth." For an instance of Darwin's speculation along these lines, see his comment made soon after the voyage in Notebook A: "In Cleavage discussion, state broadly indication of new law acting in certain directions predominantly, connection with magnetism &c counteracting gravity.—" *Notebooks,* 102 (A:62).

36. James Secord, personal communication.

37. CD to his sister Catherine, 8 November 1834, and to J. S. Henslow, 11 July 1831, in *Correspondence,* 1:418–419, 125.

38. DAR 32.1:20.

39. DAR 32.1:36^{v}.

40. Daubeny, *Description of Active and Extinct Volcanoes,* 162.

41. Lyell, *Principles of Geology,* 1:455.

42. Ibid., 1:459.

43. DAR 32.1:23^{v}. The passage appears as a footnote (signified by "(a)") to the passage on the facing page (24), where Darwin described his impression of the cliffs of São Tiago as seen from offshore and his judgment regarding the action of the force required to raise the cliffs:

> When viewing from a distance an extent of cliffs. one is struck by the great force it must have required to have raised fields 2 or 3 miles broard [*sic*] of these rocks at least 50 feet.—(Which is *supposing* the former <coast> <<bottom>> was at the surface of the sea at high water mark!)—A considerable thickness of the lower crystalline rocks. must likewise have been elevated at the same time.—["(a)" in margin] Taking this into consideration it is perfectly astonishing. that the force should have acted so uniformly that a spirit level with sights. proved the former beach to be as truly level as the present.—

44. CD to Henslow, 18 May 1832, in *Correspondence,* 1:236; *Autobiography,* 101.

45. Paul N. Pearson and Christopher J. Nicholas, "'Marks of Extreme Violence': Charles Darwin's Geological Observations on St. Jago (São Tiago), Cape Verde Islands," in Wyse Jackson, *Geological Travellers.* These authors have also redone Darwin's route and have provided photographs and excellent commentary on a number of points.

46. Lyell, *Principles of Geology,* 1:frontispiece, 449–459.

47. On the 1875 founding of the Lyell Medal, whose reverse bears the image of the Temple of Serapis, see Woodward, *History of the Geological Society of London,* 250. On Lyell's employment of the image, see Gordon L. Herries Davies, *The Earth in Decay: A History of British Geomorphology, 1578–1878* (London: Macdonald, 1969), 217.

48. *Correspondence,* 1:557; Herries Davies, *Earth in Decay,* 217.

49. Playfair, *Illustrations of the Huttonian Theory,* 446.

50. Ibid., 452.

51. Lyell, *Principles of Geology,* 1:455–456. Among his contemporaries, Fitton criticized Lyell for insufficiently acknowledging his Huttonian heritage. See Frank H. T. Rhodes, "Darwin's Search for a Theory of the Earth: Symmetry, Simplicity and Speculation," *British Journal for the History of Science* 24 (1991):211; also Roy Porter, "Charles Lyell and the Principles of the History of Geology," *British Journal for the History of Science* 9 (1976):95–96.

52. For the argument that Darwin departed from Lyell in coming to emphasize large-scale crustal movements, see Herbert, "Les divergences entre Darwin et Lyell," 69–76. Also see *Notebooks,* 114 (A:95), for Darwin's later employment of the Temple of Serapis as an exemplar of crustal motion. The Temple of Serapis is presently regarded as having been a market building rather than a temple. See the entry for Puteoli (Pozzuoli) in Richard Stillwell, ed., *The Princeton Encyclopedia of Classical Sites* (Princeton: Princeton University Press, 1976), 743–744.

53. DAR 32.1:20. The full passage, dated 17–18 January 1832, reads, "I have not mentioned a small covering of diluvium on the Western side of the Island.—At first. I thought it merely debris from the upper feldspathic rocks.—but on examining I found numerous fragments of the lower augitic rocks.—it does not appear to be of marine origin although it is the most probable explanation.—It looks to me like a part of the long disputed Diluvium.—" Sometime later Darwin struck the passage, "it does not appear . . . Diluvium.—" along with several others in his notes on the island, adding, without elaboration (fol. 20), "I have drawn my pen through those parts which appear absurd.—"

54. Sedgwick, "On the Origin of Alluvial and Diluvial Deposits," 243. On the "English school," see Rupke, *Great Chain of History,* 15–18.

55. Adam Sedgwick, "On Diluvial Formations," *Annals of Philosophy* n.s. 10 (1825), 34.

56. John Stevens Henslow, "On the Deluge," *Annals of Philosophy* 6 (1823): 344–348.

57. James Hall, "On the Revolutions of the Earth's Surface," *Transactions of the Royal Society of Edinburgh* 44 (1815):139–212. Greenough, *Critical Examination,* 151–155. Hall's connection with the Darwin-Wedgwood circle is illustrated by the fact that he used a Wedgwood pyrometer in his work. See V. A. Eyles, "James Hall," Dictionary of Scientific Biography 6:55. For CD's later use of Hall's article, see *Notebooks,* 94 (A:36), and *JR,* 621–625.

58. Laudan, *From Mineralogy to Geology,* 36–41, 110.

59. Santiago Notebook (catalogued 1.18), Down House Manuscript Collection, p. 28 (facing p. 27) in rear of notebook.

60. For an indication of Darwin's changing usage of the term "diluvium," see the following representative quotations from his geological notes: in 1832 at the Cape Verde Islands, "great beds of diluvium" (DAR 32.1:34); in 1832 at Bahía Blanca, "The whole county was elevated.—at this period <or later> the diluvium was deposited, which I have said probably owes its origin to a flood coming from the W in direction of the Andes.—" (32.1:68); in 1832 at Bahía Blanca, "superficial gravels & caverns.—or as it is sometimes called diluvial formations.—" (32.1:71); in 1834 in a note entitled "Elevations on coast of Patagonia," " . . . Valleys, . . . at S.Cruz How formed? sea or diluvial wave?—" (34.1:64^{v}); in 1834 at the Santa Cruz River,

"I suppose some would call it Diluvium" (34.2:150); in 1834 at Chiloé, "On any other supposition (putting aside *Diluvium*)" (35.1:216); in 1834 or (possibly) 1835 in Chile, "The ultimate conclusion which I <draw> <<have>> come to—Is that primarily the line of elevation determines the figure of a Continent; secondarily that a gradually *retreating* ocean models the elevated points; smooths with so called Diluvium some of its asperities, determines the directions of the great slopes—" (36.1:435); in 1835 in Chile, "Will not the arguments of [Jean André] De Luc [1727–1817] &c about the quantity of detritus in a mountain talus &c &c apply in most districts to <measure> <<ascertain>> the remoteness of that Epoch in <each case> <<place of an>> *universal* deluge?—" (36.1:459); in 1835 at Copiapò, "From the description of the valleys in this line of coast, an extent of about 400 miles, the following conclusions appear to me inevitable.—That the sea, during a long & quiet residence deposited those masses of Shingle *stratified with seams of sand & Clay.* which in Europe would be called Diluvium." (*underl*ining is *added pencil*) (37.1:675); in 1836 near Sydney in the valley of the Nepean River, "The present river could never have placed this gravel in its present position; it belongs to the substances *called* Diluvium; & which in this case probably was left by the retiring sea." (underlining is added pencil) (38.1:815); in 1836 of the forming of great valleys near Sydney, "If the agency of Debacles be attempted to be brought into play; it must be remembered that the accumulations, called Diluvium, is in this country very infrequent" (38.1:832).

61. Lyell, *Principles of Geology,* 3:145, "We defined alluvium to be such transported matter as has been thrown down, either by rivers, floods, or other causes, upon land liable to inundations, or which is not *permanently* submerged beneath the waters of lakes or seas. As examples of the *other causes* adverted to in the above definition, we might instance a wave of the sea raised by an earthquake, or a water-spout, or a glacier."

62. Ibid., 3:147.

63. Representative quotations regarding alluvium are as follows: in 1835 at Lowes Harbour in the Chonos Archipelago, "There are cliffs about 300 ft high composed of what would generally be called Alluvium.—" (DAR 35.1:233); in 1834 at Chiloé, "there was (so called) Alluvium" (35.1:294); in 1835 at Chiloé, "The materials are almost invariably arranged in horizontal lines; from their loose texture & coarse structure, they resembled exactly what is generally called Alluvium, (for instance the general covering of Shropshire).—Now the country consisting of *plains.* & the arrangement <of materials> being alternately coarse & finer sediment in horizontal lines. These present state of things. can only be accounted for by marine deposition. (lakes are here manifestly out of the question).—The deposit being in its appearance so exactly that of a river in is nature. (but not form) is interesting.—it is a step to a conclusion, which I am strongly induced to believe, that nearly all the so called grand deposits of *Alluvium* are of a marine origin.—" (35.1:301); in 1835 at the Lacuy Peninsula in Chiloé, "passage from the finer sedimentary beds to this mass (of so called) Alluvium, <which> <<such as>> has been so frequently described on the E. Coast.—We must suppose this band of gravel, traces the line of a current in the ocean, which deposited all the above Strata.—" (35.2:339–340); in 1835 in a valley of the Cordillera, "The Alluvium on each side of the uppermost torrents is irregular in outline.—the pebbles are only partially & to a different degree, rounded & are mingled with earth.—This will be seen to be essentially different from the constitution of the Shingle terraces; & properly deserve the name of Alluvium.—" (36.1:454[v]). For Darwin's later treatment of Shropshire alluvium, see "Alluvium/Shropshire/1838/July," DAR 5:21–22.

64. Herries Davies, *Earth in Decay,* 254, and, continuing, "Lyell merely took the diluvial theory, stripped it of its catastrophism, replaced the catastrophic agencies by such acceptable uniformitarian processes as wave and current action, and then adopted the refurbished theory as his own. It was thus no accident that the marine erosion theory came into vogue at the very moment when support for catastrophism and the diluvial theory was rapidly fading."

65. Patrick Armstrong, *Charles Darwin in Western Australia: A Young Scientist's Perception of an Environment* (Nedlands, Western Australia: University of Western Australia Press, 1985); Patrick Armstrong, *Darwin's Desolate Islands: A Naturalist in the Falklands, 1833 and 1834* (Chippenham, England: Picton, 1992); Patrick Armstrong, *Under the Blue Vault of Heaven: A*

Study of Charles Darwin's Sojourn in the Cocos (Keeling) Islands (Nedlands, Western Australia: Indian Ocean Centre for Peace Studies, 1991). Armstrong approaches these individual settings with camera in hand, as well as copies of relevant Darwin manuscripts. Since all three sites are infrequently visited, the resultant text and photographs are the more valuable.

66. Nicol Morton, "In the Footsteps of Charles Darwin in South America," *Geology Today* 11 (1995):190–195.

67. I will give a minor example of such an experience. On a brief unexpected stop in the Blue Mountains west of Sydney, knowing that Darwin had traveled in that region but without his writings at hand, I asked myself what Darwin might have done had he found himself at the same spot. I decided he would have looked for a good exposure. I began walking along a streambed and soon came to a path labeled "Darwin's Walk." Indeed, Darwin had come that way, and some one before me had identified his route.

68. Sandra Herbert, "From Charles Darwin's Portfolio: An Early Essay on South American Geology and Species," *Earth Sciences History* 14 (1995):23–36.

69. DAR 42: "Reflection on reading my Geological notes." Passages are quoted from fols. 6 through 10 in Darwin's pagination. The superscript "(x)" refers to a footnote that reads, "Perhaps the first opening of the N & S. crack in the crust of the globe. forming the Cordilleras" (fol. 7^{v}).

70. DAR 34.1:40–60. The title is written in pencil and, as the text is in ink, is presumably a later addition. Dating of the manuscript is approximate. From internal references (fol. 42) it is clear that Darwin had already completed the expedition up the Santa Cruz River but not yet traveled extensively in the Andes. See fol. 47 for the speculation, "If on some future day I shall be able to prove that the West coast has been elevated within the same period[.]" On 10 June 1834 the *Beagle* "Sailed for the last time from Tierra del Fuego," arriving on the west coast of South America via the Magdalen channel. *Correspondence,* 1:541.

71. Presumably William Whewell, "Essay towards a First Approximation to a Map of Cotidal Lines," *Philosophical Transactions of the Royal Society of London* 123 (1833):147–236.

72. DAR 34.1:40^{v}:

> From evidence drawn alone from these plains I do not know any proof that the land has risen in preference to the sea having subsided.—As <a> fall in the Atlantic would necessarily affect the <seas of the> whole world.—I think we may feel certain that no catastrophe has been so violent as to cause <any great &> sudden subsidence <of> such as 100 feet.—Reason will be given for supposing that one set of plains was elevated at one period more than that number of feet.—Of course the weightiest argument against the <hypothesis of> the fall of sea. is simply that it is more improbable; it requires a greater amount of change.—

73. CD to Robert FitzRoy, 10 October 1831, in *Correspondence,* 1:175.

74. FitzRoy, *Narrative,* 2:26.

75. Ibid., 2:34. The charts compiled from data gathered during the voyage are listed in *Catalogue of Charts, Plans, Views, and Sailing Directions, &c.* (London: Published by Order of the Lords Commissioners of the Admiralty, 1852 (or other editions). Charts 1324, 1288, and 1284 are helpful as providing an overview of the Patagonian coastline. A comparison with an earlier chart will suggest the scale of FitzRoy's improvements. See the British version of a Spanish chart of 1798 (British Admiralty chart 1059 south; published by W. Faden, London, 1821). While some of the place names Darwin mentioned in "Elevation of Patagonia" are obscure, they all appear in the above-cited charts.

76. DAR 34.1:48. For a reproduction of the chart of the river and its mouth, see FitzRoy, *Narrative,* 2:facing 339. Keynes suggested that FitzRoy's party probably came within a few miles of discovering Lago Argentino, out of which the river flows. *Diary,* 239. For reproductions of Conrad Martens's sketches of scenes from the expedition, see Keynes, *Beagle Record,* 200–213.

77. DAR 34.1:57.

78. *GSA,* 18. Published figures for some of the plains differ from those contained in "Elevation of Patagonia"; however, all but the 60-foot plain at Port Desire/St. George Bay are mentioned in the text.

79. DAR 34.1:57–58.

80. DAR 34.1:44.

81. DAR 34.1:44. In an estimate based on actual measurements Darwin posited a slope of 1233 feet (205.5 fathoms) in 160 miles over land at Santa Cruz, more than double the estimated slope in the ancient sea bottom (100 fathoms over 160 miles), and far greater than the estimated actual slope of the sea bottom (54 fathoms over 137 miles). However, the inland measurement Darwin thought "doubtful by Lava occurring in the above interval." He therefore added the conjecture that "the highest plain <which> (not a very regular one) which I measured is 97 miles from Coast & 1416 ft high. Now this is only 516 ft higher than the 900 plains on coast (assumed 900 between 840 & 950): accordingly as 160 <<miles>> gives 100 Fathoms:: 97 <<miles>> gives 363 <<ft>>.—So that these plains are only 153 ft higher than would be expected with a concentric elevation.—"

82. DAR 34.1:59.

83. DAR 34.1:50^{v} note "(a)" and continuing, "Perhaps from analogy a <small> sudden rise is more probable. & (if granted) its retiring waters may (perhaps?) explain some of the valleys.—"

84. DAR 34.1:57.

85. DAR 34.1:51, 40. Darwin supplied the volume number but not a page citation to the book, but the probable reference was to Lyell, *Principles of Geology,* 3:111–113.

86. DAR 34.1:58.

87. DAR 34.1:59. Regarding Lyell he wrote (fols. 58–59), "If my conclusion is granted (& in no other way I think can the coincidence in heights be accounted [for]), it appears to me, that the phenomenon of the elevation of strata is so grand, so uniform in its nature, that, the explanation offered by Mr Lyell of injection of Hypogene rocks is quite insufficient.—Can one imagine a mass of melted matter 600 miles in length <forcing> <<lifting>> up<wards> a great thickness of Strata to almost exactly the same height?—" Also see Lyell, *Principles of Geology,* 3:374–81. Lyell spoke of hypogene rocks originating from great depths (p. 380), but his hypothetical example for the source of Etnean lavas mentions only ten miles.

88. Adam Sedgwick, "Presidential Address [19 February 1830]," *Proceedings of the Geological Society of London* 1 (November 1826–June 1833):211.

89. DAR 34.1:47.

90. DAR 34.1:47.

91. DAR 34.1:60.

92. DAR 34.1:60^{v}:

> Conversing with Capt: FitzRoy, < . . . > concerning the recent elevation of the continent he suggested the following bold hypothesis: The number of distinct languages in T. del Fuego & <the *difference* of their habits from surrounding> the <<similarity in physical structure>> suggests an high antiquity to the race of the<<se>> Indians:—It seems a most strange fact, that any power could have induced a set of men to leave the <fertile> <<immense &>> <bountiful> <<fertiles>> regions of temperate America & inhabit the miserable country of the South.—May we conjecture. that this migration took place. anterior to the last 2 or 3000 ft elevation; when the greater part of America <would be> <<being>> covered with the sea. <necessity> <<want of food>> might <<well>> compel small tribes to follow to the extremity the ridge of mountains:? May we venture to <enlarge> extend this idea—the lofty plains of Mexico & Peru. <would form fertile regions> <<probably existed as dry land>> at an immensely remote epoch.—Hence <<did>> they not become th [*sic*] the two centres of <civi> aboriginal civilization?—

Also see *JR,* 409–413. On FitzRoy's interest in early human migrations, see FitzRoy, *Narrative,* 2, chap. 27. Darwin's views on elevation in relation to human habitation (*GSA,* 49) were later challenged by James Dwight Dana (1813–1895). See Daniel E. Appleman, "James Dwight Dana and Pacific Geology," in Herman J. Viola and Carolyn Margolis, eds., *Magnificent Voyagers: The U.S. Exploring Expedition, 1838–1842* (Washington, D.C.: Smithsonian Institution, 1985), 90.

93. *JR,* 201–202.

94. Ibid., 204.

95. Ibid., 205.

96. CD to Lyell, 1 September 1844, in *Correspondence,* 3:56. The first two chapters of *GSA* were entitled "On the Elevation of the Eastern Coast of South America" and "On the Elevation of the Western Coast of South America."

97. Alcide d'Orbigny to CD, 14 February 1845, in *Correspondence,* 3:143–145. *GSA,* chap. 1.

98. *GSA.* Sections of plains figured were those south of Nuevo Gulf (p. 6), in the Bay of St. George (p. 6), at Port Desire (p. 7), at Port St. Julian (p. 7), at the mouth of the Santa Cruz River (as shown, p. 8), and across terraces high up along the same river (p. 10).

99. Ibid., 14–16.

100. Ibid., 19–25. Also CD, "On the Distribution of the Erratic Boulders and on the Contemporaneous Unstratified Deposits of South America," *Transactions of the Geological Society of London,* 2d ser., 6 (1842):415–431, in *CP,* 1:145–163.

101. *GSA,* 18.

102. Ibid., 9, and, for further comment on the rise, 18. In a penciled note on the first page of "Elevation of Patagonia" Darwin remarked "All used" except the theory of gravel plains on p. 8 of the manuscript, which contains his initial interpretation of the 108-foot rise at Santa Cruz.

103. DAR 41. This volume does not bear library foliation. Citations of folio numbers refer to those in the text. Syms Covington, Darwin's servant, copied the text in a beautifully legible script. Darwin corrected Covington's copy in his own hand and, later, added numerous cross-references and notes to the whole. Darwin's original draft has not survived. The manuscript is not dated. Covington's fair copy was done on "J. Whatman 1834" paper that came into Darwin's possession in early April or early May 1836 (Sloan, "Darwin's Invertebrate Program," 118–119, n. 58). However, there is internal evidence suggesting the likelihood of an earlier date than 1836 for a first draft. The opening phrase of the extant text, which may not be solely rhetorical, reads, "Before finally leaving the shores of South America." This would correspond to the period from 19 July to 6 September 1835, when Darwin was at Callao and Lima. *Correspondence,* 1:541, and see 461 and 463 for locutions similar to "Before finally leaving the shores of South America." Also, one emendation does argue against an 1836 dating for a first draft. It appears on fol. 4^{v} and refers to the "coral paper" of late 1835. If the "coral paper" were already in existence at the time the draft for the "Recapitulation" was written, it probably would have been referred to in the original text rather than appearing as an emendation. (The emendation, in Darwin's hand, reads, "The ancient <& long continued> active condition of craters in Cordillera. harmonize. (according to deduction in coral paper) with idea of movement of elevation.—")

104. DAR 41, "Recapitulation," fols. 21, 29.

105. CD to Catherine Darwin, 31 May 1835, in *Correspondence,* 1:449: "I am lucky in having plenty of occupation for the Sea part, in writing up my journal & Geological memoranda.— I have already got two books of rough notes.—"

106. DAR 41, "Recapitulation," fols. 1–14 discuss the Cordillera. Darwin organized the formations he observed in central and northern Chile into five principal divisions: granitic, crystalline slate rocks, porphyritic breccia, and the gypseous and supergypseous formations.

107. The map is a separate sheet of tracing paper attached to the verso of fol. 1 of "Recapitulation." The date of the map is not certain, but its generality and placement suggest that it was roughly contemporary with the essay.

108. DAR 41, "Recapitulation," fol. 15.

109. DAR 41, "Recapitulation," fols. 16, 15. In fols. 14–19 the crosscurrents and contra-

dictory elements in Darwin's thinking about elevation are apparent. Thus, while maintaining a gradualist outlook, he hypothesized a dramatic history for the Cordillera beginning with "an epoch of excessive volcanic eruptions which has never since been nearly equalled" (fol. 14). Similarly, he sought to allow for instances of angular movement within a general picture of horizontal elevation (fol. 15).

110. DAR 41, "Recapitulation," fol. 15: "I conceive these views are in perfect conformity with the existence of those level remarkable basins, situated on the summit of the Andes. We may instance—Titicaca—Cuenca elevated 1350 toises—the valley of Quito from 1340 to 1490 toises—and the grand Mexican platform which between 19 and 24 1/2 of latitude, remains constantly at the height of 950 to 1200 toises (Humboldt, *Personal Narrative,* vol. VI, part II [1826:162]). Are not the strata in these plains horizontal? Can they all be thought lacustrine." Also fol. 19: "That the continental elevation is a phenomenon intimately connected with the rise in mass of the Andes, I think no one will dispute.—In Patagonia, within the recent period I have shown that its influence has been felt on the whole coast at the distance of 2 to 300 miles. . . . In Patagonia we have proofs that the coast has been elevated in a horizontal manner to a height of from 300 to 400 feet; and it may be assumed that there is no part which has not risen." The diluvium on the west coast, as on the east, Darwin believed to be of marine origin, including that present in the intervals between mountains (fol. 18).

111. DAR 34.1:40, 42.

112. DAR 41, "Recapitulation," fol. 29.

113. DAR 42, "Recapitulation," fol. 20.

114. DAR 41, "Recapitulation," fol. 11 on the simultaneity of volcanic eruptions over large areas, also fol. 24 on volcanic eruptions as "mere accidents, consequent on a more complete rupture of the strata, . . . accidents which happened during the periods, when that far more important phenomenon of injection determined the lines of elevation."

115. DAR 41, "Recapitulation," fol. 22. On Darwin's later development of the thin-crust model, see *Notebooks,* 63, 70 (RN:131, 154), 107–108, 122, 130, 131, 136 (A:77–79, 114, 133, 136, 147).

116. DAR 41, "Recapitulation," fol. 18: "I have called the gradual rise an horizontal upheaval.—Speaking with accuracy, the more probable movement is that of a curved enlargement of a narrow space of the superficies of the globe." Also fol. 19: "It is stated that in the Valparaiso earthquake of 1822, the land at the distance of 5 or 6 miles inshore, rose in the proportion of three to one, to that on the coast. On the other hand, it must be confessed, than in 1835 at Concepción, the rise of the Island of St Mary was greater than on the neighbouring mainland."

117. Beaufort as cited in FitzRoy, *Narrative,* 2:38:

> An exact geological map of the whole [coral] island should be constructed, showing its form, the greatest height to which the solid coral has risen, as well as that to which the fragments appear to have been forced. The slope of its sides should be carefully measured in different places, and particularly on the external face, by a series of soundings. . . . A modern and very plausible theory has been put forward, that these wonderful formations, instead of ascending from the bottom of the sea, have been raised from the summits of extinct volcanoes; and therefore the nature of the bottom at each of these soundings should be noted, and every means exerted that ingenuity can devise of discovering at what depth the coral formation begins, and of what materials the substratum on which it rests is composed. The shape, slope, and elevation of the coral knolls in the lagoon would also help the investigation; and no circumstances should be neglected which can render an account of the general structure clear and perspicuous.

FitzRoy did comment on coral formations (chaps. 22, 26) but referred to Darwin the question of the depth at which coral formation begins (p. 634) and, by implication, left the larger questions regarding reefs to him as well.

118. *Athenaeum* 24 (24 December 1831):834.

119. Jean René Constant Quoy and Joseph Paul Gaimard, "Mémoire sure l'accroissement des polypes lithophytes considéré géologiquement," *Annales des Sciences Naturelles* 6 (1825):273–290.

120. De la Beche, *Geological Manual,* 140–143; Frederick William Beechey, *Narrative of a Voyage to the Pacific and Beering's Strait, . . . in His Majesty's Ship Blossom . . . in the years 1825,26, 27, 28,* 2 vols. (London: Henry Colburn and R. Bentley, 1831), 1:254–264; Lyell, *Principles of Geology,* 2:283–301. De la Beche and Beechey, writing independently, accepted Quoy and Gaimard's findings. Lyell built on their views, relying particularly on the navigator Beechey for descriptions of reefs.

121. Beechey, *Narrative of a Voyage,* 169. Darwin would presumably have read this in the "Yanky Edit" he owned. *Notebooks,* 34 (RN:45).

122. *Autobiography,* 98–99. Quotations from the Santiago Book appear in Frederick Burkhardt, "Darwin's Early Notes on Coral Reef Formation," *Earth Sciences History* 3 (1984):160–163; reprinted in *Correspondence,* 1:567–571. Nora Barlow referred to the entries on reefs in the notebook but did not quote them. See Barlow, *Charles Darwin and the Voyage of the Beagle,* 243–244, noted in Stoddart, "Coral Islands," 4 n. 8.

123. Quoted in *Correspondence,* 1:568. Further along in the notebook, in a passage that postdates a reference to King George's Sound (Australia), which the *Beagle* visited from 6 to 14 March 1836, Darwin noted, "The Coral theory rests on the supposition of depressions being very slow & at small intervals[.]" Santiago Book, p. 30, Down House. On dating passages in the Santiago Book, see Sulloway, "Darwin's Spelling Habits," 369–372.

124. DAR 37.2:791. In *JR,* 453, the craters are described as being of volcanic sandstone; in *VI,* 113–114, the craters are treated in detail and described as being formed of tuff. Darwin was the first to describe these craters, now termed "tuff cones," and to connect their lowered southern lips with having been broken down by the prevailing winds. See McBirney and Williams, *Geology and Petrology of the Galápagos Islands,* 3. For Darwin's full remarks, see DAR 37.2: "All the Craters in a large Archipelago, thus having one certain side high & the opposite low or broken down, immediately calls to mind the nearly parallel fact in the Lagoon Islands in another part of the Pacific. . . . It is well known, that the opening & low parts of those Islands, is with respect to the direction of the wind, on the Leeward side; the surf however from the SW swell is nearly as great on that, as on the windward side; in this respect therefore the case of the Sandstone craters <is not entirely similar> & that of the Lagoon Is^{lds}[.]—"

125. DAR 37.2:791^{v}.

126. DAR 37.2:792. In "Coral Islands" Darwin put aside this question (Stoddart, "Coral Islands," 14). James Dwight Dana emphasized the bearing of temperature on the growth of reefs. See Appleman, "James Dwight Dana and Pacific Geology," 92–94.

127. DAR 37.2:793–793^{v}.

128. Stoddart, "Coral Islands," 5. The view that Darwin had of the reef, looking toward Moorea from Tahiti, is suggested by a photograph in Sandra Herbert, "Darwin as a Geologist," *Scientific American* 254 (May 1986):117. Moorea is presently a center for research on coral reefs. See Bernard Salvat, "Coral Reefs, Science, and Politics: Relationships and Criteria for Decisions over Two Centuries—A French Case History," in Keith R. Benson and Philip R. Rehbock, eds., *Oceanographic History: The Pacific and Beyond* (Seattle: University of Washington Press, 2002), 473–474.

129. Stoddart, "Coral Islands," 8. Beechey, *Narrative of a Voyage,* 169.

130. See the bibliography compiled by Stoddart, "Coral Islands," 18–20.

131. Ibid., 17.

132. DAR 41: "Cleavage." The manuscript is undated. Since its content is entirely South American and since there is a reference to "Coral Islands" as a "future paper," one might argue for a late 1835 dating, that is, before the coral paper was written. Yet a reference in the Santiago Book, 33, suggests otherwise: "Before concluding the Cleavage paper. consult the VI Vol of Pers. Narra." This volume of Humboldt is at the core of the essay on "Cleavage." The Humboldt reference in the Santiago Book can be dated to after Darwin's visit to King George's

Sound, Australia on 6–14 March 1836, which is referred to on p. 24 of the notebook. Sulloway, "Further Remarks," 375–376 n. 15, has suggested a May 1836 dating for the "Cleavage," which is plausible.

133. DAR 41, "Cleavage," fol. 2.

134. Secord, *Controversy in Victorian Geology,* 58–59. Secord pointed to an 1835 paper by Sedgwick as cementing a distinction between bedding and cleavage drawn informally in the 1820s by British geologists. See Adam Sedgwick, "Remarks on the Structure of Large Mineral Masses, and Especially on the Chemical Changes Produced in the Aggregation of Stratified Rocks during Different Periods after Their Deposition," *Transactions of the Geological Society of London,* 2d ser., 3 (1835):461–486. (Presumably this was the paper Darwin cited in an annotation on fol. 1v of "Cleavage": "Study Sedgwicks paper on Cleavage Geolog Transacts.")

135. DAR 41, "Cleavage," fol. 1. See also DAR 32.2:111–117. Darwin probably opened his essay with definitions to counter what he called (fol. 36) the "doubt and confusion" exhibited in "On Stratification" in Greenough, *Critical Examination,* 1–90. In a sense Darwin was pursuing a parallel course to Sedgwick in working out appropriate definitions for cleavage and stratification.

136. Humboldt, *Personal Narrative,* 6, 590–591.

137. DAR 41, "Cleavage," fols. 12–13.

138. DAR 41, "Cleavage," fol. 35.

139. DAR 41, "Cleavage," fols. 32, 33; also DAR 32.2:115v, the added comment, "From other observations I now consider it as established that there is some physical connection between lines of Elevation. metamorphic action & cleavage."

140. *GSA,* chap. 6. Also CD, "On the Geology of the Falkland Islands," *Quarterly Journal of the Geological Society of London* 2 (1846):267–274, in *CP,* 1:203–212.

141. A core set of these postvoyage notebooks (lettered A, B, C, D, E, M, and N), together with related writings, was published in 1987 as *Notebooks.* The Red Notebook was first published in 1980, and was also included in *Notebooks.* The Santiago Book, stored at Down House, remains unpublished. The Santiago Book takes its name from Darwin's stay in Santiago, Chile (28 August–6 September 1834). It was kept as a field notebook through February 1835. This section of the notebook is unpaginated. After a transitional unnumbered two-page section in which Darwin referred to his forthcoming visit to Concepción (4–6 March 1835) and Valparaiso (arrived 11 March 1835), a numbered thirty-three-page section begins. It served, like the Red Notebook and the later alphabetically lettered notebooks, as a storage place for theoretical inquiries and reading notes. Page 24 of the notebook refers to an observation Darwin made in King George's Sound, Australia, which Darwin visited from 6 to 14 March 1836. Presumably from the continuity in their style and content, Darwin opened the Red Notebook only after the Santiago Book was filled. Page 15 of the Red Notebook refers to a measurement taken off the Cape of Good Hope, presumably from the *Beagle.* The *Beagle* arrived at the Cape of Good Hope on 31 May 1836 and departed 18 June. Red Notebook entries yield a perfect progression of place names corresponding to points visited by the *Beagle* from late May through September 1836. (*Notebooks,* 17; Herbert, *Red Notebook,* 6.) On dating the Santiago Book, see Sulloway, "Darwin's Spelling Habits," 368–376. Patrick Armstrong has suggested that the opening pages of the Red Notebook may have been written as early as March. This possibility cannot be totally discounted, but Armstrong's argument does not consider the relation between the Santiago Book entries and those in the Red Notebook. See Armstrong, *Darwin in Western Australia,* 67–71. If the Red Notebook's excised pages 1e–4e are ever located, they may provide a secure answer as to its precise opening date.

142. CD to J. S. Henslow, 9 July 1836, *Correspondence* 1:499; also see pp. 158–159, 517. Darwin became a Fellow of the Geological Society on 30 November 1836. *Proceedings of the Geological Society of London* 2 (November 1833–June 1838):435.

143. DAR 41, "Cleavage," fols. 28–30. The reference is to Humboldt, *Personal Narrative,* 6:592.

144. *Notebooks,* 36 (RN:49).

145. Santiago Book, 28–29.
146. *Notebooks*, 48 (RN:85)
147. Ibid., 35 (RN:46).
148. Ibid., 26 (RN:22).
149. Ibid., 42 (RN:68–69).
150. Ibid., 25 (RN:18).
151. Santiago Book, 19.
152. *Notebooks,* 44 (RN:72).
153. Ibid., 58 (RN:117).
154. Ibid., 80 (RN:180).
155. Ibid., 56–46 (RN:77–78).
156. Alan G. Gross, *The Rhetoric of Science* (Cambridge: Harvard University Press, 1990), 145.
157. Santiago Book, 1.
158. *Correspondence,* 1:516.
159. CD (misidentified as "F. Darwin"), "Geological Notes Made during a Survey of the East and West Coasts of South America, in the Years 1832, 1833, 1834, and 1835, with an Account of a Transverse Section of the Cordilleras of the Andes between Valparaiso and Mendoza," *Proceedings of the Geological Society of London* 2 (November 1833–June 1838):210–212, in *CP,* 1:16–19. The notes, communicated by Sedgwick, were taken from the extracts of Darwin's letters to Henslow privately printed by the Cambridge Philosophical Society. See Freeman, *Annotated Bibliographical Handlist,* 24–26. Even though these letters aroused the interest of geologists in his work, Darwin was "a good deal horrified" by their unauthorized circulation. *Correspondence,* 1:469–470, 498. Later he attempted to guard against publications of this sort. On 6 September 1871 he wrote to Francis Ellingwood Abbot (DAR 139.12): "It never occurred to me that you would wish to print any extract from my notes: if it had, I would have kept a copy. I put 'private' from habit, only as yet partially acquired, from some hasty notes of mine having been printed which were not in the least degree worth printing, though otherwise unobjectionable."
160. CD, "Observations of Proofs of Recent Elevation on the Coast of Chili, Made during the Survey of His Majesty's Ship Beagle, Commanded by Capt. Fitzroy, R.N.," *Proceedings of the Geological Society of London* 2 (November 1833–June 1838):446–449, in *CP,* 1:41–43.
161. CD, "A Sketch of the Deposits containing Extinct Mammalia in the Neighbourhood of the Plata," *Proceedings of the Geological Society of London* 2 (November 1833–June 1838):542–544, in *CP,* 1:44–45; Richard Owen, "A Description of the Cranium of *Toxodon Platensis,* a Gigantic Extinct Mammiferous Species, Referrible by Its Dentition to the *Rodentia,* but with Affinities to the *Pachydermata,* and the *Herbivorous Cetacea,*" *Proceedings of the Geological Society of London* 2 (November 1833–June 1838):541–542.
162. CD, "On Certain Areas of Elevation and Subsidence in the Pacific and Indian Oceans, as Deduced from the Study of Coral Formations," *Proceedings of the Geological Society of London* 2 (November 1833–June 1838):552–554, in *CP,* 1:46–49. Lyell to Herschel, 24 May 1837, in Katharine Lyell, ed., *Life, Letters and Journals of Sir Charles Lyell, Bart,* 2 vols. (London: John Murray, 1881), 2:11.
163. CD, "On the Formation of Mould," *Proceedings of the Geological Society of London* 2 (November 1833–June 1838):574–576, in *CP,* 1:49–53.
164. Elizabeth Wedgwood to CD, 10 November 1837, in *Correspondence* 2:55, and 56 n. 1. Darwin's paper, drawn from observations made on earthworms by his uncle Josiah Wedgwood (1769–1843), included a suggestion that chalk was produced from coral by the digestive action of marine animals. Buckland, who reviewed the paper for inclusion in the *Transactions* of the Geological Society of London, recommended that this part of the paper be omitted as "very disputable." *Correspondence,* 2:76–77.
165. CD, "On the Connexion of Certain Volcanic Phænomena, and on the Formation of Mountain-chains and Volcanos, as the Effects of Continental Elevations," *Proceedings of the Ge-*

ological Society of London 2 (November 1833–June 1838):654–660. The paper was published in full in the *Transactions* under the slightly different title listed in the bibliography (*CP*, 1:53–86).

166. CD to Robert W. Darwin, 8 February–1 March 1832, in *Correspondence*, 1:205, 204; 2:431.

167. CD to Henslow, 28 May 1837, in ibid., 2:21.

168. Ibid., 2:431 for Darwin's schedule of work, and 435, 436 n. 62 for reference to the draft. Also CD to Lyell, 15 or 22 September 1843, in ibid., 2:389.

169. CD to Henslow, 14 October 1837, in ibid., 2:51; 21 January 1838, 2:69.

170. Ibid., 2:70 n. 3.

171. CD to Lyell, 14 September 1838, in ibid., 2:105, 107.

172. Ibid., 2:432.

173. CD to W. D. Fox, 24 October 1839, in ibid., 2:234; see p. 207 for an ambiguous reference (CD to Leonard Jenyns, 15 July 1839).

174. CD to Lyell, 15 February 1840, in ibid., 2:253.

175. Darwin did fieldwork at Glen Roy from 28 June to 5 July 1838. See the Glen Roy Notebook edited by Sydney Smith, Peter Gautrey, and Paul H. Barrett in *Notebooks*, 141–165. The paper appeared as "Observations on the Parallel Roads of Glen Roy, and of Other Parts of Lochaber in Scotland, with an Attempt to Prove that They Are of Marine Origin," *Philosophical Transactions of the Royal Society of London.* (1839):39–81, in *CP*, 1:89–137. Also see *JR*, 614–625.

Chapter 6. Negotiating Genesis and Geology

1. *Notebooks*, 56–57 (RN:111–112), including n. 112-1 for the Hutton passage.

2. On the mediating rather than the strictly apologetic functions of natural theological arguments, see John Hedley Brooke, "The Natural Theology of the Geologists: Some Theological Strata," in L. J. Jordanova and Roy S. Porter, eds., *Images of the Earth: Essays in the History of the Environmental Sciences* (Chalfont St. Giles, England: British Society for the History of Science, 1979), 39–64. On the "safe [i.e., nonmaterialistic] science" approach of the most important British books from the period in the natural theological tradition—the *Bridgewater Treatises on the Power, Wisdom and Goodness of God as Manifested in the Creation* (1833–1836)—see Jonathan Topham, "Science and Popular Education in the 1830s: The Role of the Bridgewater Treatises," *British Journal for the History of Science* 25 (1992):397–430. Brooke's and Topham's interpretations soften the approach, generally critical of the Bridgewater Treatises, taken by Charles Coulston Gillispie in *Genesis and Geology: A Study in the Relations of Scientific Thought, Natural Theology, and Social Opinion in Great Britain, 1790–1850* (Cambridge: Harvard University Press, 1951), 209–216.

3. Quoted in Francis C. Haber, *The Age of the World: Moses to Darwin* (Baltimore: Johns Hopkins University Press, 1959), 109.

4. Ursula B. Marvin, "Meteorites, the Moon and the History of Geology," *Journal of Geological Education* 34 (1986):141.

5. Rhoda Rappaport, "Geology and Orthodoxy: The Case of Noah's Flood in Eighteenth-Century Thought," *British Journal for the History of Science* 11 (1978): 15. For a thorough treatment of the period when naturalists sought human witnesses to past events, see Rappaport, *When Geologists Were Historians, 1665–1750.* Of singular importance on the project of integrating human and geological history is John C. Greene, *The Death of Adam: Evolution and Its Impact on Western Thought* (Ames: Iowa State University Press, 1959; rev. ed. 1996).

6. Haber, *Age of the World*, 206. Although she does not use the phrase "Cuvierian synthesis," my interpretation follows that of Dorinda Outram, *Georges Cuvier: Vocation, Science and Authority in Post-Revolutionary France* (Manchester: Manchester University Press, 1984), chap. 7. For treatment of the key texts in Cuvier's intellectual development, see Rudwick, *Georges Cuvier, Fossil Bones, and Geological Catastrophes,* and also note Rudwick's introduction (pp. 173–183) to Cuvier's "preliminary discourse." For discussion of an important predeces-

sor to Cuvier, see Martin J. S. Rudwick, "Jean-André de Luc and Nature's Chronology," in C. L. E. Lewis and S. J. Knell, eds., *The Age of the Earth: from 4004 BC to AD 2002* (London: Geological Society, 2001), 51–60.

7. Georges Cuvier, *Recherches sur les ossemens fossiles de quadrupèdes,* 4 vols. (Paris: Deterville, 1812). The "Discours préliminaire" was published separately in English as *Essay on the Theory of the Earth,* trans. Robert Kerr with mineralogical notes by Robert Jameson (Edinburgh: William Blackwood, 1813). The "Discours" appeared separately in French in 1825. While the "Discours" went through many editions in the nineteenth century, published alone or as part of the *Recherches,* the only four new versions were those of 1812, 1821, 1825, and 1830. This information is on the authority of Jean Chandler Smith (personal communication). See her *Georges Cuvier: An Annotated Bibliography of His Published Works* (Washington, D.C.: Smithsonian Institution Press, 1993). For comparison of the different editions of the "Discours," see Outram, *George Cuvier,* 239–240 n. 1. Rudwick (*Georges Cuvier, Fossil Bones, and Geological Catastrophes,* 183–252) has provided a new English translation of the "Discours," with helpful references to the French original. However, in subsequent notes I have cited the English translations current at Darwin's time.

8. Cuvier, *Essay on the Theory of the Earth* (1813), 147.

9. Ibid., 147–148.

10. Ibid., 147, 181.

11. Ibid., 148, 171.

12. On Cuvier's paleontology, see Rudwick, *Meaning of Fossils,* chap. 3, and his *Georges Cuvier, Fossil Bones, and Geological Catastrophes.* Also see William Coleman, *Georges Cuvier, Zoologist: A Study in the History of Evolution Theory* (Cambridge: Harvard University Press, 1964), chap. 5.

13. For a summary of the contents of Cuvier's work on fossil quadrupeds and for outlines of geological and paleontological successions and catastrophes as understood by Cuvier, see Coleman, *Georges Cuvier, Zoologist,* 126, 128, 133. Rudwick, *Georges Cuvier, Fossil Bones, and Geological Catastrophes,* reconstructs the development of Cuvier's research.

14. Cuvier, *Essay on the Theory of the Earth* (1813), 16, 15.

15. Ibid., 1. On Cuvier's enlarged sense of his role as a paleontologist, see Outram, *Georges Cuvier,* 150.

16. Cuvier, *Essay on the Theory of the Earth,* sects. 6, 32, 30.

17. Toby A. Appel, *The Cuvier-Geoffroy Debate: French Biology in the Decades before Darwin* (Oxford: Oxford University Press, 1987), 56–57. For a differing view, see Outram, *Georges Cuvier,* 143–147.

18. Cuvier, *Discours sur les révolutions de la surface du globe* (1825), 288–289; Cuvier, *Essay on the Theory of the Earth,* with geological illustrations by Robert Jameson, 5th ed. (Edinburgh: William Blackwood, 1827), 243–244.

19. Rudwick, *Georges Cuvier, Fossil Bones, and Geological Catastrophes,* 258–259. Rudwick notes (p. 259) that Cuvier's adult daughter, a "devout Protestant," prayed for his conversion, which would suggest her doubt of his religious convictions.

20. Robert Jameson, "Preface," in Cuvier, *Essay on the Theory of the Earth* (1813), v.

21. As an example of an author respectful toward Cuvier without engaging his historical method in regard to texts, see [W. H. Fitton], "Geology of the Deluge," *Edinburgh Review* 39 (1823):206, 229–230. Identification of Fitton is from *The Wellesley Index to Victorian Periodicals, 1824–1900.*

22. Eyles, "Robert Jameson"; Leroy Page, "The Rise of Diluvial Theory in British Geological Thought" (Ph.D. dissertation, University of Oklahoma, 1963), 148–149; and James A. Secord, "Edinburgh Lamarckians: Robert Jameson and Robert E. Grant," *Journal of the History of Biology* 24 (1991):1–18. Also see Browne, *Charles Darwin: Voyaging,* 554 n. 37, for the suggestion that a third party, rather than Grant or Jameson, may have been the author of the unsigned 1826 transmutationist article in the *Edinburgh New Philosophical Journal.*

23. Walter F. [Susan Faye] Cannon, "William Buckland," Dictionary of Scientific Biography 2:569; also, Page, "Rise of Diluvial Theory," 75–77. More generally, see also Pietro Corsi,

Science and Religion: Baden Powell and the Anglican Debate, 1800–1860 (Cambridge: Cambridge University Press, 1988), chaps. 9, 10.

24. Rupke, *Great Chain of History,* 51–63.

25. William Buckland, *Vindiciæ Geologicæ; or the Connexion of Geology with Religion* (Oxford: Oxford University Press, 1820), 35–38.

26. Buckland, *Reliquiæ Diluvianæ,* 146.

27. Buckland, *Vindiciæ Geologicæ,* 28–29.

28. On the geological usage of the term, see *Oxford English Dictionary;* also William Buckland, "On the Quartz Rock of the Lickey Hill in Worcestershire, and of the Strata Immediately Surrounding It; with Considerations on the Evidence of a Recent Deluge Afforded by the Gravel Beds of Warwickshire and Oxfordshire, and the Valley of the Thames from Oxford Downwards to London," *Transactions of the Geological Society of London* 5 (1821):533.

29. See Buckland's remarks of 1819 quoted in Rupke, *Chain of History,* 91; also Buckland, *Reliquiæ Diluvianæ,* 170. The issue of human remains in the diluvium was of greater concern for Buckland than for Cuvier since Buckland believed the land and sea held their same relative relation during the deluge as at present, while Cuvier did not. Buckland did find flints contemporaneous with fossil mammal remains at Paviland Cave, but did not recognize them as such. See F. J. North, "Paviland Cave, the 'Red Lady,' the Deluge, and William Buckland," *Annals of Science* 5 (1942):113–114.

30. Rupke, *Great Chain of History,* 203, 95.

31. A student in 1832 recorded Buckland's remarks on the cause of diluvial gravel: "whether is Mosaic inundation or not, will not say." Quoted in Rupke, *Great Chain of History,* 95. Buckland's published retraction of his former views came in his Bridgewater Treatise entitled *Geology and Mineralogy Considered with Reference to Natural Theology,* 1:95.

32. Robert Jameson, "On the Universal Deluge," in Cuvier, *Theory of the Earth* (1827), 436–437.

33. Secord, "Edinburgh Lamarckians," 9–11.

34. J. H. Ashworth, "Charles Darwin as a Student in Edinburgh, 1825–1827," *Proceedings of the Royal Society of Edinburgh* 55 (1934–1935):102.

35. The exchange between Fleming and Buckland is contained in the following series: John Fleming, "Remarks Illustrative of the Influence of Society on the Distribution of British Animals," *Edinburgh Philosophical Journal* 11 (1824):287–305; William Buckland, "Professor Buckland's Reply to Dr. Fleming," *Edinburgh Philosophical Journal* 12 (1825):304–319; John Fleming, "The Geological Deluge, as Interpreted by Baron Cuvier and Professor Buckland, Inconsistent with the Testimony of Moses and the Phenomena of Nature," *Edinburgh Philosophical Journal* 14 (1826):205–239. On Fleming, see Gillispie, *Genesis and Geology,* 123–124; Page, "Rise of the Diluvial Theory," 146–150; Leroy Page, "Diluvialism and Its Critics in Great Britain in the Early Nineteenth Century," in Cecil J. Schneer, ed., *Toward a History of Geology* (Cambridge: MIT Press, 1969), 267–271; Leroy Page, "John Fleming," Dictionary of Scientific Biography 5:31–32.

36. Fleming, "Remarks Illustrative," 305.

37. Fleming, "Geological Deluge," 213 (also see 209–210 against Cuvier's approach to biblical history); Lyell, *Principles of Geology,* 3:271–273.

38. Fleming, "Remarks Illustrative," 290.

39. George Poulett Scrope, *Memoir on the Geology of Central France* (London: Longman, 1827), 165.

40. Charles Lyell, "Analogy of Geology and History [1828]," in Wilson, *Charles Lyell,* 216. Also see Martin J. S. Rudwick, "Historical Analogies in the Geological Work of Charles Lyell," *Janus* 64 (1977):89–107.

41. Gillispie, *Genesis and Geology,* 140.

42. Lyell, *Life, Letters and Journals of Sir Charles Lyell,* 1:139.

43. Two important analyses are Michael Bartholomew, "Lyell and Evolution: An Account of Lyell's Response to the Prospect of an Evolutionary Ancestry for Man," *British Journal for the History of Science* 6 (1973):261–303, and Jan M. Ivo Klaver, *Geology and Religious Senti-*

ment: The Effect of Geological Discoveries on English Society and Literature between 1829 and 1859 (Leiden: Brill, 1997), 15–84.

44. [Sir Charles Lyell], Obituary, *Unitarian Herald* (Manchester), March 5, 1875. Also see Robert K. Webb, "The Unitarian Background," and Jean Raymond and John V. Pickstone, "The Natural Sciences and the Learning of the English Unitarians," in Barbara Smith, ed., *Truth, Liberty, Religion: Essays Celebrating Two Hundred Years of Manchester College* (Oxford: Manchester College, 1986):3–30, 129–164.

45. Dendy Agate in *The Christian Life* (London), October 5, 1912. I am indebted to Robert K. Webb for the references to Unitarian newspapers. Leonard G. Wilson kindly supplied me with information regarding Lyell's religious upbringing as well as the reference containing the following reminiscence from Frances Power Cobbe: "The Lyells regularly attended Mr. Martineau's chapel in Little Portland Street, as we did; and ere long it became a habit for us to adjourn after the service to Harley Street [Lyell's home] and spend some of the afternoon with our friends, discussing the large supply of mental food which our pastor never failed to lay before us. Those were never-to-be-forgotten Sundays." Frances Power Cobbe, *Life*, 2 vols. (Boston: Houghton Mifflin, 1894), 2:404. On Lyell's critical attitude toward the Church of England, see Leonard G. Wilson, *Lyell in America: Transatlantic Geology, 1841–1853* (Baltimore: Johns Hopkins University Press, 1998), 299–300.

46. Leonard Jenyns, *Memoir of the Reverend John Stevens Henslow* (London: J. Van Voorst, 1862), 261–262.

47. In addition to material cited earlier, see Gillispie, *Genesis and Geology*, 112–113, 142–148. For Sedgwick's views after the 1830s, particularly in the context of his advancements in his clerical career, see Klaver, *Geology and Religious Sentiment*, 102–131.

48. Henslow, "On the Deluge," 344–348.

49. CD to W. D. Fox, 9 May 1830, in *Correspondence*, 1:104.

50. *Autobiography*, 64–65.

51. John Stevens Henslow, *A Sermon on the First and Second Resurrection Preached at Great St. Mary's Church on Feb. 15, 1829* (Cambridge, 1829), viii.

52. John Stevens Henslow, *Descriptive and Physiological Botany* (London: Longman, 1836), 313–314.

53. *Autobiography*, 64.

54. Ibid., 56–57, 85–87. On Henslow's career as a clergyman, see Walters and Stow, *Darwin's Mentor*, 155–173.

55. FitzRoy, *Narrative*, 2:658.

56. Ibid., 2:666.

57. James R. Moore, "Geologists and Interpreters of Genesis in the Nineteenth Century," in David C. Lindberg and Ronald L. Numbers, eds., *God and Nature: Historical Essays on the Encounter between Christianity and Science* (Berkeley: University of California Press, 1986), 322–350. Other essays in this collection also bear on the subject at hand.

58. CD to Caroline Wedgwood, 17 October 1839, in *Correspondence*, 2:236.

59. Quotations are from a portion of a FitzRoy letter, addressee unknown, bearing a date of 16 April 1858 on its envelope, a copy of which was given to me in 1994 by Harold Krusell of Santiago, Chile. The letter is consistent with ideas expressed by FitzRoy in *Narrative*, 2:640–656 and 657–682. On the work of Müller, see Stephen G. Alter, *Darwinism and the Linguistic Image* (Baltimore: Johns Hopkins University Press, 1999), 35–56.

60. *Athenaeum* of 14 July 1860 as reported in *Correspondence*, 8:595; Browne, *Charles Darwin: Power of Place*, 123.

61. Sedgwick, "Presidential Address [1831]," 313.

62. *Autobiography*, 85.

63. Ibid., 22 n. 1. Desmond and Moore, *Darwin*, and Browne, *Charles Darwin: Voyaging*, convey a sense of that legacy.

64. See particularly the material, including Notebooks M and N, grouped under Darwin's own title "Metaphysical Enquiries" in *Notebooks*, 517–641.

65. Gillian Beer, "Darwin and the Growth of Language Theory," in John Christie and Sally

Shuttleworth, eds., *Nature Transfigured: Science and Literature, 1700–1900* (Manchester: Manchester University Press, 1989), 152.

66. Rupke, *Great Chain of History,* 106–107.

67. John McPhee as quoted in Joe D. Burchfield, "The Age of the Earth and the Invention of Geological Time," in Derek J. Blundell and Andrew C. Scott, eds., *Lyell: The Past Is the Key to the Present* (London: Geological Society, 1998), 137.

68. John Phillips, *A Treatise on Geology,* 2 vols. (London: Longman, 1837 [vol. 1], 1839 [vol. 2]), 1:11.

69. Phillips as quoted in Burchfield, " Age of the Earth and the Invention of Geological Time," 139. Burchfield identified this usage of "geological time" as the first he has found.

70. Phillips, *Treatise on Geology,* 1:154–157, 18.

71. Ibid., 1:11.

72. Ibid., 1:266.

73. On Sedgwick's defense of geology against scriptural geologists, see Klaver, *Geology and Religious Sentiment,* 106–117.

74. William Whewell, "[Review of] *Principles of Geology,* Vol. I," *British Critic, Quarterly Theological Review and Ecclesiastical Record* 19 (1831):202.

75. See the richly detailed account in James A. Secord, *Victorian Sensation: The Extraordinary Publication, Reception, and Secret Authorship of Vestiges of the Natural History of Creation* (Chicago: University of Chicago Press, 2000).

76. George W. Featherstonehaugh to Sedgwick, 16 November 1844, in Secord, *Victorian Sensation,* 222, also 222–247.

77. Alvar Ellegård, *Darwin and the General Reader: The Reception of Darwin's Theory of Evolution in the British Periodical Press, 1859–1872* (Göteborg: Göteborg University, 1958).

78. CD to Leonard Horner, 20 March 1861, in *Correspondence,* 9:62 and n. 5 for the Horner quotation. The dates remained in the edition of the Bible printed for British schools until 1885. See John G. C. M. Fuller, "Before the Hills in Order Stood: The Beginning of the Geology of Time in England," in Lewis and Knell, *Age of the Earth,* 15–23. Cherry Lewis tells the story of the geologist Arthur Holmes (1890–1965) as a young boy puzzling, at the turn of the twentieth century, over the 4004 b.c. date in his parents' Bible. See her *Dating Game: One Man's Search for the Age of the Earth* (Cambridge: Cambridge University Press, 2000), 27–28.

79. Charles Wycliffe Goodwin, "Mosaic Cosmogony," in *Essays and Reviews,* 7th ed. (London: Longman, 1861), 252–253.

80. Baden Powell, "On the Study of the Evidences of Christianity," in *Essays and Reviews,* 139. Also see Corsi, *Science and Religion.*

81. Quoted in John Hedley Brooke, *Science and Religion: Some Historical Perspectives* (Cambridge: Cambridge University Press, 1991), 274.

82. On Darwin's participation in defense of the *Essays and Reviews* authors, see *Correspondence,* 9:416–419 (and see 12:35–36 on the later similar case of the biblical scholar John William Colenso [1814–1883]). On controversy overall, see Ieuan Ellis, *Seven against Christ: A Study of "Essays and Reviews"* (Leiden: E. J. Brill, 1980). On Darwin's religious trajectory, see Desmond and Moore, *Darwin,* and Browne, *Charles Darwin: Voyaging* and *Charles Darwin: Power of Place.*

Chapter 7. Toward Simplicity

1. Charles Darwin, *The Structure and Distribution of Coral Reefs,* foreword by H. W. Menard (Berkeley: University of California Press, 1962), vi.

2. Martin J. S. Rudwick, "Charles Lyell, F.R.S. (1797–1875) and His London Lectures on Geology, 1832–33," *Notes and Records of the Royal Society of London* 29 (1975):256–257.

3. Herbert, " Place of Man in the Development of Darwin's Theory of Transmutation: Part II," 158–178.

4. *Notebooks,* 44 (RN:72–73); next quotation p. 45 (RN:77).
5. Herschel, *Preliminary Discourse,* 270. Italics added
6. *Correspondence,* 1:238.
7. Stoddart, "Coral Islands," 17.
8. See *Notebooks,* 44 n. 73–1, for the original French (translation by S. Herbert). Humboldt expressed a similar thought in his letter to Darwin of 18 September 1839. The translation provided in *Correspondence,* 2:427, reads, "I have long believed that primitive vegetation had an additional source of heat beyond that available to present-day vegetation. I have believed that our earth . . . has received much of its climate . . . for a long time not so much from its position relative to a central star (the sun), but from within. At every latitude there is a fissuring of the crust of a planet. Volcanicity is nothing but the reaction produced by the fluid part of the Interior, near the oxidized, hardened surface losing heat through radiation."
9. Ronald Numbers, *Creation by Natural Law: Laplace's Nebular Hypothesis in American Thought* (Seattle: University of Washington Press, 1977), 20. Also see Brush, *Nebulous Earth;* Charles Coulston Gillispie with the collaboration of Robert Fox and Ivor Grattan-Guinness, *Pierre-Simon Laplace: A Life in Exact Science* (Princeton, Princeton University Press, 1997); Michael Hoskin, "William Herschel," *Dictionary of Scientific Biography* 6:328–336; and Stanley L. Jaki, "The Five Forms of Laplace's Cosmogony," *American Journal of Physics* 44 (1976):4–11.
10. Simon Schaffer, "Herschel in Bedlam: Natural History and Stellar Astronomy," *British Journal for the History of Science* 13 (1980):227. Stephen Brush adds (personal communication): "The metaphor [of development of individuals of the same species at different stages in their life cycles] survives in the early twentieth century concept of 'stellar evolution' as 'sliding down the main sequence' in the Hertzsprung-Russell diagram; it helped to support the 'long' time scale (10^{12} to 10^{13} years) but proved fallacious in the 1930s."
11. Pierre-Simon Laplace, *Exposition du système du monde,* 2 vols. (Paris: Cercle-Social, 1796), 2:298–299; also cited in Brush, *Nebulous Earth,* 27. Gillispie has emphasized the theme of stability, rather than nascent evolutionism, in Laplace and prefers the phrase "atmospheric hypothesis" to "nebular hypothesis" as a description of Laplace's intent. Gillispie, *Pierre-Simon Laplace,* 173.
12. Laplace in an English translation (1830) from Appendix 2 in Numbers, *Creation by Natural Law,* 126. This was from Laplace's fifth edition. Jaki also referred to Laplace's fourth edition (which also mentioned Cuvier) as being the first to show a departure from an "unqualified assertion of the stability of the solar system." Jaki, "Five Forms of Laplace's Cosmogony," 9. For key documents from Cuvier, see Rudwick, *Georges Cuvier, Fossil Bones, and Geological Catastrophes,* 13–32.
13. Quoted in Rudwick, *Georges Cuvier, Fossil Bones, and Geological Catastrophes,* 166–169.
14. Robert Chambers, *Vestiges of the Natural History of Creation and Other Evolutionary Writings,* ed. James A. Secord (Chicago: University of Chicago Press, 1994). Secord, *Victorian Sensation,* 9–10, 90–91, 102, 386–387.
15. Quoted in Martin J. S. Rudwick, "Charles Lyell Speaks in the Lecture Theatre," *British Journal for the History of Science* 11 (1976):154.
16. Lyell, *Principles of Geology,* 1:143.
17. Ibid., 1:141, 153.
18. Series of passages from Herschel, *Preliminary Discourse,* 211–212.
19. Ibid., 109.
20. Quotations from John Herschel, "On the Astronomical Causes Which May Influence Geological Phænomena," *Transactions of the Geological Society of London,* 2d ser., 3 (1835):298–299. The report of Herschel's oral presentation of 15 December 1830 appears in *Proceedings of the Geological Society of London* 1 (1826–1833):244–245, but the version in the *Transactions* is taken as expressing Herschel's meaning.
21. Herschel to Lyell, 20 February 1836, cited by Cannon, "The Impact of Uniformitarianism," 307–308.

22. Charles Lyell, *Principles of Geology,* 5th ed., 4 vols. (London: John Murray, 1837), 2:308.

23. *Notebooks,* 125 (A:121).

24. Ibid., 85 (A:inside front cover).

25. Brush, *Nebulous Earth,* 106.

26. Marvin, "Meteorites," 141–144, and Ursula B. Marvin, "Impacts from Space: The Implications for Uniformitarian Geology," in G. Y. Craig and J. H. Hull, eds., *James Hutton—Present and Future* (London: Geological Society, 1999), 89–98.

27. *Notebooks,* 90 (A:22–24), for all quotations in this paragraph. For the sake of comparison, note Marvin, "Impacts," 92: "Today the Earth gains, on average about 10 000 tonnes of meteoritic and cometary dust each year."

28. Whewell, "[Review of] *Principles of Geology,* Vol. II," 125.

29. The Glen Roy case will be discussed in greater detail in chapter 8. The standard treatment of Darwin's paper is Martin R. S. Rudwick, "Darwin and Glen Roy: A 'Great Failure' in Scientific Method?" *Studies in the History and Philosophy of Science* 5 (1974):128–129.

30. *CP,* 1:128–131.

31. All quotations in this section are from ibid., 1:128–131.

32. CD, "On the Connexion of Certain Volcanic Phenomena," in ibid., 1:60.

33. *Notebooks,* 126 (A:123), all of A:121–126 relevant.

34. Ibid., 127 (A:126).

35. Greene, *Geology in the Nineteenth Century,* 69–92. *CP,* 1:82.

36. Herschel to Lyell as quoted in Charles Babbage, *The Ninth Bridgewater Treatise. A Fragment,* 2d ed. (London: John Murray, 1838), 227, 230, 233.

37. *Notebooks,* 118–119 (A:104–105). The article cited is William Beer and Johann Heinrich Madler, "Survey of the Surface of the Moon," *Edinburgh New Philosophical Journal* 25 (1838):38–69.

38. *Notebooks,* 124 (A:118).

39. On Hopkins, see Cannon, *Science in Culture,* 29–71; Brush, *Nebulous Earth,* 76–92; Robert P. Beckinsale, "William Hopkins," Dictionary of Scientific Biography 6:502–504; Crosbie Smith, "William Hopkins and the Shaping of Dynamical Geology: 1830–1860," *British Journal for the History of Science* 22 (1989):27–52; and David S. Kushner, "The Emergence of Geophysics in Nineteenth Century Britain" (Ph.D. dissertation, Princeton University, 1990), 33–58, 117–125, 260–261. As indicated by the string of citations from Kushner, he weaves Hopkins's story in with that of other figures, whom I do not discuss.

40. William Hopkins, "Researches in Physical Geology [1835]," *Transactions of the Cambridge Philosophical Society* 6 (1836):1–84; and *An Abstract of a Memoir on Physical Geology* (Cambridge: Pitt Press, 1836). Darwin's copies are held at Cambridge University Library.

41. *Notebooks,* 132 (A:137).

42. William Hopkins to CD, 3 March 1845, 27 April 1846, 5 May 1846, in *Correspondence,* 3:151–152, 314–316, 317–318.

43. William Hopkins, "On the Phenomena of Precession and Nutation, Assuming the Fluidity of the Interior of the Earth," *Philosophical Transactions of the Royal Society of London* 129 (1839):381–423; "On Precession and Nutation, Assuming the Interior of the Earth to Be Fluid and Heterogeneous," *Philosophical Transactions of the Royal Society of London* 130 (1840):193–208; "On the Thickness and Constitution of the Earth's Crust," *Philosophical Transactions of the Royal Society of London* 132 (1842):43–55.

44. Hopkins, "Researches in Physical Geology" (1842), 51.

45. Ibid., 51.

46. William Hopkins, "Report on the Geological Theories of Elevation and Earthquakes," *Report of the British Association for the Advancement of Science* [Oxford 1847] (1848):33–92.

47. Ibid., 73, 69; Hopkins, *Abstract of a Memoir on Physical Geology,* 44–45. In the *Abstract* Hopkins mentioned Élie de Beaumont by name; in "Report on the Geological Theories of Elevation and Earthquakes" he did not.

48. On the development of seismology, see Oldroyd, *Thinking about the Earth,* 224–247, and Brush, *Nebulous Earth,* 144–147, 166–169, 178–202.

49. Robert Mallet, "Earthquakes," in J. F. W. Herschel, ed., *A Manual of Scientific Enquiry* (London: John Murray, 1849), 208.

50. William Hopkins, Review of the *Origin* [1860] "Physical Theories of the Phenomena of Life," reprinted with commentary in David Hull, *Darwin and His Critics: The Reception of Darwin's Theory of Evolution by the Scientific Community* (Cambridge: Harvard University Press, 1973), 266, 267.

51. CD to Asa Gray, 3 July 1860, in *Correspondence,* 8:274.

52. CD to Asa Gray, 22 July 1860, in ibid., 8:299.

53. George Howard Darwin, "On the Influence of Geological Changes on the Earth's Axis of Rotation [1877]," reprinted in George Howard Darwin, *Scientific Papers,* 5 vols. (Cambridge: Cambridge University Press, 1907–1916), 3:1–46. For George Howard Darwin's discussions of his father's work, see p. 27: "[M]y father, who has especially attended to the subject of the subsidence of the Pacific islands, has marked for me, on the map given in his work on Coral Reefs, a large area which he believes to have undergone a general subsidence motion. . . . I marked these areas on a globe, and cut out of number of pieces of paper to fit them, and then weighed them." The slips represented about 6.5 percent of the globe's area. He concluded from this that between 5 and 7 percent of the globe's area had undergone subsidence in a late geological period. For an early example of CD's use of paper slips molded on a globe to aid in thinking about the earth, see *Notebooks,* 35–36 (A:48–49).

54. On George Howard Darwin's support for Thomson's argument for a solid earth, see "On the Stresses Caused in the Interior of the Earth by the Weight of Continents and Mountains [1882]," in G. Darwin, *Scientific Papers,* 2:459–514, particularly p. 514.

55. Quotations are from Kushner, "Emergence of Geophysics in Nineteenth Century Britain," 547, 548; and for an authoritative treatment of George Howard Darwin's career, 379–585. On the deference of the children of Charles Darwin toward their father, see the account by George Darwin's daughter, Gwen Raverat, *Period Piece: A Cambridge Childhood* (London: Faber and Faber, 1960), 153, 175–209. Also see David Kushner, "Sir George Darwin and a British School of Geophysics," *Osiris* 8 (1993):196–223.

56. Brush, *Nebulous Earth,* 151.

57. The current view concerning the core of the earth suggests that it is of two parts, with the outer core, liquid and mostly iron, surrounding a small inner core that is solid and partly iron. There is still some controversy about the nature of the inner core since the only methods for measuring it are rather indirect. Stephen Brush, personal communication. Also see Oldroyd, *Thinking about the Earth,* 239–240, and Brush, *Nebulous Earth,* 175–202.

58. Geological and Geographical Committee, *Report of the First and Second Meetings of the BAAS* [York 1831, Oxford 1832] (1833):54.

59. William Whewell, "Account of a Level Line, Measured from the Bristol Channel to the English Channel, during the Year 1837–8, by Mr. Bunt, under the Direction of a Committee of the British Association," *Report of the British Association for the Advancement of Science* [Newcastle 1838] (1839):1–11 (passages quoted from p. 1); for the instruments used, see Thomas G. Bunt, "Account of the Leveling Operations between the Bristol Channel and the English Channel," *Report of the British Association for the Advancement of Science* [Newcastle 1838] (1839):11–18. On Whewell's role, see Morrell and Thackray, *Gentlemen of Science,* 505.

60. Greenough, "Presidential Address [1834]," 54.

61. Herschel, *Preliminary Discourse,* 215.

62. Lyell, *Principles of Geology,* 1:80. The argument in this section is taken from Herbert, "Les divergences entre Darwin et Lyell," 70–71.

63. Graham, "An Account of Some Effects of the late Earthquakes in Chili," 415.

64. Lyell, "Presidential Address [1837]," 505–506, for all passages quoted from the address.

65. Robert Edward Alison to CD, 25 June 1835, in *Correspondence,* 1:451.

66. *GSA*, 36.

67. Basil Hall, *Extracts from a Journal Written on the Coasts of Chili, Peru and Mexico for the Years 1820, 1821, 1822,* 2 vols. (Edinburgh: A. Constable, 1824), 2:9.

68. Lyell, *Principles of Geology*, 3:132.

69. Hall, *Extracts from a Journal Written on the Coasts of Chili, Peru, and Mexico,* 2:6. Hall referred to "Sir Thomas Lauder Dick," a variant of the present standard use.

70. Alexander Caldcleugh, *Travels in South America during the Years 1819–20–21, Containing an Account of the Present State of Brazil, Buenos Ayres, and Chile,* 2 vols. (London: John Murray, 1825), 1:298.

71. Quoted in Nicol Morton, "In the Footsteps of Charles Darwin in South America," 191. I wish to express appreciation to Dr. Morton for supplying me with the text of Victor Ramous's unpublished "Field Guide to the Geology of the Principal Cordillera" prepared for the Fourth International Congress on Jurassic Stratigraphy and Geology held in Argentina in October 1994.

72. CD, "Extracts of Letters Addressed to Professor Henslow."

73. CD to Henslow, March 1834, in *Correspondence,* 1:370.

74. CD to Henslow, 18 April 1835, in ibid., 1:443.

75. CD to Susan Darwin, 23 April 1835, in ibid., 1:446. See also p. 403 for a letter from a British diplomat, writing from Rio de Janeiro, recommending Caldcleugh to Darwin.

76. Works by the authors named in this paragraph are listed in the bibliography. (For an additional listing on Gay, see the entry under "Brongniart.") The volume by Molina, held by the Cambridge University Library, is inscribed "Charles Darwin Valparaiso 1834." For a color reproduction of "Map of the Country between Buenos Ayres and the Pacific Ocean, with a Specification of the different Geological Formations" included in the Caldcleugh volumes, see Herbert, "From Charles Darwin's Portfolio," 25. The map conveys both Caldcleugh's interest and knowledge, and the highly generalized nature of topographical features represented. There are five formations indicated on the map: "Primitive," "a very new Stalactiform Limestone," "Red Marl," "Pebbles & Sand," and "Clay." The map of the pass over the Andes published by W. Eaden is stored in DAR 44, along with other maps belonging to Darwin.

77. CD to Lyell, 1 September 1844, in *Correspondence,* 3:56.

78. *Diary,* 321.

79. CD to J. S. Henslow, 12 August 1835, in *Correspondence,* 1:461.

80. Herbert, "From Charles Darwin's Portfolio," 32.

81. See "Map of the Country between the Rio de La Plata and the Pacific Ocean between the Parallels of 29°45′ and 36° South Latitude," in John Miers, *Travels in Chile and La Plata,* 2 vols. (London: Baldwin, 1826), 1:frontispiece. Miers labeled the western chain "The Central Ridge of the Cordillera of Los Andes" and the eastern chain "El Paramillo de Cordillera."

82. *GSA,* third sketch in fold-out section at rear of book. In drawing beds, Darwin generally used straight lines (one recalls William Smith's rulelike edges). Darwin's intention was schematic rather than pictorial, though his results could be consistent with later, more pictorially drawn sections. See Morton, "In the Footsteps of Charles Darwin in South America," 192.

83. DAR 36.1:501ᵛ. The Portillo chain on the east is separated from the Peuquenes line on the west by the valley of Tenuyan. On the two chains, see *GSA,* 175–187 and Sketch 1 of the fold-out section drawings.

84. As an instance of this point, see the exchange: Lyell to CD, 6 and 8 September 1838, and CD to Lyell, 14 September 1838, in *Correspondence,* 2:100, 104–105. As an indication of Darwin's regard for Élie de Beaumont, he sent him a presentation copy of the *Journal of Researches* (*Correspondence,* 2:212).

85. DAR 36.1:422.

86. DAR 34.1:60ᵛ.

87. CD to Henslow, 18 April 1835, in *Correspondence,* 1:442.

88. CD to Henslow, 12 August 1835, in ibid., 1:461.

89. CD, "On the Connexion of Certain Volcanic Phenomena in South America," 609, in *CP,* 1:60–61.

90. Ibid., 615, in *CP,* 1:68.

91. Quotations in this paragraph from ibid., 626–627, in *CP,* 1:76–78.

92. Quotations in this paragraph from ibid., 630–631, in *CP,* 1:80–82.

93. The key interpretation is Greene, *Geology in the Nineteenth Century,* 69–143. Also see Robert H. Dott Jr., "Recognition of the Tectonic Significance of Volcanism in Ancient Orogenic Belts," in Nicoletta Morello, ed., *Volcanoes and History* (Genova: Brigati, 1998), 123–131. Dott's article underlines a point made by Greene that interpretations of mountain chains were often dependent on *which* mountains the observer studied. On the secure understanding (by Darwin's time) of volcanism as being a regular, as opposed to an accidental, feature of nature, see Kenneth L. Taylor, "Volcanoes as Accidents: How 'Natural' Were Volcanoes to 18th-Century Naturalists?" in Morello, *Volcanoes and History,* 595–618.

94. Lyell quoted in Rhodes, "Darwin's Search for a Theory of the Earth," 211.

95. Ibid., 206.

96. William B. Rogers and Henry D. Rogers, "On the Physical Structure of the Appalachian Chain, As Exemplifying the Laws Which Have Regulated the Elevation of Great Mountain Chains Generally," *Reports of the First, Second, and Third Meetings of the Association of American Geologists and Naturalists* 1 (1843):474–531. In a note referring to an 1842 publication, Darwin reminded himself to consider the "curious folds" of the Falkland Islands in relation to the Rogers's work (DAR 40:37), though his later article did not cite them ("On the Geology of the Falkland Islands").

97. *Correspondence,* 3:396–397.

98. Passages in this paragraph quoted from *GSA,* 246–248.

99. "Summary on the Geological History of the Chilian Cordillera, and of the southern parts of S. America," in *GSA,* 237–248.

100. Alcide Charles Victor Dessalines d'Orbigny to CD, 14 February 1845, in *Correspondence,* 3:143–145. *GSA,* 181. Challinor, *Dictionary of Geology,* 201: "The first three stages [of the Cretaceous] are (or at least were) . . . grouped together as the Neocomian ["Neocomium" referring to Neuchâtel, Switzerland]." See the *Oxford English Dictionary* on the history of the usage of the term, which dates to the 1830s.

101. Edward Forbes to CD, after 14 February 1845, in *Correspondence,* 3:146.

102. *GSA,* 234. Also p. 238: "The fossil shells in the Cordillera of central Chile, in the opinion of all the palaeontologists who have examined them, belong to the earlier stages of the cretaceous system; whilst in northern Chile there is a most singular mixture of cretaceous and oolitic forms: from the geological relations, however, of these two districts, I cannot but think that they all belong to nearly the same epoch, which I have provisionally called cretaceo-oolitic."

103. *JR,* 406; see also *GSA,* 202.

104. Morton, "In the Footsteps of Charles Darwin in South America," 195.

105. See *Correspondence,* vol. 3 for the exchanges. For example, d'Orbigny challenged an identification by Sowerby (p. 143). In turn, d'Orbigny's list of fossils (p. 145) was annotated by both Sowerby and CD. Forbes also commented on Sowerby's identifications (p. 151).

106. *CR,* 119.

107. *VI,* 128–129. In this passage CD also cited "On the Connexion of Certain Volcanic Phenomena in South America."

108. *CR,* 140.

109. *VI,* 126. As time went on, Darwin sharpened his views on the geological age and permanence of continents and oceans. In an 1877 article his son George Howard Darwin captured the direction of Darwin's shift: "There is only one area of large extent in which we possess fairly well-marked evidence of a general subsidence; and this is the area embracing the Coral islands of the Pacific Ocean. The evidence is derived from the structure of the Coral islands, and is confirmed in certain points by the geographical distribution of plants and animals. Some naturalists are of [the] opinion that there is evidence of the existence of a previous continent; others (and amongst them my father, Mr. Charles Darwin) that there existed there an archipelago of islands." G. H. Darwin, *Scientific Papers,* 3:26–27.

110. William Montgomery, "Charles Darwin's Theory of Coral Reefs and the Problem of

the Chalk," *Earth Sciences History* 7 (1988):111–120. The quotations from the Santiago Book in this paragraph, with the exception of the last one, are from *Correspondence,* 1:568. The last quotation is from Santiago Book (Down Notebook 1.18):30.

111. See, for example, *CR,* 194: "I cannot . . . avoid suspecting that [Christian] Ehrenberg has rather under-rated the influence of corals . . . on the formation of the tertiary deposits of the Red Sea." On Christian Ehrenberg (1795–1876), who traced chalk formations of the Cretaceous to fossilized microscopic remains of Foraminifera, see Montgomery, "Charles Darwin's Theory of Coral Reefs," 112–118. Ehrenberg's theory opposed, and eventually helped to replace, a notion defended by Lyell that the chalk had largely developed from coral reefs. On the petrology of the chalk, see Challinor, *Dictionary of Geology,* 47.

112. Armstrong, *Under the Blue Vault of Heaven,* 85, 9.

113. FitzRoy, *Narrative,* 2:629.

114. John Clunies Ross quoted in Armstrong, *Under the Blue Vault of Heaven,* 13. Ross, who never met FitzRoy or Darwin in person, opposed Darwin's theory of the origin of coral reefs in the 1840s.

115. The chart was issued as number 2510. The several copies, of varying dates, of this chart that I consulted at the Library of Congress bear what appears to be a publication date of 8 November 1856 for the original version. Armstrong, who consulted charts at the Ministry of Defence Hydrographic Office's Archives at Taunton, Somerset, reported a chart from 1845 that used data FitzRoy had gathered. Armstrong, *Under the Blue Vault of Heaven,* 11. In another variation, Stoddart listed an issue date of 1855 for the chart. David R. Stoddart, "Darwin and the Seeing Eye: Iconography and Meaning in the Beagle Years," *Earth Sciences History* 14 (November 1995):21 n. 62.

116. Darwin's geological notes for the Keeling Islands are contained in DAR 41. Stoddart ("Darwin and the Seeing Eye") reproduces and analyzes the important series of field sketches and finished drawings from Darwin's Keeling notes. Armstrong quotes extensively from them.

117. *Diary,* 418.

118. FitzRoy, *Narrative,* 2:629. For a number of photographs of Keeling Island, taken to illustrate Darwin's work there, see Armstrong, *Under the Blue Vault of Heaven.*

119. *Diary,* 417.

120. The version produced in Figure 7.8 is from *CR,* 5. For comparison, see Darwin's field sketch and his hand-drawn fair sketch depicted in Stoddart, "Darwin and the Seeing Eye," 12, 13.

121. *CR,* 6.

122. Ibid., 8.

123. Ibid. The location of the measure is taken from Admiralty chart 2510.

124. Helen Rozwadowski, "Fathoming the Ocean: Discovery and Exploration of the Deep Sea, 1840–1880" (Ph.D. dissertation, University of Pennsylvania, 1996), 58. Rozwadowski argues that it was explorers, rather than natural historians, who led the way in the study of ocean depths.

125. Stoddart, "Darwin and the Seeing Eye," 11.

126. See Lyell, *Principles of Geology,* 3:359–361 (plus separately paginated glossary, 70, 82); Challinor, *Dictionary of Geology,* 142. Stoddart supplied "basalt" as a working equivalent.

127. *CP,* 1:47. Italics added.

128. *CR,* 54. The geological notes (DAR 38.1:882–891) refer to reef shapes at Mauritius that might be explained by elevatory movement, but not in the words quoted from *CR*. For a transcription of the Mauritius notes, with discussion, see Patrick Armstrong, "Charles Darwin's Geological Notes on Mauritius," *Indian Ocean Review* 1, no. 2 (1988):1–20. Darwin did not use the phrase "fringing reef" in his voyage notes.

129. The *Calendar* to the Darwin Correspondence project is useful for evaluating Darwin's lifelong pattern of communication. See Frederick Burkhardt and Sydney Smith, *A Calendar of the Correspondence of Charles Darwin, 1821–1882, with Supplement* (Cambridge: Cambridge University Press, 1994). Stoddart ("Darwin and the Seeing Eye," 15) has expressed surprise

that Darwin did not establish a correspondence with Scrope, with whom he shared a great deal, but the same could be said for a number of other geologists, as, for example, Sedgwick. Leonard Wilson has suggested that "in writing to Darwin, Lyell discussed scientific questions more fully and frankly than he did in his letters to anyone else." Wilson, *Lyell in America,* 395. On the subject of Darwin as correspondent in his later years, see Janet Bell Garber, "Charles Darwin as a Laboratory Director" (Ph.D. dissertation, University of California Los Angeles, 1989).

130. CD to Henslow, 30–31 October 1836, in *Correspondence,* 1:512.

131. Lyell to John Herschel, 24 May 1837, and Lyell to CD, 26 December 1836, in Lyell, *Life, Letters and Journals of Sir Charles Lyell,* 2:12, 1:474.

132. Stoddart notes that Lyell included some discussion of Darwin's theory in *Elements of Geology* (London: John Murray, 1838) and more in the sixth edition of the *Principles of Geology* (1840). David R. Stoddart, "Darwin, Lyell, and the Geological Significance of Coral Reefs," *British Journal for the History of Science* 9 (1976):206. Using large-scale illustrations, Lyell lectured on Darwin's coral reef theory in the United States. See Robert H. Dott Jr., "Charles Lyell in America—His Lectures, Field Work, and Mutual Influences, 1841–1853," *Earth Sciences History* 15 (1996):105; also see Wilson, *Lyell in America,* 89. On Dana's support of Darwin's coral reef hypothesis, see Appleman, "James Dwight Dana and Pacific Geology," 91–95, and David R. Stoddart, "This Coral Episode: Darwin, Dana and the Coral Reefs of the Pacific," in Roy MacLeod and Philip F. Rehbock, eds., *Darwin's Laboratory: Evolutionary Theory and Natural History in the Pacific* (Honolulu: University of Hawai'i Press, 1994), 21–48. Dana learned of Darwin's theory in 1839 while participating in the U.S. Exploring Expedition of 1838–1842.

133. The issue at hand was one raised by Dana, who asserted that when Mauna Loa erupted, there was no sympathetic eruption from Kilauea, a crater on its flanks. To Dana this argued against the "so-called *elevation* theory." Darwin believed Dana's case "*proved too much*" and defended his view that a common source of lava fed both ("it really seems monstrous to suppose that the lava within the same crater is not connected at no very great depth"). To make his case, Darwin invoked the analogy of a "high-pressure Boiler" with cracks at various points. Darwin also urged his views, consistent with those he had expressed in 1838, on Lyell, who indeed supported them in the ninth edition of the *Principles of Geology* (1853). See CD to Lyell, 24 March 1853, in *Correspondence,* 5:124–126.

134. Alcide Charles Victor Dessalines d'Orbigny, *Voyage dans l'Amérique Méridionale (le Brésil, la République orientale de l'Uruguay, la République Argentine, la Patagonie, la République du Chili, la République de Bolivia, la République de Pérou), exécuté pendant les années 1826 . . . 1833,* 9 vols. (Paris and Strasbourg: Pitous-Levrault, 1835–1847).

135. CD to Lyell, 1 September 1844, in *Correspondence,* 3:56, and n. 7 for the citation from d'Orbigny 1835–1847, vol. 3, pt. 3: *Géologie,* 85–86.

136. *GSA,* 98–99. Darwin listed a third possibility, also rejected, that the formation was thrown down by rivers before they assumed their present course.

137. Lyell, *Principles of Geology,* 1:386–395; see Wilson, *Charles Lyell: The Years to 1841,* 420, for Mary Lyell's notice of her husband's "famous fight with v. Buch & de Beaumont" over the issue, which took place in Bonn in 1835.

138. Leopold von Buch, "On Volcanos and Craters of Elevation," *Edinburgh New Philosophical Journal* 21 (1836):189. For background, see Dennis R. Dean, "Graham Island, Charles Lyell, and the Craters of Elevation," *Isis* 71 (1980):571–588. For an example of an 1850s adoption of von Buch's classification of volcanic phenomena as applied to mapping, see Barbara Mae Christy and Paul D. Lowman Jr., "Global Maps of Volcanism: Two Maps from Two Centuries," in Morello, *Volcanoes and History,* 65–90.

139. DAR 38.1:883; also in Armstrong, "Charles Darwin's Geological Notes on Mauritius," 16.

140. *VI,* 93. While Darwin published his ideas in 1844, he noted in his journal that he "Began craters of Elevation Theory" in September 1838. *Correspondence,* 2:432.

141. *CP,* 1:243. Years earlier Sedgwick had suggested that too much fuss was being made

about "elevation craters." He also suggested "local elevatory forces"—something like Darwin's solution—to explain their appearance. Sedgwick to Lyell, 20 September 1835, in Clark and Hughes, *Life and Letters of Adam Sedgwick,* 1:446.

142. Leonard G. Wilson, "Lyell: the Man and His Times," in Blundell and Scott, *Lyell,* 30. For Darwin's comments on the "craters of elevation" controversy and his ultimate full support of Lyell's field-drawn conclusions from the 1850s, see *Correspondence,* 3:54, 378–379; 4:137, 298; 6:83–84; 7:137–138; 8:183.

143. CD to Caroline Darwin, 29 April 1836, in *Correspondence,* 1:495.

144. *Notebooks,* 58 (RN:117).

145. CD to J. D. Hooker, 23 October 1859, in *Correspondence,* 7:356, and 264 for a similar remark.

146. CD to Leonard Horner, 23 December 1846–January 1847, in ibid., 3:379.

Chapter 8. Simplicity Challenged

1. Louis Agassiz, *Studies on Glaciers Preceded by the Discourse of Neuchâtel,* ed., trans., and with an introduction by Albert V. Carozzi (New York: Hafner, 1967).

2. Playfair, *Illustrations of the Huttonian Theory of the Earth,* 388–389.

3. Herries Davies, *Earth in Decay,* 265–266.

4. Quoted in Barlow, *Charles Darwin and the Voyage of the "Beagle,"* 175.

5. Patrick Syme, *Werner's Nomenclature of Colours,* 2d ed. (London: Blackwood, 1821).

6. Conrad Martens's painting "Mount Sarmiento from Warp Bay." A notation on the pencil sketch from which the painting derived reads, "The grand glacier, Mount Sarmiento. The mountain rises to about 3 times the height here seen, but all is here hidden by dark misty clouds—a faint sunny gleam lights the upper part of the glacier, giving its snowy surface a tinge which appears almost of a rose colour by being contrasted with the blue of its icy crags—a faint rainbow was likewise visible to the right of the glacier, but the whole was otherwise very grey & gloomy. June 9 1834." Initialed top right RF [Robert FitzRoy]. From Keynes, Beagle Record, 398. Also see FitzRoy, Narrative 2, facing p. 359.

7. FitzRoy, *Narrative,* 2:216–217.

8. As a side point, Venetz's first publication in English described an avalanche that occurred on 27 December 1819. He described the mass that descended from the avalanche as "nearly 150 feet high; so that the whole body of ice precipitated is about 360,000,000 cubic feet." Ignace Venetz, "Account of the Descent of the Glacier of the Weisshorn, on the 27th of December 1819, and the Destruction of the Village of Randa," *Edinburgh New Philosophical Journal* 3 (1820):276.

9. References in this paragraph are to FitzRoy, *Narrative,* 1:337; DAR 32.2:122; CD, "On the Distribution of the Erratic Boulders," in *CP,* 158; DAR 32.2:121; DAR 32.2:122; DAR 32.2:122^{v}; DAR 32.2:122 and FitzRoy, *Narrative,* 2:216; *Diary,* 139–140; and von Buch, *Travels through Norway and Lapland,* 305.

10. CD, "On the Distribution of the Erratic Boulders," in *CP,* 1:151.

11. Phillip Parker King, "Some Observations upon the Geography of the Southern Extremity of South America, Tierra del Fuego, and the Strait of Magalhaens," *Journal of the Royal Geographical Society* 1 (1831):155–175.

12. DAR 32.2:111–117; DAR 34.2:177; *GSA,* 151–156.

13. One contrary example should be cited from DAR 34.2:143^{v}: "Glaciers & floating Icebergs have frequently been brought forward as the means of Transportal." This comment, however, appears as a later addition, of uncertain date, to Darwin's original run of manuscript notes.

14. DAR 34.2:171

15. DAR 34.2:169.

16. DAR 34.2:150^{v}; William Buckland, "On the Excavation of Valleys by Diluvial Action, as Illustrated by a Succession of Valleys Which Intersect the South Coast of Dorset and Devon," *Transactions of the Geological Society of London,* 2d ser., 1 (1824):95–102.

17. DAR 34.2:171^{v}.

18. *Correspondence,* 2:431. The notation was made in what Darwin termed his "Journal," a small notebook opened in August 1838 that he used until 1881 to record the major events in his life.

19. CD to W. D. Fox, 28 August [1837], in *Correspondence,* 2:39.

20. Lyell, *Principles of Geology,* 3:148.

21. Lyell, "Presidential Address [1836]," 382–383.

22. Lyell, "Presidential Address [1837]," 507.

23. *JR,* 288. The theory of transport by icebergs also solved the problem presented by the often-great depths of water in channels at Tierra del Fuego, depths that argued against transportal of rocks by ordinary tides and currents, or by sudden rushes of water (DAR 34.2:171).

24. *Notebooks,* 26 (RN:20).

25. Alexander von Humboldt, "On Isothermal Lines," *Edinburgh New Philosophical Journal* 3 (1820):1–20, 256–274; 4 (1821):23–37, 262–281; 5 (1821):28–39.

26. Lyell, *Principles of Geology,* 1:106.

27. The quotations in this paragraph are from *JR,* 291–292.

28. Ibid., 286.

29. Quotations in this paragraph from ibid., 606.

30. Bruce Hevly, "The Heroic Science of Glacier Motion," *Osiris,* 2d ser., 11 (1996):66–86.

31. FitzRoy, *Narrative,* 1:574.

32. Edward Lurie, *Louis Agassiz: A Life in Science* (Baltimore: Johns Hopkins University Press, 1988), 371–377.

33. Quotations in this paragraph are from *JR,* 616, 619, 621, and Lyell, *Principles of Geology,* 1:112.

34. Adam Sedgwick to Louis Agassiz, 5 March 1838, in Clark and Hughes, *Life and Letters of Adam Sedgwick,* 1:503–504.

35. CD, "Observations on the Parallel Roads of Glen Roy," 72, in *CP,* 1:122.

36. Elizabeth Wedgwood to CD, 10 November 1837, in *Correspondence,* 2:55. It was at Maer Hall, the Wedgwood home in Staffordshire, where Darwin's investigation of the action of earthworms took place at the instigation of his uncle, Josiah Wedgwood II (1769–1843).

37. *Correspondence,* 2:443. "Europe, yes" and "If I travel . . . Zoological" are circled in pencil. This note appears to have been written after 7 April but before July 1838. The key document summarizing Darwin's assessment of his problems with health that were to soon foreclose geological travel appears in *Correspondence,* 13:481–484. According to this document, his great problems with health arose in 1840–1841. He commented (p. 482) that he "cannot walk above 1/2 mile."

38. See notes entitled "Alluvium/Shropshire/July 1837." DAR 5:23–29.

39. References in this paragraph to CD's calendar are to *Correspondence,* 2:29, 84, 431–432.

40. CD to W. D. Fox, 15 June 1838, in ibid., 2:91.

41. Ibid., 2:432. DAR 5:19–22; "Salisbury Craigs/June 1838/Geological Notes" are in DAR 5:33b-37. The Glen Roy notebook is catalogued as DAR 130 and has been published in *Notebooks,* 141–165.

42. *Correspondence,* 2:435.

43. DAR 42:206: "Any person <merely> look at <Secondary or supermedial formation> Geolog: Map of England. would believe that the different formations had successively silted up an ocean. so that the margins successively cropped out towards the <old> coast lines.—" Darwin then argued against this view on the grounds that the animals, present in fossilized form in many beds, could not have lived in so deep an ocean. He preferred to credit oscillations in "elevation, upheaval and subsidence acting with remarkable equality over whole England."

44. DAR 5:B33–B37: "Salisbury Craigs/June 1838/Geological Notes"; also DAR 5:B38: notes on Thomas Allan, "On the Rocks in the Vicinity of Edinburgh," *Transactions of the Royal*

Society of Edinburgh 6 (1812):405–433. For a guide to the Salisbury Crags, see G. P. Black, "Arthurs's Seat," in A. D. McAdam and E. N. K. Clarkson, eds., *Lothian Geology: An Excursion Guide* (Edinburgh: Scottish Academic Press, 1986), 38–41.

45. Quotations in paragraph are from Louis Agassiz, "Upon Glaciers, Moraines and Erratic Blocks: Being the Address Delivered at the Opening of the Helvetic Natural History Society, at Neuchatel [*sic*], on 24th July 1837," *Edinburgh New Philosophical Journal* 24 (1838):378, 382.

46. Herries Davies, *Earth in Decay,* 182–196, 237, 254.

47. DAR 36.1:435. On 434[v] Darwin referred to Playfair.

48. Playfair, *Illustrations of the Huttonian Theory of the Earth,* 351.

49. DAR 5:19.

50. John MacCulloch, "On the Parallel Roads of Glen Roy." *Transactions of the Geological Society of London* 4, part 2 (1817):314, 392.

51. Thomas Dick Lauder, "On the Parallel Roads of Lochaber," *Transactions of the Royal Society of Edinburgh* 9 (1823):1. Lauder's paper was read 2 March 1818. He reported (p. 62) that he read MacCulloch's paper in January 1818, after completing the major portion of his own paper.

52. MacCulloch, "On the Parallel Roads of Glen Roy," 314, 319.

53. George Fennell Robson, *Scenery of the Grampian Mountains* (1814; repr. London: Longman, 1819), preface.

54. References in this paragraph are to Lauder, "On the Parallel Roads of Lochaber," 1, 2, 4, 14, 16, 17, 27, 40.

55. Albert Way to CD, 9 December 1838, in *Correspondence,* 2:140.

56. CD, "Observations on the Parallel Roads of Glen Roy," 39. For MacCulloch's primary list of measurements of the elevation of the roads, see his "On the Parallel Roads of Glen Roy," 327. From his own barometrical measurements, Darwin suspected that MacCulloch's measurements were too high, which was justified, though it did not affect the argument (Rudwick, "Darwin and Glen Roy," 184).

57. MacCulloch, "On the Parallel Roads of Glen Roy," 379–385.

58. Lauder, "On the Parallel Roads of Lochaber," 51.

59. *Notebooks,* 146–148 (Glen Roy Notebook, 13, 16–17, 27–28). For a list of features both for and against the marine interpretation, see Rudwick, "Darwin and Glen Roy," 138–139.

60. Lauder, "On the Parallel Roads of Lochaber," 12.

61. Darwin, "Observations on the Parallel Roads of Glen Roy," 55, in *CP,* 106.

62. Darwin, "Observations on the Parallel Roads of Glen Roy," 63–64, in *CP,* 113–114.

63. Darwin, "Observations on the Parallel Roads of Glen Roy," 68–71, in *CP,* 1:118–122.

64. Murchison, *Silurian System,* quoted portions from pp. 509–510. Murchison's role in establishing the term "drift" needs to be emphasized since others, such as Lyell, are sometimes credited with this term.

65. See DAR 5:B31–B32.

66. Passages referred to are from ibid., 523, 543–544, 532–533, 546–547, and from Darwin, "Observations on the Parallel Roads of Glen Roy," 63–64, in *CP,* 113–114. Darwin also suggested that shells in Lochaber might have more easily dissolved owing to the gravel and sand being derived from granitic rocks, and thus not containing any free carbonate of lime.

67. Charles Lyell, "On the Boulder Formation or Drift, and Associated Freshwater Deposits Composing the Mud Cliffs of Eastern Norfolk [read 22 January 1840]," *Proceedings of the Geological Society of London* 3 (November 1838–June 1842):176.

68. Louis Agassiz, *Études sur les glaciers* (Neuchâtel: Jent et Gassmann, 1840). See also English translation by Carozzi (1967). In this chapter I will cite the English version as Agassiz, *Studies on Glaciers.*

69. CD, "Geology," 183, in *CP,* 1:229.

70. Agassiz, *Studies on Glaciers,* 32.

71. William Buckland, "Evidences of Glaciers in Scotland and the North of England [read 4 and 18 November and 2 December 1840]," *Proceedings of the Geological Society of London* 3 (November 1838–June 1842):332–337, 345–348, and Charles Lyell, "On the Geological Evidence of the Former Existence of Glaciers in Forfarshire [read 18 November and 2 December 1840]," *Proceedings of the Geological Society of London* 3 (November 1838–June 1842):337–345. Boylan has provided an excellent field guide to the joint work of Agassiz, Buckland, and Lyell, which is available as a document at www.city.ac.uk/artspol/glacloc.html.

72. Darwin attended the council meetings on 8 and 22 January but none later in the year. Geological Society Archives CM 1/5 (1837–1842). I am grateful to Wendy Cawthorne, assistant librarian of the Geological Society Library, for rechecking this information for me.

73. CD to Asa Gray, 4 August 1863, in *Correspondence,* 11:582 and n. 23.

74. Louis Agassiz to Robert Jameson, 3 October 1840, printed in "Discovery of the Former Existence of Glaciers in Scotland, Especially in the Highlands, by Professor Agassiz," *Scotsman* 66 (7 October 1840).

75. Louis Agassiz, "On Glaciers, and the Evidence of Their Having Once Existed in Scotland, Ireland, and England." *Proceedings of the Geological Society of London* 3 (November 1838–June 1842):331–332.

76. Buckland, "Evidences of Glaciers in Scotland and the North of England," 333.

77. Agassiz, "On Glaciers, and the Evidence of Their Having Once Existed in Scotland, Ireland, and England," 328, 329.

78. Citations from the 18 November debate are taken from Woodward, *History of the Geological Society of London,* 141.

79. Roderick Murchison to Adam Sedgwick, 26 September 1840, in Archibald Geikie, *Memoir of Sir Roderick Murchison,* 2 vols. (London, John Murray, 1875), 1:307.

80. CD to Louis Agassiz, 1 March 1841, in *Correspondence,* 2:284.

81. CD to Lyell, 9 March 1841, in ibid., 2:285.

82. CD to Lyell, 12 March 1841, in ibid., 2:86.

83. CD to William Lonsdale, 14 April 1841, in ibid., 2:289. An expanded version of the paper was printed in the *Transactions* of the society in 1842.

84. CD, "On the Distribution of the Erratic Boulders and on the Contemporaneous Unstratified Deposits of South America," (*Transactions*), 425, in *CP,* 1:156.

85. CD, "On a Remarkable Bar of Sandstone off Pernambuco, on the Coast of Brazil," *London, Edinburgh and Dublin Philosophical Magazine and Journal of Science,* 3d ser., 19 (1841):257–260, in *CP,* 1:139–142.

86. Wilson, *Lyell in America,* 6–7, 128–129.

87. Joshua Trimmer, "On the Diluvial Deposits of Caernarvonshire, between the Snowdon Chain of Hills and the Menai Strait, and on the Discovery of Marine Shells in Diluvial Sand and Gravel on the Summit of Moel Tryfane, near Caernarvon, 1000 ft above the Level of the Sea," *Proceedings of the Geological Society of London* 1 (November 1826–June 1833):331–332 (quotation, 331). Bernard Smith and T. Neville George, *British Regional Geology: North Wales,* 3d ed. (London: Her Majesty's Stationery Office, 1961), 84: "Although the Irish Sea Ice was not able to override completely the mountains of North Wales, the great force of its movement may be gauged by the occurrence of patches of such ice-dredged marine sand at heights of nearly 1,400 ft on Moel Tryfaen between Carnarvon and Snowdon." For a map of the area presumed affected by Irish Sea Ice, see p. 83. Trimmer's "1000'" was clearly an estimate, made before the Ordnance Survey entered the region.

88. Herries Davies, *Earth in Decay,* 248, 249, 288, 289, 297, 298, 301, 312.

89. John E. Bowman, "On the Question, Whether There Are Any Evidences of the Former Existence of Glaciers in North Wales?" *Philosophical Magazine* 19 (1842):469–479. The passage quoted is from pp. 477–478.

90. Quotations in this and the following paragraph are from William Buckland, "On the Glacia-Diluvial Phænomena in Snowdonia and the Adjacent Parts of North Wales [read 15 December 1841]," *Proceedings of the Geological Society of London* 3 (November 1838–June

1842):579–584. For discussion of Buckland's fieldwork, I am indebted to Michael Roberts, "Buckland, Darwin and the Ice Age in Wales and the Marches, 1841–1842" (paper presented at the International Conference on the History of Geology in Neuchâtel, September 1998).

91. *Correspondence,* 2:435.

92. CD, "Notes on the Effects Produced by the Ancient Glaciers of Caernarvonshire, and on the Boulders Transported by Floating Ice," *Edinburgh New Philosophical Journal* 23 (1842):352, in *CP,* 1:163.

93. *CD,* "Notes on the Effects Produced by the Ancient Glaciers of Caernarvonshire," 358.

94. Kenneth Addison, *Classic Glacial Landforms of Snowdonia* (Sheffield: Geographical Association, 1997), 11, and 57 for glossary.

95. DAR 27.1:7.

96. Quotations in this paragraph are from CD, "Notes on the Effects Produced by the Ancient Glaciers of Caernarvonshire," 354.

97. Jean de Charpentier, "Account of One of the Most Important Results of the Investigations of M. Venetz, Regarding the Present and Earlier Condition of the Glaciers of the Canton Vallais," *Edinburgh New Philosophical Journal* 21 (1836): 216–217. Agassiz also referred to blocks disappearing by falling into crevasses, in *Studies on Glaciers,* 59–61, 66, 69.

98. CD, "Notes on the Effects Produced by the Ancient Glaciers of Caernarvonshire," 354. Examples of Agassiz's plates showing boss- or dome-formed rocks are shown in Figures 8.6 and 8.7.

99. Agassiz, *Studies on Glaciers,* 144. The word "cirque" does appear in Agassiz's work (p. 69) but simply to mean a circular-shaped moraine.

100. CD, "Notes on the Effects Produced by the Ancient Glaciers of Caernarvonshire," 361–363, for quotations in this paragraph.

101. DAR 27.1.

102. CD to W. H. Fitton, ca. 28 June 1842, in *Correspondence,* 2:322.

103. *Autobiography,* 70.

104. CD to W. D. Fox, 4 September 1843, in *Correspondence,* 2:387.

105. CD to Lyell, 5 and 7 October 1842, in ibid., 2:338.

106. CD to W. D. Fox, 9 December 1842 and 4 September 1843, in ibid., 2:345, 387.

107. CD to W. H. Fitton, ca. 28 June 1842, in ibid., 2:322.

108. Quotations in this paragraph from CD to James Geikie, 16 November 1876, in F. Darwin, *Life and Letters of Charles Darwin,* 1888, 3:213–214. Darwin's "Rough notes on Southampton gravel, 1874" are contained in DAR 52.

109. James Geikie, *Prehistoric Europe: A Geological Sketch* (London: E. Stanford, 1881), 141–142. Included in *CP,* 2:233–235.

110. CD to John Phillips, November 1840, in *Correspondence,* 2:273; CD to Adolf von Morlot, 10 October 1844, in ibid., 3:65; CD to Charles Woodd, 4 March 1850, in ibid., 4:317.

111. CD to T. H. Huxley, 2 June 1859, in ibid., 7:301; CD to Lyell, 23November 1859, in ibid., 7:392. See also 7:359 and 7:385 for Darwin's use of the term "monomaniac."

112. *Autobiography,* 84. Italics added.

113. On eliminative induction, see David L. Hull, "Charles Darwin and Nineteenth-Century Philosophies of Science," in Ronald N. Giere and Richard S. Westfall, eds., *Foundations of Scientific Method* (Bloomington: Indiana University Press, 1974), 128. For a previous interpretation of the passage from the *Autobiography,* see Rudwick, "Darwin and Glen Roy," 167–169. I concur with Rudwick that it was Darwin's degree of trust in the method of eliminative induction that was at issue, but I differ from him in associating Darwin's use of the method with Whewell. Darwin used the "principle of exclusion" during the *Beagle* voyage, which ended before Whewell had published his historical and philosophical works. In contrast, Herschel's work, the importance of which Rudwick also stressed, was fully available to Darwin in time to be useful.

114. CD to T. F. Jamieson, 6 September 1861, in *Correspondence,* 9:256. Italics added.

115. CD to Buckland, November 1840–17 February 1841, in ibid., 13:356. I thank Michael Roberts for the original reference.

116. CD to Henry Walter Bates, 22 November 1860, in ibid., 8:484–485.

117. CD to W. D. Fox, 4 September 1843, in ibid., 2:387

118. CD to Leonard Horner, 17 August–7 September 1846, in ibid., 3:333–334.

119. David Milne, "On the Parallel Roads of Lochaber, with Remarks on the Change of Relative Levels of Sea and Land, and on the Detrital Deposits in that Country," *Transactions of the Royal Society of Edinburgh* 16 (1849):399.

120. For Darwin's draft of his letter, see *Correspondence,* 7:76–80.

121. Robert Chambers, *Ancient Sea-Margins, As Memorials of Changes in the Relative Level of Sea and Land* (Edinburgh: W. and R. Chambers, 1848), 108, 113–114, 270–280, 318–319. The passage quoted is from p. 108. See also Secord, *Victorian Sensation,* 391–399, 426–436.

122. George Mackenzie, "An Attempt to Classify the Phenomena in the Glens of Lochaber with those of the Diluvium, or Drift, Which Covers the Face of the Country," *Edinburgh New Philosophical Journal* 44 (1848):2.

123. James Thomson, "On the Parallel Roads of Lochaber," *Edinburgh New Philosophical Journal* 14 (1848):52. Rudwick, "Darwin and Glen Roy," 147–148.

124. CD To A. C. Ramsay, 26 June 1859, in *Correspondence,* 7:311. In this letter Darwin also praised Andrew Crombie Ramsay's "The Old Glaciers of Switzerland and North Wales," in John Ball, ed., *Peaks, Passes, and Glaciers: A Series of Excursions by Members of the Alpine Club,* ser. 1 (London: Longman, 1859–60), 400–466. After De la Beche's death in 1855, Ramsay "became nearly as Lyellian as his friend Darwin." Secord, "Geological Survey," 258.

125. CD to J. D. Hooker, 20 January 1859, in *Correspondence,* 7:236.

126. T. F. Jamieson to CD, 24 March 1862, in ibid., 10:132–133. The Ordnance Survey surveyed the area in May 1858. For the relevant list of benchmarks, see Henry James, *Abstracts of the Principal Lines of Spirit Levelling in Scotland* (London: Eyre and Spottiswoode, 1861), 254–255. Also see Henry James, *Notes on the Parallel Roads of Lochaber . . . with Illustrative Maps and Sketches from the Ordnance Survey of Scotland* (Southampton: Ordnance Survey Office, 1874).

127. T. F. Jamieson to Lyell, 5 October 1861, in *Correspondence,* 9:452. *Correspondence,* 9:429–459, reprints "Letters between Thomas Francis Jamieson and Charles Lyell on the Geology of Glen Roy, Scotland."

128. Bowdoin A. Van Riper, *Men Among the Mammoths: Victorian Science and the Discovery of Human Prehistory* (Chicago: University of Chicago Press, 1993).

129. T. V. Jamieson to CD, 3 September 1861, in *Correspondence,* 9:247.

130. T. V. Jamieson to Lyell, 22 and 27 October and 25 September 1861, in ibid., 9:456–457, 446–447.

131. CD to Lyell, 1 April 1862, in ibid., 10:144; for background, see pp. 136–137.

132. T. F. Jamieson to Lyell, 25 September 1861, in ibid., 9:448.

133. Thomas Francis Jamieson, "On the Parallel Roads of Glen Roy, and Their Place in the History of the Glacial Period," *Quarterly Journal of the Geological Society of London* 19 (1863):247. Also see his "On the Ice-Worn Rocks of Scotland," *Quarterly Journal of the Geological Society of London* 18 (1862):164–184, particularly 178–180 on the weight of moving land-ice. The quoted passage is from his "On the History of the Last Geological Changes in Scotland," *Quarterly Journal of the Geological Society of London* 21 (1865):165.

134. CD to Lyell, 11 August 1862, in *Correspondence,* 10:378.

135. Charles Lyell, *The Geological Evidences of the Antiquity of Man* (London: John Murray, 1863), 252–264.

136. Quoted in *Correspondence,* 11:218 n.1.

137. CD to C. Lyell, 6 March 1863, in ibid., 11:208.

138. "Charles Darwin's Essay of 1844," in Charles Darwin and Alfred Russel Wallace, *Evolution by Natural Selection* (Cambridge: Cambridge University Press, 1958), 180–182. Darwin regretted that Edward Forbes "forestalled" him when he published ideas in 1845. See CD to Hooker, 27 June 1845, in *Correspondence,* 3:207; *Autobiography,* 124–125. Darwin further explored the effects of glacial cold on the distribution of species in Robert C. Stauffer, ed., *Charles*

Darwin's Natural Selection: Being the Second Part of His Big Species Book Written from 1856 to 1858 (Cambridge: Cambridge University Press, 1975):534–566; also see *Origin,* 365–382.

139. The passages cited in this paragraph are from CD to J. D. Hooker, 15 March, 30 March, and 23 October 1859, in *Correspondence,* 7:264, 271–272, 356.

140. CD to A. C. Ramsay, 5 September 1862, in *Correspondence,* 10:397.

141. James Croll, *Climate and Time in Their Geological Relations: A Theory of Secular Changes of the Earth's Climate* (London: Daldy, Tsbister, 1875). For discussion of Croll and a history of subsequent developments of glacial theory, see John Imbrie and Katherine Palmer Imbrie, *Ice Ages: Solving the Mystery* (Cambridge: Harvard University Press, 1979).

Chapter 9. Geology and Species

1. CD to John Murray, 31 May 1859, in *Correspondence,* 7:300.

2. Quoted in Maureen McNeil, *Under the Banner of Science: Erasmus Darwin and His Age* (Manchester: Manchester University Press, 1987), 32.

3. Griffith's edition of Cuvier's *Animal Kingdom* referred at points to transmutationist views, not always to refute them. See Herbert, "From Charles Darwin's Portfolio," 28. In addition, the *Dictionnaire classique d'histoire naturelle,* edited by J. B. Bory de Saint-Vincent (17 vols.; Rey and Gravier, 1822–1833), reflected some of the diversity of opinion then current in French natural history. For example, vol. 4 (pp. 378–379) had two tables of classification for mollusks, one "after the method" of Lamarck, the other "after the method" of Cuvier. Interestingly, Étienne Geoffroy Saint-Hilaire (1772–1844), who was open to evolutionist theorizing, was a contributor to the *Dictionnaire.* On the climate of opinion in French biology during the 1820s and 1830s, see Appel, *Cuvier-Geoffroy Debate,* 105–174. On Grant, see Desmond and Moore, *Darwin,* 33–40.

4. *Autobiography,* 49.

5. Richard Darwin Keynes, ed., *Charles Darwin's Zoology Notes and Specimen Lists from H.M.S. Beagle* (Cambridge: Cambridge University Press, 2000).

6. Duncan M. Porter, ed., "Darwin Notes on *Beagle* Plants," *Bulletin of the British Museum (Natural History) Historical Series* 14, no. 2 (1987):145–233, and "Charles Darwin's Vascular Plant Specimens from the Voyage of H.M.S. *Beagle,*" *Botanical Journal of the Linnean Society* (1986):1–172. Kenneth G. V. Smith, ed., "Darwin's Insects," *Bulletin of the British Museum (Natural History) Historical Series* 14, no. 1 (1987):1–143. Museum collections are also of interest. See, for example, Gordon Chancellor, Angelo DiMauro, Ray Ingle, and Gillian King, "Charles Darwin's *Beagle* Collections in the Oxford University Museum," *Archives of Natural History* 15:197–231.

7. *Autobiography,* 118.

8. Keynes, *Charles Darwin's Zoology Notes,* 300, my reading substituting "their" for "the."

9. Ibid., 188.

10. Ibid., 291.

11. [Marsha Richmond], "Darwin's Study of the Cirripedia," in *Correspondence,* 4:388. Richmond's essay is an excellent introduction to Darwin's treatment of invertebrates generally.

12. Keynes, Charles Darwin's *Zoology Notes,* 116 n. 1, 139.

13. *CR,* 9. Porter in "Darwin's Notes on *Beagle* Plants" includes photographs of several of Darwin's specimens of coralline algae.

14. Mary P. Winsor, "Barnacle Larvae in the Nineteenth Century," *Journal of the History of Medicine and Allied Sciences* 24 (July 1969):304–305. See also Alan C. Love, "Darwin and *Cirripedia* Prior to 1846: Exploring the Origins of Barnacle Research," *Journal of the History of Biology* 35 (2002):251–289; and Stott, *Darwin and the Barnacle,* 53–54, 62–63, 66, 72, 98–109, 122–124, 155–156, 162, 179–180, 182–183, 197–200, 207–221, 245, 253.

15. M. J. S. Hodge, "Darwin as a Lifelong Generation Theorist," in David Kohn, ed., *The Darwinian Heritage* (Princeton: Princeton University Press, 1985), 207–243, and Phillip R. Sloan, "Darwin's Invertebrate Program, 1826–1836: Preconditions for Transformism," 71–

120. Also see Sandra Herbert, "The Logic of Darwin's Discovery" (Ph.D. dissertation, Brandeis University, 1968), 103–137.

16. Quotations in the paragraph are from Keynes, *Charles Darwin's Zoology Notes,* 48, 201, 209, 210, 245, 138, 28, 130.

17. Ibid., 283. Another interesting comment on adaptation made in 1833 reads (p. 175), "This absence of Coprophagous beetles appears to me to be a very beautiful fact; as showing a connection in the creation between animals as widely apart as Mammalia & . . . Insects. Coleoptera, which when one of them is removed out of its original zone, can scarcely be produced by a length of time & the most favourable circumstances.—" I read "creation" rather than "creating" (manuscript DAR 30.2:200).

18. Cuvier, *Animal Kingdom,* 1:1.

19. Darwin and Wallace, *Evolution by Natural Selection,* 67, 69, 76, 84, 85. In searching for Darwin's use of the word I have relied on "Concordance to Charles Darwin's 'Sketch of 1842' and 'Essay of 1844'," produced by James R. Fleming.

20. Lyell, *Principles of Geology,* 2:324, and index p. 76; Cuvier, *Animal Kingdom,* 5:26.

21. DAR 32.1:48.

22. DAR 32.1:72.

23. DAR 32.1:125–125[v].

24. *Notebooks,* 66 (RN:142). Much later Murchison queried his own judgment, referring the specimens possibly to Upper Silurian(?) and Devonian. Roderick Impey Murchison, *Siluria: The History of the Oldest Fossiliferous Rocks and Their Foundations,* 3d ed.(London: John Murray, 1859), 455. The same query also occurred in the 1867 and 1872 editions. Also see Stafford, *Scientist of Empire,* 88 n. 177.

25. DAR 32.2:69.

26. DAR 32.2:119.

27. Lyell, *Principles of Geology,* 2:127, 126. For an overview of the subject of geographical distribution, see Janet Browne, *The Secular Ark: Studies in the History of Biogeography* (New Haven: Yale University Press, 1983). On the Linnaeus connection, see Robert C. Stauffer, "Ecology in the Long Manuscript Version of Darwin's *Origin of Species* and Linnaeus' *Oeconomy of Nature,*" *Proceedings of the American Philosophical Society* 104 (1960):235–241. On the Lyell-Darwin link, see M. J. S. Hodge, "Darwin and the Laws of the Animate Part of the Terrestrial System (1835–1837): On the Lyellian Origins of His Zoonomical Explanatory Program," *Studies in History of Biology* 6 (1983):1–106.

28. DAR 32.1:64.

29. DAR 32.1:67.

30. Cuvier, *Recherches sur les ossemens fossiles.* At the time the *Beagle* departed, there had also been a second (1821–1824) and a third (1825) edition. Volume 11 of Griffith's edition of Cuvier's *Animal Kingdom* was entitled "The Fossil Remains of the Animal Kingdom" and was almost certainly part of the *Beagle*'s library.

31. Rudwick, *Meaning of Fossils,* 104–107.

32. Henslow to CD, 31 August 1833, in *Correspondence,* 1:326–327.

33. DAR 32.1:59[v]. The newspaper carrying the account is not known to me, but the context of Darwin's note suggests that he read the account after he had done his own collecting. There is a marginal date of 20 November, which may indicate when he read a newspaper report. Relevant journal publications are Woodbine Parish, "An Account of the Discovery of Portions of Three Skeletons of the Megatherium in the Province of Buenos Ayres in South America [read 13 June 1832]," *Proceedings of the Geological Society of London* 1 (November 1826–June 1833):403–404, and Buckland, "On the Fossil Remains of the Megatherium [read 23 June 1832]," 104–107. For the full account of Parish's specimen, which also discusses Darwin's *Megatherium* find, and which includes a plate of what was then interpreted as dermal armor plates of the animal, see William Clift, "Some Account of the Remains of the Megatherium Sent to England from Buenos Ayres by Woodbine Parish, Jun.," *Transactions of the Geological Society of London,* 2d ser., 3 (1835):437–460.

34. Cuvier, *Animal Kingdom,* 11:132.

35. Buckland, "On the Fossil Remains of the Megatherium," 106–107. For Darwin's comment that he was reading the volume containing Buckland's report, see *Correspondence,* 1:370.

36. DAR 32.1:65–66.

37. DAR 236.

38. In DAR 32.1:66^{v} Darwin cited Thomas Falkner, *A Description of Patagonia, and the Adjoining Parts of South America* (Hereford: C. Pugh, 1774), 55. A "span" was formerly an English unit of measure equal to about nine inches.

39. DAR 32.1:74^{v}, 66. The "Megatherium hide" phrase appeared in an appendix to the original notes dated 1833. DAR 32.1:66^{v}; *Correspondence,* 1:276, 301, 312; *JR,* 96; *Notebooks,* 37 (A:9), 188–189 (B:69–70), 195 (B:9).

40. Woodbine Parish, *Buenos Ayres and the Provinces of the Rio de la Plata* (London: John Murray, 1838):178b–e plus frontispiece, including reduced drawing of original sketch of *Glyptodon* with the feet added conjecturally, a side view of a molar tooth, and the grinding surface of the same tooth. The alphabetically identified pages (178b-e) and the frontispiece were added after the book had already been set in print; the *Glyptodon* was clearly a newsworthy find.

41. Richard Owen, "Description of a Tooth and Part of the Skeleton of the *Glyptodon clavipes,* a Large Quadruped of the Edentate Order, to which Belongs the Tesselated Bony Armour Described and Figured by Mr. Clift with a Consideration of the Question Whether the Megatherium Possessed an Analogous Dermal Armour [read 23 March 1839]," *Transactions of the Geological Society of London* 6 (1842):97, 105.

42. *A Descriptive and Illustrated Catalogue of the Fossil Organic Remains of Mammalia and Aves Contained in the Museum of the Royal College of Surgeons of England* (London: Richard and John E. Taylor, 1845), 107–120 plus plates. In the catalogue *Glyptodon* is listed as a genus in the order Edentata under the class Mammalia.

43. The genus *Hoplophorus* had been established by the Danish naturalist Peter Wilhelm Lund (1801–1880) in 1839 on the basis of specimens from an area in Brazil. Lund stressed that several families of mammals were exclusively found in America formerly, as they are at present. Peter Wilhelm Lund, "List of Fossil Mammifera from the Basin of the Rio das Velhas, with an Extract of Some of Their Distinguishing Characters," *Annals of Natural History* 3 (1839):422–427.

44. *Fossil Mammalia,* 106–108.

45. On the osseous plates of an armadillo-like animal, see *JR* (1845), 82, 155, 173, 371.

46. *Notebooks,* 175, 184 (B:20, 53–54).

47. Darwin and Wallace, *Evolution by Natural Selection,* 188.

48. Buckland, *Geology and Mineralogy Considered with Reference to Natural Theology,* 1:159–162 plus 2:Plate 5, Figures 12, 13.

49. Owen, "Description of a Tooth and Part of the Skeleton of the *Glyptodon clavipes,*" 96.

50. DAR 205.9:65.

51. The essay has been published in Herbert, "From Charles Darwin's Portfolio," 29–36.

52. Quotations in paragraph are from Darwin's text as transcribed in Herbert, "From Charles Darwin's Portfolio," 31–32. The word in bold script is a later annotation. Also see *Notebooks,* 25 (RN:19), where CD, citing material from De la Beche's *Geological Manual,* stated his view that life is rare at the bottom of the sea. For investigation of these ideas by De la Beche, William Broderip (1789–1859), and Edward Forbes, see Philip F. Rehbock, "The Early Dredgers: 'Naturalizing' in British Seas, 1830–1850," *Journal of the History of Biology* 12 (1979):311–323.

53. Quotation from Darwin is in Herbert, "From Charles Darwin's Portfolio," 32–33 and note 41 for references to Darwin's probable use of Cuvier, *Animal Kingdom,* vol. 5.

54. Lyell, *Principles of Geology,* 3:58. On this point, see Rudwick, "Charles Lyell's Dream of a Statistical Palaeontology."

55. Herbert, "From Charles Darwin's Portfolio," 33.

56. DAR 42:97–99. All quotations referred to "February 1835" are from this document.

The first three folios are published in Hodge, "Darwin and the Laws of the Animate Part of the Terrestrial System," 19–20.

57. Buckland, "On the Occurrence of the Remains of Elephants, and Other Quadrupeds," 345.

58. Keynes, Charles Darwin's *Zoology Notes,* 236–237; *JR,* 363–364.

59. Lindley Darden, *Theory Change in Science: Strategies from Mendelian Genetics* (Oxford: Oxford University Press, 1991), 105, 116–117.

60. Darwin's contemporary remarks are in DAR 33:246 and *Diary,* 213–214; his later remarks in *Fossil Mammalia,* 11. Owen's treatment of the fossil is in *Fossil Mammalia,* 35–56. On a beach near Monte Hermoso footprints of *Macrauchenia* dating from approximately 12,000 years ago were identified by Silvia A. Aramayo and Teresa M. de Bianco in 1986. For a photograph of the prints, see Keynes, *Fossils, Finches and Fuegians,* 112.

61. Edward J. Larson, *Evolution's Workshop: God and Science on the Galápagos Islands* (New York: Basic Books, 2001); Estes, Grant, and Grant, "Darwin in Galápagos," 343–368.

62. CD to Caroline Darwin, 19 July–12 August 1835, in *Correspondence,* 1:458.

63. *Diary,* 356.

64. Cuvier, *Animal Kingdom,* 9:63–64. In the 1820s names for the Galápagos tortoise had already been proposed, though Darwin did not cite any scientific name in his contemporary notes. Van Denburgh and Pritchard suggested that Quoy and Gaimard offered the first proposal for scientific names for the Galápagos tortoise in 1824 from a specimen they believed came from California. John Van Denburgh, "The Gigantic Land Tortoises of the Galapagos Archipelago," *Proceedings of the California Academy of Sciences,* 4th ser., 1, pt. 1 (1914):244–245. Peter C. H. Pritchard, *The Galápagos Tortoises: Nomenclatural and Survival Status* (Lunenberg, MA: Chelonian Research Monographs, 1996), 21. On Darwin's changing treatment of the tortoises, see Frank J. Sulloway, "Darwin and the Galapagos," *Biological Journal of the Linnean Society* 21 (1984):33–42. Sulloway speculates, plausibly, that as of 1835 Darwin might have presumed that the Galápagos tortoises were of the same species as the Aldabra tortoise.

65. *JR,* 466. For the current view that tortoises arrived at the islands already as giant forms, see Pritchard, *Galápagos Tortoises,* 18–19.

66. Keynes, *Charles Darwin's Zoology Notes,* 297; *JR,* 474.

67. Quoted in Barlow, *Charles Darwin and the Voyage of the Beagle,* 247. I have dated the passage from a photocopy of the manuscript (1.17). The entry on the mockingbirds occurs directly following an entry containing material that appears in the diary entry for 21 September (*Diary,* 354). Shortly following the field notebook entry is a passage comparing St. Jago in the Cape Verde Island to the landscape at hand, which is presumably Chatham Island (for the published version of this point, see *GSA,* 99).

68. Abel Dupetit-Thouars, *Voyage autour du monde sur la frégate la Vénus pendant les années 1836–1839, Relation,* 4 vols. (Paris: Gide, 1841–1845), 2:284, 309; Browne, *Charles Darwin: Voyaging,* 304. Dupetit-Thouars described Lawson as "interim" governor of the islands as of 1835. This is consistent with FitzRoy's characterization of Lawson as "acting for the governor of this archipelago," who was at that time José de Villamil (1788–1866). FitzRoy, *Narrative,* 2:490.

69. Ecuador, *Manual de información . . . del Ecuador,* 3 vols. (Quito: Cientifica Latina Editores, 1980), 3:678, 602; the useful article on the Galápagos, which appears in English as well as Spanish, runs from pp. 638–735. Charles was the only inhabited island in 1835; FitzRoy (*Narrative,* 2:490) estimated its population as 200, Darwin as between 200 and 300 (*JR,* 456).

70. Keynes, *Charles Darwin's Zoology Notes,* 291.

71. FitzRoy, *Narrative,* 2:492. Saddle-back tortoises exist only on the Galápagos Islands. Since their carapace does not protect them completely, it is presumably the lack of predators that permits their presence there. With their longer necks, the saddle-backs are able to reach higher to feed than are the domed tortoises. They are, or were, present on the smaller, dryer islands. See Thomas H. Fritts, "Morphometrics of Galapagos Tortoises: Evolutionary Implica-

tions," in Robert I. Bowman, Margaret Berson, and Alan E. Leviton, eds., *Patterns of Evolution in Galapagos Organisms* (San Francisco: Pacific Division, American Association for the Advancement of Science, 1983), 107–122. Sulloway claimed that Darwin and FitzRoy did not see living tortoises on Charles Island. Sulloway, "Darwin and the Galapagos," 36; Sulloway, "Darwin's Conversion," 341–342. However, this would seem to be an overstatement since Darwin simply did not say whether or not he saw tortoises on the island. Further, of the Charles Island tortoises he did comment that "such numbers yet remain that it is calculated two days hunting will find food for the other five in the week." *Diary,* 356. David W. Steadman reports the existence of some tortoises on the island in 1835 and 1837. See his *Holocene Vertebrate Fossils from Isla Floreana, Galápagos* (Washington, D.C.: Smithsonian Institution Press, 1986), 65. For a photograph of a mounted specimen owned by the Natural History Museum, London, said to be of the extinct Charles Island tortoise, see Pritchard, *Galápagos Tortoise,* 64.

72. Godfrey Merlen, *Restoring the Tortoise Dynasty* (Quito: Charles Darwin Foundation for the Galapagos Islands, 1999), 2. Charles Haskins Townsend, "The Galapagos Tortoises in Their Relation to the Whaling Industry: A Study of Old Logbooks," *Zoologica* 4 (1925):63, 77–80, 82, 104–111. Townsend suggested that the tortoises reported collected on Charles Island in the 1840s may have originated elsewhere.

73. David Porter, *Journal of a Cruise* (1815; repr. Annapolis: Naval Institute Press, 1986), 244. By the time of the second edition of *JR* Darwin cited Porter. Sulloway, "Darwin and the Galapagos," 39.

74. *JR,* 465. After CD's initial draft of *JR* was complete, the French zoologist Gabriel Bibron (1806–1848) informed him "that he has seen full-grown animals, brought from this Archipelago, which he considers undoubtedly to be distinct species." CD added this information as an addendum (*JR,* 628). With Bibron's information on the tortoises to encourage him, CD wrote to Henslow of the plants to inquire "how many cases . . . there are of *near* species, of the same genus,;—one species coming from one island, & the other from a second island.—" CD hoped that the plants would "support my case of the birds & Tortoises in a *glorious* manner." CD to Henslow, 3 November 1838, in *Correspondence,* 13:350. See *JR,* 629, for Henslow's affirmative reply.

75. Herman Heinzel and Barnaby Hall, *Galápagos Diary: A Complete Guide to the Archipelago's Birdlife* (Berkeley: University of California Press, 2000), 234–235. For the handsome lithograph of the Charles Island mockingbird in the Beagle's *Zoology,* see Plate 16 (facing p. 62). For photographs by Robert Curry of the four presently recognized species, see Peter R. Grant, *Ecology and Evolution of Darwin's Finches* (Princeton: Princeton University Press, 1999), 270–271. Also see the discussion of the mockingbirds, with color plates, in David W. Steadman and Steven Zousmer, *Galápagos: Discovery on Darwin's Islands* (Washington, D.C.: Smithsonian Institution, 1988), 188–191.

76. *Zoology,* 63.

77. Keynes, *Charles Darwin's Zoology Notes,* 298.

78. John Gould, "Three Species of the Genus Orpheus, from the Galapagos, in the collection of Mr. Darwin," *Proceedings of the Zoological Society of London* 5 (1837):27. See Herbert, *Red Notebook,* 117.

79. For discussion of the taxonomic issues, see Robert Curry, "Evolution and Ecology of Cooperative Breeding in Galapagos Mockingbirds (*Nesomimus* SPP.)" (Ph.D. dissertation, University of Michigan, 1987), 4–6.

80. CD to J. S. Henslow, 28–29 January 1836, in *Correspondence,* 1:485.

81. *Diary,* 429; *JR,*454.

82. Nora Barlow, ed., "Darwin's Ornithological Notes," *Bulletin of the British Museum (Natural History) Historical Series* 2, no. 7 (1963):262. Barlow initially dated the notes as contemporary with Darwin's visit to the Galápagos Island in 1835. Later she argued for an 1836 dating. See Barlow, *Charles Darwin and the Voyage of the Beagle,* 245–247, and the "Ornithological Notes," 203–209. In my treatment of the notes, I argued for a date "well into 1836" and provided evidence that the notes were in existence before January 1837. I also argued more

generally that the Ornithological Notes should be judged, as with similar lists in other fields, for their intended use by individual taxonomists. Herbert, "Place of Man in the development of Darwin's Theory of Transmutation: Part I," 237–239. Frank J. Sulloway has further narrowed the date of composition of the various notes on specialties to the period from mid-June to mid-August 1836, with the Ornithological Notes having been completed by 19 July. See his "Darwin's Conversion: The *Beagle* Voyage and Its Aftermath," *Journal of the History of Biology* 15 (1982):337.

83. Frank J. Sulloway, "Darwin and His Finches: The Evolution of a Legend," *Journal of the History of Biology* 15 (1982):1–53; but see also Estes, Grant, and Grant, "Darwin in Galápagos," 367 n. 204, for the claim that Darwin collected the finches systematically. Also see Jonathan Weiner, *The Beak of the Finch: The Story of Evolution in Our Time* (New York: Knopf, 1994), 54–56, 134–135, 151. Darwin emphasized the differences among the beaks of the various species of finches in *JR,* 462, and, with greater detail, in *JR* (1845), 379–380. However, as Weiner (pp. 144–146) and others have suggested, it was the twentieth-century ornithologist David Lack (1910–1973) who drew attention to the "adaptive radiation" shown by the finches. See David Lack, *Darwin's Finches* (Cambridge: Cambridge University Press, 1947), 146–148. For current work on finch speciation see Peter Grant, *Ecology and Evolution of Darwin's Finches.*

84. Cannon, "The Impact of Uniformitarianism," 305.

85. Ibid., 305. If Herschel gave Darwin the vocabulary of origins and origination, it should be noted that the terms posed some problems for Darwin later in that they suggest both a beginning in time as well as a process. Interestingly, when Lyell took up the subject again in the 1850s, he usually referred to Darwin's theory in terms "of . . . variety-making, viz. . . . species-making" rather than origination. Leonard G. Wilson, ed., *Sir Charles Lyell's Scientific Journals on the Species Question* (New Haven: Yale University Press, 1970), 288, 458, though see 163 for "origination." In the *Principles of Geology* (2:123) Lyell had spoken of the "introduction" of new species.

86. *Notebooks,* 413 (E:59). While Herschel's letter was published in 1837, Darwin does not seem to have read it until 1838. He was, however, clearly familiar with at least some of the views expressed in the letter in early 1837. See *Correspondence,* 2:8–9, 4:319.

Chapter 10. Geology and the *Origin of Species*

1. Herbert, *Red Notebook,* 7–12; *Notebooks,* 17–20 (RN:127, 130).
2. Herbert, *Red Notebook,* 11–12.
3. *Notebooks,* 62–63 (RN:130).
4. Herbert, *Red Notebook,* 107. On Darwin's collecting of the specimen, see *Diary,* 212.
5. *Notebooks,* 175, 180, 196, 277, 304–305, 340–341; (B:16, 37–38, 105), (C:126, 208–209), (D:29–30).
6. For a listing of Darwin's *Myothera* specimens, see Herbert, *Red Notebook,* 115.
7. *JR,* 208.
8. Quoted in Wilson, *Charles Lyell: The Years to 1841,* 437.
9. Lyell, "Presidential Address [1837]," 511.
10. *Notebooks,* 17, 19; Stan P. Rachootin, "Owen and Darwin Reading a Fossil: *Macrauchenia* in a Boney Light," in Kohn, *Darwinian Heritage,* 155–183. While Rachootin's argument that Darwin did not believe that the fossil *Macrauchenia* was ancestor to the guanaco at the point he wrote his earliest transmutationist entries (RN:127–130) is incorrect (as substantiated by the later notebook entry A:9), there is still much of value in his important and original discussion of the anatomy of the fossils.
11. *Notebooks,* 19 n. 16, which includes reference to Owen's notebooks. See also *Fossil Mammalia,* 35–56, for discussion of "Description of Parts of the Skeleton of *Macrauchenia Patachonica;* A large extinct Mammiferous Animal, referrible to the Order Pachydermata; but with affinities to the Ruminantia, and especially to the Camelidæ."

12. Armadillos and glyptodonts are grouped, together with sloths and anteaters, under the placental mammalian order Xenarthra (the term "edentates" being in disuse). See Timothy J. Gaudin, "Xenarthrans," in Ronald Singer, ed., *Encyclopedia of Paleontology,* 2 vols. (Chicago: Fitzroy Dearborn, 1999), 2:1347–1353. On Litopterna, see Darin A. Croft, "Placentals: Endemic South American Ungulates," in Singer, *Encyclopedia of Paleontology,* 2:891. On the evolution of placental mammals generally, see Michael Novacek, "Placentals: Overview," in Singer, *Encyclopedia of Paleontology,* 2:884–890. The complicated history of South American mammals, which was not known during Darwin's lifetime, is described in Simpson, *Splendid Isolation: The Curious History of South American Mammals* (New Haven: Yale University Press, 1980). As Simpson stated (p. 3), during most of the age of mammals, the Cenozoic, South America was an island continent, and its land mammals evolved in almost complete isolation. For drawings and discussion of the fossil mammals collected by Darwin and described by Owen, see Alfred S. Romer, "Darwin and the Fossil Record," *Natural History* 58 (October 1958):456–469. The present-day South American llamas and guanacos are regarded as closely related to the Old World camels. Christine Janis, "Artiodactyls," in Singer, *Encyclopedia of Paleontology,* 1:125–135. Also see Ernst Mayr, *What Evolution Is* (New York: Basic Books, 2001), 31: "Camels and their relatives are found on two different continents: the true camels in Asia and Africa, and their close relatives the llamas in South America. If we believe in continuous evolution there should be a connection between the two now isolated areas; in other words, camels should occur in North America, but they are absent. This situation led to the inference that camels had indeed at one time existed in North America, serving as a connecting link between the Asian and South American camels, but then had become extinct. In due time, this conjecture was indeed confirmed by the discovery in North America of a large fossil fauna of Tertiary camels."

13. *Notebooks,* 87 (A:9). On the basis of this entry it would seem that Darwin did not surrender his idea of a "fossil llama" even after Owen revised his initial judgment. Of course, if one is a believer in descent, the branching-off may simply be viewed as having occurred earlier.

14. *Notebooks,* 183–184 (B:53–54).

15. CD to Hooker, 10 April 1846, in *Correspondence,* 3:311.

16. CD, *A Monograph on the Fossil Balanidæ and Verrucidæ of Great Britain* (London: Palæontographical Society, 1854), Figure 1.

17. Phillip Reid Sloan, ed., *The Hunterian Lectures in Comparative Anatomy, May–June 1837 by Richard Owen* (Chicago: University of Chicago Press, 1992), 71. Owen also discussed the idea of succession of forms in his 11 May lecture (p. 69). For the Darwin quotation, see *Notebooks,* 210 (B:161). On the Owen-Darwin relationship, see Robert J. Richards, *The Meaning of Evolution: The Morphological Construction and Ideological Reconstruction of Darwin's Theory* (Chicago: University of Chicago Press, 1992), 126–136.

18. *Notebooks,* 391, 382 (D:174, 158).

19. Richard Owen to CD, 17 September 1841?, in *Correspondence,* 13:357; CD to Hooker, 10 May 1848, in ibid., 4:140.

20. The "Sketch" and "Essay" appear in Darwin and Wallace, *Evolution by Natural Selection;* "Natural Selection," in Stauffer, *Charles Darwin's Natural Selection* (1975).

21. On Darwin's activities during the hiatus, see Browne, *Charles Darwin: Voyaging,* 448–543; Stott, *Darwin and the Barnacle;* and James A. Secord, "Nature's Fancy: Charles Darwin and the Breeding of Pigeons," *Isis* 72 (1981):163–186.

22. As an entry point to the literature on Darwin's development of his species theory, see Browne, *Charles Darwin: Voyaging* and *Charles Darwin: The Power of Place;* Desmond and Moore, *Darwin;* the bibliography compiled by Malcolm J. Kottler in Kohn, *Darwinian Heritage,* 1023–1099; and the bibliography in Jonathan Hodge and Gregory Radick, *The Cambridge Companion to Darwin* (Cambridge: Cambridge University Press, 2003), 424–460. Darwin recorded his adoption of a transmutationist view in the Red Notebook and his reading of Malthus on population in Notebook D. *Notebooks,* 61–63 (RN:127–130), 374–376 (D:134–145). Also see Sandra Herbert, "Darwin, Malthus, and Selection," *Journal of the History of Biology* 4 (1971):209–217. The main run of the theoretical notebooks are A (geology), B through E (species), M and N (metaphysical enquiries).

23. Sedgwick, "Presidential Address [1830]," 205. In the same paragraph Sedgwick took a swipe at the concept of the transmutation of species in the course of observing that formations do "pass" into each other. Sedgwick discussed the first volume of Lyell's *Principles of Geology* in his next (1831) "Presidential Address."

24. Cannon, "Impact of Uniformitarianism," 308.

25. *Notebooks,* 352 (D:60).

26. Ibid., 352 n. 60–1. Also see Notebook E:5–6. On the language analogy, which derived from Herschel, see Cannon, "Impact of Uniformitarianism," 308, and Alter, *Darwinism and the Linguistic Image,* 12–14.

27. *Notebooks,* 398–399 (E:6).

28. Ibid., 433 (E:126–127), also n. 126–1 for relevant passages from Adam Sedgwick and Roderick Impey Murchison, "Classification of the Older Stratified Rocks of Devonshire and Cornwall," *Philosophical Magazine* 14 (1839):241–260, 317, 354–358.

29. *Notebooks,* 227 (B:225).

30. Ibid., 443 (E:155–156).

31. Ibid., 285 (C:152).

32. Ibid., 180 (B:37).

33. Ibid., 284 (C:147); also see p. 419 (E:85) for reference to an "absolute quantity of vitality <in the World>."

34. Ibid., 231 (B:243).

35. For a concise statement of Lyell's views on species, see William Coleman, "Lyell and the Reality of Species," *Isis* 53 (1962):325–338. On Darwin's reaction to Lyell, see Sydney Smith, "The 'Origin' of the *Origin* as Discerned from Charles Darwin's Notebooks and His Annotations in the Books He Read between 1837 and 1842," *Advancement of Science* 16 (1960):392, 397–398.

36. Wilson, *Sir Charles Lyell's Scientific Journals.*

37. *Notebooks,* 209 (B:155). Darwin added the line in bold at a later date from his original entry. On the points raised in Darwin's remark, see Malcolm J. Kottler, "Charles Darwin's Biological Species Concept and Theory of Geographic Speciation: the Transmutation Notebooks," *Annals of Science* 35 (1978):275–297.

38. Ernst Mayr, *The Growth of Biological Thought: Diversity, Evolution, and Inheritance* (Cambridge: Harvard University Press, 1982), 265–269.

39. *Notebooks,* 277, 285 (C:126, 152).

40. Ibid., 431 (C:122).

41. Richard W. Burkhardt Jr., *The Spirit of System: Lamarck and Evolutionary Biology,* 2d ed. (Cambridge: Harvard University Press, 1995), 115–142.

42. Quotations and figures from *Notebooks,* 177 (B:26).

43. *Origin,* 116–117, 420–421. It is debatable to what extent Darwin intended to convey a commitment to change that might now be described as "phyletic gradualism" or "anagenesis" in these sketches. On this point, see Mayr, *What Evolution Is,* 11, 62–63.

44. Jean Baptiste de Lamarck, *Zoological Philosophy: An Exposition with Regard to the Natural History of Animals,* trans. Hugh Elliot (Chicago: University of Chicago Press, 1984), 179. For a reproduction of the original diagram, labeled in French, from *Philosophie zoologique* (1809), see Burkhardt, *Spirit of System,* 176.

45. Sloan, *Hunterian Lectures,* 22.

46. *Notebooks,* 487–516.

47. Jack Morrell, "Genesis and Geochronology: The Case of John Phillips (1800–1874)," in Lewis and Knell, *Age of the Earth: From 4004 BC to AD 2002,* 85. "Cenozoic" is an alternative spelling to "Cainozoic."

48. On Huxley's "lack of acquaintance with practical paleontology in the 1850s," see Adrian Desmond, *Archetypes and Ancestors: Palaeontology in Victorian London, 1850–1875* (London: Blond and Briggs, 1982), 58. On the history of nineteenth-century paleontology as related to Darwin, see Peter J. Bowler, *Fossils and Progress: Paleontology and the Idea of Progressive Evolution in the Nineteenth Century* (New York: Science History Publications, 1976), and his *Life's*

Splendid Drama: Evolutionary Biology and the Reconstruction of Life's Ancestry 1860–1940 (Chicago: University of Chicago Press, 1996); Rudwick, *Meaning of Fossils;* Karl Alfred von Zittel, *History of Geology and Palæontology to the End of the Nineteenth Century,* trans. Maria M. Ogilvie-Gordon (London: Walter Scott, 1901). Also important as relating to the theme of progressive development are Greene, *Death of Adam;* Richards, *Meaning of Evolution;* and Michael Ruse, *Monad to Man: The Concept of Progress in Evolutionary Biology* (Cambridge: Harvard University Press, 1996). On the interpretive issue, debated among the scholars cited above, as to whether Darwin himself believed in, to use Richards's phrase, "progressive evolution" (p. 91), I lean to the view that it was usually, though not always, a peripheral rather than a central theme for him. Most often the appearance of progressive evolution functioned as something to be explained rather than as a tool to be used as part of an explanation.

49. Charles Darwin, *A Monograph on the Sub-Class Cirripedia,* 2 vols. (London: Ray Society, 1851, 1854), 2:34.

50. Stott, *Darwin and the Barnacle,* 100; also Dennis J. Crisp, "Extending Darwin's Investigations on the Barnacle Life-history," *Biological Journal of the Linnean Society* 20 (1983):74.

51. *Origin,* 27, 23.

52. CD to James Dwight Dana, 7 January 1863, in *Correspondence,* 11:23.

53. CD to Andrew Murray, 28 April 1860, in ibid., 8:176. On Salter's difficult career, see James A. Secord, "John W. Salter: The Rise and Fall of a Victorian Palaeontological Career," in Alwynne Wheeler and James H. Price, eds., *From Linnaeus to Darwin: Commentaries on the History of Biology and Geology* (London: Society for the History of Natural History, 1985), 61–75.

54. CD to Robert Chambers, 30 April 1861, in *Correspondence,* 9:106.

55. CD to Thomas William St Clair Davidson, 26 April 1861, in ibid., 9:104; also see 107–108, 111–114.

56. CD to Lyell, 23 September 1860, in ibid., 8:377–378, 379–380, for the diagrams in Figure 10.5.

57. *Notebooks,* 286 (C:155).

58. Sydney Smith, "The Darwin Collection at Cambridge with One Example of Its Use: Charles Darwin and *Cirripedes,*" *Actes du XIe congrès international d'histoire des sciences* 5 (1965):97. It remains an ongoing subject of discussion among historians how far Darwin embraced Quinarianism at any time in his life. As an editor, having run into difficulty squaring a literal Quinarian reading of terms relating to "inosculation" with the meaning of Darwin's notebook entries (RN:130, "distinct species inosculate"), and being aware that Darwin could use the word broadly (A:76, "Three inosculating rivers in Southern America"), I drew on a meaning for "inosculation" that predated MacLeay, that is, the medical term referring to the joining of one blood vessel to another (Herbert, *Red Notebook,* 7). While this was an overstatement—for Darwin had used "inosculation" in connection with MacLeay's name earlier (*Correspondence,* 1:280, MacLeay "never imagined such an inosculating creature"—said of a bird looking like "a happy mixture of a lark, pidgeon [*sic*] & snipe") and therefore probably wished to evoke MacLeay in RN:130 (and B:8)—I remain persuaded that one need not make Darwin into a practicing Quinarian simply on the basis of his using the term "inosculation." It was a term with more layers of meaning and wider applicability than that (*Notebooks,* 19 n. 17). Rachootin ("Owen and Darwin Reading a Fossil," 171–174) also struggles with the question of how far to read Darwin's intention in RN:130 as Quinarian. Figure 10.6 is drawn from William Sharp MacLeay, *Horæ Entomologicæ* (London: S. Bagster, 1819), 318.

59. *Notebooks,* 286 (C:155), also 302 (C:201–202). In tracing Darwin's diminishing use of Quinarian terminology over the course of the notebooks, the following work is useful: Donald J. Windshank, Stephan J. Ozminski, Paul Ruhlen, and Wilma M. Barrett, *A Concordance to Charles Darwin's Notebooks, 1836–1844* (Ithaca: Cornell University Press, 1990).

60. *Notebooks,* 281 (C:139). Dov Ospovat made this point in *The Development of Darwin's Theory: Natural History, Natural Theology, and Natural Selection, 1838–1859* (Cambridge: Cambridge University Press, 1981), 108.

61. CD to G. R. Waterhouse, August 1838–1840 (first quotation, exact date uncertain), and 26 July 1843 (second quotation), in *Correspondence*, 2:93, 376. For an indication of why Darwin might have approached Waterhouse, see *Notebooks*, 463 (T:55).

62. G. R. Waterhouse to CD, 9 August 1843, in *Correspondence*, 2:381.

63. CD to G. R. Waterhouse, 26 and 31 July 1843, in ibid., 3:376, 378.

64. George Robert Waterhouse, "Observations on the Classification of the Mammalia," *Annals and Magazine of Natural History* 12 (1843):399–412. The article is devoted to explicating a diagram (p. 399) depicting the class Mammalia as containing ten orders represented as contiguous circles. Waterhouse did not reproduce the diagram in his *Natural History of the Mammalia,* 2 vols. (London: Billière, 1846–1848). As regard to the close, but sometimes obscure, ties connecting Darwin to museum-based zoologists, consider the difficulties faced by editors in interpreting a letter to Owen in Waterhouse's handwriting but signed "yrs Ch. Darwin." Frederick H. Burkhardt, "A Troublesome Letter Signed 'yrs Ch. Darwin,'" *Documentary Editing* 23 (2001):73–81, and Phillip Reid Sloan, "A Plea for Caution: A Response to Frederick Burkhardt," *Documentary Editing* 23 (2001):82–84. The editors of the *Correspondence* are clearly right in not taking this to be a genuine Darwin letter; I share Sloan's opinion that Waterhouse intended the false signature to be a joke.

65. CD to George Robert Waterhouse, 8 July 1843, in *Correspondence,* 2:373.

66. Herbert, "Place of Man in the Development of Darwin's Theory: Part I," 240, 244, 255; also 245 where Darwin, showing similar deference to naturalists, backed Gould's "experience" as a ornithologist in making the call between species and varieties.

67. John Beatty, "Speaking of Species: Darwin's Strategy," in Kohn, *Darwinian Heritage,* 265–281.

68. Hugh E. Strickland, "Report of a Committee Appointed to Consider of the Rules by Which the Nomenclature of Zoology May Be Established on a Uniform and Permanent Basis," *Report of the British Association for the Advancement of Science* [Manchester 1842] (1843):105–121.

69. Gordon R. McOuat, "Species, Rules and Meaning: The Politics of Language and the Ends of Definitions in 19th Century Natural History," *Studies in History and Philosophy of Science* 27 (1996):473–519.

70. Gordon McOuat, "Cataloguing Power: Delineating 'Competent Naturalists' and the Meaning of Species in the British Museum," *British Journal for the History of Science* 34 (2001):1–28, especially 25 n. 102.

71. McOuat, "Cataloguing Power," 24 (quoted from Gray's cataloguing rules), 21.

72. Hugh E. Strickland, "Report of a Committee Appointed to Print and Circulate a Report on Zoological Nomenclature," *Report of the British Association for the Advancement of Science* [Cork 1843] (1844):120.

73. Ernst Mayr has noted the change between the notebooks and the *Origin* on the treatment of species, with the *Origin* depending more on morphological standards, which were the province of museum specialists, and the notebooks leaning more toward the biological species concept. See his *Growth of Biological Thought,* 265–269, and McOuat, "Cataloguing Power," 1, 27. Also useful in outlining a number of current positions on systematics is Elliott Sober, *Philosophy of Biology* (Boulder, CO: Westview Press, 1993), 143–183. Present-day debates resonate with those contemporaneous with publication of the *Origin,* as, for example, the interest of the botanist Hewett Cottrell Watson in the subject of convergence/divergence of species. See H. C. Watson to CD, 3? January 1860, in *Correspondence,* 8:10–13.

74. *Notebooks,* 222 (B:202).

75. Ibid., 423 (E:96).

76. Lyell's most dramatic statement of his cyclicism regarding species occurs in a letter to Gideon Mantell of 15 February 1830: "All these changes [of climate] are to happen in the future again, and iguanodons . . . must as assuredly live again in the latitude of Cuckfield as they have done so." Lyell, *Life, Letters and Journals of Sir Charles Lyell,* 1:262. *Notebooks,* 425 (E:105). The reference is to Lyell, *Elements of Geology,* 275.

77. See, example, the comment of J. D. Hooker to CD, 4 August 1856, in *Correspondence,*

6:199: "I do not find that repetition of species, or of forms, under closely similar physical conditions, that I should have expected transmutation to have effected."

78. *Notebooks,* 446 (E:160). On the views regarding spontaneous generation, particularly after 1860, see James E. Strick, *Sparks of Life: Darwinism and the Victorian Debates over Spontaneous Generation* (Cambridge: Harvard University Press, 2000).

79. CD to J. D. Hooker, 10 June 1855, in *Correspondence,* 5:350.

80. *Notebooks,* 427 (E:108). This is the "first filling of the ecological barrel" argument. See Stephen Jay Gould, *Wonderful Life: The Burgess Shale and the Nature of History* (New York: W. W. Norton, 1990), 228–230.

81. *Notebooks,* 283–284 (C:144–147). Note that in C:147 Darwin did suggest how the number of forms (or the "subdivision of stations & diversity"—depending how one read his sentence) "perhaps on long average equal." The "Sketch" of 1842 employed the vocabulary of diversity; the "Essay" of 1844 used the vocabulary of diversity but added the term "diverging" though not "divergence." Darwin and Wallace, *Evolution by Natural Selection,* 52, 57, 219, 97, 129, 201, 205 (sample entries). In the *Origin* Darwin frequently employed the term "divergence" as well as retaining the use of "diversity." Paul H. Barrett, Donald J. Weinshank, and Timothy T. Gottleber, *A Concordance to Darwin's Origin of Species, First Edition* (Ithaca: Cornell University Press, 1981), 209–210. The phrase "principle of divergence" appears in *Origin,* 118. The slow emergence of the vocabulary of divergence is consistent with Darwin's recollection in his *Autobiography* (pp. 120–121) that it was "long after" he came to Down, which was in September 1842, that the principle came to him. On the timing of Darwin's arrival at what would now be called an ecological understanding of divergence, see David Kohn, "Darwin's Principle of Divergence as Internal Dialogue," in Kohn, *Darwinian Heritage,* 245–257. For a critique of the validity of Darwin's concept of divergence, see Ernst Mayr, "Darwin's Principle of Divergence," *Journal of the History of Biology* 25 (1992):357–358.

82. Ospovat, *Development of Darwin's Theory,* 192. Ospovat credited Darwin's appreciation of the concept of the division of physiological labor held by Henri Milne-Edwards (1800–1885) with securing the principle of divergence (p. 210). Darwin was also indebted to the experiments of George Sinclair (1786–1834), head gardener at Woburn Abbey. See Andy Hector and Rowan Hooper, "Darwin and the First Ecological Experiment," *Science* 295 (25 January 2002):639–640.

83. CD to Lyell, 25 June 1856, in *Correspondence,* 6:154.

84. David Forbes to CD, November ? 1860, in ibid., 8:457.

85. The key letter, with Forbes's own map, is Forbes to CD, 25 February 1846, in ibid., 3:290–293. Darwin's immediate reply is lost, but for his judgment regarding Forbes's ideas, see his comments to Hooker (p. 290, 300). See Edward Forbes, "On the Distribution of Endemic Plants, More Especially Those of the British Islands, Considered with Regard to Geological Changes," *Report of the British Association for the Advancement of Science* [Cambridge 1845] (1846):67–68. The expanded version appeared as "On the Connexion between the Distribution of the Existing Fauna and Flora of the British Isles, and the Geological Changes Which Have Affected Their Area, Especially during the Epoch of the Northern Drift," *Memoirs of the Geological Survey of Great Britain and of the Museum of Economic Geology in London* 1 (1846):336–432. Also see Browne, *Secular Ark,* 114–136, and Philip F. Rehbock, *The Philosophical Naturalists: Themes in Early Nineteenth-Century British Biology* (Madison: University of Wisconsin Press. 1983), 150–191.

86. CD to Hooker, 19 July 1856, in *Correspondence,* 6:191.

87. CD to Hooker, 30 July 1856, in ibid., 6:195.

88. Forbes, "On the Distribution of Endemic Plants," 67.

89. CD to Charles Lyell, 25 June 1856, in *Correspondence,* 6:155.

90. CD to Hooker, 6 September 1860, in ibid., 8:345.

91. *Notebooks,* 30 (RN:34–35). Mocha Island lies twenty miles off the coast of south central Chile.

92. Edward William Binney, "On the Origin of Coal," *Memoirs of the Literary and Philo-*

sophical Society of Manchester, 2d ser., 8 (1848):180–181. In 1847 Binney's paper had been privately printed as a pamphlet of which Darwin owned a copy. *Correspondence,* 4:38 n. 2.

93. CD to Hooker, 19 May 1846, in *Correspondence,* 3:320. Presently *Sigillaria* is defined as a genus of fossil arborescent club mosses (Lycopsida) with a time range from the Middle Carboniferous to near the end of the Permian.

94. CD to Hooker, 1 May 1847, in ibid., 4:38.

95. CD to Hooker, 6 May 1847, in ibid., 4:40 ("you are right indeed in thinking me mad, if you suppose that I would class any ferns as marine plants"); *Autobiography,* 105.

96. Joseph Dalton Hooker, "On the Vegetation of the Carboniferous Period, As Compared with that of the Present Day," *Memoirs of the Geological Survey of Great Britain and of the Museum of Practical Geology in London* 2, pt. 2 (1848):417.

97. Andrew Sclater of the Darwin Correspondence Project has provided me with the text of a letter from January 1866 in which Hooker noted that Darwin had once offered five-to-one odds that coal was formed by submarine plants, and asked Darwin whether he would still offer the same odds. By the mid-1860s the notion of a peat bog as a model for the mechanism of coal formation had been suggested, thus altering the terms of Hooker's and Darwin's discussion. According to a personal communication from the paleobotanist Dr. Stephen Scheckler, present-day views hold that *Sigillaria* lived in the drier parts of coastal swamps in warm temperate to subtropical areas of the world. *Sigillaria* fossils from the early Carboniferous are found mixed with other plants to form peat, but those from the late Carboniferous do not occur in peaty deposits, perhaps indicating that they had moved to the edges of the swamps. In any case they did not grow in water. These would have been freshwater swamps. If a saltwater incursion occurred, it was due to a rare storm surge.

98. CD to Hooker, 6 May 1847, in *Correspondence.,* 4:40. Italics added.

99. *Notebooks,* 175 (B:17); *Origin,* 159.

100. *Notebooks,* 172 (B:7).

101. Darwin and Wallace, *Evolution by Natural Selection,* 65–72 (1842 "Sketch"), 91–116 (1844 "Essay").

102. Stauffer, *Charles Darwin's Natural Selection,* 173–174; *Origin,* 43.

103. Darwin, *Monograph on the Sub-Class Cirripedia,* 2:155.

104. CD to W. D. Fox, 19 March 1855, in *Correspondence,* 5:288–289.

105. Stauffer, *Charles Darwin's Natural Selection,* 173; also see *Origin,* 43, for a similar assertion in different words.

106. Quoted in Herbert, "From Charles Darwin's Portfolio," 28.

107. *Notebooks,* 63 (RN:130).

108. Hooker to CD, 22 November 1856, in *Correspondence,* 6:280.

109. CD to Hooker, 23 November 1856, in ibid., 6:282, quoted in Browne, *Secular Ark,* 136. The entire passage is of interest:

> When I sent my M.S. I felt strongly that some preliminary questions on causes of variation ought to have been sent you. Whether I am right or wrong in these points is quite a separate question, but the conclusion which I have come to, quite independently of geographical distribution, is that external conditions (to which naturalists so often appeal) do by themselves *very little.* How much they do is the point of all others on which I feel myself very weak.—I judge from facts of variation under domestication, & I may not yet get more light. But at present, after drawing up a rough copy of this subject, my conclusion is that external conditions do *extremely* little, except in causing mere variability. This mere variability (causing the child *not* to resemble its parent) I look at as *very* different from the formation of a marked variety or new species.—(No doubt the variability is governed by laws, some of which I am endeavouring very obscurely to trace).—The formation of a strong variety or species, I look at as almost wholly due to the selection of what may be incorrectly called *chance* variations or variability. This

> power of selection stands in the most direct relation to time, & state of nature can be only excessively slow.—Again the slight differences selected, by which a race or species is at last formed, stands, as I think can be shown (even with plants & obviously with animals) in far more important relation to its associates than to external conditions.—Therefore, according to my principles, whether right or wrong, I cannot agree with your proposition that Time + altered conditions + altered associates are 'convertible terms'. I look at first & last as *far* more important;—time being important only so far as giving scope to selection.—

110. *Origin,* 313.

111. Lyell to CD, 28 August 1860, in *Correspondence,* 8:336, 338 n. 4. The main run of letters between Darwin and Lyell on the issue of rates of change and percentages of species remaining constant appears in ibid., 8:336–341, 348–349, 354–357, 366–369, 377–381 (the diagram reproduced as Figure 9 in this chapter appears on 379), 383–386, 391–393, 396–404, 406–407, 410–411, 412–413, 421–423.

112. CD to Lyell, 1 September 1860, in ibid., 8:339, 340.

113. Lyell to CD, 8 September 1860, in ibid., 8:348: "The only speculation I have been able to make is this that by far the majority 9/10ths perhaps are immutable, & when changes come, must die rather than yield. The hereditary power has got too strong a set, after a million of years, in one way. It cannot go in any other."

114. CD to Lyell, 12 September 1860, in ibid., 8:355: "I entirely agree with what you say about only one species of many becoming modified. . . . It is absolutely implied on my ideas of classification & divergence that only one [or] two species of even large genera give birth to new species."

115. Lyell to CD, 30 September 1860, in ibid., 8:398–399.

116. CD to Lyell, 3 October 1860, in ibid., 8:402.

117. For Grant's observation, see Weiner, *Beak of the Finch,* 127. For Darwin's description, see *Origin,* 68, and also Stauffer, *Charles Darwin's Natural Selection,* 185.

118. Niles Eldredge and Stephen Jay Gould, "Punctuated Equilibria: An Alternative to Phyletic Gradualism," in Thomas J. M. Schopf, ed., *Models in Paleobiology* (San Francisco: Freeman, Cooper, 1972), 82–115; most recently elaborated in Stephen Jay Gould, *The Structure of Evolutionary Theory* (Cambridge: Harvard University Press, 2002).

119. Eldredge and Gould, "Punctuated Equilibria," 84; Frank H. T. Rhodes, "Darwinian Gradualism and Its Limits: The Development of Darwin's Views on the Rate and Pattern of Evolutionary Change," *Journal of the History of Biology* 20 (1987):139–157.

120. Hooker to CD, 29 June 1854, in *Correspondence,* 5:199.

121. CD to Hooker, 7 July 1854, in ibid., 5:201.

122. CD to William Erasmus Darwin, 6 October 1858, in ibid., 7:163.

123. CD to Lyell, 23 February 1860. The first Wallace letter, dated 1 September 1860, was to George Silk; the second Wallace letter, dated 24 December 1860, was to Henry Walter Bates. See *Correspondence,* 8:102 and 221 n. 1. Italics added.

124. *Origin,* 282–287. On the phrase "lapse of time," see Henry Thomas De la Beche, *Researches in Theoretical Geology* (London: Charles Knight, 1834): "[Man] feels it difficult to conceive that great lapse of time which geology teaches us has been necessary to produce the present condition of the earth's surface." CD used "lapse [of time]" and "lapse of ages" in *Notebooks,* 63 (RN:130, 132).

125. *Origin,* 287.

126. Ibid., 282.

127. G. Brent Dalrymple, "The Age of the Earth in the Twentieth Century: A Problem (Mostly) Solved," in Lewis and Knell, *Age of the Earth,* 207. Kenneth L. Taylor, "Buffon, Desmarest and the Ordering of Geological Events in *Époques,*" in Lewis and Knell, *Age of the Earth,* 40. Eyles, "De la Beche," 4:10.

128. Morrell, "Genesis and Geochronology," 86.

129. *Origin,* 283.

130. Hugh Miller, *The Testimony of the Rocks; or, Geology in Its Bearings on the Two Theologies, Natural and Revealed* (Edinburgh: T. Constable, 1857). On Miller's *Testimony,* see David R. Oldroyd, "The Geologist from Cromarty," in Michael Shortland, ed., *Hugh Miller and the Controversies of Victorian Science* (Oxford: Clarendon Press, 1996), 76–121.

131. *Origin,* 282. Joe D. Burchfield, "Darwin and the Dilemma of Geological Time," *Isis* 65 (1974):303.

132. John Fuller and Hugh Torrens of the History of Geology Group have written a useful but unpublished field guide ("A Wealden Geology and English History Excursion: Ships, Guns, Fuel," 2000) to the history of geological exploration of the Weald from John Farey's work of 1806–1808 forward. It was Farey who coined the term "denudation." William Topley, *The Geology of the Weald* (London: Longman, 1875), 11. Also see J. F. Kirkaldy, "William Topley and 'The Geology of the Weald,'" *Proceedings of the Geologists' Association* 86 (1975):373–388, and, on the history of the discovery of the spectacular fossils of southeastern England, Dean, *Gideon Mantell.*

133. Lyell, *Principles of Geology,* 3:289 and frontispiece map.

134. *Origin,* 285–287, quotation on 287. Darwin's computation was as follows: 22 miles equals "1393920—inches in whole distance." This multiplied by 220 (100 [1-inch erosion per century] times 2.2 [the ratio of 1100 to 500 feet]) equals "306,662,400 years for denudation of Weald." Darwin's computation appears in DAR 205.9:341, dated 23 December 1858. Also see DAR 209.9:246, dated March 1857, which includes measurements of thicknesses of formations and in which Darwin notes, "In about year will know more."

135. Andrew Crombie Ramsay, "On the Denudation of South Wales and the Adjacent Counties of England," *Memoirs of the Geological Survey of Great Britain, and of the Museum of Economic Geology in London* 1 (1846):297–335.

136. Richard J. Chorley, Antony J. Dunn, and Robert P. Beckinsale, *The History of the Study of Landforms; or, The Development of Geomorphology,* vol. I: *Geomorphology before Davis* (London: Methuen, 1964), 307–308.

137. CD to A. C. Ramsay, 10 October 1846, in *Correspondence,* 3:352–353.

138. Entry from the diary of A. C. Ramsay, 13 February 1848, in ibid., 4:108 n. 2.

139. A. C. Ramsay to CD, 29 December 1858, in ibid., 7:225.

140. *Origin,* 284.

141. Ramsay, "On the Denudation of South Wales," 335.

142. A. C. Ramsay to CD, 21 February 1860, in *Correspondence,* 8:97–98, also see 116. Herries Davies, *Earth in Decay,* 317–353. On the support for the *Origin* within the Geological Survey, see M. J. S. Hodge, "England," in Thomas F. Glick, ed., *The Comparative Reception of Darwinism* (Austin: University of Texas Press), 11–13.

143. Burchfield, "Darwin and the Dilemma," 303. The geological critique first appeared in the anonymous review of the *Origin* in *Saturday Review* (24 December 1859):775–776. While the author of the review has long been listed as anonymous, Frederick Burkhardt has suggested to me that John Phillips was its author, which I believe to be correct, though the prominent weekly's discerning editor John Douglas Cook (1808?–1868) probably had a hand in shaping the review as well. With Owen and Huxley both writing for the *Saturday Review,* and Huxley appealing to Cook not to allow Owen to review the *Origin,* Phillips would have seemed a good compromise, suggested possibly by Ramsay, who was a member of Cook's inner circle. See Merle Mowbray Bevington, *The "Saturday Review" 1855–1868: Representative Educated Opinion in Victorian England* (New York: Columbia University Press, 1941), 277–286.

144. Morse Peckham, *The Origin of Species by Charles Darwin: A Variorum Text* (Philadelphia: University of Pennsylvania Press, 1959), 484. Also CD to Asa Gray, 1 February 1860, in *Correspondence,* 8:61–63.

145. CD to Asa Gray, 3 April 1860, in ibid., 8:141 and n. 9 and n. 10.

146. CD to Lyell, 25 November 1860, in ibid., 8:495.

147. CD to Asa Gray, 3 April 1860, in ibid., 8:141.

148. Key sources include Burchfield, *Lord Kelvin and the Age of the Earth* (New York: Science History Publications, 1975); Lewis and Knell, *Age of the Earth;* and G. Brent Dalrymple, *The Age of the Earth* (Stanford: Stanford University Press, 1991). On twentieth-century developments, see the biography of Arthur Holmes: Lewis, *The Dating Game.* For the connection to cosmology, see Martin J. Rees, "Understanding the Beginning and the End," in Lewis and Knell, *Age of the Earth,* 275–283.

149. John Phillips, *Manual of Geology: Practical and Theoretical* (London: Richard Griffin, 1855), 621: "The nearly vertical position of certain fossil plants . . . affords good ground for caution in assigning very great extension of years to geological periods." Also Morrell, "Genesis and Geochronology," 88, and Simon J. Knell and C. L. E. Lewis, "Celebrating the Age of the Earth," in Lewis and Knell, *Age of the Earth,* 6.

150. John Phillips, "Presidential Address [17 February 1860]," *Quarterly Journal of the Geological Society of London* 16, pt. 1 (1860):li–lii.

151. Phillips, *Manual of Geology,* 594–598, 616–634.

152. John Phillips, *Life on the Earth: Its Origin and Succession* (Cambridge, England: Macmillan, 1860), 126: "[T]he Ganges . . . delivers annually to the Bay of Bengal 6,368,044,440 cubic feet of sediment, which is equal to 1/111th of an inch in a year. The maximum thickness of the strata is supposed to be about 72,000 feet = 864,000 inches, and dividing this by 1/111th we have the calculated antiquity of the base of the stratified rocks = 95,904,000 years." See pp. 134–135 for an estimate of the age of stratified deposits as 63,936,000 years based on an assumption of higher temperatures, and higher "atmospheric power in wasting the earth's surface" in the past. Also see Burchfield, "Age of the Earth and the Invention of Geological Time," 140.

153. John H. Callomon, "Fossils as Geological Clocks," in Lewis and Knell, *Age of the Earth,* 237–252.

154. Burchfield, *Lord Kelvin and the Age of the Earth,* 21–22; Jed Z. Buchwald, "William Thomson," *Dictionary of Scientific Biography* 13:382–383. On the philosophical import of Thomson's work, see Crosbie Smith and M. Norton Wise, *Energy and Empire: A Biographical Study of Lord Kelvin* (Cambridge: Cambridge University Press, 1989), 497–523 ("The Irreversible Cosmos").

155. William Thomson, "On the Age of the Sun's Heat," *Macmillan's Magazine* 5 (1862):388–393, quotation on 393, which concludes, "As for the future, we may say, with equal certainty, that inhabitants of the earth cannot continue to enjoy the light and heat essential to their life, for many million years longer, unless sources now unknown to us are prepared in the great storehouse of creation." His earlier treatment was "Physical Considerations Regarding the Possible Age of the Sun's Heat," *Report of the British Association for the Advancement of Science* [Manchester 1861] (1862):27–28 (Transactions of the Sections).

156. Morrell, "Genesis and Geochronology," 88–89. William Thomson's figure appeared in his "On the Secular Cooling of the Earth," *Philosophical Magazine* 25 (1863):5, where it was linked to an estimated temperature of melting rock as 7000° F. Thomson suggested that other values for melting temperatures would produce different ages for consolidation of the crust. A 10,000° F melting temperature would yield a figure of 200 million years before the present for the consolidation of the crust.

157. George H. Darwin to William Thomson, 1 November 1878, quoted in Smith and Wise, *Energy and Empire,* 579.

158. Nineteenth-century estimates in the range of 100 million years for the age of the earth's strata differ from present values by an approximate factor of 6. Current values are in the range of 570 million years before the present to the bottom of the Cambrian (Dalrymple, *Age of the Earth,* 60). Darwin's initial value of 300 million years for the denudation of the Weald was too long by an approximate factor of 4 for the date of the strata he described (Fuller and Torrens, "Wealden Geology," Figure 2, "Stratigraphic table for the Mesozoic rocks exposed in the Weald").

Conclusion

1. CD, "On the Formation of Mould," and *The Formation of Vegetable Mould, through the Action of Worms, with Observations on Their Habits* (London: John Murray, 1881). On Darwin's late work with earthworms, see Browne, *Charles Darwin: Power of Place,* 446–449, 478–480, 489–490.

2. *Origin,* 487.

3. *Notebooks,* 83–84, introduction to Notebook A.

4. Geikie, *Charles Darwin as Geologist,* 82.

5. Judd, "Darwin and Geology," 369–370. Roy MacLeod, "Imperial Reflections in the Southern Seas: The Funafuti Expeditions, 1896–1904," in Roy MacLeod and Philip F. Rehbock, eds., *Nature in Its Greatest Extent: Western Science in the Pacific* (Honolulu: University of Hawaii Press, 1988), 168–181 and n. 49 describing Judd's work analyzing boring samples from the Funafuti expeditions.

6. Joshua I. Tracey Jr. (1915–2004) as quoted in Daphne G. Fautin, "Before Darwin: Coral Reef Research in the Twentieth Century," in Benson and Rehbock, *Oceanographic History,* 447. See also Frank C. Whitmore Jr. and Joshua I. Tracey Jr., "Memorial to Harry Stephen Ladd, 1899–1982," *Geological Society of America Memorials* 14 (1984):1–7.

7. Young, *Mind over Magma,* 128.

8. Hewett Cottrell Watson to CD, 21 November 1859, in *Correspondence,* 7:385.

9. Judd, "Darwin and Geology," 377.

10. Falconer to CD, 25 October 1859, in *Correspondence,* 7:360; Hooker to CD, 24 May 1863, in ibid., 11:443. On this point, also see Stephen Jay Gould, "Agassiz's Marginalia in Lyell's *Principles,* or the Perils of Uniformity and the Ambiguity of Heroes," *Studies in History of Biology* 3 (1979):119–138.

BIBLIOGRAPHY

This bibliography is organized into four sections: (1) manuscripts, (2) works by Charles Darwin, (3) works published before 1900 by authors other than Charles Darwin, and (4) works published after 1900. Titles and names of publishers have been shortened in some cases.

(1) Manuscripts

Cambridge University Library (CUL). Darwin Papers. Manuscripts are referred to by the volume number assigned in the *Handlist of Darwin Papers* published by CUL in 1960, or by the catalogue number assigned in the list under preparation by Nicholas W. Gill.

Cambridge University Library. University Archive.
Geological Society of London Archive.
Mitchell Library, Sydney. Syms Covington Papers.
Natural History Museum, London. Richard Owen Papers.
Royal Society of London Archive.
University of Keele Archive. Darwin-Wedgwood Papers.

(2) Works by Charles Darwin

Barlow, Nora, ed. *The Autobiography of Charles Darwin*. London: Collins, 1958.
——, ed. *Charles Darwin and the Voyage of the Beagle*. London: Pilot Press, 1945.
——, ed. *Charles Darwin's Diary of the Voyage of H.M.S. "Beagle."* Cambridge: Cambridge University Press, 1933.
——, ed. "Darwin's Ornithological Notes." *Bulletin of the British Museum (Natural History) Historical Series* 2, no. 7 (1963):203–278.
Barrett, Paul H. *The Collected Papers of Charles Darwin*. Chicago: University of Chicago Press, 1977.
Barrett, Paul H., Peter J. Gautrey, Sandra Herbert, David Kohn, and Sydney Smith, eds. *Charles*

Darwin's Notebooks, 1836–1844: Geology, Transmutation of Species, Metaphysical Enquiries. London and Ithaca: British Museum (Natural History) and Cornell University Press, 1987.
Barrett, Paul H., Donald J. Weinshank, and Timothy T. Gottleber. *A Concordance to Darwin's Origin of Species, First Edition.* Ithaca: Cornell University Press, 1981.
Burkhardt, Frederick, and Sydney Smith. *A Calendar of the Correspondence of Charles Darwin, 1821–1882 with Supplement.* Cambridge: Cambridge University Press, 1994.
Burkhardt, Frederick, et al., eds. *The Correspondence of Charles Darwin.* 13 vols. Cambridge: Cambridge University Press, 1985–2002.
Darwin, Charles. "Extracts of Letters Addressed to Professor Henslow." Printed for Private Distribution by the Cambridge Philosophical Society, 16 November 1835.
——. "Extracts from Letters to the General Secretary, on the Analogy of the Structure of Some Volcanic Rocks with That of Glaciers." *Proceedings of the Royal Society of Edinburgh* 2 (1845):17–18. *CP,* 1:193–195.
——. *The Formation of Vegetable Mould, through the Action of Worms, with Observations on Their Habits.* London: John Murray, 1881.
——. "Geological Notes Made during a Survey of the East and West Coasts of South America, in the Years 1832, 1833, 1834, and 1835, with an Account of a Transverse Section of the Cordilleras of the Andes between Valparaiso and Mendoza." *Proceedings of the Geological Society of London* 2 (November 1833–June 1838):210–212. *CP,* 1:16–19.
——. *Geological Observations on South America. Being the Third Part of the Geology of the Voyage of the Beagle.* London: Smith, Elder and Co., 1846.
——. *Geological Observations on the Volcanic Islands Visited During the Voyage of H.M.S. Beagle, together with Some Brief Notices of the Geology of Australia and the Cape of Good Hope. Being the Second Part of the Geology of the Voyage of the Beagle.* London: Smith, Elder and Co., 1844.
——. "Geology." In *A Manual of Scientific Enquiry; Prepared for the Use of Her Majesty's Navy: and Adapted for Travellers in General,* edited by John F. W. Herschel, 156–195. London: John Murray, 1849.
——. "The Geology of the Falkland Islands." *Quarterly Journal of the Geological Society of London* 2 (1846):267–274. *CP,* 1:203–212.
——. *The Geology of the Voyage of the Beagle.* With a critical introduction to each work by John W. Judd. London: Ward, Lock and Co., 1890.
——. *Journal of Researches into the Natural History and Geology of the Countries Visited during the Voyage of H.M.S. Beagle round the World, under the Command of Capt. Fitz Roy, R.N.* 2d ed. London: John Murray, 1845.
——. *A Monograph on the Fossil Lepadidæ, or Pedunculated Cirripedes of Great Britain.* [Vol. I] London: Palæontographical Society, 1851.
——. *A Monograph on the Fossil Balanidaeæ and Verrucidæ of Great Britain.* [Vol. II] London: Palæontographical Society, 1854. Index to Vol. 11, 1858.
——. *A Monograph on the Sub-Class Cirripedia.* Vol. 1: *The Lepadidæ; or, Pedunculated Cirripedes.*Vol. 2: *The Balanidæ, or Sessile Cirripedes; The Verrucidæ.* London: Ray Society, 1851, 1854.
——. "Notes on the Effects Produced by the Ancient Glaciers of Caernarvonshire, and on the Boulders Transported by Floating Ice." *Edinburgh New Philosophical Journal* 23 (1842): 352–363. *CP,* 1:163–171.
——. "Observations on the Parallel Roads of Glen Roy, and of Other Parts of Lochaber in Scotland, with an Attempt to Prove that They Are of Marine Origin." *Philosophical Transactions of the Royal Society of London* (1839):39–81. *CP,* 1:89–137.
——. "Observations of Proofs of Recent Elevation on the Coast of Chili, Made during the Survey of His Majesty's Ship Beagle, commanded by Capt. Fitzroy, R.N." *Proceedings of the Geological Society of London* 2 (November 1833–June 1838): 446–449. *CP,* 1:41–43.
——. "On Certain Areas of Elevation and Subsidence in the Pacific and Indian Oceans, as De-

duced from the Study of Coral Formations." *Proceedings of the Geological Society of London* 2 (November 1833–June 1838):552–554. *CP*, 1:46–49.

——. "On the Analogy of the Structure of some Volcanic Rocks with That of Glaciers [1845]." *Proceedings of the Royal Society of Edinburgh* 2 (1851):17–18.

——. "On the Connexion of Certain Volcanic Phænomena, and on the Formation of Mountain-chains and Volcanos, as the Effects of Continental Elevations." *Proceedings of the Geological Society of London* 2 (November 1833–June 1838):654–660.

——. "On the Connexion of Certain Volcanic Phænomena in South America; and on the Formation of Mountain Chains and Volcanos, as the Effect of the Same Power by which Continents Are Elevated." *Transactions of the Geological Society of London*, 2d ser., 5 (1840):601–631. *CP*, 1:53–86.

——. "On the Distribution of the Erratic Boulders and on the Contemporaneous Unstratified Deposits of South America." *Proceedings of the Geological Society of London* 3 (November 1838–June 1842):426–430. *Transactions of the Geological Society of London*, 2d ser., 6 (1842):415–431. *CP*, 1:145–163.

——. "On the Formation of Mould." *Proceedings of the Geological Society of London* 2 (November 1833–June 1838):574–576. *CP*, 1:49–53.

——. "On the Geology of the Falkland Islands." *Quarterly Journal of the Geological Society of London* 2 (1846):267–274. *CP*, 1:203–212.

——. *On the Origin of Species by Means of Natural Selection, or the Preservation of Favoured Races in the Struggle for Life*. London: John Murray, 1859.

——. "On the Power of Icebergs to Make Rectilinear, Uniformly-Directed Grooves Across a Submarine Undulatory Surface." *Philosophical Magazine* 10 (1855):96–98. *CP*, 1:252–255.

——. "On a Remarkable Bar of Sandstone Off Pernambuco, on the Coast of Brazil." *London, Edinburgh and Dublin Philosophical Magazine and Journal of Science*, 3d ser., 19 (1841):257–260. *CP*, 1:139–142.

——. *On the Structure and Distribution of Coral Reefs; Also Geological Observations on Volcanic Islands and Parts of South America*. With a Critical Introduction to Each Work by John W. Judd. London: Ward, Lock, and Co., 1890.

——. "On the Transportal of Erratic Boulders from a Lower to a Higher Level." *Quarterly Journal of the Geological Society of London* 4 (1848):315–323. *CP*, 1:218–227.

——. "A Sketch of the Deposits containing Extinct Mammalia in the Neighbourhood of the Plata." *Proceedings of the Geological Society of London* 2 (November 1833–June 1838):542–544. *CP*, 1:44–45.

——. *The Structure and Distribution of Coral Reefs. Being the First Part of the Geology of the Voyage of the Beagle*. London: Smith, Elder and Co., 1842; 2d ed. 1874.

——. *The Structure and Distribution of Coral Reefs*. Forward by H. W. Menard. Berkeley: University of California Press, 1962.

——, ed. *The Zoology of the Voyage of H.M.S. Beagle, under the Command of Captain Fitzroy, during the Years 1832 to 1836*. Part I. *Fossil Mammalia*, by Richard Owen. London, 1838.

Darwin, Charles, and Alfred Russel Wallace. *Evolution by Natural Selection*. Cambridge: Cambridge University Press, 1958.

Darwin, Francis, ed. *The Life and Letters of Charles Darwin*. 3 vols. London: John Murray, 1887.

De Beer, Gavin, ed. "Darwin's Journal." *Bulletin of the British Museum (Natural History) Historical Series* 2, no. 1 (1959):1–21.

Di Gregorio, Mario A., and N. W. Gill, eds. *Charles Darwin's Marginalia*. New York: Garland, 1990.

FitzRoy, Robert, ed. *Narrative of the Surveying Voyages of His Majesty's Ships Adventure and Beagle;* 3 vols. + appendix to vol. 2. Vol. 1: *Proceedings of the First Expedition, 1826–1830, under the Command of Captain P. Parker King;* vol. 2: *Proceedings of the Second Expedition, 1831–1836 under the Command of Captain Robert Fitz-Roy* + appendix; vol. 3: *Journal and*

Remarks. 1832–1836, by Charles Darwin. London: Henry Colburn, 1839. Vol. 3 also issued separately as *Journal of Researches into the Geology and Natural History of the Various Countries Visited by H.M.S. 'Beagle' . . . 1832–1836*. London: Henry Colburn, 1839.
Freeman, Richard B. *The Works of Charles Darwin: An Annotated Bibliographical Handlist*. 2d ed. Folkestone: Dawson, 1977.
Herbert, Sandra, ed. *The Red Notebook of Charles Darwin*. London and Ithaca: British Museum (Natural History) and Cornell University Press, 1980.
Keynes, Richard Darwin, ed. *Charles Darwin's Zoology Notes and Specimen Lists from H.M.S. Beagle*. Cambridge: Cambridge University Press, 2000.
——, ed. *The Beagle Record: Selections from the Original Pictorial Records and Written Accounts of the Voyage of H.M.S. Beagle*. Cambridge: Cambridge University Press, 1979.
——, ed. *Charles Darwin's 'Beagle' Diary*. Cambridge: Cambridge University Press, 1988.
Stauffer, Robert C., ed. *Charles Darwin's Natural Selection: Being the Second Part of His Big Species Book Written from 1856 to 1858*. Cambridge: Cambridge University Press, 1975.
Stoddart, David R., ed. "Coral Islands by Charles Darwin." *Atoll Research Bulletin* 88 (December 1962):1–20.

(3) Works Published before 1900 by Authors Other than Charles Darwin

An asterisk preceding an entry signifies a work known to be part of H.M.S. *Beagle*'s library (1831–1836). For a complete listing of such works, see Appendix IV, "Books on board the *Beagle*," in *Correspondence*, 1:553–557.

Agate, Dendy. [On Charles Lyell.] *The Christian Life* (London), October 5, 1912.
[Agassiz, Louis.] "Discovery of the Former Existence of Glaciers in Scotland, Especially in the Highlands, by Professor Agassiz." *Scotsman* 66 (7 October 1840).
Agassiz, Louis. *Études sur les glaciers*. Neuchâtel: Jent et Gassmann, 1840.
——. "On Glaciers and Boulders in Switzerland." *Report of the British Association for the Advancement of Science* [Glasgow 1840] (1841):113–114 (Transactions of the Sections).
——. "On Glaciers, and the Evidence of Their Having Once Existed in Scotland, Ireland, and England." *Proceedings of the Geological Society of London* 3 (November 1838–June 1842):327–332.
——. "On the Polished and Striated Surfaces of the Rocks Which Form the Beds of Glaciers in the Alps." *Proceedings of the Geological Society of London* 3 (November 1838–June 1842):320–321.
——. *Studies on Glaciers Preceded by the Discourse of Neuchâtel*. Edited, translated and with an introduction by Albert V. Carozzi. New York: Hafner, 1967.
——. "Upon Glaciers, Moraines and Erratic Blocks: Being the Address Delivered at the Opening of the Helvetic Natural History Society, at Neuchatel [*sic*], on 24th July 1837." *Edinburgh New Philosophical Journal* 24 (1838):364–383.
Allan, Thomas. "On the Rocks in the Vicinity of Edinburgh," *Transactions of the Royal Society of Edinburgh* 6 (1812):405–433.
Anonymous. *The Wonders of the World*. Dublin: B. Smith, 1825.
*Arrowsmith, Aaron. *A New General Atlas, Constructed from the Latest Authorities*. London and Edinburgh: A. Constable, 1823.
Athenæum 24 (24 December 1831):834.
*Aubuisson de Voisins, Jean François d'. *Traité de géognosie*. 2 vols. Strassbourg: F. G. Levrault, 1819.
Austen, Robert Alfred Cloyne (afterward Godwin-Austen). "Account of the Raised Beach Near Hope's Nose, in Devonshire, and Other Recent Disturbances in That Neighbourhood." *Proceedings of the Geological Society of London* 2 (November 1833–June 1838):102–103.
Babbage, Charles. *The Ninth Bridgewater Treatise. A Fragment*. 2d ed. London: John Murray, 1838.
Bakewell, Robert. *An Introduction to Geology: Comprising the Elements of the Science in Its Pre-*

sent Advanced State, and All the Recent Discoveries; with an Outline of the Geology of England and Wales. 3d ed. London: Longman, 1828.
*Beechey, Frederick William. *Narrative of a Voyage to the Pacific and Beering's Strait, . . . in His Majesty's Ship Blossom . . . 1825, 26, 27, 28.* 2 vols. London: Henry Colburn and R. Bentley, 1831.
*——. *Narrative of a Voyage to the Pacific and Beering's Strait . . . 1825.* Philadelphia: Carey and Lea, 1832.
Beer, William, and Johann Heinrich Madler. "Survey of the Surface of the Moon." *Edinburgh New Philosophical Journal* 25 (1838):38–69.
Berzelius, Jakob J. *The Use of the Blowpipe in Chemical Analysis, and in the Examination of Minerals.* Translated by J. G. Children. London: Baldwin, 1822.
Binney, Edward William. "On the Origin of Coal." *Memoirs of the Literary and Philosophical Society of Manchester,* 2d ser., 8 (1848):148–194.
*Bory de Saint-Vincent, J. B., ed., *Dictionnaire classsique d'histoire naturelle.* 17 vols. Paris: Rey and Gravier, 1822–1831.
Bowman, John E. "On the Question, Whether There Are Any Evidences of the Former Existence of Glaciers in North Wales?" *Philosophical Magazine* 19 (1842):469–479.
Brande, William Thomas. *A Manual of Chemistry.* 2d ed., 3 vols. London: John Murray, 1821.
Brongniart, Alexandre. "Notice Concerning the Method of Collecting, Labelling, and Transmitting Specimens of Fossil Organized Bodies, and of the Accompanying Rocks." *American Journal of Science and Arts* 1 (1819):71–74.
*——. "On the Relative Position of the Serpentines (Ophiolites), Diallage Rocks (Euphotides), Jasper, &c., in Some Parts of the Apennines." In *A Selection of the Geological Memoirs Contained in the Annales des Mines,* edited by Henry De la Beche, 161–207. London: William Phillips, 1824.
*——. "On the Zoological Characters of Formations, with the Application of These Characters to the Determination of Some Rocks of the Chalk Formation." In *A Selection of the Geological Memoirs Contained in the Annales des Mines,* edited and translated by Henry De la Beche, 235–261. London: William Phillips, 1824.
*Brongniart, Alexandre. "Rapport fait à l'Académie Royale des Sciences, sur les travaux de M. Gay." *Annales des Sciences Naturelles* 28 (1833):26–35.
Brooke, Henry James. *A Familiar Introduction to Crystallography; Including an Explanation of the Principle and Use of the Goniometer.* London: W. Phillips, 1823.
*Buch, Leopold von. *Description physique des îles Canaries, suivie d'une indication des principaux volcans du globe.* With atlas. Paris: F. G. Levrault, 1836.
——. "On Volcanos and Craters of Elevation." *Edinburgh New Philosophical Journal* 21 (1836):189–206.
*——. *Travels through Norway and Lapland.* Translated by John Black. Notes by Robert Jameson. London: H. Colburn, 1813.
Buckland, William. "Evidences of Glaciers in Scotland and the North of England." *Proceedings of the Geological Society of London* 3 (November 1838–June 1842): 332–337, 345–348.
——. *Geology and Mineralogy Considered with Reference to Natural Theology.* 2 vols. London: Pickering, 1836.
——. *Geology and Mineralogy, Considered with Reference to Natural Theory.* 2d ed., 2 vols. in 1. London: Pickering, 1837.
——. "Instructions for Conducting Geological Investigations, and Collecting Specimens." *American Journal of Science and Arts* 3 (1821):249–251.
——. "On the Excavation of Valleys by Diluvial Action, as Illustrated by a Succession of Valleys Which Intersect the South Coast of Dorset and Devon." *Transactions of the Geological Society of London,* 2d ser., 1(1824):95–102.
*——. "On the Fossil Remains of the Megatherium, Recently Imported into England from South America [read 23 June 1832]." *Report of the British Association for the Advancement of Science* [York 1831, Oxford 1832] (1833):104–107.

——. "On the Glacia-Diluvial Phænomena in Snowdonia and the Adjacent Parts of North Wales." *Proceedings of the Geological Society of London* 3 (November 1838–June 1842):579–584.

*——. "On the Occurrence of Remains of Elephants, and Other Quadrupeds, in the Cliffs of Frozen Mud, in Eschscholtz Bay, within Beering's Strait, and in Other Distant Parts of the Shores of the Arctic Seas." In *Narrative of a Voyage to the Pacific and Beering's Strait . . . 1825, 26, 27, 28,* by F. W. Beechey, 2:331–356. London: Henry Colburn and R. Bentley, 1831.

——. "On the Quartz Rock of the Lickey Hill in Worcestershire, and of the Strata Immediately Surrounding It; with Considerations on the Evidence of a Recent Deluge Afforded by the Gravel Beds of Warwickshire and Oxfordshire, and the Valley of the Thames from Oxford Downwards to London," *Transactions of the Geological Society of London* 5 (1821):506–544.

——. "Presidential Address [21 February 1840]." *Proceedings of the Geological Society of London* 3 (November 1838–June 1842):210–267.

——. "Presidential Address [19 February 1841]." *Proceedings of the Geological Society of London* 3 (November 1838–June 1842):469–540.

——. "Professor Buckland's Reply to Dr. Fleming." *Edinburgh Philosophical Journal* 12 (1825):304–319.

——. *Reliquiae Diluvianae; or, Observations on the Organic Remains Contained in Caves, Fissures, and Diluvial Gravel, and on Other Geological Phenomena, Attesting the Action of an Universal Deluge.* London: John Murray, 1823.

——. *Vindiciae Geologicae; or the Connexion of Geology with Religion.* Oxford: Oxford University Press, 1820.

Bunt, Thomas G. "Account of the Leveling Operations between the Bristol Channel and the English Channel." *Report of the British Association for the Advancement of Science* [Newcastle 1838] (1839):11–18.

Catalogue of Charts, Plans, Views, and Sailing Directions, &c. London: Published by Order of the Lords Commissioners of the Admiralty, 1852 (or other editions).

*Caldcleugh, Alexander. *Travels in South America, during the Years 1819–20–21, Containing an Account of the Present State of Brazil, Buenos Ayres, and Chile.* 2 vols. London: John Murray, 1825.

Chambers, Robert. *Ancient Sea-Margins, As Memorials of Changes in the Relative Level of Sea and Land.* Edinburgh: W. and R. Chambers, 1848.

——. *Vestiges of the Natural History of Creation.* Original work published anonymously. London: John Churchill, 1844.

Charpentier, Jean de. "Account of One of the Most Important Results of the Investigations of M. Venetz, Regarding the Present and Earlier Condition of the Glaciers of the Canton Vallais." *Edinburgh New Philosophical Journal* 21 (1836):210–220.

The Charter of the Geological Society of London. London: Richard Taylor, 1836.

Clark, John Willis, and Thomas McKenny Hughes, eds. *Life and Letters of the Reverend Adam Sedgwick.* 2 vols. Cambridge: Cambridge University Press, 1890.

Clift, William. "Some Account of the Remains of the Megatherium Sent to England from Buenos Ayres by Woodbine Parish, Jun." *Transactions of the Geological Society of London,* 2d ser., 3 (1835):437–460.

Cobbe, Frances Power. *Life.* 2 vols. Boston: Houghton Mifflin, 1894.

*Conybeare, William Daniel. "Report on the Progress, Actual State, and Ulterior Prospects of Geological Science." *Report of the British Association for the Advancement of Science* [York 1831, Oxford 1832] (1833):365–414.

*Conybeare, William Daniel, and William Phillips. *Outlines of the Geology of England and Wales, with an Introductory Compendium of the General Principles of that Science and Comparative Views of the Structure of Foreign Countries.* Part I. London: William Phllips, 1822.

Croll, James. *Climate and Time in Their Geological Relations: A Theory of Secular Changes of the Earth's Climate.* London: Daldy, Isbister, 1875.

*Cuvier, Georges. *The Animal Kingdom Arranged in Conformity with Its Organization . . . with Additional Descriptions of All the Species Hitherto Named, and of Many Not Before Noticed.* By Edward Griffith and others. 16 vols. London: Whittaker, 1827–1835.

——. *Discours sur les Révolutions de la Surface du Globe.* 3d French ed. Paris: G. Dufour, 1825.

——. *Essay on the Theory of the Earth.* Translated by Robert Kerr. Mineralogical notes by Robert Jameson. Edinburgh: William Blackwood, 1813. Also 5th ed., 1827.

——. *Recherches sur les ossemens fossiles de quadrupèdes.* 4 vols. Paris: Deterville, 1812.

Darwin, Erasmus. *The Botanic Garden, A Poem, in Two Parts.* Part 1: *The Economy of Vegetation.* Part 2: *The Loves of the Plants.* (Part 1, 1791 1st ed.; Part 2, 1790 2d ed.) London: J. Johnson, 1791–1790.

——. *The Temple of Nature; or, The Origin of Society:* a Poem, with *Philosophical Notes.* London: J. Johnson, 1803.

*Daubeny, Charles. *A Description of Active and Extinct Volcanos; with Remarks on Their Origin, Their Chemical Phænomena, and the Character of Their Products, As Determined by the Condition of the Earth during the Period of Their Formation.* London: William Phillips, 1826.

Dawson, L. S. *Memoirs of Hydrography.* 2 vols. Eastbourne: Henry W. Keyay, 1885.

*De la Beche, Henry Thomas. *A Geological Manual.* London: Treuttel and Würtz, 1831.

——. *Researches in Theoretical Geology.* London: Charles Knight, 1834.

——. *Sections and Views Illustrative of Geological Phaenomena.* London: Treuttel and Würtz, 1830.

*——, ed. *A Selection of the Geological Memoirs Contained in the Annales des Mines, Together with a Synoptical Table of Equivalent Formations, and M. Brongniart's Table of the Classification of Mixed Rocks.* London: William Phillips, 1824.

——. *A Tabular and Proportional View of the Superior, Supermedial, and Medial (Tertiary and Secondary) Rocks.* 2d ed. London: Truettel, 1828.

——. *A Descriptive and Illustrated Catalogue of the Fossil Organic Remains of Mammalia and Aves Contained in the Museum of the Royal College of Surgeons of England.* London: Richard and John E. Taylor, 1845.

Dupetit-Thouars, Abel. *Voyage autour du monde sur la frégate la Vénus pendant les années 1836–1839, Relation.* 4 vols. Paris: Gide, 1841–1845.

**Encyclopædia Britannica.* 6th ed., 20 vols. + *Supplement.* Edinburgh: Archibald Constable, 1823.

Essays and Reviews. 7th ed. London: Longman, 1861. (lst ed. 1860.)

Falkner, Thomas. *A Description of Patagonia, and the Adjoining Parts of South America.* Hereford: C. Pugh, 1774.

[Fitton, William.] "Geology of the Deluge." *Edinburgh Review,* 39 (1823):196–234.

*Fitton, William Henry. "Instructions for Collecting Geological Specimens." In *Narrative of a Survey of the Intertropical and Western Coasts of Australia . . . 1818–1822,* by Phillip Parker King, 2:623–629. London: John Murray, 1827.

——. "Presidential Address [15 February 1828]." *Proceedings of the Geological Society of London* 1 (November 1826–June 1833):50–62.

——. "Presidential Address [20 February 1829]." *Proceedings of the Geological Society of London* 1 (November 1826–June 1833):112–134.

FitzRoy, Robert, ed. *Narrative of the Surveying Voyages of His Majesty's Ships Adventure and Beagle;* 3 vols. + appendix to vol. 2. Vol. 1: *Proceedings of the First Expedition, 1826–1830, under the Command of Captain P. Parker King;* vol. 2: *Proceedings of the Second Expedition, 1831–1836 under the Command of Captain Robert Fitz-Roy* + appendix; vol. 3: *Journal and Remarks. 1832–1836,* by Charles Darwin. London: Henry Colburn, 1839. Vol. 3 also issued separately as *Journal of Researches into the Geology and Natural History of the Various Countries Visited by H.M.S. 'Beagle' . . . 1832–1836.* London: Henry Colburn, 1839.

Fleming, John. "The Geological Deluge, as Interpreted by Baron Cuvier and Professor Buckland, Inconsistent with the Testimony of Moses and the Phenomena of Nature." *Edinburgh Philosophical Journal* 14 (1826):205–239.

——. "Remarks Illustrative of the Influence of Society on the Distribution of British Animals." *Edinburgh Philosophical Journal* 11 (1824):287–305.

Forbes, Edward. "On the Connexion between the Distribution of the Existing Fauna and Flora of the British Isles, and the Geological Changes Which Have Affected Their Area, Especially during the Epoch of the Northern Drift." *Memoirs of the Geological Survey of Great Britian and of the Museum of Economic Geology in London* 1 (1846):336–432.

——. "On the Distribution of Endemic Plants, More Especially Those of the British Islands, Considered with Regard to Geological Changes." *Report of the British Association for the Advancement of Science* [Cambridge 1845] (1846):67–68.

Forchhammer, Johann Georg. "On Some Changes of Level Which Have Taken Place during the Historical Period in Denmark." *Proceedings of the Geological Society of London* 2 (November 1833–June 1838) 554–556.

——. "On Some Changes of Level Which Have Taken Place in Denmark During the Present Period." *Transactions of the Geolgoical Society of London,* 2d ser., 6 (1842): 157–160.

*Gay, Claude. "Aperçu sur les recherches d'historie naturelle faites dans l'Amérique du Sud . . . 1830 et 1831." *Annales des Sciences Naturelles* 18 (1833):369–373.

Geikie, Archibald. *Memoir of Sir Roderick Murchison.* 2 vols. London: John Murray, 1875.

Geikie, James. *The Great Ice Age, and Its Relation to the Antiquity of Man.* 2d ed. London: W. Isbister, 1877.

——. *Prehistoric Europe: A Geological Sketch.* London: E. Stanford, 1881.

Geological and Geographical Committee. *Report of the First and Second Meetings of the BAAS* [York 1831, Oxford 1832] (1833):54.

Geological Society of London. *The Charter of the Geological Society of London: Instituted 1807; Incorporated 1826, With the Bye-Laws Adopted at the General Meeting on the 1st of May, 1827.* London: Richard Taylor, 1836.

Gould, John. "Three Species of the Genus Orpheus, from the Galapagos, in the collection of Mr. Darwin." *Proceedings of the Zoological Society of London* 5 (1837):27.

Graham, Maria. "An Account of Some Effects of the Late Earthquakes in Chili [*sic*]. Extracted from a Letter to Henry Warburton, Esq." *Transactions of the Geological Society of London,* 2d ser., 1 (1824):413–415.

——. *A Letter to the President and Members of the Geological Society of London, in Answer to Certain Observations Contained in Mr. Greenough's Anniversary Address of 1834.* London: T. Brettel, 1834.

*Greenough, George. *A Critical Examination of the First Principles of Geology.* London: Longman, 1819.

*——. "Presidential Address [21 February 1834]." *Proceedings of the Geological Society of London* 2 (November 1833–June 1838):42–70.

——. "Presidential Address [20 February 1835]." *Proceedings of the Geological Society of London* 2 (November 1833–June 1838):145–175.

Griffin, John. *Chemical Recreations: A Compendium of Experimental Chemistry.* 8th ed. Glasgow: R. Griffin, 1838.

——. *A Practical Treatise on the Use of the Blowpipe, in Chemical and Mineral Analysis.* Glasgow: Richard Griffin, 1827.

Griffith, Edward. *See* Cuvier, Georges.

*Hall, Basil. *Extracts from a Journal Written on the Coasts of Chili, Peru and Mexico for the Years 1820, 1821, 1822.* 2 vols. Edinburgh: A. Constable, 1824.

*Hall, Basil, and John Playfair. "Account of the Structure of the Table Mountain, and Other Parts of the Peninsula of the Cape." *Transactions of the Royal Society of Edinburgh* 7 (1815):269–278 + plates 13–15.

Hall, James. "On the Revolutions of the Earth's Surface." *Transactions of the Royal Society of Edinburgh* 44 (1815):139–212.

Henry, William. *The Elements of Experimental Chemistry.* 9th ed., 2 vols. London: Baldwin, 1823.
Henslow, John Stevens. *Descriptive and Physiological Botany.* London: Longman, 1836.
*——. "Geological Description of Anglesea." *Transactions of the Cambridge Philosophical Society* 1 (1822):359–452.
——. "On the Deluge." *Annals of Philosophy* 6 (1823):344–348.
——. "On Typical Objects in Natural History." *Report of the British Association for the Advancement of Science* [Glasgow 1855] (1856):108–110.
——. *A Sermon on the First and Second Resurrection Preached at Great St. Mary's Church on Feb. 15, 1829.* Cambridge, 1829.
Herschel, John F. W. "On the Astronomical Causes Which May Influence Geological Phænomena." *Transactions of the Geological Society of London,* 2d ser., 3 (1835):293–300. Also *Proceedings of the Geological Society of London* 1 (1826–1833):244–245.
——. *A Preliminary Discourse on the Study of Natural Philosophy.* London: Longman, 1831.
——. *Results of Astronomical Observations Made during the Years 1834, 5, 6, 7, 8, at the Cape of Good Hope; Being the Completion of a Telescopic Survey of the Whole Surface of the Visible Heavens, Commenced in 1825.* London: Smith, Elder and Co., 1847.
Hooker, Joseph Dalton. "On the Vegetation of the Carboniferous Period, As Compared with that of the Present Day." *Memoirs of the Geological Survey of Great Britain and of the Museum of Practical Geology in London* 2, pt. 2 (1848):387–430.
Hopkins, William. *An Abstract of a Memoir on Physical Geology.* Cambridge: Pitt Press, 1836.
——. "On the Phenomena of Precession and Nutation, Assuming the Fluidity of the Interior of the Earth." *Philosophical Transactions of the Royal Society of London* 129 (1839):381–423.
——. "On Precession and Nutation, Assuming the Interior of the Earth to Be Fluid and Heterogeneous." *Philosophical Transactions of the Royal Society of London* 130 (1840):193–208.
——. "On the Thickness and Constitution of the Earth's Crust." *Philosophical Transactions of the Royal Society of London* 132 (1842):43–55.
——. "Report on the Geological Theories of Elevation and Earthquakes." *Report of the British Association for the Advancement of Science* [Oxford 1847] (1848):33–92.
——. "Researches in Physical Geology [1835]." *Transactions of the Cambridge Philosophical Society* 6 (1836):1–84.
——. Review of the *Origin* [1860] "Physical Theories of the Phenomena of Life." In *Darwin and His Critics: The Reception of Darwin's Theory of Evolution by the Scientific Community,* by David Hull, 229–275. Cambridge: Harvard University Press, 1973.
*Humboldt, Alexander von. *Fragmens de géologie et de climatologie asiatiques.* Paris: Gide, A. Phian Delaforest, Delaunay, 1831.
*——. *A Geognostical Essay on the Superposition of Rocks, in Both Hemispheres.* London: Longman, 1823.
——. "On Isothermal Lines." *Edinburgh New Philosophical Journal* 3 (1820):1–20, 256–274; 4 (1821):23–37, 262–281; 5 (1821):28–39.
*——. *Personal Narrative of Travels to the Equinoctial Regions of the New Continent, during the Years 1799–1804.* Translated by Helen Maria Williams. 7 vols. in 9. London: Longman, 1814–1829.
Iddings, Joseph P. "The Origin of Igneous Rocks." *Bulletin of the Philosophical Society of Washington* 12 (1892–1894):89–214.
James, Henry. *Abstracts of the Principal Lines of Spirit Levelling in Scotland.* London: Eyre and Spottiswoode, 1861.
——. *Notes on the Parallel Roads of Lochaber . . . with Illustrative Maps and Sketches from the Ordnance Survey of Scotland.* Southampton: Ordnance Survey Office, 1874.
Jameson, Robert. *Manual of Mineralogy: Containing an Account of Simple Minerals, and Also a Description and Arrangement of Mountain Rocks.* Edinburgh: Archibald Constable, 1821.
——. *A System of Mineralogy.* 3 vols. Edinburgh: Archibald Constable, 1820.

Jamieson, Thomas Francis. "On the History of the Last Geological Changes in Scotland." *Quarterly Journal of the Geological Society of London* 21 (1865):161–203.
——. "On the Ice-Worn Rocks of Scotland." *Quarterly Journal of the Geological Society of London* 18 (1862):164–184.
——. "On the Parallel Roads of Glen Roy, and Their Place in the History of the Glacial Period." *Quarterly Journal of the Geological Society of London* 19 (1863): 235–259.
Jenyns, Leonard. *Memoir of the Rev. John Stevens Henslow.* London: J. Van Voorst, 1862.
Judd, John W. "Critical Introduction" to Charles Darwin's *Geological Observations on South America.* In *The Works of Charles Darwin,* edited by Paul H. Barrett and R. B. Freeman, 9:7–14. New York: New York University Press, 1987.
——. "Critical Introduction" to Charles Darwin's *Geological Observations on the Volcanic Islands.* In *The Works of Charles Darwin,* edited by Paul H. Barrett and R. B. Freeman, 8:7–15. New York: New York University Press, 1987.
——. "Critical Introduction" to Charles Darwin's *Structure and Distribution of Coral Reefs.* In *The Works of Charles Darwin,* edited by Paul H. Barrett and R. B. Freeman, 7:7–15. New York: New York University Press, 1987.
*King, Phillip Parker. "Some Observations upon the Geography of the Southern Extremity of South America, Tierra del Fuego, and the Strait of Magalhaens." *Journal of the Royal Geographical Society* 1 (1831):155–175.
Labillardière, Jacques Julien Houton de. *Relation du voyage à la recherche de 'La Pérouse' . . . pendant les années 1791, 1792, et pendant la 1ère et la 2ème année de la République françoise.* 2 vols. + atlas. Paris: H. J. Jansen, 1800.
Lamarck, Jean Baptiste de. *Zoological Philosophy: An Exposition with Regard to the Natural History of Animals.* Translated by Hugh Elliot. Chicago: University of Chicago Press, 1984.
Laplace, Pierre-Simon. *Exposition du système du monde.* 2 vols. Paris: Cercle-Social, 1796.
Lauder, Thomas Dick. "On the Parallel Roads of Lochaber." *Transactions of the Royal Society of Edinburgh* 9 (1823):1–64.
Lund, Peter Wilhelm. "List of Fossil Mammifera from the Basin of the Rio das Velhas, with an Extract of Some of Their Distinguishing Characters." *Annals of Natural History* 3 (1839):422–427.
Lyell, Charles. *Elements of Geology.* London: John Murray, 1838.
——. *The Geological Evidences of the Antiquity of Man.* London: John Murray, 1863.
[——.] Obituary. *Unitarian Herald* (Manchester), March 5, 1875.
——. "On the Boulder Formation or Drift, and Associated Freshwater Deposits Composing the Mud Cliffs of Eastern Norfolk." *Proceedings of the Geological Society of London* 3 (November 1838–June 1842):171–179.
——. "On the Geological Evidence of the Former Existence of Glaciers in Forfarshire." *Proceedings of the Geological Society of London* 3 (November 1838–June 1842):337–345.
——. "Presidential Address [19 February 1836]." *Proceedings of the Geological Society of London* 2 (November 1833–June 1838):357–390.
——. "Presidential Address [17 February 1837]." *Proceedings of the Geological Society of London* 2 (November 1833–June 1838):479–523.
*——. *Principles of Geology, Being an Attempt to Explain the Former Changes of the Earth's Surface, by Reference to Causes Now in Operation.* 3 vols. London: John Murray, 1830 (vol. 1), 1832 (vol. 2), 1833 (vol.3). (Reprint, with bibliography of sources by Martin J. S. Rudwick. Chicago: University of Chicago Press, 1990–1991.)
——. *Principles of Geology.* 5th ed., 4 vols. London: John Murray, 1837.
Lyell, Katharine M., ed. *Life, Letters and Journals of Sir Charles Lyell, Bart.* 2 vols. London: John Murray, 1881.
MacLeay, William Sharp. *Horæ Entomologicæ: Or Essay on the Annulose Animals.* 1 vol. in 2 pts. London: S. Bagster, 1819–1821.
MacCulloch, John. "On the Forms of Geological Hammers." *Quarterly Journal of Science, Literature, and the Arts* 11 (1821):1–10.

——, "On the Parallel Roads of Glen Roy." *Transactions of the Geological Society of London* 4, pt. 2 (1817):314–392.

Mackenzie, George. "An Attempt to Classify the Phenomena in the Glens of Lochaber with those of the Diluvium, or Drift, Which Covers the Face of the Country." *Edinburgh New Philosophical Journal* 44 (1848):1–12.

Malaspina, Alessandro. *Viaje Pólitico-Científico Alrededor del undo por las Corbetas Descubierta y Atrevide . . . desde 1789–1794.* Madrid: Impr. de la viuda e hijos de Abienzo, 1885.

Mallet, Robert. "Earthquakes." In *A Manual of Scientific Enquiry,* edited by J. F. W. Herschel, 196–223. London: John Murray, 1849.

Martin, A. Patchett, ed. *Life and Letters of the Right Honourable Robert Lowe, Viscount Sherbrooke.* 2 vols. London: Longman, 1893.

Mawe, John. *Familiar Lessons on Mineralogy and Geology.* London: Longman, 1829.

*——. *Travels in the Gold and Diamond Districts of Brazil.* London: Longman, 1825.

*Michell, John, "Conjectures Concerning the Cause, and Observations on the Phænomena of Earthquakes; Particularly of That Great Earthquake of the First of November 1755, Which Proved So Fatal to the City of Lisbon, and Whose Effects Were Felt As Far As Africa, and More or Less throughout Almost All Europe," *Philosophical Transactions* 51 (1760):566–634.

*Miers, John. *Travels in Chile and La Plata.* 2 vols. London: Baldwin, 1826.

Miller, Hugh. *The Testimony of the Rocks; or, Geology in Its Bearings on the Two Theologies, Natural and Revealed.* Edinburgh: T. Constable, 1857.

Milne, David. "On the Parallel Roads of Lochaber, with Remarks on the Change of Relative Levels of Sea and Land, and on the Detrital Deposits in that Country." *Transactions of the Royal Society of Edinburgh* 16 (1849):395–418 + map.

*Molina, Juan Ignacio. *Compendio de la historia geografica natural y civil del Reyno de Chile.* 2 vols. Madrid: A. de Sancha, 1788–1795.

Morris, John, and Daniel Sharpe. "Description of Eight Species of Brachiopodus Shells from the Palaeozoic Rocks of the Falkland Islands." *Quarterly Journal of the Geological Society of London* 2 (1846):274–278.

Mulhall, Michael G. *The English in South America.* Buenos Ayres: Standard Office, 1878.

Murchison, Roderick Impey. "Presidential Address [17 February 1832]." *Proceedings of the Geological Society of London* 1 (November 1826–June 1833):362–386.

——. "Presidential Address [15 February 1833]." *Proceedings of the Geological Society of London* 1 (November 1826–June 1833):438–464.

——. "Presidential Address [18 February 1842]." *Proceedings of the Geological Society of London* 3 (November 1838–June 1842):637–687.

——. "Presidential Address [17 February 1843]." *Proceedings of the Geological Society of London* 4 (November 1843–April 1845):65–151.

——. *Siluria: The History of the Oldest Fossiliferous Rocks and Their Foundations.* 3d ed. London: John Murray, 1859.

——. *The Silurian System, Founded on Geological Researches in the Counties of Salop, Hereford, Radnor, Montgomery, Caermarthen, Brecon, Pembroke, Monmouth, Gloucester, Worcester, and Stafford; with Descriptions of the Coal-Fields and Overlying Formations.* London: John Murray, 1839.

*d'Omalius d'Halloy, Jean-Baptiste. "Memoir on the Geological Extent of the Formation of the Environs of Paris." In *A Selection of the Geological Memoirs Contained in the Annales des Mines, Together with a Synoptical Table of Equivalent Formations, and M. Brongniart's Table of the Classification of Mixed Rocks.* London: William Phillips, 1824, edited and translated by Henry De la Beche, 9–35. London: William Phillips, 1824.

Orbigny, Alcide Charles Victor Dessalines d. *Voyage dans l'Amérique Méridionale (le Brésil, la République orientale de l'Uruguay, la République Argentine, la Patagonie, la République du Chili, la République de Bolivia, la République de Pérou), exécuté pendant les années 1826 . . . 1833.* 9 vols. Paris and Strasbourg: Pitous- Levrault, 1835–1847.

Owen, Richard. "A Description of the Cranium of *Toxodon Platensis,* a Gigantic Extinct Mammiferous Species, Referrible by Its Dentition to the *Rodentia,* but with Affinities to the *Pachydermata,* and the *Herbivorous Cetacea.*" *Proceedings of the Geological Society of London* 2 (November 1833–June 1838):541–542.

——. "Description of a Tooth and Part of the Skeleton of the *Glyptodon clavipes,* a Large Quadruped of the Edentate Order, to which Belongs the Tesselated Bony Armour Described and Figured by Mr. Clift with a Consideration of the Question Whether the *Megatherium* Possessed an Analogous Dermal Armour." *Transactions of the Geological Society of London* 6 (1842):81–106.

Parish, Woodbine. "An Account of the Discovery of Portions of Three Skeletons of the Megatherium in the Province of Buenos Ayres in South America [read 13 June 1832]." *Proceedings of the Geological Society of London* 1 (November 1826–June 1833):403–404.

——. *Buenos Ayres and the Provinces of the Rio de la Plata.* London: John Murray, 1838.

Parkes, Samuel. *The Chemical Catechism.* 10th ed. London: Baldwin, 1822.

Pattinson, Hugh L. "On a New Process for the Extraction of Silver from Lead." *Report of the British Association for the Advancement of Science* [Newcastle 1838] (1839):50–55 (Transactions of the Sections).

[Phillips, John.] "Darwin's Origin of Species." *Saturday Review,* 24 December 1859, 775–776.

——. *Life on the Earth: Its Origin and Succession.* Cambridge, England: Macmillan, 1860.

——. *Manual of Geology: Practical and Theoretical.* London: Richard Griffin, 1855.

——. "Presidential Address [17 February 1860]." *Quarterly Journal of the Geological Society of London.* 16, pt. 1 (1860):xxv–lv.

——. *A Treatise on Geology.* 2 vols. London: Longman, 1837 (vol. 1), 1839 (vol. 2).

*Phillips, William. *An Elementary Introduction to the Knowledge of Mineralogy: Comprising Some Account of the Characters and Elements of Minerals; Explanations of Terms in Common Use; Descriptions of Minerals, with Accounts of the Places and Circumstances in Which They Are Found; and Especially the Localities of British Minerals.* 3d ed. London: William Phillips, 1823.

*Playfair, John. *Illustrations of the Huttonian Theory of the Earth.* Edinburgh: Cadell and Davies, 1802.

Porter, David. *Journal of a Cruise* 1815; repr. Annapolis: Naval Institute Press, 1986.

Prévost, Constant."*Diluvion.*" In *Dictionnaire classique d'histoire naturelle,* edited by J. B. Bory de Saint-Vincent, 5:508–509. 17 vols. Paris: Rey and Gravier, 1822–1831.

Quoy, Jean René Constant, and Joseph Paul Gaimard. "Mémoire sure l'accroissement des polypes lithophytes considéré géologiquement." *Annales des Sciences Naturelles* 6 (1825): 273–290.

Ramsay, Andrew Crombie. "The Old Glaciers of Switzerland and North Wales." In *Peaks, Passes, and Glaciers: A Series of Excursions by Members of the Alpine Club,* ser. 1, edited by John Ball, 400–466. London: Longman, 1859–1860.

——. "On the Denudation of South Wales and the Adjacent Counties of England." *Memoirs of the Geological Survey of Great Britain, and of the Museum of Economic Geology in London* 1 (1846):297–335.

Robson, George Fennell. *Scenery of the Grampian Mountains.* 1814; repr. London: Longman, 1819.

Rogers, William B., and Henry D. Rogers. "On the Physical Structure of the Appalachian Chain, As Exemplifying the Laws Which Have Regulated the Elevation of Great Mountain Chains Generally." *Reports of the First, Second, and Third Meetings of the Association of American Geologists and Naturalists* 1 (1843):474–531.

Rose, Henry. *A Manual of Analytical Chemistry.* Translated by John Griffin. London: Thomas Tegg, 1831.

Sabine, Edward. "Report on the Variations of the Magnetic Intensity Observed at Different

Points of the Earth's Surface." *Report of the British Association for the Advancement of Science* [Liverpool 1837] (1838):1–85, 497–500.
*Scrope, George Poulett. *Considerations on Volcanos: The Probable Causes of Their Phenomena, the Laws Which Determine Their March, the Disposition of Their Products, and Their Connexion with the Present State and Past History of the Globe; Leading to the Establishment of a New Theory of the Earth.* London: William Phillips, 1825.
——. *The Geology and Extinct Volcanos of Central France.* 2d ed. London: John Murray, 1858.
——. *Memoir on the Geology of Central France.* London: Longman, 1827.
——. [Review of] "Principles of Geology . . . by Charles Lyell." *Quarterly Review* 86 (October 1830):411–469.
Sedgwick, Adam. "On Diluvial Formations." *Annals of Philosophy* n.s. 10 (1825): 18–37.
——. "On the Origin of Alluvial and Diluvial Formations." *Annals of Philosophy* n.s. 9 (1825):241–257.
——. "Presidential Address [19 February 1830]." *Proceedings of the Geological Society of London* 1 (November 1826–June 1833):187–212.
——. "Presidential Address [18 February 1831]." *Proceedings of the Geological Society of London* 1 (November 1826–June 1833):270–316.
——. "Remarks on the Structure of Large Mineral Masses, and Especially on the Chemical Changes Produced in the Aggregation of Stratified Rocks during Different Periods after Their Deposition." *Transactions of the Geological Society of London,* 2d ser., 3 (1835):461–486.
——. *A Syllabus of a Course of Lectures on Geology.* Cambridge: J. Hodson, 1821; 2d ed., 1832; 3d ed., at the University Printers, 1837.
Sedgwick, Adam, and Roderick Impey Murchison. "Classification of the Older Stratified Rocks of Devonshire and Cornwall." *Philosophical Magazine* 14 (1839):241–260, 317, 354–358.
Strickland, Hugh E. "Report of a Committee Appointed to Consider of the Rules by Which the Nomenclature of Zoology May Be Established on a Uniform and Permanent Basis." *Report of the British Association for the Advancement of Science* [Manchester 1842] (1843):105–121.
——. "Report of a Committee Appointed to Print and Circulate a Report on Zoological Nomenclature." *Report of the British Association for the Advancement of Science* [Cork 1843] (1844):119–120.
*Syme, Patrick. *Werner's Nomenclature of Colours.* 2d ed. London: Blackwood, 1821.
Taylor, J. E., ed. *Notes on Collecting and Preserving Natural-History Objects.* London: Hardwicke and Bogue, 1876.
Thomson, James. "On the Parallel Roads of Lochaber." *Edinburgh New Philosophy Journal* 14 (1848):49–61.
Thomson, William. "On the Age of the Sun's Heat." *Macmillan's Magazine* 5 (1862):388–393.
——. "On the Secular Cooling of the Earth." *Philosophical Magazine* 25 (1863):1–14.
——. "Physical Considerations Regarding the Possible Age of the Sun's Heat." *Report of the British Association for the Advancement of Science* [Manchester 1861] (1862):27–28 (Transactions of the Sections).
Todhunter, Isaac. *William Whewell: An Account of His Writings with Selections from His Literary and Scientific Correspondence.* 2 vols. London: Macmillan, 1876.
Topley, William. *The Geology of the Weald.* London: Longman, 1875.
Trimmer, Joshua. "On the Diluvial Deposits of Caernarvonshire, between the Snowdon Chain of Hills and the Menai Strait, and on the Discovery of Marine Shells in Diluvial Sand and Gravel on the Summit of Moel Tryfane, near Caernarvon, 1000 ft above the Level of the Sea." *Proceedings of the Geological Society of London* 1 (November 1826–June 1833):331–332.
*Ure, Andrew. *A Dictionary of Chemistry.* 2d ed. London: Thomas Tegg, 1823.

Venetz, Ignace. "Account of the Descent of the Glacier of the Weisshorn, on the 27th of December 1819, and the Destruction of the Village of Randa." *Edinburgh New Philosophical Journal* 3 (1820):274–277.

Waterhouse, George Robert. *A Natural History of the Mammalia*. 2 vols. London: Billière, 1846–1848.

——. "Observations on the Classification of the Mammalia." *Annals and Magazine of Natural History* 12 (1843):399–412.

Webster, Thomas. "On Some Freshwater Formations on the Isle of Wight, With Some Observations on the Strata over the Chalk in the Southeast Part of England." *Transactions of the Geological Society of London* 2 (1814):161–254.

*Webster, William. *Narrative of a Voyage to the Southern Atlantic Ocean, in the Years 1828, 29, 30, Performed in H.M. Sloop Chanticleer, under the Command of the Late Captain Henry Foster.* 2 vols. London: R. Bentley, 1834.

Werner, Abraham Gottlob. *Short Classification and Description of the Various Rocks*. Translated and edited by Alexander M. Ospovat. 1786; repr. New York: Hafner, 1971.

Whewell, William. "Account of a Level Line, Measured from the Bristol Channel to the English Channel, during the Year 1837–8, by Mr. Bunt, under the Direction of a Committee of the British Association." *Report of the British Association for the Advancement of Science* [Newcastle 1838] (1839):1–11.

*——. "Essay towards a First Approximation to a Map of Cotidal Lines." *Philosophical Transactions of the Royal Society of London* 123 (1833):147–236.

——. *History of the Inductive Sciences from the Earliest to the Present Times*. 3 vols. London: J. W. Parker, 1837.

——. "Presidential Address [16 February 1838]." *Proceedings of the Geological Society of London* 2 (November 1833–June 1838):624–649.

——. "Presidential Address [15 February 1839]." *Proceedings of the Geological Society of London* 3 (November 1838–June 1842):61–98.

——. "[Review of] *Principles of Geology*, Vol. I." *British Critic, Quarterly Theological Review and Ecclesiastical Record* 19 (1831):180–206.

——. "[Review of] *Principles of Geology*, Vol. II, by Charles Lyell." *Quarterly Review* 47 (1832):103–132.

Willis, Robert, and John Willis Clark. *The Architectural History of the University of Cambridge*. 4 vols. Cambridge: Cambridge University Press, 1886.

Witham, Henry T. M. *The Internal Structure of Fossil Vegetables Found in the Carboniferous and Oolitic Deposits of Great Britain*. Edinburgh: Adam and Charles Black, 1833.

Wordsworth, William. *The Excursion*. 1814; facs. Oxford: Woodstock Books, 1991.

(4) Works Published after 1900

Addison, Kenneth. *Classic Glacial Landforms of Snowdonia*. Sheffield: Geographical Association, 1997.

Albury, W. R., and D. R. Oldroyd. "From Renaissance Mineral Studies to Historical Geology, in the Light of Michel Foucault's *The Order of Things*." *British Journal for the History of Science* 10 (1977):187–215.

Aldrich, Michele L. "Women in Paleontology in the United States, 1840–1860." *Earth Sciences History* 1 (1982):14–22.

Allen, David. *The Naturalist in Britain: A Social History*. 2d ed. Princeton: Princeton University Press, 1994.

Alter, Stephen G. *Darwinism and the Linguistic Image*. Baltimore: Johns Hopkins University Press, 1999.

Appel, Toby A. *The Cuvier-Geoffroy Debate: French Biology in the Decades before Darwin*. Oxford: Oxford University Press, 1987.

Appleman, Daniel E. "James Dwight Dana and Pacific Geology." In *Magnificent Voyagers: The*

U.S. Exploring Expedition, 1838–1842, edited by Herman J. Viola and Carolyn Margolis, 88–117. Washington, D.C.: Smithsonian Institution, 1985.
Armstrong, Patrick. "Charles Darwin's Geological Notes on Mauritius." *Indian Ocean Review* 1, no. 2 (1988):1–20.
——. *Charles Darwin in Western Australia: A Young Scientist's Perception of an Environment.* Nedlands, Western Australia: University of Western Australia Press, 1985.
——. *Darwin's Desolate Islands: A Naturalist in the Falklands, 1833 and 1834.* Chippenham, England: Picton, 1992.
——. *Under the Blue Vault of Heaven: A Study of Charles Darwin's Sojourn in the Cocos (Keeling) Islands.* Nedlands, Western Australia: Indian Ocean Centre for Peace Studies, 1991.
Ashworth, J. H. "Charles Darwin as a Student in Edinburgh, 1825–1827." *Proceedings of the Royal Society of Edinburgh* 55 (1934–1935):97–113.
Baker's Map of the University and Town of Cambridge 1830. Cambridge: Cambridgeshire Records Society, 1998.
Banks, Maxwell R. "A Darwin Manuscript on Hobart Town." *Papers and Proceedings of the Royal Society of Tasmania* 105 (1971):5–19.
Banks, Maxwell R., and David Leaman. "Charles Darwin's Field Notes on the Geology of Hobart Town—A Modern Appraisal." *Papers and Proceedings of the Royal Society of Tasmania* 133, no. 1 (1999):29–50.
Barnes, John. *Basic Geological Mapping.* 3d ed. Chichester: John Wiley, 1995.
Barrett, Paul H. "Darwin's 'Gigantic Blunder.'" *Journal of Geological Education* 21 (January 1977):19–28.
——. "The Sedgwick-Darwin Geological Tour of North Wales." *Proceedings of the American Philosophical Society* 118, no. 2 (1974):146–164.
Barrett, Paul H., and Richard B. Freeman, eds. *The Works of Charles Darwin.* 29 vols. New York: New York University Press, 1987–1989.
Bartholomew, Michael. "Lyell and Evolution: An Account of Lyell's Response to the Prospect of an Evolutionary Ancestry for Man." *British Journal for the History of Science* 6 (1973):261–303.
Beaglehole, John C. *Exploration of the Pacific.* 3d ed. Stanford: Stanford University Press, 1966.
——. "James Cook." *Dictionary of Scientific Biography* 3:396–397.
——. *The Life of Captain James Cook.* London: Adam and Charles Black, 1974.
Beatty, John. "Speaking of Species: Darwin's Strategy." In *The Darwinian Heritage,* edited by David Kohn, 265–281. Princeton: Princeton University Press, 1985.
Beckinsale, Robert P. "William Hopkins." *Dictionary of Scientific Biography* 6:502–504.
Bedell, Rebecca. *The Anatomy of Nature: Geology and American Landscape Painting, 1825–1875.* Princeton: Princeton University Press, 2001.
Beer, Gillian. "Darwin and the Growth of Language Theory." In *Nature Transfigured: Science and Literature, 1700–1900,* edited by John Christie and Sally Shuttleworth, 152–170. Manchester: Manchester University Press, 1989.
——. "Travelling the Other Way." In *Cultures of Natural History,* edited by Nicholas Jardine, James Secord, and Emma Spary, 322–327. Cambridge: Cambridge University Press, 1996.
Benson, Keith R., and Philip F. Rehbock, eds. *Oceanographic History: The Pacific and Beyond.* Seattle: University of Washington Press, 2002.
Bevington, Merle Mowbray. *The "Saturday Review," 1855–1868: Representative Educated Opinion in Victorian England.* New York: Columbia University Press, 1941.
Biermann, Kurt-R. "Alexander von Humboldt." *Dictionary of Scientific Biography* 6:549–555.
Birembaut, Arthur. "Jean-François d'Aubuisson de Voisin." *Dictionary of Scientific Biography* 1:327–328.
Black, G. P. "Arthurs's Seat." In *Lothian Geology: An Excursion Guide,* edited by A. D. McAdam and E. N. K. Clarkson, 33–48. Edinburgh: Scottish Academic Press, 1986.
Blundell, Derek J., and Andrew C. Scott. *Lyell: The Past Is the Key to the Present.* London: Geological Society, 1998.

Bowen, Norman L. *The Evolution of the Igneous Rocks*. Princeton: Princeton University Press, 1928.

Bowlby, John. *Charles Darwin: A New Life*. New York: W. W. Norton, 1990.

Bowler, Peter J. *Fossils and Progress: Paleontology and the Idea of Progressive Evolution in the Nineteenth Century*. New York: Science History Publications, 1976.

——. *Life's Splendid Drama: Evolutionary Biology and the Reconstruction of Life's Ancestry 1860–1940*. Chicago: University of Chicago Press, 1996.

Boylan, Patrick J. "English and Scottish Glacial Localities of Agassiz, Buckland and Lyell, 1840." www.city.ac.uk/artspol/glaclocs.html (Accessed 1999).

——. "Lyell and the Dilemma of Quaternary Glaciation." In *Lyell: The Past Is the Key to the Present*, edited by D. J. Blundell and A. C. Scott, 145–159. London: Geological Society, 1998.

Brock, W. H. "Humboldt and the British: A Note on the Character of British Science." *Annals of Science* 50 (July 1993):365–372.

Brooke, John Hedley. "The Natural Theology of the Geologists: Some Theological Strata." In *Images of the Earth: Essays in the History of the Environmental Sciences*, edited by L. J. Jordanova and Roy Porter, 39–64. Chalfont St. Giles, England: British Society for the History of Science, 1979.

——. *Science and Religion: Some Historical Perspectives*. Cambridge: Cambridge University Press, 1991.

Browne, Janet. *Charles Darwin: The Power of Place*. New York: Alfred A. Knopf, 2002.

——. *Charles Darwin: Voyaging*. New York: Alfred A. Knopf, 1995.

——. *The Secular Ark: Studies in the History of Biogeography*. New Haven: Yale University Press, 1983.

Brush, George J. *Manual of Determinative Mineralogy*. Revised by Samuel L. Penfield. New York: John Wiley, 1911.

Brush, Stephen G. *Nebulous Earth: The Origin of the Solar System and the Core of the Earth from Laplace to Jeffreys*. Cambridge: Cambridge University Press, 1996.

——. *Transmuted Past: the Age of the Earth and Evolution of the Elements from Lyell to Patterson*. Cambridge: Cambridge University Press, 1996.

Buchwald, Jed Z. "William Thomson." *Dictionary of Scientific Biography* 13:374–388.

Burchard, Ulrich. "Blowpipe." In *Instruments of Science*, edited by Robert Bud and Deborah Jean Warner, 68–69. New York: Garland, 1998.

——. "The History and Apparatus of Blowpipe Analysis." *Mineralogical Record* 25, no. 4 (1994):251–277.

——. "History of the Development of the Crystallographic Goniometer." *Mineralogical Record* 29 (1998):517–583.

Burchfield, Joe D. "The Age of the Earth and the Invention of Geological Time." In *Lyell: The Past Is the Key to the Present*, edited by Derek J. Blundell and Andrew C. Scott, 137–143. London: Geological Society, 1998.

——. "Darwin and the Dilemma of Geological Time," *Isis* 65 (1974):301–321.

——. *Lord Kelvin and the Age of the Earth*. New York: Science History Publications, 1975.

Burke, John G. "Friedrich Mohs." *Dictionary of Scientific Biography* 9:447–449.

Burkhardt, Frederick H. "Darwin's Early Notes on Coral Reef Formation." *Earth Sciences History* 3 (1984):160–163.

——. "A Troublesome Letter Signed 'yrs. Ch. Darwin.'" *Documentary Editing* 23 (2001):73–81.

Burkhardt, Richard W., Jr. *The Spirit of System: Lamarck and Evolutionary Biology*. 2d ed. Cambridge: Harvard University Press, 1995.

Bushnell, O. A. "Aftermath: Britons' Response to News of the Death of Captain James Cook." *Hawaiian Journal of History* 25 (1991):1–20.

Cain, P. J., and A. G. Hopkins. *British Imperialism: Innovation and Expansion, 1688–1914*. London: Longman, 1993.

Callomon, John H. "Fossil as Geological Clocks." In *The Age of the Earth: From 4004 BC to*

AD 2002, edited by C. L. E. Lewis and S. J. Knell, 237–252. London: Geological Society, 2001.

Cannon, Susan Faye. *Science in Culture: The Early Victorian Period.* New York: Dawson and Science History Publications, 1978.

Cannon, Walter F. [Susan Faye]. "The Impact of Uniformitarianism: Two Letters from John Herschel to Charles Lyell, 1836–1837." *Proceedings of the American Philosophical Society* 105 (1961):301–314.

——. "John Herschel and the Idea of Science." *Journal of the History of Ideas* 22 (April–June 1961):215–239.

——. "William Buckland." *Dictionary of Scientific Biography* 2:566–572.

Cassirer, Ernst. *The Philosophy of the Enlightenment.* Translated by Fritz C. A. Koelln and James P. Pettegrove. Princeton: Princeton University Press, 1951 (original German edition 1932).

Cawood, John. "The Magnetic Crusade: Science and Politics in Early Victorian Britain." *Isis* 70 (1979):493–518.

Challinor, John. *A Dictionary of Geology.* 5th ed. Cardiff: University of Wales Press, 1978.

——. *The History of British Geology: A Bibliographical Study.* New York: Barnes and Noble, 1971.

——. "Thomas Webster." *Dictionary of Scientific Biography* 14:210–211.

Chancellor, Gordon, Angelo DiMauro, Ray Ingle, and Gillian King. "Charles Darwin's *Beagle* Collections in the Oxford University Museum." *Archives of Natural History* 15:197–231.

Chorley, Richard J., Antony J. Dunn, and Robert P. Beckinsale. *The History of the Study of Landforms; or, The Development of Geomorphology.* Vol. I: *Geomorphology before Davis.* London: Methuen, 1964.

Christy, Barbara Mae, and Paul D. Lowman Jr. "Global Maps of Volcanism: Two Maps from Two Centuries." In *Volcanoes and History,* edited by Nicoletta Morello, 65–90. Genova: Brigati, 1998.

Chubb, Lawrence John. "Geology of Galapagos, Cocos, and Easter Islands." *Bernice P. Bishop Museum Bulletin* 110 (1933):1–44.

Cohen, I. Bernard. *Revolution in Science.* Cambridge: Harvard University Press, 1985.

Coleman, William. *Georges Cuvier, Zoologist: A Study in the History of Evolution Theory.* Cambridge: Harvard University Press, 1964.

——. "Lyell and the Reality of Species." *Isis* 53 (1962):325–338.

Corsi, Pietro. *Science and Religion: Baden Powell and the Anglican Debate, 1800–1860.* Cambridge: Cambridge University Press, 1988.

Cowan, Charles F. "Notes on Griffith's *Animal Kingdom of Cuvier* (1824–1835)." *Journal of the Society for the Bibliography of Natural History* 5 (1969):137–140.

Cox, L. R. "New Light on William Smith and His Work." *Proceedings of the Yorkshire Geological Society* 25 (1942):1–99.

Crisp, Dennis J. "Extending Darwin's Investigations on the Barnacle Life-history." *Biological Journal of the Linnean Society* 20 (1983):73–83.

Croft, Darin A. "Placentals: Endemic South American Ungulates." In *Encyclopedia of Paleontology,* edited by Ronald Singer, 2:290–906. 2 vols. Chicago: Fitzroy Dearborn, 1999.

Crossley, Louise. *Explore Antarctica.* Cambridge: Cambridge University Press, 1995.

Curry, Robert. "Evolution and Ecology of Cooperative Breeding in Galapagos Mockingbirds (*Nesomimus* SPP.)." Ph.D. dissertation, University of Michigan, 1987.

Dalrymple, G. Brent. *The Age of the Earth.* Stanford: Stanford University Press, 1991.

——. "The Age of the Earth in the Twentieth Century: A Problem (Mostly) Solved." In *The Age of the Earth: From 4004 BC to AD 2002,* edited by C. L. E. Lewis and S. J. Knell, 205–221. London: Geological Society, 2001.

Darden, Lindley. *Theory Change in Science: Strategies from Mendelian Genetics.* Oxford: Oxford University Press, 1991.

Darwin, George Howard. *Scientific Papers.* 5 vols. Cambridge: Cambridge University Press, 1907–1916.

Daston, Lorraine, and Peter Galison. "The Image of Objectivity." *Representations* 40 (1992): 81–128.

Davies, G. L. *See* Herries Davies.

Davies, John. *The Making of Wales*. Cardiff: Cadw, 1996.

Day, Archibald. *The Admiralty Hydrographic Service, 1795–1919*. London: Her Majesty's Stationery Office, 1967.

Deacon, Margaret. *Scientists and the Sea 1650–1900*. Aldershot, England: Ashgate, 1997.

Dean, Dennis R. *Gideon Mantell and the Discovery of Dinosaurs*. Cambridge: Cambridge University Press, 1999.

——. "Graham Island, Charles Lyell, and the Craters of Elevation." *Isis* 71 (1980):571–588.

De Beer, Gavin. *The Sciences Were Never at War*. London: Thomas Nelson, 1960.

Desmond, Adrian. *Archetypes and Ancestors: Palaeontology in Victorian London, 1850–1875*. London: Blond and Briggs, 1982.

——. *The Politics of Evolution: Morphology, Medicine, and Reform in Radical London*. Chicago: University of Chicago Press, 1989.

Desmond, Adrian, and James Moore. *Darwin: The Life of a Tormented Evolutionist*. New York: Warner Books, 1991.

Dettelbach, Michael. "Global Physics and Aesthetic Empire: Humboldt's Physical Portrait of the Tropics." In *Visions of Empire: Voyages, Botany and Representations of Nature*, edited by David Philip Miller and Peter Hanns Reill, 258–292. Cambridge: Cambridge University Press, 1996.

——. "Humboldtian Science." In *Cultures of Natural History*, edited by N. Jardine, J. A. Secord, and E. C. Spary, 287–304. Cambridge: Cambridge University Press, 1996.

Dolan, Brian P. "Governing Matters: The Values of English Education in the Earth Sciences, 1790–1830." Ph.D. dissertation, University of Cambridge, 1995.

Dott, Robert H., Jr. "Charles Lyell in America—His Lectures, Field Work, and Mutual Influences, 1841–1853." *Earth Sciences History* 15 (1996):101–140.

——. "Recognition of the Tectonic Significance of Volcanism in Ancient Orogenic Belts." In *Volcanoes and History*, edited by Nicoletta Morello, 123–131. Genova: Brigati, 1998.

Douglas, Norman, and Ngaire Douglas, eds. *Pacific Islands Yearbook*. 16th ed. North Ryde, Australia: Augus and Robertson, 1989.

Ecuador. *Manual de información . . . del Ecuador*. 3 vols. Quito: Cientifica Latina Editores, 1980.

Eldredge, Niles, and Stephen Jay Gould, "Punctuated Equilibria: An Alternative to Phyletic Gradualism." In *Models in Paleobiology*, edited by Thomas J. M. Schopf, 82–115. San Francisco: Freeman, Cooper, 1972.

Elena, Alberto. "The Imaginary Lyellian Revolution." *Earth Sciences History* 7 (1988):126–133.

Ellegård, Alvar. *Darwin and the General Reader: The Reception of Darwin's Theory of Evolution in the British Periodical Press, 1859–1872*. Göteborg: Göteborg University, 1958.

Ellis, Elizabeth. *Conrad Martens: Life and Art*. Sydney: State Library of New South Wales Press, 1994.

Ellis, Ieuan. *Seven against Christ: A Study of "Essays and Reviews."* Leiden: E. J. Brill, 1980.

Estes, Gregory, K. Thalia Grant, and Peter R. Grant. "Darwin in Galápagos: His Footsteps through the Archipelago." *Notes and Records of the Royal Society of London* 54 (2000):343–368.

Etherington, Norman. "Reconsidering Theories of Imperialism." *History and Theory* 21, no. 1 (1982):1–36.

Eyles, Joan M. "Robert Jameson." *Dictionary of Scientific Biography* 7:69–71.

——. "William Smith." *Dictionary of Scientific Biography* 12:486–492.

Eyles, V. A. "George Bellas Greenough." *Dictionary of Scientific Biography* 5:518–519.

——. "Henry De la Beche." *Dictionary of Scientific Biography* 4:9–11.

——. "James Hall." *Dictionary of Scientific Biography* 6:53–56.

Fautin, Daphne G. "Beyond Darwin: Coral Reef Research in the Twentieth Century." In *Oceanographic History: The Pacific and Beyond*, edited by Keith R. Benson and Philip F. Rehbock, 446–449. Seattle: University of Washington Press, 2002.

Ferns, H. S. *Argentina*. London: Ernest Benn, 1969.

——. "Britain's Informal Empire in Argentina, 1806–1814." *Past and Present* 4 (November 1953):60–75.

Fleming, James R. "Charles Lyell and Climatic Change: Speculation and Certainty." In *Lyell: The Past Is the Key to the Present*, edited by D. J. Blundell and A. C. Scott, 161–169. London: Geological Society, 1997.

——. "Concordance to Charles Darwin's 'Sketch of 1842' and 'Essay of 1844.'" Privately printed, n.d.

Friendly, Alfred. *Beaufort of the Admiralty: The Life of Sir Francis Beaufort, 1774–1847*. London: Hutchinson, 1977.

Friis, Herman R., ed. *The Pacific Basin: A History of Its Geographical Exploration*. New York: American Geographical Society, 1967.

Fritts, Thomas H. "Morphometrics of Galapagos Tortoises: Evolutionary Implications." In *Patterns of Evolution in Galapagos Organisms*, edited by Robert I. Bowman, Margaret Berson, and Alan E. Leviton, 107–122. San Franciso: Pacific Division, American Association for the Advancement of Science, 1983.

Frost, Alan. "Captain James Cook and the Early Romantic Imagination." In *Captain James Cook: Image and Impact*, edited by Walter Veit, 90–106. Melbourne: Hawthorn, 1972.

Fuller, John G. C. M. "Before the Hills in Order Stood: The Beginning of the Geology of Time in England." In *The Age of the Earth: From 4004 BC to AD 2002*, edited by C. L. E. Lewis and S. J. Knell, 15–23. London: Geological Society, 2001.

Fuller, John, and Hugh Torrens. "A Wealden Geology and English History Excursion: Ships, Guns, Fuel." Unpublished fieldguide, 2000.

Gallagher, John, and Ronald Robinson. "The Imperialism of Free Trade." *Economic History Review*, 2d ser., 1 (1953):1–15.

Garber, Janet Bell. "Charles Darwin as a Laboratory Director." Ph.D. dissertation, University of California Los Angeles, 1989.

Gascoigne, John. *Science in the Service of Empire: Joseph Banks, the British State and the Uses of Science in the Age of Revolution*. Cambridge: Cambridge University Press, 1998.

Gaudin, Timothy J. "Xenarthrans." In *Encyclopedia of Paleontology*, edited by Ronald Singer, 2:1347–1353. 2 vols. Chicago: Fitzroy Dearborn, 1999.

Geikie, Archibald. *Charles Darwin as Geologist*. Cambridge: Cambridge University Press, 1909.

Geschwind, Carl-Henry. "Becoming Interested in Experiments: American Igneous Petrologists and the Geophysical Laboratory, 1905–1965." *Earth Sciences History* 14 (1995):47–61.

Ghiselin, Michael T. *The Triumph of the Darwinian Method*. Chicago: University of Chicago Press, 1969.

Gill, Stephen. *Wordsworth and the Victorians*. Oxford: Clarendon Press, 1998.

Gillispie, Charles Coulston. *Genesis and Geology: A Study in the Relations of Scientific Thought, Natural Theology, and Social Opinion in Great Britain, 1790–1850*. Cambridge: Harvard University Press, 1951.

——; with the collaboration of Robert Fox and Ivor Grattan-Guinness. *Pierre-Simon Laplace: A Life in Exact Science*. Princeton: Princeton University Press, 1997.

Glick, Thomas F., ed. *The Comparative Reception of Darwinism*. Austin: University of Texas Press, 1974.

Goetzmann, William H. *Exploration and Empire: The Explorer and the Scientist in the Winning of the American West*. New York: W. W. Norton, 1966.

Gohau, Gabriel. *A History of Geology*. Revised and translated by Albert V. Carozzi and Marguerite Carozzi. New Brunswick: Rutgers University Press, 1990.

Golinski, Jan. *Science as Public Culture: Chemistry and Enlightenment in Britain, 1760–1820*. Cambridge: Cambridge University Press, 1992.

Goodman, D. C. "William Hyde Wollaston." *Dictionary of Scientific Biography* 14: 486–494.
Gould, Stephen Jay. "Agassiz's Marginalia in Lyell's *Principles,* or the Perils of Uniformity and the Ambiguity of Heroes." *Studies in History of Biology* 3 (1979):119–138.
——. *The Structure of Evolutionary Theory.* Cambridge: Harvard University Press, 2002.
——. *Time's Arrow, Time's Cycle: Myth and Metaphor in the Discovery of Geological Time.* Cambridge: Harvard University Press, 1987.
——. *Wonderful Life: The Burgess Shale and the Nature of History.* New York: W. W. Norton, 1990.
Graham, Richard. *Independence in Latin America: A Comparative Approach.* 2d ed. New York: McGraw-Hill, 1994.
——. "Robinson and Gallagher in Latin America: The Meaning of Informal Imperialism." In *Imperialism: The Robinson and Gallagher Controversy,* edited by William Roger Louis, 217–221. New York: Franklin Watts, 1976.
Grant, Peter R. *Ecology and Evolution of Darwin's Finches.* Princeton: Princeton University Press, 1999.
Greene, John C. *The Death of Adam: Evolution and Its Impact on Western Thought.* Ames: Iowa State University Press, 1959; rev. ed. 1996.
——. "The Kuhnian Paradigm and the Darwinian Revolution in Natural History." In *Perspectives in the History of Science and Technology,* edited by Duane H. D. Roller, 3–25. Norman: University of Oklahoma Press, 1971).
Greene, Mott T. *Geology in the Nineteenth Century: Changing Views of a Changing World.* Ithaca: Cornell University Press, 1982.
Gross, Alan G. *The Rhetoric of Science.* Cambridge: Harvard University Press, 1990.
Gruber, Howard E., and Valmai Gruber. "The Eye of Reason: Darwin's Development during the *Beagle* Voyage." *Isis* 53 (1962):186–200.
Haber, Francis. *The Age of the World: Moses to Darwin.* Baltimore: Johns Hopkins University Press, 1959.
Hall, Elizabeth. "Ethology's Warning: A Conversation with Niko Tinbergen." *Psychology Today* 7, no. 10 (March 1974):65–80.
Harker, Alfred. *The Natural History of Igneous Rocks.* London: Methuen, 1909.
——. "Notes on the Rocks of the 'Beagle' Collection.—I." *Geological Magazine* n.s. 4 (March 1907):100–106.
Hector, Andy, and Rowan Hooper. "Darwin and the First Ecological Experiment." *Science* 295 (25 January 2002):639–640.
Heinzel, Herman, and Barnaby Hall. *Galápagos Diary: A Complete Guide to the Archipelago's Birdlife.* Berkeley: University of California Press, 2000.
Herbert, Sandra. "An 1830s View from Outside Switzerland: Charles Darwin on the "Beryl Blue" Glaciers of Tierra del Fuego." *Eclogae Geologicae Helvetiae* 92 (1999):339–346.
——. "Between Genesis and Geology: Darwin and Some Contemporaries in the 1820s and 1830s." In *Religion and Irreligion in Victorian Society: Essays in Honor of R.K. Webb,* edited by R. W. Davis and R. J. Helmstadter, 68–84. London: Routledge, 1992.
——. "Charles Darwin as a Prospective Geological Author." *British Journal for the History of Science* 24 (1991):159–192.
——. "Darwin as a Geologist." *Scientific American* 254 (May 1986):116–123.
——. "Darwin, Malthus, and Selection." *Journal of the History of Biology* 4 (1971): 209–217.
——. "Darwin the Young Geologist." In *The Darwinian Heritage,* edited by David Kohn, 483–510. Princeton: Princeton University Press, 1985.
——. "Les divergences entre Darwin et Lyell sur quelques questions géologiques." In *De Darwin au darwinisme: science et idéologie,* edited by Yvette Conry, 69–76. Paris: J. Vrin, 1983.
——. "Doing and Knowing: Charles Darwin and Other Travellers." In *Geological Travellers,* edited by Patrick Wyse Jackson. New York: Pober Publications, in press.
——. "Essay Review." *Isis* 84 (1993):113–127.
——. "From Charles Darwin's Portfolio: An Early Essay on South American Geology and Species." *Earth Sciences History* 14 (1995):23–36.

——. "The Logic of Darwin's Discovery." Ph.D. dissertation, Brandeis University, 1968.
——. "The Place of Man in the Development of Darwin's Theory of Transmutation." *Journal of the History of Biology.* Part I: 7 (1974):217–258; Part II: 10 (1977):155–227.
——. "Remembering Charles Darwin as a Geologist." In *Charles Darwin, 1809–1882: A Centennial Commemorative,* edited by Roger G. Chapman and Cleveland T. Duval, 231–258. Wellington, New Zealand: Nova Pacifica, 1982.
Herbert, Sandra, and Michael B. Roberts. "Charles Darwin's Notes on His 1831 Geological Map of Shrewsbury." *Archives of Natural History* 29 (2002):27–29.
Herries Davies, Gordon L. *The Earth in Decay: A History of British Geomorphology, 1578–1878.* London: Macdonald, 1969.
——. *North from the Hook: 150 Years of the Geological Survey of Ireland.* Dublin: Geological Survey of Ireland, 1995.
Hevly, Bruce. "The Heroic Science of Glacier Motion." *Osiris,* 2d ser., 11 (1996): 66–86.
Hoare, Michael E. "The Forsters and Cook's Second Voyage 1772–1775." In *Captain James Cook: Image and Impact,* edited by Walter Veit, 107–116. Melbourne: Hawthorn Press, 1972.
Hodge, Jonathan, and Gregory Radick. *The Cambridge Companion to Darwin,* Cambridge: Cambridge University Press, 2003.
Hodge, M. J. S. "Darwin and the Laws of the Animate Part of the Terrestrial System (1835–1837): On the Lyellian Origins of His Zoonomical Explanatory Program." *Studies in History of Biology* 6 (1983):1–106.
——. "Darwin as a Lifelong Generation Theorist." In *The Darwinian Heritage,* edited by David Kohn, 207–243. Princeton: Princeton University Press, 1985.
——. "England." In *The Comparative Reception of Darwinism,* edited by Thomas F. Glick. Austin: University of Texas Press, 1974.
Hooykaas, Reijer. *The Principle of Uniformity in Geology, Biology and Theology.* 2d impression. Leiden: E. J. Brill, 1963.
Hoskin, Michael. "William Herschel." *Dictionary of Scientific Biography* 6:328–336.
Hoyningen-Huene, Paul. *Reconstructing Scientific Revolutions: Thomas S. Kuhn's Philosophy of Science.* Chicago: University of Chicago Press, 1989.
Hull, David L. "Charles Darwin and Nineteenth-Century Philosophies of Science." In *Foundations of Scientific Method,* edited by Ronald N. Giere and Richard S. Westfall, 115–132. Bloomington: Indiana University Press, 1974.
——. *Darwin and His Critics: The Reception of Darwin's Theory of Evolution by the Scientific Community.* Cambridge: Harvard University Press, 1973.
Humphries, D. W. *The Preparation of Thin Sections of Rocks, Minerals, and Ceramics.* Oxford: Oxford University Press, 1992.
Imbrie, John, and Katherine Palmer Imbrie. *Ice Ages: Solving the Mystery.* Cambridge: Harvard University Press, 1979.
Jaki, Stanley L. "The Five Forms of Laplace's Cosmogony." *American Journal of Physics* 44 (1976):4–11.
Janis, Christine. "Artiodactyls." In *Encyclopedia of Paleontology,* edited by Ronald Singer, 1:125–135. 2 vols. Chicago: Fitzroy Dearborn, 1999.
Jardine, Nicholas, James A. Secord, and Emma Spary, eds., *Cultures of Natural History.* Cambridge: Cambridge University Press, 1996.
Jensen, William B. "The Development of Blowpipe Analysis." In *The History and Preservation of Chemical Instrumentation,* edited by John T. Stock and Mary Orna, 123–149. Dordrecht: D. Reidel, 1986.
Johannsen, Albert. *Manual of Petrographic Methods.* New York: McGraw-Hill, 1918.
Judd, John Wesley. "Darwin and Geology." In *Darwin and Modern Science,* edited by Albert Charles Seward, 337–384. Cambridge: Cambridge University Press, 1909.
Kelly, Gary. *Women, Writing, and Revolution, 1790–1827.* Oxford: Clarendon Press, 1993.
Keynes, Richard. *Fossils, Finches and Fuegians: Charles Darwin's Adventures and Discoveries on the Beagle, 1832–1836.* London: HarperCollins, 2002.

Kirkaldy, J. F. "William Topley and 'The Geology of the Weald.'" *Proceedings of the Geologists' Association* 86 (1975):373–388.

Klaver, Jan M. Ivo. *Geology and Religious Sentiment: The Effect of Geological Discoveries on English Society and Literature between 1829 and 1859.* Leiden: Brill, 1997.

Knell, Simon J. *The Culture of English Geology, 1815–1851: A Science Revealed through Its Collecting.* Aldershot, England: Ashgate, 2000.

Knell, Simon J., and C. L. E. Lewis. "Celebrating the Age of the Earth." In *The Age of the Earth: From 4004 BC to AD 2002,* edited by C. L. E. Lewis and S. J. Knell, 1–14. London: Geological Society of London, 2001.

Kölbl-Ebert, Martina. "British Geology in the Early Nineteenth Century: A Conglomerate with a Female Matrix." *Earth Sciences History* 21 (2002):3–25.

——. "Charlotte Murchison (née Hugonin) 1788–1869." *Earth Sciences History* 16 (1997): 39–43.

——. "Mary Buckland (née Morland), 1797–1857)." *Earth Sciences History* 16 (1997):33–38.

Koebner, Richard, and Helmut Dan Schmidt. *Imperialism: The Story and Significance of a Political Word, 1840–1960.* Cambridge: Cambridge University Press, 1964.

Kohn, David. "The Aesthetic Construction of Darwin's Theory." In *The Elusive Synthesis: Aesthetics and Science,* edited by Alfred I. Tauber, 13–48. Dordrecht: Kluwer, 1996.

——. "Darwin's Ambiguity: The Secularization of Biological Meaning." *British Journal for the History of Science* 22 (1989):215–239.

——. "Darwin's Principle of Divergence as Internal Dialogue." In *The Darwinian Heritage,* edited by David Kohn, 245–257. Princeton: Princeton University Press, 1985.

Kohn, David, ed., with bibliographic assistance from Malcolm J. Kottler. *The Darwinian Heritage.* Princeton: Princeton University Press, 1985.

Kottler, Malcolm J. "Charles Darwin's Biological Species Concept and Theory of Geographic Speciation: the Transmutation Notebooks." *Annals of Science* 35 (1978):275–297.

Kuhn, Thomas S. "The History of Science." *International Encyclopedia of the Social Sciences* 14 (1968):74–83.

——. *The Structure of Scientific Revolutions.* 2d ed. Chicago: University of Chicago Press, 1970.

Kushner, David S. "The Emergence of Geophysics in Nineteenth Century Britain." Ph.D. dissertation, Princeton University, 1990.

——. "Sir George Darwin and a British School of Geophysics." *Osiris* 8 (1993):196–223.

Lakatos, Imre. *Philosophical Papers: The Methodology of Scientific Research Programmes,* edited by John Worrall and Gregory Currie. Cambridge: Cambridge University Press, 1978.

Larsen, Anne. "Equipment for the Field." In *Cultures of Natural History,* edited by N. Jardine, J. A. Secord, and E. C. Spary, 358–377. Cambridge: Cambridge University Press, 1996.

Larson, Edward J. *Evolution's Workshop: God and Science on the Galápagos Islands.* New York: Basic Books, 2001.

Latour, Bruno. *Science in Action: How to Follow Scientists and Engineers through Society.* Cambridge: Harvard University Press, 1987.

Laudan, Rachel. *From Mineralogy to Geology: The Foundations of a Science, 1650–1830.* Chicago: University of Chicago Press, 1987.

Lenzen, Victor F., and Robert P. Multhauf. "Development of Gravity Pendulums in the Nineteenth Century." *Smithsonian Institution National Museum Bulletin* 240 (1966):301–347.

Lewis, Cherry. *The Dating Game: One Man's Search for the Age of the Earth.* Cambridge: Cambridge University Press, 2000.

Lewis, C. L. E., and S. J. Knell, eds. *The Age of the Earth: From 4004 BC to AD 2002.* London: Geological Society of London, 2001.

Lindberg, David C., and Ronald L. Numbers, eds. *God and Nature: Historical Essays on the Encounter between Christianity and Science.* Berkeley: University of California Press, 1986.

Litchfield, Henrietta, ed. *Emma Darwin: A Century of Family Letters.* 2 vols. London: John Murray, 1915.

Lopes, Maria Margaret, and Irina Podgorny. "The Shaping of Latin American Museums of Natural History, 1850–1990." *Osiris* 15 (2000):108–118.

Lord, Ruth. *Henry F. du Pont and Winterthur: A Daughter's Portrait.* New Haven: Yale University Press, 1999.

Love, Alan C. "Darwin and *Cirripedia* Prior to 1846: Exploring the Origins of Barnacle Research." *Journal of the History of Biology* 35 (2002):251–289.

Lucas, Peter. "Jigsaw with Pieces Missing: Charles Darwin with John Price at Bodnant, the Walking Tour of 1826 and the Expeditions of 1827." *Archives of Natural History* 29 (2002):359–370.

——. "'A Most Glorious Country': Charles Darwin and North Wales, Especially His 1831 Geological Tour." *Archives of Natural History* 29 (2002):1–26.

——. "'Three Weeks Which Now Appears like Three Months': Charles Darwin at Plas Edwards, 1819." *National Library of Wales Journal* 32 (2001):133–146.

Lurie, Edward. *Louis Agassiz: A Life in Science.* 1960; repr. Baltimore: Johns Hopkins University Press, 1988.

Lynch, John. *The Spanish American Revolutions, 1808–1826.* 2d ed. New York: W. W. Norton, 1986.

MacLeod, Roy. "Imperial Reflections in the Southern Seas: The Funafuti Expeditions, 1896–1904." In *Nature in Its Greatest Extent: Western Science in the Pacific,* edited by Roy MacLeod and Philip F. Rehbock, 159–191. Honolulu: University of Hawaii Press, 1988.

MacLeod, Roy, ed. *Nature and Empire: Science and the Colonial Enterprise. Osiris* 15 (2000).

Marvin, Ursula B. "Impacts from Space: The Implications for Uniformitarian Geology." In *James Hutton—Present and Future,* edited by G. Y. Craig and J. H. Hull, 89–117. London: Geological Society, 1999.

——. "Meteorites, the Moon and the History of Geology." *Journal of Geological Education* 34 (1986):140–165.

Mayr, Ernst. "Darwin's Principle of Divergence." *Journal of the History of Biology* 25 (1992): 343–359.

——. *The Growth of Biological Thought: Diversity, Evolution, and Inheritance.* Cambridge: Harvard University Press, 1982.

——. *One Long Argument: Charles Darwin and the Genesis of Modern Evolutionary Thought.* Cambridge: Harvard University Press, 1991.

——. *What Evolution Is.* New York: Basic Books, 2001.

McBirney, Alexander R., and Howel Williams. *Geology and Petrology of the Galápagos Islands.* Boulder, CO: Geological Society of America, 1969.

McNeil, Maureen. *Under the Banner of Science: Erasmus Darwin and His Age.* Manchester: Manchester University Press, 1987.

McOuat, Gordon. "Cataloguing Power: Delineating 'Competent Naturalists' and the Meaning of Species in the British Museum." *British Journal for the History of Science* 34 (2001):1–28.

——. "Species, Rules and Meaning: The Politics of Language and the Ends of Definitions in 19th Century Natural History." *Studies in History and Philosophy of Science* 27 (1996):473–519.

Merlen, Godfrey. *Restoring the Tortoise Dynasty.* Quito: Charles Darwin Foundation for the Galápagos Islands, 1999.

Mitch, David. *The Rise of Popular Literacy in Victorian England.* Philadelphia: University of Pennsylvania Press, 1992.

Montgomery, William. "Charles Darwin's Theory of Coral Reefs and the Problem of the Chalk." *Earth Sciences History* 7 (1988):111–120.

Moore, D. T., J. C. Thackray, and D. L. Morgan. "A Short History of the Museum of the Geological Society of London, 1807–1911, with a Catalogue of the British and Irish Accessions, and Notes on Surviving Collections." *Bulletin of the British Musueum (Natural History) Historical Series* 19 (1):51–160.

Moore, James R. "Geologists and Interpreters of Genesis in the Nineteenth Century." In *God and Nature: Historical Essays on the Encounter between Christianity and Science,* edited by David C. Lindberg and Ronald L. Numbers, 322–350. Berkeley: University of California Press, 1986.

Morrell, Jack. "Genesis and Geochronology: The Case of John Phillips (1800–1874)." In *The Age of the Earth: From 4004 BC to AD 2002,* edited by C. L. E. Lewis and S. J. Knell, 85–90. London: Geological Society of London, 2001.

Morrell, Jack, and Arnold Thackray. *Gentlemen of Science: Early Years of the British Association for the Advancement of Science.* Oxford: Clarendon Press, 1981.

Morris, Solene, and Louise Wilson. *Down House: The Home of Charles Darwin.* London: English Heritage, 1998.

Morton, Nicol. "In the Footsteps of Charles Darwin in South America." *Geology Today* 11 (1995):190–195.

Moses, Alfred J., and Charles Lathrop Parsons. *Elements of Mineralogy, Crystallography and Blowpipe Analysis.* New York: Van Nostrand, 1916.

Multhauf, Robert P., and Gregory Good. *A Brief History of Geomagnetism and a Catalog of the Collections of the National Museum of American History.* Washington, D.C.: Smithsonian Institution Press, 1987.

Newcomb, Sally. "Contributions of the British Experimentalists to the Discipline of Geology: 1780–1820." *Proceedings of the American Philosophical Society* 134 (1990):161–225.

Nicholas, F. W., and J. M. Nicholas. *Charles Darwin in Australia.* Cambridge: Cambridge University Press, 1989.

North, F. J. "Paviland Cave, the 'Red Lady,' the Deluge, and William Buckland." *Annals of Science* 5 (1942):91–128.

Novacek, Michael. "Placentals: Overview." In *Encylopedia of Paleontology,* edited by Ronald Singer, 2:884–890. 2 vols. Chicago: Fitzroy Dearborn, 1999.

Numbers, Ronald L. *Creation by Natural Law: Laplace's Nebular Hypothesis in American Thought.* Seattle: University of Washington Press, 1977.

Obeyesekere, Gananath. *The Apotheosis of Captain Cook: European Mythmaking in the Pacific.* Princeton: Princeton University Press, 1992.

Oldroyd, David R. "Adam Sedgwick and Lakeland Geology (1822–24)." *Proceedings Thirtieth International Geological Congress* 26 (1997):197–204.

——. "Charles Darwin (1809–1882)." In *Encyclopedia of Geology,* edited by Richard C. Selley, L. Robin M. Cocks, and Ian R. Plimer, vol. 2, 184–187. Amsterdam: Elsevier, 2005.

——. "Edward Daniel Clarke, 1769–1822, and His Rôle in the History of the Blowpipe." *Annals of Science* 29 (October 1972):214–234.

——. "The Geologist from Cromarty." In *Hugh Miller and the Controversies of Victorian Science,* edited by Michael Shortland, 76–121. Oxford: Clarendon Press, 1996.

——. *The Highlands Controversy: Constructing Geological Knowledge through Fieldwork in Nineteenth-Century Britain.* Chicago: University of Chicago Press, 1990.

——. "A Note on the Status of A.F. Cronstedt's Simple Earths and His Analytical Methods." *Isis* 65 (1974):506–512.

——. *Sciences of the Earth: Studies in the History of Mineralogy and Geology.* Aldershot, England: Variorum, 1998.

——. *Thinking about the Earth: A History of Ideas in Geology.* Cambridge: Harvard University Press, 1996.

Ospovat, Alexander M. "Reflections on A.G. Werner's 'Kurze Klassifikation.'" In *Toward a History of Geology,* edited by Cecil J. Schneer, 242–256. Cambridge: MIT Press, 1969.

Ospovat, Dov. *The Development of Darwin's Theory: Natural History, Natural Theology, and Natural Selection, 1838–1859.* Cambridge: Cambridge University Press, 1981.

——. "Lyell's Theory of Climate." *Journal of the History of Biology* 10 (Fall 1977):317–339.

Ottone, Eduardo G. "The French Botanist Aimé Bonpland and Paleontology at Cuenca del Plata." *Earth Sciences History* 21 (2002):150–165.

Outram, Dorinda. *Georges Cuvier: Vocation, Science and Authority in Post-Revolutionary France*. Manchester: Manchester University Press, 1901.

Owen, Roger, and Bob Sutcliffe, *Studies in the Theory of Imperialism*. London: Longman, 1972.

Page, Leroy. "Diluvialism and Its Critics in Great Britain in the Early Nineteenth Century." In *Toward a History of Geology*, edited by Cecil J. Schneer, 257–271. Cambridge: MIT Press, 1969.

——. "John Fleming." *Dictionary of Scientific Biography* 5:31–32.

——. "The Rise of the Diluvial Theory in British Geological Thought." Ph.D. dissertation, University of Oklahoma, 1963.

Palladino, Paolo, and Michael Worboys. "Science and Imperialism." *Isis* 84 (1993):91–102.

Paradis, James. "Darwin and Landscape." In *Victorian Science and Victorian Values: Literary Perspectives,* edited by James Paradis and Thomas Postlewait, 85–110. New York: New York Academy of Sciences, 1981.

Pearson, Paul N. "Charles Darwin on the Origin and Diversity of Igneous Rocks." *Earth Sciences History* 15 (1996):49–67.

Pearson, Paul N., and Christopher J. Nicholas. "Defining the Base of the Cambrian: The Hicks-Geikie Confrontation of April 1883." *Earth Sciences History* 11 (1992):70–80.

——. "'Marks of Extreme Violence': Charles Darwin's Geological Observations on St. Jago (São Tiago), Cape Verde Islands." In *Geological Travellers,* edited by Patrick Wyse Jackson. New York: Pober Publications, in press.

Peckham, Morse. *The Origin of Species by Charles Darwin: A Variorum Text.* Philadelphia: University of Pennsylvania Press, 1959.

Phillips, John A., and Charles Wetherell. "The Great Reform Act of 1832 and the Political Modernization of England," *American Historical Review* 100 (April 1995):411–436.

Platt, D. C. M. "Economic Imperialism and the Businessman: Britain and Latin America before 1914." In *Studies in the Theory of Imperialism,* edited by Roger Owen and Bob Sutcliffe, 295–311. London: Longman, 1972.

Porter, Duncan M. "The *Beagle* Collector and His Collections." In *The Darwinian Heritage,* edited by David Kohn, 973–1019. Princeton: Princeton University Press, 1985.

——. "Charles Darwin's Vascular Plant Specimens from the Voyage of HMS *Beagle.*" *Botanical Journal of the Linnean Society* (1986):1–172.

——. "Darwin's Missing Notebooks Come to Light." *Nature* 291 (7 May 1981):13.

——, ed. "Darwin Notes on *Beagle* Plants." *Bulletin of the British Museum (Natural History) Historical Series* 14, no. 2 (1987):145–233.

Porter, Roy. "Charles Lyell and the Principles of the History of Geology." *British Journal for the History of Science* 9 (1976):91–103.

——. "Gentlemen and Geology: The Emergence of a Scientific Career, 1660–1920." *Historical Journal* 21 (1978):809–836.

——. *The Making of Geology: Earth Science in Britain, 1660–1815.* Cambridge: Cambridge University Press, 1977.

Pritchard, Peter C. H. *The Galápagos Tortoises: Nomenclatural and Survival Status.* Lunenberg, MA: Chelonian Research Monographs, 1996.

Pyenson, Lewis. "Cultural Imperialism and Exact Sciences Revisited." *Isis* 84 (1993):103–108.

Rachootin, Stan P. "Owen and Darwin Reading A Fossil: *Macrauchenia* in a Boney Light." In *The Darwinian Heritage,* edited by David Kohn, 155–184. Princeton: Princeton University Press, 1985.

Rappaport, Rhoda. "Geology and Orthodoxy: The Case of Noah's Flood in Eighteenth-Century Thought." *British Journal for the History of Science* 11 (1978):1–18.

——. *When Geologists Were Historians, 1665–1750.* Ithaca: Cornell University Press, 1997.

Raverat, Gwen. *Period Piece: A Cambridge Childhood.* London: Faber and Faber, 1960.

Raymond, Jean, and John V. Pickstone. "The Natural Sciences and the Learning of the English Unitarians." In *Truth, Liberty, Religion: Essays Celebrating Two Hundred Years of Manchester College,* edited by Barbara Smith, 129–164. Oxford: Manchester College, 1986.

Rees, Martin J. "Understanding the Beginning and the End." In *The Age of the Earth: From 4004 BC to AD 2002,* edited by C. L. E. Lewis and S. J. Knell, 275–283. London: Geological Society, 2001.

Rehbock, Philip F. "The Early Dredgers: 'Naturalizing' in British Seas, 1830–1850." *Journal of the History of Biology* 12 (1979):293–368.

——. *The Philosophical Naturalists: Themes in Early Nineteenth-Century British Biology.* Madison: University of Wisconsin Press, 1983.

Reingold, Nathan. "Edward Sabine." *Dictionary of Scientific Biography* 12:49–53.

Rhodes, Frank H. T. "Darwinian Gradualism and Its Limits: The Development of Darwin's Views on the Rate and Pattern of Evolutionary Change." *Journal of the History of Biology* 20 (1987):139–157.

——. "Darwin's Search for a Theory of the Earth: Symmetry, Simplicity and Speculation." *British Journal for the History of Science* 24 (1991):193–229.

Richards, Robert J. *The Meaning of Evolution: The Morphological Construction and Ideological Reconstruction of Darwin's Theory.* Chicago: University of Chicago Press, 1992.

——. *The Romantic Conception of Life: Science and Philosophy in the Age of Goethe.* Chicago: University of Chicago Press, 2002.

Richardson, Constance. "Petrology of the Galapagos Islands." *Bernice P. Bishop Museum Bulletin* 110 (1933):45–67.

[Richmond, Marsha.] "Darwin's Study of the Cirripedia." In *The Correspondence of Charles Darwin, 1847–1850,* edited by Frederick Burkhardt et al., 4:388–409. Cambridge: Cambridge University Press, 1988.

Ritchie, George Stephen. *The Admiralty Chart: British Naval Hydrography in the Nineteenth Century.* London: Hollis and Carter, 1967.

Roberts, Michael B. "Buckland, Darwin and the Ice Age in Wales and the Marches, 1841–1842." Paper presented at the International Conference on the History of Geology, Neuchâtel, September 1998.

——. "Darwin at Llanymynech: The Evolution of a Geologist." *British Journal for the History of Science* 29 (1996):469–478.

——. "Darwin's Dog-leg: The Last Stage of Darwin's Welsh Field Trip of 1831." *Archives of Natural History* 25 (1998):59–73.

——. "I Coloured a Map: Darwin's Attempts at Geological Mapping in 1831." *Archives of Natural History* 27 (2000):69–79.

——. "Just before the *Beagle:* Charles Darwin's Geological Fieldwork in Wales, Summer 1831." *Endeavour* 25 (2001):33–37.

Romer, Alfred S. "Darwin and the Fossil Record." *Natural History* 58 (October 1958):456–469.

Rothblatt, Sheldon. *The Revolution of the Dons: Cambridge and Society in Victorian England.* London: Faber and Faber, 1968.

Rozwadowski, Helen. "Fathoming the Ocean: Discovery and Exploration of the Deep Sea, 1840–1880." Ph.D. dissertation, University of Pennsylvania, 1996.

Rudwick, Martin J. S. "Caricature as a Source for the History of Science: De la Beche's Anti-Lyellian Sketches of 1831." *Isis* 66 (1975):534–560.

——. "Charles Darwin in London: The Integration of Public and Private Science." *Isis* 73 (1982):186–206.

——. "Charles Lyell, F.R.S. (1797–1875) and His London Lectures on Geology, 1832–33." *Notes and Records of the Royal Society of London* 29 (1975):231–263.

——. "Charles Lyell Speaks in the Lecture Theatre." *British Journal for the History of Science* 11 (1976):147–155.

——. "Charles Lyell's Dream of a Statistical Palaeontology." *Palaeontology* 21, pt. 2 (May 1978):225–244.

——. "Cuvier and Brongniart, William Smith, and the Reconstruction of Geohistory." *Earth Sciences History* 15 (1996):25–36.

——, "Darwin and Glen Roy: A 'Great Failure' in Scientific Method?" *Studies in History and Philosophy of Science* 5 (1974):97–185.

——. "Darwin and the World of Geology (Commentary)." In *The Darwinian Heritage,* edited by David Kohn, 511–518. Princeton: Princeton University Press, 1985.

——. "The Foundation of the Geological Society of London: Its Scheme for Co-operative Research and Its Struggle for Independence." *British Journal for the History of Science* 1 (1963):325–355.

——. *Georges Cuvier, Fossil Bones, and Geological Catastrophes: New Translations and Interpretations of Primary Texts.* Chicago: University of Chicago Press, 1997.

——. *The Great Devonian Controversy: The Shaping of Scientific Knowledge among Gentlemanly Specialists.* Chicago: University of Chicago Press, 1985.

——. "Historical Analogies in the Geological Work of Charles Lyell." *Janus* 64 (1977):89–107.

——. "Introduction" and "Bibliography." *The Principles of Geology* by Charles Lyell. 3 vols. Chicago: University of Chicago Press, 1990–1991.

——. "Jean-André de Luc and Nature's Chronology." In *The Age of the Earth: From 4004 BC to AD 2002,* edited by C. L. E. Lewis and S. J. Knell, 51–60. London: Geological Society, 2001.

——. *The Meaning of Fossils: Episodes in the History of Palaeontology.* London: Macdonald, 1972.

——. "The Strategy of Lyell's *Principles of Geology.*" *Isis* 61 (1970):5–33.

——. "A Year in the Life of Adam Sedgwick and Company." *Archives of Natural History* 15 (1988):243–268.

Ruggles, Richard I. "Geographical Exploration by the British." In *The Pacific Basin: A History of Its Geographical Exploration,* edited by Herman R. Friis, 221–255. New York: American Geographical Society, 1967.

Rupke, Nicolaas A. *The Great Chain of History: William Buckland and the English School of Geology, 1814–1849.* Oxford: Clarendon Press, 1983.

——. "Humboldt's Fame." Paper presented at the History of Science Society annual meeting, 29 October 1995, Minneapolis.

——. *Richard Owen: Victorian Naturalist.* New Haven: Yale University Press, 1994.

Ruse, Michael. "Darwin's Debt to Philosophy: An Examination of the Influence of the Philosophical Ideas of John F.W. Herschel and William Whewell on the Development of Charles Darwin's Theory of Evolution." *Studies in History and Philosophy of Science* 6 (June 1975):159–181.

——. *Monad to Man: The Concept of Progress in Evolutionary Biology.* Cambridge: Harvard University Press, 1996.

Sahlins, Marshall. *How 'Natives' Think: About Captain Cook, for Example.* Chicago: University of Chicago Press, 1995.

Salvat, Bernard. "Coral Reefs, Science, and Politics: Relationships and Criteria for Decisions over Two Centuries—A French Case History." In *Oceanographic History: The Pacific and Beyond,* edited by Keith R. Benson and Philip R. Rehbock, 468–478. Seattle: University of Washington Press, 2002.

Schaer, Jean-Paul. *Les géologues et le développement de la géologie en pays de Neuchâtel.* Neuchâtel: Muséum d'histoire naturelle de Neuchâtel, 1998.

Schaffer, Simon. "Herschel in Bedlam: Natural History and Stellar Astronomy." *British Journal for the History of Science* 13 (1980):211–239.

Schlee, Susan. *The Edge of an Unfamiliar World: A History of Oceanography.* New York: Dutton, 1973.

Schofield, Robert E. *The Lunar Society of Birmingham: A Social History of Provincial Science and Industry in Eighteenth-Century England.* Oxford: Clarendon Press, 1963.

Searby, Peter. *A History of the University of Cambridge.* Vol. III: *1750–1870.* Cambridge: Cambridge University Press, 1997.

Secord, James A. *Controversy in Victorian Geology: The Cambrian-Silurian Dispute.* Princeton: Princeton University Press, 1986.

——. "The Discovery of a Vocation: Darwin's Early Geology." *British Journal for the History of Science* 24 (1991):133–158.

——. "Edinburgh Lamarckians: Robert Jameson and Robert E. Grant." *Journal of the History of Biology* 24 (1991):1–18.

——. "The Geological Survey of Great Britain as a Research School, 1839–1855." *History of Science* 24 (1986):223–275.

——. "John W. Salter: The Rise and Fall of a Victorian Palaeontological Career." In *From Linnaeus to Darwin: Commentaries on the History of Biology and Geology,* edited by Alwynne Wheeler and James H. Price, 61–75. London: Society for the History of Natural History, 1985.

——. "King of Siluria: Roderick Murchison and the Imperial Theme in Nineteenth-Century British Geology." *Victorian Studies* 25 (1982):413–442.

——. "Nature's Fancy: Charles Darwin and the Breeding of Pigeons." *Isis* 72 (1981):163–186.

——. *Victorian Sensation: The Extraordinary Publication, Reception, and Secret Authorship of Vestiges of the Natural History of Creation.* Chicago: University of Chicago Press, 2000.

Simpson, George Gaylord. *Discoverers of the Lost World: An Account of Some of Those Who Brought Back to Life South American Mammals Long Buried in the Abyss of Time.* New Haven: Yale University Press, 1984.

——. *Splendid Isolation: The Curious History of South American Mammals.* New Haven: Yale University Press, 1980.

Sloan, Phillip Reid. "Darwin's Invertebrate Program, 1826–1836: Preconditions for Transformism." In *The Darwinian Heritage,* edited by David Kohn, 71–120. Princeton: Princeton University Press, 1985.

——. "A Plea for Caution: A Response to Frederick Burkhardt." *Documentary Editing* 23 (2001):82–84.

——, ed. *The Hunterian Lectures in Comparative Anatomy, May–June 1837 by Richard Owen.* Chicago: University of Chicago Press, 1992.

Smith, Bernard. *Imagining the Pacific in the Wake of the Cook Voyages.* New Haven: Yale University Press, 1992.

Smith, Bernard, and T. Neville George. *British Regional Geology: North Wales.* 3d ed. London: Her Majesty's Stationery Office, 1961.

Smith, Crosbie. "Geologists and Mathematicians: The Rise of Physical Geology." In *Wranglers and Physicists: Studies on Cambridge Physics in the Nineteenth Century,* edited by P. M. Harman, 49–83. Manchester: Manchester University Press, 1985.

——. "William Hopkins and the Shaping of Dynamical Geology: 1830–1860." *British Journal for the History of Science* 22 (1989):27–52.

Smith, Crosbie, and M. Norton Wise. *Energy and Empire: A Biographical Study of Lord Kelvin.* Cambridge: Cambridge University Press, 1989.

Smith, Jean Chandler. *Georges Cuvier: An Annotated Bibliography of His Published Works.* Washington, D.C.: Smithsonian Institution Press, 1993.

Smith, Kenneth G. V., ed. "Darwin's Insects." *Bulletin of the British Museum (Natural History) Historical Series* 14, no. 1 (1987):1–143.

Smith, Sydney. "The Darwin Collection at Cambridge with One Example of Its Use: Charles Darwin and *Cirripedes.*" *Actes du XIe congrès international d'histoire des sciences* 5 (1965):96–100.

——. "The 'Origin' of the *Origin* as Discerned from Charles Darwin's Notebooks and His Annotations in the Books He Read between 1837 and 1842." *Advancement of Science* 16 (1960):391–401.

Sober, Elliott. *Philosophy of Biology.* Boulder, CO: Westview Press, 1993.

Sörlin, Sverker. "Ordering the World for Europe: Science as Intelligence and Information as Seen from the Northern Periphery. *Osiris* 15 (2000):51–69.

Stafford, Robert A. "Geological Surveys, Mineral Discoveries, and British Expansion, 1835–1871." *Journal of Imperial and Commonwealth History* 12 (1984):5–32.

——. "Scientific Exploration and Empire." In *The Oxford History of the British Empire,* edited by William Roger Louis. Vol. 3: *The Nineteenth Century,* edited by Andrew Porter, 294–319. Oxford: Oxford University Press, 1999.

——. *Scientist of Empire: Sir Roderick Murchison, Scientific Exploration and Victorian Imperialism.* Cambridge: Cambridge University Press, 1989.

Stauffer, Robert C. "Ecology in the Long Manuscript Version of Darwin's *Origin of Species* and Linnaeus' *Oeconomy of Nature.*" *Proceedings of the American Philosophical Society* 104 (1960):235–241.

Steadman, David W. *Holocene Vertebrate Fossils from Isla Floreana, Galápagos.* Washington, D.C.: Smithsonian Institution Press, 1986.

Steadman, David W., and Steven Zousmer. *Galápagos: Discovery on Darwin's Islands.* Washington, D.C.: Smithsonian Institution, 1988.

Stillwell, Richard, ed. *The Princeton Encyclopedia of Classical Sites.* Princeton: Princeton University Press, 1976.

Stoddart, David R. "Darwin, Lyell, and the Geological Significance of Coral Reefs." *British Journal for the History of Science* 9 (1976):199–218.

——. "Darwin and the Seeing Eye: Iconography and Meaning in the Beagle Years," *Earth Sciences History* 14 (November 1995):3–22.

——. "This Coral Episode: Darwin, Dana and the Coral Reefs of the Pacific." In *Darwin's Laboratory: Evolutionary Theory and Natural History in the Pacific,* edited by Roy MacLeod and Philip F. Rehbock, 21–48. Honolulu: University of Hawai'i Press, 1994.

Stott, Rebecca. *Darwin and the Barnacle.* New York: W. W. Norton, 2003.

Strick, James E. *Sparks of Life: Darwinism and the Victorian Debates over Spontaneous Generation.* Cambridge: Harvard University Press, 2000.

Sulloway, Frank J. "Darwin and His Finches: The Evolution of a Legend." *Journal of the History of Biology* 15 (1982):1–53.

——. "Darwin and the Galapagos." *Biological Journal of the Linnean Society* 21 (1984):29–59.

——. "Darwin's Conversion: The *Beagle* Voyage and Its Aftermath." *Journal of the History of Biology* 15 (1982):327–398.

——. "Further Remarks on Darwin's Spelling Habits." *Journal of the History of Biology* 16 (1983):361–390.

Sutherland, Gillian. "Education." In *The Cambridge Social History of England 1750–1950,* edited by Francis Michael Longstreth Thompson, 3:119–169. Cambridge: Cambridge University Press, 1990.

Taylor, Kenneth L. "Buffon, Desmarest and the Ordering of Geological Events in *Époques.*" In *The Age of the Earth: From 4004 BC to 2002 AD,* edited by C. L. E. Lewis and S. J. Knell, 39–49. London: Geological Society of London, 2002.

——. "Nicholas Desmarest." *Dictionary of Scientific Biography* 4:70–73.

——. "Volcanoes as Accidents: How 'Natural' Were Volcanoes to 18th-Century Naturalists?" In *Volcanoes and History,* edited by Nicoletta Morello, 595–618. Genova: Brigati, 1998.

Terra, Helmut de. *Humboldt: The Life and Times of Alexander von Humboldt, 1769–1859.* New York: Alfred A. Knopf, 1955.

——. "Studies of the Documentation of Alexander von Humboldt." *Proceedings of the American Philosophical Society* 102, no. 2 (December 1958):560–589.

Tesnière, Marie-Hélène, and Prosser Gifford, eds. *Creating French Culture: Treasures from the Bibliothèque nationale de France.* New Haven: Yale University Press, 1995.

Thackray, John C. "The Archives of the Geological Society of London." *Earth Sciences History* 3 (1984):3–8.

——. "Charles Lyell and the Geological Society." In *Lyell: The Past Is the Key to the Present,* edited by D. J. Blundell and A. C. Scott, 17–20. London: Geological Society, 1998.

——. "Mineral and Fossil Collections." In *Sir Hans Sloane: Collector, Scientist, Antiquary,* edited by Arthur MacGregor, 123–135. London: British Museum Press, 1994.

——. *To See the Fellows Fight: Eye Witness Accounts of Meetings of the Geological Society of London and Its Club, 1822–1868.* Stanford in the Vale, England: British Society for the History of Science, 2003.

Thrower, Norman J. W. *Maps and Civilization: Cartography in Culture and Society.* Chicago: University of Chicago Press, 1996.

Tilley, Cecil Edgar. "The Dunite-Mylonites of St. Paul's Rocks (Atlantic)." *American Journal of Science* 245 (1947):483–491.

Tomkeieff, Sergei Ivanovich. *Dictionary of Petrology.* Chichester: John Wiley, 1983.

Topham, Jonathan. "Science and Popular Education in the 1830s: The Role of the Bridgewater Treatises." *British Journal for the History of Science* 25 (1992):397–430.

Torrens, Hugh S. "Geology in Peace Time: An English Visit to Study German Mineralogy and Geology (and Visit Goethe, Werner and Raumer) in 1816." In *Toward A History of Mineralogy, Petrology, and Geochemistry,* edited by Bernhard Fritscher and Fergus Henderson, 147–178. Munich: Institut für Geschichte der Naturwisssenschaften, 1998.

——. "Patronage and Problems: Banks and the Earth Sciences." In *Sir Joseph Banks: A Global Perspective,* edited by R. E. R. Banks, B. Elliott, J. G. Hawkes, D. King-Hele, and G. Ll. Lucas, 49–75. Kew: Royal Botanic Gardens, 1994.

——. *The Practice of British Geology, 1750–1850.* Aldershot, England: Ashgate, 2002.

——. "Timeless Order: William Smith (1769–1839) and the Search for Raw Materials 1800–1820." In *The Age of the Earth: From 4004 BC to AD 2002,* edited by C. L. E. Lewis and S. J. Knell, 61–83. London: Geological Society, 2001.

Toulmin, Stephen. *Human Understanding.* Oxford: Clarendon Press, 1972.

Townsend, Charles Haskins. "The Galapagos Tortoises in Their Relation to the Whaling Industry: A Study of Old Logbooks." *Zoologica* 4 (1925):55–135.

Turner, Steven C. "Goniometer." In *Instruments of Science,* edited by Robert Bud and Deborah Jean Warner, 290–292. New York: Garland, 1998.

Uglow, Jenny. *The Lunar Men: Five Friends Whose Curiosity Changed the World.* New York: Farrar, Straus and Giroux, 2002.

Van Denburgh, John. "The Gigantic Land Tortoises of the Galapagos Archipelago." *Proceedings of the California Academy of Sciences,* 4th ser., 1, pt. 1 (1914):203–374.

Van Riper, Bowdoin A. *Men Among the Mammoths: Victorian Science and the Discovery of Human Prehistory.* Chicago: University of Chicago Press, 1993.

Walters, Stuart Max, and E. Anne Stow. *Darwin's Mentor: John Stevens Henslow, 1796–1861.* Cambridge: Cambridge University Press, 2001.

Washington, Henry S. "The Petrology of St. Paul's Rocks (Atlantic)." In *Report on the Geological Collections Made during the Voyage of the "Quest" on the Shackleton-Rowett Expedition to the South Atlantic & Weddell Sea in 1921–1922,* 126–144. London: British Museum (Natural History), 1930.

Webb, Robert K. *Modern England from the Eighteenth Century to the Present.* 2d ed. New York: Harper and Row, 1980.

——. "The Unitarian Background." In *Truth, Liberty, Religion: Essays Celebrating Two Hundred Years of Manchester College,* edited by Barbara Smith, 3–30. Oxford: Manchester College, 1986.

Weiner, Jonathan. *The Beak of the Finch: A Story of Evolution in Our Time.* New York: Knopf, 1994.

Westman, Robert. "The Copernicans and the Churches." In *God and Nature: Historical Essays on the Encounter between Christianity and Science,* edited by David C. Lindberg and Ronald L. Numbers, 76–113. Berkeley: University of California Press, 1986.

Whitaker, Arthur P. "Alexander von Humboldt and Spanish America." *Proceedings of the American Philosophical Society* 104, no. 3 (June 1960):317–322.

S

Whitmore, Frank C., Jr., and Joshua I. Tracey Jr. "Memorial to Harry Stephen Ladd, 1899–1982." *Geological Society of America Memorials* 14 (1984):1–7.

Williams, Glyndwr. "New Holland to New South Wales: The English Approaches." In *Terra Australis to Australia,* edited by Glyndwr Williams and Alan Frost, 117–159. Melbourne: Oxford University Press Australia, 1988.

Williams, Glyndwr, and Alan Frost. "*Terra Australis:* Theory and Speculation." In *Terra Australis to Australia,* edited by Glyndwr Williams and Alan Frost, 1–37. Melbourne: Oxford University Press Australia, 1988.

Wilson, Leonard G. *Charles Lyell: The Years to 1841.* New Haven: Yale University Press, 1972.

——. *Lyell in America: Transatlantic Geology, 1841–1853.* Baltimore: Johns Hopkins University Press, 1998.

——. "Lyell: The Man and His Times." In *Lyell: The Past Is the Key to the Present,* edited by D. J. Blundell and A. C. Scott, 21–37. London: Geological Society, 1998.

——, ed. *Sir Charles Lyell's Scientific Journals on the Species Question.* New Haven: Yale University Press, 1970.

Wilton, Andrew, and Anne Lyles. *The Great Age of British Watercolours, 1750–1880.* Munich: Prestel, 1993.

Windshank, Donald J., Stephan J. Ozminski, Paul Ruhlen, and Wilma M. Barrett. *A Concordance to Charles Darwin's Notebooks, 1836–1844.* Ithaca: Cornell University Press, 1990.

Winsor, Mary P. "Barnacle Larvae in the Nineteenth Century." *Journal of the History of Medicine and Allied Sciences* 24 (July 1969):294–309.

Winstanley, D. A. *Early Victorian Cambridge.* Cambridge: Cambridge University Press, 1940.

Wolfe, Patrick. "Imperialism and History: A Century of Theory, from Marx to Postcolonialism." *American Historical Review* 102 (April 1997):338–420.

Woodward, Horace B. *The History of the Geological Society of London.* London: Geological Society, 1907.

Woolf, Harry. *The Transits of Venus: A Study of Eighteenth-Century Science.* Princeton: Princeton University Press, 1959.

Wyse Jackson, Patrick, ed. *Geological Travellers—On Foot, Bicycle, Sledge or Camel: The Search for Geological Knowledge.* New York: Pober Publications, in press.

Yoder, Hatten S., Jr. "Timetable of Petrology." *Journal of Geological Education* 41 (1993): 447–489.

Young, Davis A. *Mind over Magma: The Story of Igneous Petrology.* Princeton: Princeton University Press, 2003.

——. *N. L. Bowen and Crystallization-Differentiation: The Evolution of a Theory.* Washington, D.C.: Mineralogical Society of America, 1998.

Young, George Malcolm. *Portrait of an Age: Victorian England.* Edited by G. K. Clark. London: Oxford University Press, 1977.

Zeller, Suzanne. "The Colonial World as a Geological Metaphor: Strata(gems) of Empire in Victorian Canada." *Osiris* 15 (2000):85–107.

Zittel, Karl Alfred, von. *History of Geology and Palæontology to the End of the Nineteenth Century.* Translated by Maria M. Ogilvie-Gordon. London: Walter Scott, 1901.

INDEX

Note: Page numbers with an *f* indicate figures; *pl.* indicates color plate.